P9-CJY-071

Presidential Wives

Presidential Wives

PAUL F. BOLLER, JR.

New York Oxford
OXFORD UNIVERSITY PRESS
1988

For My Beloved Friends

Bengt and Julia
Jim and Martha
Phil and Subie
George and Elaine

Oxford University Press

Oxford New York Toronto
Delhi Bombay Calcutta Madras Karachi
Petaling Jaya Singapore Hong Kong Tokyo
Nairobi Dar es Salaam Cape Town
Melbourne Auckland

and associated companies in
Beirut Berlin Ibadan Nicosia

Published by Oxford University Press, Inc.,
200 Madison Avenue, New York, New York 10016

Library of Congress Cataloging-in-Publication Data
Boller, Paul F.
Presidential wives.
Includes index.
1. Presidents—United States—Wives—Anecdotes. I. Title.
E176.2.B65 1988 973'.09'92 87-31573
ISBN 0-19-503763-4

1 3 5 7 9 8 6 4 2

Printed in the United States of America
on acid-free paper

PREFACE

Presidential Wives completes the trilogy of books about the Presidents that began with *Presidential Anecdotes* (1981) and continued with *Presidential Campaigns* (1984). The format is the same: essays and anecdotes. The essays focus on the backgrounds, characters, outlooks, and personalities of the women the Presidents married, the lives they lived, public and private, as wives of political leaders, and the conceptions held by those who got to the White House of the duties and responsibilities of the First Lady of the Land. The stories accompanying the essays—some of them amusing, others dramatic, and still others on the sad side—illustrate, underline, and expand on the points made in the essays.

Stories about the Presidents' wives are harder to come by than those about the Presidents themselves. Until the twentieth century the wives of the Presidents, with a few notable exceptions, maintained low profiles and attracted little public attention. It did not occur to James Monroe or Martin Van Buren to mention their families in their autobiographies; nor did biographers of the Presidents, until recent years, feel obliged to spend more than a perfunctory page or paragraph or two on the President's wife. "A lady's name should appear in print only three times," said Edith Kermit Roosevelt, First Lady from 1901 to 1909; "at her birth, marriage, and death."[1] The explosion of the mass media during the twentieth century, together with new notions about the place of women in American society, changed all that. It gradually got so that the wife of a President couldn't stay out of the news even if she wanted to.

Five of the women appearing here died before their husbands became President: Martha Jefferson, Rachel Jackson, Hannah Van Buren, Ellen Arthur, and Alice Lee Roosevelt. Three of them—Letitia Tyler, Caroline Harrison, and Ellen Axson Wilson—died while in the White House, and their husbands went on to remarry: John Tyler (Julia Gardiner) and Woodrow Wilson (Edith Bolling Galt) while still President, and Benjamin Harrison (Mary Scott Dimmock) after leaving office. Five women were widows at the time of their marriages (Martha Washington, Martha Jefferson, Dolley Madison, Mary Scott Harrison, and Edith

1. Sylvia Jukes Morris, *Edith Kermit Roosevelt: Portrait of a First Lady* (New York, 1980), 525–26.

Bolling Wilson) and three were divorcées (Rachel Jackson, Florence Harding, and Betty Ford). Ronald Reagan was the first divorced man to become President; he broke up with film star Jane Wyman in 1948, and in 1952 he married Nancy Davis, who became First Lady in 1981. Two bachelors—James Buchanan and Grover Cleveland—became President, but the latter married Frances Folsom during his second year in office. Eight First Ladies became widows while in the White House—Anna Harrison, Margaret Taylor, Mary Todd Lincoln, Lucretia Garfield, Ida McKinley, Florence Harding, Eleanor Roosevelt, Jacqueline Kennedy—but only Mrs. Kennedy later remarried (Aristotle Onassis). Of the fifteen First Ladies who became widows after leaving the White House—Martha Washington, Dolley Madison, Louisa Catherine Adams, Julia Gardiner Tyler, Sarah Polk, Eliza Johnson, Julia Dent Grant, Frances Cleveland, Edith Kermit Roosevelt, Helen Taft, Edith Bolling Wilson, Grace Coolidge, Bess Truman, Mamie Eisenhower, Lady Bird Johnson—only one remarried: Frances Folsom Cleveland (Thomas J. Preston, Jr.). Six First Ladies died before their husbands did after leaving the White House—Abigail Adams, Elizabeth Monroe, Abigail Fillmore, Jane Pierce, Lucy Webb Hayes, Lou Hoover—but Millard Fillmore was the only widower to remarry (Carolyn C. McIntosh, an Albany widow).

In the nineteenth century many of the Presidents' wives were practically Invisible Women. They moved into the family quarters of the Executive Mansion right after their husbands took the oath of office and either stayed there, pleading poor health, or appeared dutifully at White House functions a few times with as little to-do as possible. Others, even in the early years of the Republic, made something of their role as Hostesses. Martha Washington won praise for her amiability when entertaining guests in the President's house, and Dolley Madison became celebrated for her grace and glamor as the nation's premier Hostess. Like Jacqueline Kennedy many years later, Mrs. Madison also shone as a Leader of Fashion, and, like Mrs. Kennedy, influenced women's fashions abroad as well as in this country.

Some of the women who became First Ladies achieved fame neither as Hostesses nor as Leaders of Fashion, but played crucial roles as their husbands' helpmates all the same. They brought wealth and social status to their marriages; or they added learning and culture to their husbands' interest in politics. And while they made no great stir in the public prints, they energetically managed the household finances, supervised the children's education, kept their husbands properly fed, clad, and rested, and probably told them off when they thought they were full of baloney. Occasionally, too, they gave informal advice on presidential speeches, policies, and appointments. Sarah Childress Polk had a lot to say about her husband's work, both before and after he became President, but she never dreamed of making public pronouncements on the issues facing him. Bess W. Truman was about as uncommunicative as it was possible for a First Lady to be when dealing with the press, but she expected her husband to keep her abreast of all his political activities and was incensed whenever he neglected to do so. The effect of family discussions—even marital tensions—on Presidential decisions is incalculable but indubitable.

The Trumans regarded themselves as Partners. So did Jimmy and Rosalynn Carter. But Mrs. Carter, like several other twentieth-century First Ladies, was a diligent Activist as well while her husband was President. She sat in on Cabinet meetings, toured Central and South America as the President's representative, shared in the decisions he made as Chief Executive, and made campaign speeches for him when he ran for President. Jacqueline Kennedy and Pat Nixon were far less activistic than Mrs. Carter, but they, too, helped out their husbands on campaigns and also sponsored special programs of their own; and so did Betty Ford and Nancy Reagan. But the greatest of all the Activists was Eleanor Roosevelt. Not only did she advise her husband on policies and appointments, she also helped shape some of the New Deal projects he championed and came up with ideas of her own for which she won his approval. At the same time she had a successful career of her own as a journalist. She started writing for newspapers and magazines in the 1920s, expanded her writing assignments while in the White House, and continued with them after her husband's death. She also served as U.S. delegate to the United Nations after World War II. Unlike most First Ladies, she never seriously entertained the possibility of retiring after leaving the White House.

Once, in the early 1920s, a judge asked an applicant for naturalization whether he knew who would become Chief Executive if the President died suddenly, and the latter responded promptly: "His wife." The judge was amused by the answer and at once issued final papers.[2] But most Presidents' wives (probably all of them) would have been appalled by the thought of stepping into first place. Even Edith Bolling Wilson, who screened people and papers for her husband after his stroke in 1919, indignantly denied charges that she became a "Presidentress" during Wilson's illness. And Eleanor Roosevelt firmly rejected the idea that she ever really acted on her own while in the White House. Despite her vigorous and varied activities as First Lady, Mrs. Roosevelt always thought of her work as being ancillary to that of her husband. "On the whole . . . ," she once wrote of her twelve years in the White House, "I think I lived those years very impersonally. It was almost as though I had erected someone a little outside of myself who was the president's wife. I was lost somewhere deep down inside myself. That is the way I felt and worked until I left the White House."[3]

2. Frederick L. Collins, "Electing a President's Wife," *Woman's Home Companion,* 55 (April 1928), 9.

3. Eleanor Roosevelt, *This I Remember* (New York, 1949), 350–51.

Some day the President's wife may feel free to pursue a career of her own while helping out in the White House. And some day, surely, as Gerald R. Ford once predicted (to his wife's delight), the American voters will put a woman in the White House as Chief Executive. But one can't help wondering what it will be like for the husband of an American President. Will he turn out to be an Invisible Man? Or will he be a Great Host, a Fashion Leader, a Partner, or an Activist? Or will the First Gentleman (if that is what he is called) carry on with an independent career of his own while residing in the Executive Mansion? Whatever role he chooses to play, one thing seems certain: there will be an outpouring of press and television coverage of his comings and goings the like of which the country has never seen before.

I am deeply grateful to those perspicacious editors, Sheldon Meyer and Leona Capeless, for their thoughtful suggestions while I was working on this study of the wives of our Presidents.

Texas Christian University Paul F. Boller, Jr.
Fort Worth

CONTENTS

Presidential Wives

CHAPTER 1

MARTHA WASHINGTON

1731–1802

There was nothing high-falutin' about America's very first First Lady. She was, said an acquaintance, "simple, easy, and dignified."[1] When she arrived at her husband's headquarters in Morristown, New Jersey, in February 1779, some people took her for a servant when she stepped from the carriage and gasped when General Washington rushed up to greet her. A little later some of the ladies in town paid a formal call and were taken aback to find "Lady Washington," as she was called, plainly dressed in homespun garb and busily knitting stockings for the soldiers. They were even more surprised when she went on with her knitting as she chatted amiably with them. She was just as affable after her husband became President. As the wife of the President of the United States, she was, noted Abigail Adams, without a trace of arrogance. She was probably too self-assured to be supercilious.

Martha Washington (born Martha Dandridge in June 1731) came up the easy way. The daughter of a wealthy Virginia planter, as a child she did the things rich girls did—studied with a tutor, embroidered, played the spinet, and rode horseback—and began moving in high society when she was fifteen. In 1749, when she was seventeen, she married Colonel Daniel Parke Custis, another wealthy planter, almost twice her age, took up residence in the "White House," as the Custis mansion was called, and embarked on her duties as a planter's wife with confidence and competence. Eight years later Custis died, leaving her a widow—one of Virginia's wealthiest—at twenty-five with two small children. For a year or two she managed the Custis estate herself. Then she met George Washington.

George, a bit younger, was no plebeian when he met Martha. He had already inherited his fine Mount Vernon plantation from his half-brother Lawrence; and, as colonel in the Virginia militia who had proven himself in action, he moved with ease himself among Virginia's high and mighty. But his marriage to Martha in 1759 boosted his wealth and prestige. Years later, when he got into an argument with David Burns, an outspoken Scotsman who refused to sell him some land he wanted, Burns exclaimed: "I suppose you think people are going to take every grist that comes from you as pure grain. What would you have been, if you hadn't married the Widow Custis?"[2] Washington, who had just retired as President, told of the encounter with some chagrin. But Martha was amused. She had probably asked herself the reverse question from time to time: "What would you have been, if you hadn't married George?"

Did the Washingtons live happily ever after? Apparently they did. Before meeting Martha, Washington had fallen hard for Sally Fairfax, beautiful young wife of one of his best friends, and he seems to have retained a sentimental attachment to her all of his life. But his devotion to Martha was deep and abiding. "Love is a mighty pretty thing," he once wrote, but it "is too dainty a food to live on *alone,* and ought not to be considered further than as a necessary ingredient for that matrimonial happiness which results from a combination of causes: none of which are of greater importance than that the partner should have good sense, a good disposition, a good reputation, and financial means."[3] Both George and Martha had all of these in abundance.

Shortly before marrying Martha, George resigned his commission in the army and made plans to devote himself, with Martha's help, to supervising his vast holdings, entertaining friends and relatives, and serving, as rich planters usually did, as vestryman and as a member of the Virginia legislature. "I am now, I believe, fixed in this spot with an agreeable partner for life," he wrote a relative in London right after the marriage; "and I hope to find more happiness in retirement than I ever experienced in the wide and bustling world."[4] The Washingtons spent a lot of time hankering for retirement when the American Revolution and then the Presidency took them away from Mount Vernon. But they probably enjoyed the excitement of public life more than they acknowledged. They certainly took it for granted that George should be called out of private life whenever there was a crisis. There was a great deal of aristocratic *noblesse*—and self-esteem—in the two of them.

When the Revolution came there were rumors that Martha was a Tory. But the idea of Martha's supporting the British when George was a Whig was preposterous. The two were as one when it came to politics.

When a relative deplored George's "folly" in joining the resistance movement, Martha exclaimed: "My heart is made up; my heart is in the cause; George is right; he is always right."[5] In August 1774, Patrick Henry and Edmund Pendleton stopped to pick up Washington for the trip to Philadelphia, where the Second Continental Congress was meeting, and they were impressed with his wife's fighting spirit. "She seemed ready to make any sacrifice, and was very cheerful, though I know she felt very anxious," Pendleton wrote a friend afterward. "She talked like a Spartan mother to her son on going to battle. 'I hope you will all stand firm— I know George will,' she said. . . . When we set off in the morning, she stood in the door and cheered us with good words. 'God be with you, gentlemen.' "[6] George professed reluctance when the Continental Congress offered him command of the Continental forces in June 1775. He told Martha he would "enjoy more real happiness in one month with you at home" than the glory of holding the highest military post in the country.[7] Still, both George and "Patsy" (his nickname for her) were pleased by the honor and probably not much surprised by it.

In November 1775, Martha journeyed to Massachusetts to join her husband at his headquarters in Cambridge. Thereafter, until the war was over, she traveled to his camp whenever he went into winter quarters and remained until the fighting was resumed in the spring. Her companionship meant much to him; she took long rides with her "Old Man," as she called him, and bolstered his spirits when things looked dark. Her presence also boosted the morale of the soldiers. And her activities—mending clothes, knitting socks, visiting the sick—encouraged other women to do war work too. "Whilst our husbands and brothers are examples of patriotism," she told them, "we must be patterns of industry."[8]

Martha was herself a model of diligence at Washington's various headquarters. "I never in my life knew a woman so busy from early morning until late at night as was Lady Washington, providing comforts for the sick soldiers," reported a woman in Valley Forge. "Every day, excepting Sunday, the wives of the officers in camp, and sometimes other women, were invited . . . to assist her in knitting socks, patching garments, and making shirts for the poor soldiers, when materials could be procured. Every fair day she might be seen, with basket in hand, and with a single attendant, going among the huts seeking the keenest and most needy sufferer, and giving all the comforts to him in her power."[9] Another woman noted that Martha "talked much of the sufferings of the poor soldiers, especially of the sick ones. Her heart seemed to be full of compassion for them."[10] In 1780, when a Philadelphia paper printed an essay entitled, "The Sentiments of an American Woman," some people thought she was the author and liked to quote from it.[11] Martha was no writer (though the few letters she wrote that survive contain some nice phrases); but she wholeheartedly subscribed to the article's patriotic views.

When Washington became the first President of the United States after the war, "Lady Washington" from the outset took her new position as seriously as she had her position as wife of the Revolutionary Commander-in-Chief. The American people looked to her, she realized, as well as to her husband, to envelop the new executive branch of government with the kind of decorum that would win it respect among the older countries of the world. Martha, as always, combined dignity with geniality in her new position. Abigail Adams saw her in New York a few weeks after Washington's inauguration in April 1789 and liked what she saw. "She is plain in her dress," observed Mrs. Adams, "but that plainness is the best of every art. . . . Her manners are modest and unassuming, dignified and feminine, not the tincture of hauteur about her."[12] A second visit confirmed the initial good impression. "Mrs. Washington is one of those characters which create Love & Esteem," Mrs. Adams decided. "A most becoming pleasentness sits upon her countanance & an unaffected deportment which renders her the object of veneration and respect. . . . I found myself much more deeply impressed than I ever did before their Majesties of Britain."[13]

While Washington was President there were weekly levees on Tuesday afternoons, weekly dinners on Thursdays, and evening receptions by the First Lady, which Washington also attended, on Fridays. There were also dinner parties for government officials and foreign visitors. Martha was solicitous of her husband on all social occasions. She never permitted guests to talk politics and when visitors tried to draw her out on public issues she tactfully steered them to non-controversial topics. She also saw to it that her husband did not stay up too late. When the clock struck nine, she was in the habit of telling guests: "The general always retires at nine o'clock, and I usually precede him."[14] Some people—Alexander Hamilton, for one—thought there should be more ceremony at the Washington receptions; but others accused the First Family of attempting "an awkward imitation of royalty."[15] Most people, though, applauded Martha's "ease and simplicity of manners."[16]

Martha was not especially happy in New York, the first capital, and found Philadelphia, to which the Federal Government moved in 1790, more to her liking. On the whole, though, she felt oppressed by her position as First Lady. She called her days as the President's wife her "lost days." To her niece she wrote frankly: "I live a very dull life here, and know nothing that passes in town. I never go to any public place—indeed, I am more like a State prisoner than anything else. There are certain bounds set for me which I must not depart from, and as I cannot do as I like, I am obstinate and stay home a great deal."[17] To Mercy Warren she complained of the "ceremonies of mere etiquette" that went with her "new and unwished-for situation," and added: "I sometimes think that arrangement is not quite as it ought to have been, that I, who had much rather be at home, should occupy a place with which a great many younger and gayer women would be extremely pleased. . . . I have learned too much of the vanity of human affairs to expect felicity from the scenes of public life." Still, she resolved to do her duty just as her husband was doing his. "I am still determined to be cheerful and happy in whatever situation I may be," she told Mrs. Warren; "for I have also learned from experience that the greater part of our happiness or misery depends on our dispositions and not on our circumstances. We carry the seeds of the one or the other about with us in our minds, wherever we go."[18] Mrs. Warren wrote back to reassure her. "Your observation may be true, that many younger and gayer ladies consider your situation as enviable; yet I know not one who, by general consent, would be more likely to obtain the suffrages of the sex, even were they to canvass an election, for this elevated station, than the lady who now holds the first rank in the United States."[19]

Martha was overjoyed—so was George—when she was at long last able to return to Mount Vernon for good after Washington left office in 1797. "I cannot tell you, my dear friend," she wrote Mrs. Knox, "how much I enjoy home after having been deprived of one so long, for our dwelling in New York and Philadelphia was not *home,* only a sojourning. The General and I feel like children just released from school or from a hard taskmaster, and we believe that nothing can tempt us to leave the sacred roof-tree again, except on private business or pleasure. We are so penurious with our enjoyment that we are loath to share it with any one but dear friends, yet almost every day some stranger claims a portion of it, and we cannot refuse." But she adds: "I am again fairly settled down to the pleasant duties of an old-fashioned Virginia house-keeper, steady as a clock, busy as a bee, and cheerful as a cricket."[20]

The joys of retirement were short-lived. In 1798, during the crisis with France, Washington accepted President John Adams' invitation to head a provisional army in case of war but fortunately never had to return to the field. The following year he took a chill while out riding in the rain and died at sixty-five. "I shall soon follow him," Martha said quietly at his death bed; "I have no more trials to pass through."[21] She spoke prematurely. When Congress made plans to move her husband's body from its grave at Mount Vernon to the new national capital in Washington, she was deeply disappointed. But she was as dutiful as ever. "Taught by the great example which I have so long before me never to oppose my private wishes to the public will—," she wrote, "I must consent to the request made by Congress . . . and in doing this I need not—I cannot say what a sacrifice of individual feeling I make of a sense of public duty."[22] In the end, however, Congress abandoned its plan and George's body remained at Mount Vernon.

Martha spent her last two years quietly at Mount Vernon keeping busy with her knitting and entertaining visitors. Her face was "very little wrinkled," her guests observed, "and remarkably fair for a person of her years." In the spring of 1802 some Boston Federalists traveled to Mount Vernon to pay their respects and had breakfast with her. They listened with glee as she made some remarks about "the new order of things" with Thomas Jefferson as President which were "frequently pointed, and sometimes very sarcastic." She still resented Jefferson's criticisms of her husband's policies as President and said she regarded him as "one of the most detestable of mankind, the greatest misfortune our country had ever experienced." She talked about "the General" a great deal, "viewing herself as left alone, and her life protracted, until she had become a stranger in the world." She "longed for the time," she said, "to follow her departed friend."[23]

A few weeks later Mrs. Washington fell ill with a fever, burned all of George's letters to her, had a clergyman administer her last communion, ordered her funeral dress laid out, and awaited the end. On May 22, 1802, after seventeen days of illness, she died at the age of seventy-one. Observed the *Port Folio* of her passing: "To those amiable and Christian virtues which adorn the female character, she added the dignity of manners, superiority of understanding, a mind intelligent and elevated. The silence of respectful grief is our best eulogy."[24]

✲ ✲ ✲ ✲

How She Can Ride!

As a young woman Martha preferred riding horses to reading books and soon became a skillful horsewoman. Once she even rode her horse Fatima up the stairs and down again at her Uncle William's house. Both her stepmother and her aunt were horrified, but when they began scolding her, her father came to her defense. "Let Patsy alone. . . !" he cried. "She's not harmed William's staircase. And, by heavens, how she can ride!"[25]

Inseparable

In May 1758 Colonel and Mrs. Richard Chamberlayne invited Martha Custis, who lived nearby, to spend a few days at their home on the Pamunkey River. Shortly after her arrival Chamberlayne took a walk down to the river, where he ran into his friend George Washington and invited him to dinner. Washington, the twenty-six-year-old commander of the Virginia militia, was on his way to Williamsburg to confer with the Governor and said he was in a hurry. But when Chamberlayne promised to introduce him to Virginia's prettiest and wealthiest widow, Washington decided to stay. He not only stayed for dinner; he also spent the night there, for, as his host told him, "No guest ever leaves my house after sunset." George and Martha, who were about the same age, seem to have hit it off at once and by the time George resumed his journey they had reached an understanding. From Fort Cumberland, Washington wrote her in July: ". . . I embrace the Opportunity to send a few words to one whose life is now inseparable from mine. Since that happy hour when we made our Pledges to each other, my thoughts have been continually going to you as to another Self." They were married in January 1759 and Washington assumed the guardianship of her children, John Parke (Jacky) and Martha Parke (Patsy) Custis. Later on, when Jacky died, he assumed the guardianship of his two children.[26]

Plucky Little Woman

In October 1775, shortly after Washington assumed command of the Continental forces at Cambridge, Massachusetts, there was a rumor that Lord Dunmore and a British crew were on their way up the Potomac to capture Mrs. Washington and burn Mount Vernon. But Mrs. Washington refused to leave. When George Mason came by to take her away, she told him: "No, I will not desert my post." But when a British ship was reported in sight of the Mount Vernon landing she finally yielded and agreed to go. But she "did so with reluctance," Mason wrote Washington, "rode only a few miles, and, plucky little woman she is, stayed away only one night." The British never came to Mount Vernon, but after the Dunmore scare Washington decided his wife would be safer with him and sent a courier to escort her to Cambridge. It was what she had wanted all along.[27]

Pattern of Industry

When Washington was stationed in Morristown, New Jersey, Mrs. Troupe and some of the other women there decided to call on his wife and carefully donned their "best bibbs and bands" for the occasion. Afterward Mrs. Troupe cornered a friend. "Well," she said, "what do you think . . . I have been to see Lady Washington!" "Have you indeed?" cried her friend. "Then tell me all about her ladyship, how she appeared, and what she said." "Well," said Mrs. Troupe, "I will honestly tell you. I never was so ashamed in all my life." She went on to say that when she and three other women called on the General's wife, they were wearing elegant ruffles and silks. "And don't you think we found her *knitting and with a speckled apron on!*" Mrs. Washington received the women graciously, Mrs. Troupe said, and then went back to her knitting. "There we were without a stitch of work. . . ," sighed Mrs. Troupe, "but General Washington's lady with her own hands was knitting stockings for herself and husband! And that was not all. In the afternoon her ladyship took occasion to say, in a way that we could not be offended at, that at this time it was very important that American ladies should be patterns of industry to their countrymen, because the separation from the mother country will dry up the sources whence many of our comforts have been derived. We must become independent by our determination to do without what we cannot make ourselves. Whilst our husbands and brothers are examples of patriotism, we must be patterns of industry." Like Mrs. Troupe, one of the other callers "felt rebuked," but was consoled by Mrs. Washington's geniality. "She seems very wise in experience, kind-hearted and winning in all her ways," she reported. "She talked much of the suffering of the poor soldiers, especially of the sick ones. Her heart seemed to be full of compassion for them."[28]

Mere Lads

When Washington established his winter quarters in Middlebrook, New Jersey, in December 1778, he engaged two young carpenters to finish off one of the rooms in the top story of the building he was occupying for his wife to use. But he let Martha supervise the work and the carpenters soon found they had an amiable boss. "She came into the place," one of the men reported years later, "a portly-looking, agreeable woman of forty-five," and told them what she wanted. "Now, young men," she said, "I care for nothing but comfort here, and should like you to fit me up a beaufet [closet] on one side of the room, and some shelves, and places for hanging clothes on the other." So the carpenters went to work, adding a new floor, filling up the holes in the walls, putting up pegs, and constructing a closet. Every morning at eleven Mrs. Washington came upstairs with some whiskey for the men, and, later in the day, after she and Washington had finished dining, she invited them down for a meal. On the fourth day she went up to see how things were going and found they had converted the old garret into a comfortable new apartment for her. "Madam," said one of the carpenters, "we have endeavored to do the best we could; I hope we have suited you." "I am astonished," cried Mrs. Washington, looking about with a big smile. "Your work would do honor to an old master, and you are mere lads. I am not only satisfied, but highly gratified by what you have done for my comfort." The two men left in a glow.[29]

Heavenly Sight

When John Hunter, an Englishman touring the United States and Canada after the Revolution, visited Mount Vernon in 1785, he found Mrs. Washington "a most agreeable woman about 50," but was surprised by her enthusiasm for things military. "It's astonishing," he wrote, "with what raptures Mrs. Washington spoke about the discipline of the army, the excellent order they were in, superior to any troops she said upon the face of the earth towards the close of the war; even the English acknowledged it, she said. What pleasure she took in the sound of the fifes and drums, preferring it to any music that was ever heard; and then to see them reviewed a week or two before the men were disbanded, when they were all well clothed was she said a most heavenly sight. . . ."[30]

Early to Rise

After Washington became President, he commissioned Charles Willson Peale to do a miniature of his wife. Peale was scheduled to arrive for the first sitting at seven in the morning, but though he got there on time he was afraid to sound the brass knocker at the Washington home so early in the day. He walked the streets for a time before making his presence known. When he finally appeared, Mrs. Washington, a stickler for punctuality, looked at the clock and asked him why he was so late. When he said he hated to disturb her at that early hour, she laughed and told him she had already attended family worship, given Nelly Custis a music lesson, and read the newspaper while waiting for him.[31]

Remain as You Are

At Mount Vernon the Washingtons dressed for dinner and expected everyone else to do the same. But Nelly Custis, Mrs. Washington's granddaughter, who was staying with them, didn't always conform. One afternoon she and her friend Martha Dandridge, Mrs. Washington's niece, appeared for dinner carelessly dressed, and though Washington said nothing his wife was deeply offended. Just as the meal ended a coach arrived in the driveway and young Charles Carroll, Jr., of Carrollton and some French officers got out. At once the two girls asked to be excused so they could change their dresses and fix their hair. "No," said Mrs. Washington sternly, "remain as you are. What is good enough for General Washington is good enough for any guest of his."[32]

Locks of Hair

In March 1797, Washington attended the inauguration of his successor, John Adams, as President, and afterward he and Martha went to a little retirement party in their honor. At one point Mrs. Oliver Wolcott came by to congratulate them and Mrs. Washington asked if she would like a memento of the first President. "Yes," said Mrs. Wolcott. "I should like a lock of his hair." Mrs. Washington at once took out a pair of scissors, it is said, and cut a large lock from her husband's head for Mrs. Wolcott. Then, with a big smile, she cut off a lock of her own hair to add to the gift.[33]

Filthy Democrat

While Washington was President, his wife carefully avoided politics and at receptions treated critics of the administration with as much graciousness as she did its supporters. But some observers thought she was thoroughly Federalist in her sympathies and held Jeffersonian Republicans in contempt. During a reception at the Executive Mansion, according to one tale, she happened to enter the parlor just as an anti-Federalist had left and asked her granddaughter Nelly who the caller was. When Nelly said she didn't know, Mrs. Washington looked around the room, noticed a dirty spot on the wall just above the settee and cried: "Ah, that was no Federalist. None but a filthy democrat would mark a place with his good-for-nothing head in that manner!" But her grandson, George Washington Parke Custis, denied the story; he said it was out of character for her to talk that way.[34]

CHAPTER 2

ABIGAIL ADAMS

1744–1818

When Abigail Adams was asked late in life whether she would have wanted her husband to go into politics had she known it would mean years of separation from him, she replied in the affirmative. "I feel a pleasure," she explained, "in being able to sacrifice my selfish passions to the general good, and in imitating the example which has taught me to consider myself and family but as the small dust of the balance, when compared with the great community."[1] Her answer was vintage Adams: a bit solemn, perhaps, but utterly sincere in its insistence on the primacy of public duty over personal concerns. Both Abigail and John Adams tended to minimize the pleasure they took in winning public recognition, though it was real enough; but both, surely, had a generous measure of what the eighteenth century called "public virtue."

Abigail Adams was the first—and, so far, only—woman to be both the wife and mother of a President. She enjoyed her husband John's eminence and found it stimulating to be the wife of a President for a term. She also took satisfaction in her son John Quincy's steady rise in public life; and, though she didn't live to see him become President, she was delighted when President Monroe made him Secretary of State, a stepping stone to the Presidency, in 1817. Abigail even thought she could take some credit for her husband's and her son's achievements. Women were, she knew, expected to be good wives and mothers; and Abigail, for all her "saucyness" (John's term), was conscientious about her womanly responsibilities. But she also struck off on her own at times; and both John and John Quincy were well aware of her unusual abilities.[2] John regarded her as his intellectual peer and depended a great deal on her counsel; and John Quincy attributed much of his own success in life to her tutelage when he was a boy. "How shall I offer you consolation for your loss," he asked his father when Abigail died in 1818, "when I feel that my own is irreparable?" "Any consolations are more than I can number," John assured his son, for he expected to join her soon. "The separation cannot be so long as twenty separations heretofore."[3]

There were far more than twenty separations. From almost the beginning, John's absences from home were frequent, at times lengthy, and always painful for him and his wife. Abigail once complained that she had been a "widow" for much of her married life. "Who shall give me back Time?" she wailed at one point, "who shall compensate to me those *years* I cannot recall?"[4] John began courting Abigail Smith, a Congregational minister's daughter, in 1761 when he was twenty-six and she only seventeen, and they were married in her father's parsonage in Weymouth, not far from Boston, in 1764, and settled down in Braintree to begin a family. But during their early married years John was away a lot on the court circuit trying to get ahead as a lawyer and Abigail was left alone as a young wife to run the house and raise the children. When the Stamp Act crisis came in 1765 and Adams joined the resistance movement, separation gradually became a way of life for the young couple. The American Revolution brought excitement, adventure, and patriotic fervor into their lives, but it also tore them apart. First it was Philadelphia, where Adams served in the First and the Second Continental Congress from 1774 until 1778. Then it was Europe, where he served, first, as Commissioner to France in 1778 and then, with Benjamin Franklin and John Jay, as negotiator of the Paris peace treaty ending the war in 1783. It wasn't until 1784, just before he became the first U.S. Minister to Great Britain that Abigail finally went to London to join him in his work.

Separation was harder on Abigail than on John. The "one left behind," she once remarked, was always "the greatest sufferer."[5] She also thought women suffered more than men did from being separated from loved ones and said she "never wondered at the philosopher who thanked the gods that he was created a Man rather than a Woman."[6] Once, when she hadn't heard from her husband for five weeks, she exclaimed: "I had rather give a dollar for a letter by the post, tho the consequence should be that I Eat but one meal a day for these three weeks to come."[7] Another time, when he was in Europe, she cried: "Alass my dear I am much afflicted with a disorder call'd the *Heartach*, nor can any remedy be found in America. . . ."[8] But Adams missed her keenly, too, and both of them lamented the way public responsibilities kept them apart. "If I were to tell you all the tenderness of my heart," he once told her, "I should do nothing but write to you. . . ."[9] From Amsterdam he wrote her in 1781: "What a fine Affair it would be if We could flit across the Atlantic as they say Angels do from Planet to Planet. I would dart to Penns Hill and bring you over on my Wings."[10] John's and Abigail's was one of America's great love stories.

Abigail wrote far more often than John did during the separations. John was immersed in his work, both in Congress and when he was in Europe, and sometimes sent her only short and rather uncommunicative notes. He was afraid, he told her, that his letters might fall into the hands of the British and be put to embarrassing use if he said personal things in them. When he was abroad, moreover, the letters he wrote were sometimes lost at sea and the result was that months went by without any news from him. Abigail understood these things, but it didn't help. Sometimes her loneliness became excruciating. In October 1778 she finally wrote to ask whether her husband "could have changed Hearts with some frozen Laplander or made a voyage to a region that has chilld every Drop of your Blood."[11] Why should he conceal his feelings, she wanted to know, simply because there was a danger the British might capture his letters? "The affection I feel for my Friend is of the tenderest kind," she wrote, "matured by years, sanctified by choise and approved by Heaven. Angels can witness to its purity, what care I then for the Ridicule of Britain should this testimony of it fall in their hands?"[12] She had no sooner finished writing her reproach than she received three letters from John all at once. But she decided to send her letter anyway, though she apologized for "harboring an Idea so unjust, to your affection." "Were you not dearer to me than all this universe contains beside," she explained, "I could not have suffered as I have done."[13] In his reply, John told her again that he didn't feel he could write her anything "that one is not willing should go to all the Newspapers of the world," but added: "For Heaven's Sake, my dear dont indulge a Thought that it is possible for me to neglect, or forget all that is dear to me in this world."[14]

While John was away, Abigail managed the household, ran the farm, saw to their four children's education, handled the family's finances, and even expanded the family holdings. At first she sought John's advice on these matters; but he was willing to trust her judgment and before long she was making most of the decisions herself. In time she also became John's informant on political developments at home and eventually his confidante on principles and policies in general. Abigail and John enjoyed discussing men and measures; they both had a wry way of looking at things and they also found they agreed on most matters. Politically the two had been in accord from the outset. Abigail heartily supported John's work for the resistance movement just before the American Revolution and soon after the war began agreed with him in thinking that the goal of the American rebels should be independence, not reconciliation with the British.

The fight for independence sharpened Abigail's critical powers. After the conflict with Britain got under way she became keenly aware of how far short America itself fell from the libertarian goals it professed. Like many other Patriots she became a strong foe of slavery. "I wish most sincerely there was not a Slave in the province," she told John in September 1774. "It allways appeared a most iniquitous Scheme to me—fight ourselfs for what we are daily robbing and plundering from those who have as good a right to freedom as we have."[15] She also began thinking seriously about the status of women in America. She began calling herself "Mrs. Delegate" when her husband became a member of the Continental Congress. "Why," she asked, "should we not assume your titles when we give you up our names?"[16] Her exchange with John on women's rights in March 1776, when Congress began debating the question of independence, has become a classic. "I long to hear," she wrote her husband, "that you have declared an independency. And, by the way, in the new code of laws which I suppose it will be necessary for you to make, I desire you would remember the ladies and be more generous and favorable to them than your ancestors. Do not put unlimited power into the hands of the husbands. Remember, all men would be tyrants if they could. If particular care and attention is not paid to the ladies, we are determined to foment a rebellion, and will not hold ourselves bound by any laws in which we have no voice or representation. That your sex are naturally tyrannical is a truth so thoroughly established as to admit of no dispute. . . ."[17]

John was amazed—and impressed—by what she said about women, but in his response decided on the light touch. Calling her "saucy," he wrote back: "Depend upon it, we know better than to repeal our masculine systems. Although they are in full force, you know they are little more than theory. We dare not exert our power in its full latitude. We are obliged to go fairly and softly, and, in practice, you know we are the subjects. We have only the name of masters, and rather than give this up, which would completely subject us to the despotism of the petticoat, I hope George Washington and all our brave heroes would fight. . . ."[18] Abigail refused to be put off by teasing. "I cannot say that I think you are very generous to the ladies," she replied; "for, whilst you are proclaiming peace and good-will to men, emancipating all nations, you insist upon retaining an absolute power over wives. But you must remember that arbitrary power is, like most other things which are very hard, very liable to be broken; and, notwithstanding all your wise laws and maxims, we have it in our power, not only to free ourselves, but to subdue our masters, and, without violence, throw both your natural and legal authority at our feet. . . ."[19]

Abigail was by no means as militant as her letters in 1776 sounded. She demanded neither votes for women nor the right to hold office. Her demands were mainly legal and educational. She wanted a separate legal existence for married women which would give them property rights and protection against abusive husbands. She also called for equal educational opportunities. Education was, in fact, her major concern. She continually deplored the "trifling narrow contracted Education" of women in America and the way people ridiculed "Female learning." How could America produce "Heroes, Statesmen and Philosophers," she wanted to know, if it didn't also produce "Learned women"?[20] Abigail herself learned her three Rs at home and was soon a voracious reader. But though she became one of the best-read women in the country, she was bothered by her educational deficiencies and lamented the lack of formal schooling.

Abigail never doubted that women were men's intellectual equals. She thought their minds "might with propriety receive the highest possible cultivation."[21] When she was in London in 1787 she enrolled in a series of lectures on natural science and lamented the fact that women were not supposed to have any interest in the subject. But she was no radical like Mary Wollstonecraft and Catherine Macaulay, though she admired their work. Unlike the radicals, she believed that women found their highest fulfillment within marriage and the family. With a better education, she said repeatedly, a woman would be a better wife and mother and contribute more in the long run to the well-being of the new nation than if she were uninformed. Well-educated women, she insisted, could help their husbands safeguard republican liberty; they could also rear boys qualified for leadership in the young republic and girls who in turn could become the devoted wives and mothers of patriots. "I will never consent to have our Sex considered in an inferiour point of light," she declared. "Let each planet shine in their own orbit, God and nature designed it so. If man is Lord, woman is *Lordess*—that is what I contend for, and if a woman does not hold the Reins of Government, I see no reason for her not judging how they are conducted."[22]

John was willing to let Abigail be "Lordess." He fully shared her views on the role of women in American society. He once told her that "upon examining the biography of illustrious men, you will generally find some female about them, in the relation of mother or wife or sister, to whose instigation a great part of their merit is to be ascribed." He singled out Aspasia, the wife of Pericles, as an example of what he had in mind. "She was a woman of the greatest beauty and the first genius," he told Abigail. "She taught him, it is said, his refined maxims of policy, his lofty imperial eloquence, nay, even composed the speeches on which so great a share of his reputation was founded." Adams said he doubted whether men could accomplish much in this world without "very great women for wives." He never doubted that Abigail was in her own way a great woman.[23]

Abigail took her duties as John's "help-meet" seriously.[24] She was timorous about appearing at the Court of St. James's when her husband became minister to Britain, but succeeded in bringing it off nicely. When Adams became George Washington's Vice President in 1789, she was also conscientious about her social obligations as the first wife of a Vice President; she dutifully returned formal calls, scheduled weekly receptions, open to everyone, for Monday nights, and held formal dinners on Wednesdays for members of Congress. But John thought the Vice-Presidency was the most insignificant office ever invented by the mind of man and Abigail came to feel the same way about it. She soon became bored with her routine and began wondering whether the position of Vice-President's wife was worth all the trouble and expense. When Washington and Adams were elected to a second term in 1792, she decided to remain home in Quincy (as Braintree was now called) managing the family estate. But she lost none of her zest for politics. The letters she exchanged with John at this time were filled with comments on public issues and, as always, the two agreed on most matters. Both favored a strong central government, accepted most of the policies Alexander Hamilton sponsored as Secretary of the Treasury to strengthen the nation's economy, and took an exceedingly dim view of France after the French Revolution. This brought them into a clash with Thomas Jefferson, with whom they had once been close, and with other old friends from the days of the American Revolution. Jefferson thought the Adamses—and the Federalists generally—were going royalist on him; the Adamses thought Jefferson—and the Republicans clustered around him—had become dangerous radicals.

Being the wife of a President turned out to be far more to Abigail's liking than being the wife of a Vice President. Though illness forced her to miss her husband's inauguration as President on March 4, 1797, she soon joined him in Philadelphia and, with some trepidation, stepped into Martha Washington's shoes. Abigail doubted she had the "patience, prudence, discretion" to carry off her new role as deftly as Martha had and she even wrote the latter for advice.[25] But she may well have done better; at least she was a more stimulating presidential wife than Mrs. Washington. Martha's entertainments tended to be dreary, partly because people were tongue-tied around George and partly because his wife tried to keep things low-keyed. Abigail's were lively, mainly because she was an intelligent and witty conversationalist and loved good talk. But as President Adams's wife, Abigail did more than hold levees and dinners for the great and near-great. She also discussed policies and measures with her husband, read over his speeches beforehand, and gave him advice on occasion. And she became a vigorous champion of his policies, seeing to it that material favorable to his administration got into the hands of newspaper editors. Adams certainly needed her help. The Republicans were merciless in their assault on the second President. Not only did they call him a monarchist and a warmonger; they also made vicious *ad hominem* attacks. The opposition press, Abigail told her sister, almost in disbelief, "calls the President old, querilous, Bald, blind, crippled Toothless Adams."[26] Like John, Abigail came to favor restrictions on the freedom of the press, and she heartily supported the Alien and Sedition Acts of 1798.

Some people thought Abigail carried great weight in her husband's administration. Abigail always denied this and when she received letters from office-seekers answered them tactfully but firmly in the negative. She was influential, of course, in the only way she had ever been influential: as a sounding board for her husband and as the one person in whom he felt he could really confide. On only one major issue did they ever disagree: on policy toward France. When relations between the United States and France reached the breaking point after the XYZ Affair in 1797–98, Abigail—and many of the Federalists—favored war and John himself seemed headed in that direction. Then, to the surprise of everyone, including Abigail, Adams suddenly announced the appointment of William Vans Murray to head a peace mission to France. Federalists like Hamilton were outraged. "Oh how they lament Mrs. Adams's absence!" Adams noted with amusement. "She is a good Counsellor! If she had been here Murray would never have been named, nor his Mission instituted."[27] From Boston, Abigail herself learned that "some of the Feds *[sic]* who did not like being taken so by surprise said they wisht the old Woman had been there; they did not believe it would have taken place." Abigail assured John that it didn't "in the least flatter my vanity, to have the public Imagine that I am not equally pacific with my Husband, or that the same Reasons & Motives, which led him to take upon his own shoulders the weight, of a measure, which he knew must excite a Clamour, would not have equally operated upon my mind, if I had been admitted a partner in the Counsel." But she added that she had "not any very sanguine expectation of success."[28] Adams's policy succeeded in keeping the peace and Abigail came to acknowledge its wisdom. John thought it was his greatest achievement as President.

Abigail thought John deserved a second term (so did John) and was tremendously disappointed when Thomas Jefferson nosed him out in the election of 1800. Still, retirement had its compensations. "I have commenced my operations of dairy-woman," she wrote gleefully from Quincy a couple of months after Jefferson's inauguration in March 1801. "Tell Nabby she might see me, at five o'clock in the morning skimming milk! And in July, you will find your father in the fields attending to his haymakers." Still, she continued to keep up with the "busy world" with unflagging interest.[29]

After 1801 there were no more searing separations for "Darby and Joan" (the "Happy Old Couple" of the English ballad), as an opposition newspaper laughingly dubbed the Adamses when John was still President.[30] Abigail enjoyed being mistress of a large household, having her children and grandchildren around her much of the time, and watching John Qunicy continue to move up in his public career. Jefferson's Presidency turned out better than she had expected and within a few years she and John were reconciled to their old Revolutionary friend. In 1808, almost to her own surprise, she supported Jefferson's friend, James Madison, a Republican, for President. John did too. And in 1816 they both supported James Monroe, another Republican, for the office Adams had once held. By this time Abigail and John had long since abandoned the Federalists because of their slavishly pro-British policies and their refusal to support the War of 1812.

Abigail's last years were not all pleasant. She was ill a great deal of the time, sometimes seriously so, and confined to her bedroom for weeks at a time. She was also saddened by the loss of her only daughter, Nabby, in 1814, as well as by the deaths, one after another, of relatives and close friends. But she was never dejected for long. Hers was an essentially cheerful temperament, and her serene religious faith also helped keep her on keel. "I am determined to be very well pleased with the world and wish well to all its inhabitants," she declared toward the end of her life. "Altho in my journey through it, I meet with some who are too selfish, others too ambitious, some uncharitable, others malicious and envious, yet these vices are counterbalanced by opposite virtues . . . and I always thought the laughing philosopher a much wiser man than the sniveling one."[31]

In October 1818 Abigail took sick with typhoid fever, lingered a few days, her mind as clear as ever, and then died a few days before her seventy-fourth birthday. Shortly before her death, John, who lived until 1826, told his son John Quincy that "the affectionate participation and cheering encouragement" of his wife had been "his never-failing support" and that he doubted whether he could have made it without her by his side. His son felt the same way.[32] "My mother was an angel upon earth," he said. "She was a minister of blessing to all human beings within her sphere of action." Her life, he declared, "gave the lie to every libel on her sex that was ever written."[33]

* * * *

Catalogue

When John Adams was courting Abigail Smith he was playfully critical in one of his letters; she blushed too much, he wrote, hung her head "like a Bulrush," sat "with the Leggs across," and walked with "Toes bending inward." Abigail took the criticism in good humor. "I thank you for your Catalogue," she wrote back, "but must confess I was so hardned as to read over most of my Faults with as much pleasure, as an other person would have read their perfections." But she added that she thought a gentleman had no business "to concern himself about the Leggs of a lady." Abigail's parents weren't sure that John, a farmer's son, was good enough for her, a parson's daughter, even though he was a Harvard graduate, for the Smiths had deep roots in Massachusetts history. They wondered, too, about his quick temper. At the wedding on October 25, 1764, Abigail's father, a Congregational minister, took his text from Luke vii.33: "For John came neither eating bread nor drinking wine, and ye say, *He hath a devil.*"[34]

Saw It

About 3:00 a.m., Saturday morning, June 17, 1775, Abigail was awakened by the sound of cannon fire in the distance. She at once roused her seven-year-old son, John Quincy, and took him to the top of Penn's Hill, behind the Adams farm, so he could see the battle being fought between American and British soldiers over Breed's Hill and Bunker's Hill in Charlestown. Some day, she told little Johnny, schoolboys would memorize the date, June 17, and Johnny would be able to tell his children and grandchildren: "Yes, I *saw* it." Johnny never forgot watching what came to be called the Battle of Bunker Hill. In a speech in 1843, recalling the episode, he exclaimed: "Do you wonder that a boy of seven who witnessed this scene would be a patriot?"[35]

Wrong Sweetheart

When a smallpox epidemic hit Massachusetts in the summer of 1776, Abigail Adams took the children to Boston for inoculation and the effects were so severe that everyone in the family, including Abigail, felt terrible for a long time afterward. Her husband, then in Philadelphia, heard about her illness and decided to send a little gift by way of comfort. So he purchased some tea, entrusted it to a man named Garry who was traveling to Boston, and felt happy to think that Abigail would soon have "the poor relief of a dish of good tea." But Mrs. Adams never received it. When Garry returned to Philadelphia, Adams asked: "You delivered the tea?" "Yes," said Garry brightly, "to Mr. Samuel Adams' lady." Adams was thunderstruck.

Meanwhile Abigail wrote to say that "the herbs you mentioned I never received." But she got to taste the tea anyway. "I was upon a visit to Mrs. Samuel Adams about a week after Garry returned," she told her husband, "when she entertained me with a very fine dish of green tea. The scarcity of the article made me ask where she got it. She replied that her *sweetheart* sent it to her by Mr. Garry. I said nothing, but thought my sweetheart might have been equally kind, considering the disease I was visited with, and that It was recommended as a bracer."

Adams was so upset by Garry's blunder that he ordered more tea and sent it on by a more trustworthy messenger; he also urged his wife to tell Samuel Adams's wife what had happened and try to get back what was left of the original gift. The tea was "amazingly dear," he told Abigail; "nothing less than forty shillings, lawful money, a pound." Abigail said she would do what she could; but she reminded her husband that if he had "mentioned a single word" of Garry's mission in his letters, "I should have immediately found out the mistake."[36]

On the Active

On June 20, 1784, Abigail boarded the little trading ship *Active* with her daughter Nabby and headed for Britain to join John in London. She hadn't been aboard long before she recalled her husband's remark some years before that "no being in nature was so disagreeable as a lady at sea." For ten days she and the other passengers, about ten in all, were deathly ill, and it was all she could do to crawl to the deck for a few minutes each day to get some fresh air. When she finally threw off her sea-sickness, she was overwhelmed by the "horrid dirtiness" of the ship and the "slovenliness" of the steward. "I soon exerted my authority," she wrote her sister Mary, "with scrapers, mops, brushes, infusions of vinegar, &c., and in a few hours you would have thought yourself in a different ship." She also went to work on the cook, persuading him to abandon his "higgledy-piggedly" way of handling the food and even doing a bit of cooking herself. She toured the ship with the captain, too, learning the names and locations of all the masts and sails and before long the latter was telling her "he is sure I know well enough how to *steer,* to take a trick at the helm."

In her spare time Mrs. Adams sewed, read, and wrote letters. She also engaged the other passengers in conversation. But one of the passengers, a Mr. Green from Scotland, rubbed her the wrong way. He was haughty, imperious, and, as she put it, "a high prerogative man." Tired, at length, of his aristocratic arrogance, she finally told him that "merit, not title, gave a man preeminence" in the United States, and that she "did not doubt it was a mortifying circumstance to the British nobility to find themselves so often conquered by mechanics and mere husbandmen; but that we esteemed it our glory to draw such characters not only into the field, but into the Senate; and I believed no one would deny that they had shone in both." The other passengers applauded her put-down.

Storms bothered Mrs. Adams; she was amused when the sailors called one storm a mere "breeze" and "prayed, if this was only a breeze, to be delivered from a storm." But when the ship encountered a calm, she found she liked it even less. "I begin to think," she wrote, "that a calm is not desirable in any situation in life. Every object is beautiful in motion; a ship under sail, trees gently agitated with the wind, and a fine woman dancing, are three instances in point. Man was made for action and for bustle too, I believe. I am quite out of conceit with calms." She tried to relax—"Patience, patience, patience," she reminded herself, "is the first, second, and third virtue of a seaman"—but rejoiced when she sighted the white cliffs of Dover. Soon she was going ashore at Deal in a pilot-boat, stepping out, sinking into the sand, and with the other passengers, looking like a "parcel of Naiads, just rising from the sea." Two days later she was in London, eagerly awaiting the arrival of husband John and son John Quincy.[37]

Reunion in London

In London Mrs. Adams checked into the Adelphi Hotel late in July 1784 and awaited to hear news of her husband and son, now in Paris, whom she had not seen in over four years. While waiting, she received callers, mainly Americans who heard she was in town, did some sightseeing with her daughter Nabby, and looked over British women with a critical eye. "They paint here nearly as much as in France, but with more art," she wrote her sister Mary. "The head-dress disfigures them in the eye of an American. I have seen many ladies, but not one elegant one since I came. . . ." Then came a Puritan plea. "O, my country, my country!" she exclaimed, "Preserve, preserve the little purity and simplicity of manners you yet possess. Believe me, they are jewels of inestimable value; the softness, peculiarly characteristic of our sex, which is so pleasing to the gentlemen, is wholly laid aside here for the masculine attire and manners of Amazonians."

One afternoon, when Mrs. Adams was busy writing letters, a servant suddenly came rushing into the room huffing and puffing. "Young Mr. Adams is come!" he announced. "Where, where is he?" cried Mrs. Adams. "In the other house, Madam," said the servant; "he stopped to get his hair dressed." A few minutes later the seventeen-year-old John Quincy walked into the room and Mrs. Adams, hardly believing her eyes, drew back. "O," cried Johnny, "my mamma and my dear sister!" At first Mrs. Adams didn't recognize the tall, slim fellow standing before her; he had been only twelve when she saw him off to Europe with his father back in 1779. But the eyes were the same; and, feeling quite matronly, she was soon having a happy reunion with him.

Johnny brought her a letter from his father, who was answering her latest. "Your letter," Adams told her, "has made me the happiest man upon earth. I am twenty years younger than I was yesterday. It is a cruel mortification to me that I cannot go to meet you in London." Adams suggested she buy clothes in London, urged her to watch her health, promised to show her around Paris when she and Nabby got there, and signed his letter, "with more ardor than ever." During the next few days Johnny purchased a carriage, took Nabby to the theater, and made arrangements for his mother's journey to the Continent. One day, while Mrs. Adams was packing for the trip, the door opened and there was John. He had become impatient, finished his work posthaste, and headed for London. The reunion of John and Abigail was doubtless ecstatic. Mrs. Adams couldn't bring herself to give details. "You will chide me for not recording the event," she wrote her sister a few months later, "but you know my dear Sister, that poets and painters wisely draw a veil over those Scenes which surpass the pen of the one and the pencil of the other; we were indeed a very very happy family once more met together after a Seperation of 4 years." The journey to Paris, though, she acknowledged, was like a second honeymoon.[38]

Barber

In London, Mrs. Adams was appalled by the attention to fashion, but as wife of the American minister she couldn't wholly eschew it herself. She and Nabby shopped for clothes, fretted over expenses, and then decided not to buy anything until they got to Paris, where the fashions might be different. She finally broke down, however, and made plans to have her hair dressed in the style of London ladies. But when she sent a servant to hire a barber for her, she soon learned there were differences between the American and the English language. "The fellow stared," she recalled, "and was loth to ask for what purpose I wanted him." At last he said, "You mean a hair-dresser, Madam, I believe?" "Aye," she said, "I want my Hair dressed." "Why, barbers, Madam, in this country," he told her, "do nothing but shave." She quickly rectified her "bad mistake" by getting a real hair stylist.[39]

Madame Helvétius

A couple of months after the Adamses arrived in Paris, Benjamin Franklin invited them to have dinner with Madame Helvétius, sixty-year-old widow of the noted French philosopher, who was one of his best French friends.

Mrs. Adams was not impressed. Madame Helvétius entered the room "with a careless, jaunty air," she later wrote, and bawled out: "Ah! mon Dieu, where is Franklin? Why did you not tell me there were ladies here?" She looked abominably dirty to Mrs. Adams, and her manners seemed even worse. When Franklin entered the room, Madame Helvétius ran up, caught his hand, and cried: "Helas! Franklin!" Then, to Mrs. Adams's shock, she kissed him on both cheeks and on his forehead too. She went on to monopolize the conversation during dinner and held hands with Franklin throughout, frequently throwing her arms around him as well. "I own I was highly disgusted," Mrs. Adams wrote later, "and never wish for an acquaintance with any ladies of this cast."

The evening got worse for Mrs. Adams. After dinner, Madame Helvétius threw herself on the settee, "where she showed more than her feet." And when her little lap-dog, whom she continually kissed, wet the floor, she calmly wiped the mess up with her chemise. "You see that manners differ exceedingly in different Countries," Mrs. Adams told her niece. "I hope however to find amongst the French Ladies manners more consistent with my ideas of decency, or I shall be a mere recluse."[40]

Delicacy Wounded

Mrs. Adams enjoyed the theater when she was in Paris with her husband, but as a Puritan found it hard to get used to seeing women dancing on the stage in short skirts. "I can never look upon a Woman in such Situations," she said, "without conceiving all that adorns and Beautifies the female Character, delicacy, modesty and difference, as wholly laid aside, and nothing of the Woman but the Sex left." After attending her first opera, she reported that the "dresses and beauty of the performers were enchanting," but "no sooner did the dance commence," she added, "than I felt my delicacy wounded and I was ashamed to be seen to look at them. Girls, clothed in the thinnest silk and gauze, with their petticoats short, springing two feet from the floor, poising themselves in the air, with their feet flying, and as perfectly showing their garters and drawers as though no petticoat had been worn, was a sight altogether new." Still, she admitted, she had to "speak a truth, and say that repeatedly seeing these dances has worn off that disgust . . . and that I see them now with pleasure."[41]

Desdemona

Although the Adamses began their married life owning two slaves, they soon freed them, for they became convinced that slavery was incompatible with the ideals of the American Revolution. Blacks, they insisted, had as much right to freedom as anyone else.

But like most whites Mrs. Adams found it hard to suppress her revulsion against race-mixing. When she was in London in 1785 she saw Mrs. Sarah Siddons, the most famous actress of the age, play Desdemona in Shakespeare's *Othello* and was bothered by the fact that a black took the part of Othello. She "lost much of the play," she confessed, "from the sooty appearance of the Moor. Perhaps it may be early prejudice; but I could not separate the African color from the man, nor prevent that disgust and horror which filled my mind every time I saw him touch the gentle Desdemona; nor did I wonder that Brabantio thought some love potion or some witchcraft had been practised to make his daughter fall in love with what she scarcely dared to look upon."

But back in Massachusetts, Mrs. Adams enrolled one of her two black servants in the local schools, and when a townsman objected, she said: "The Boy is a Freeman as much as any of the Young Men, and merely because his Face is Black, is he to be denied instruction, how is he to be qualified to procure a livelihood? Is this the Christian principle of doing to others as we would have others to do us? . . . I have not thought it any disgrace to my self to take him into my parlour and teach him both to read and write. . . . I hope we shall all go to Heaven together."[42]

The Virgin and the Abbé

In December 1786, John Adams was so busy with his work in London that Abigail decided to take a trip into the English countryside. She was gone about two weeks but exchanged letters with her husband nearly every day. She worried a great deal about his health in the damp, cold weather of London, and Adams finally promised to use extra blankets at night to keep him warm. But if they weren't sufficient, he added, he would take a "Virgin" to bed with him. "Ay a Virgin," he repeated; then he explained what he meant: a stone bottle filled with hot water placed between the sheets to warm the bed and popularly known as a virgin. "An old Man you see may comfort himself with such a Virgin . . .," he wrote his wife, "and not give the least Jealousy even to his Wife." To this Mrs. Adams replied that she missed her "bedfellow" badly, but promised not to resort to an "Abbé." Then she explained. "You recollect in France," she wrote, "that they are so polite to the Ladies as to accommodate them with an *Abbé* when they give the gentlemen a Nun as an escort."[43]

Abigail at Court

In June 1785, Mrs. Adams was presented to the Court on St. James's a few days after her husband had appeared there as American minister to Britain. In preparing for the occasion she told her dressmaker she wanted no "foil or tincil about me" and that the dress should be "elegant, but plain as I could possibly appear, with decency." As she headed for the Court with Nabby she thought she looked "very tasty."

The event itself was a gigantic bore. After being escorted into a large reception room in the palace, where about two hundred people eagerly awaited the royal sovereigns, Mrs. Adams had to wait four hours before King George III and Queen Charlotte, making the rounds, finally got around to her. The King was surprisingly friendly. He asked Mrs. Adams if she had taken a walk that day and she said, "No, Sire," instead of saying she spent the morning preparing for him. "Why," exclaimed the King, "don't you love walking?" Mrs. Adams murmured something about being rather indolent and the King bowed and moved on. The Queen was less friendly. "Mrs. Adams," she said, "have you got into your house? Pray how do you like the situation?" Mrs. Adams thought the Queen, with her dump figure, froglike mouth, and pug nose, singularly unattractive, but she also thought the ladies of the court in general were all "very plain, ill-shaped, and ugly." She took great pleasure later on in presenting several attractive American women at Court, including the dazzling Mrs. Bingham, and was delighted when an English lord told her: "You have one of the finest ladies to present that I ever saw."

When Mrs. Adams got home after her first Court appearance, she was exhausted. "What a fool do I look like," she sighed, "to be thus accutored & stand here for 4 hours together, only for to be spoken to by 'royalty.' " She wished she never had to appear there again, but knew that diplomatic etiquette demanded it. She also knew that the "smile of royalty is bestowed as a mighty boon," but that she couldn't look at it that way. "I consider myself as complimenting the power before which I appear," she told her sister, "as much as I am complimented by being noticed by it." She simply refused to be awed. "I found the Court like the rest of mankind, mere Men & women & not of the most personable kind neither,'" she told Johnny; "I had vanity enough to come away quite self-satisfied. I saw many who were vastly richer drest . . . but I will venture to say that I saw none neater or more elegant."[44]

Swims

In September 1787, just before returning to the United States, the Adamses took a trip throughout western England, visiting cathedrals, villages, castles, and seaside resorts. Mrs. Adams had lived near the ocean all her life but had never gone swimming and she was fascinated by the seaside resorts she visited on the trip. At Southampton she decided to go in the water. She went into one of the beachfront dressing rooms, put on an ankle-length flannel gown, socks, and an oil-cloth cap, and managed to take a dip. Delighted by the "experiment," she wrote her sister Mary saying she wished Americans would start constructing seaside resorts and suggested Boston, Braintree, and Weymouth as likely locations. But after observing the uninhibited way people behaved at English watering places, she suddenly changed her mind. The "rage" for seaside resorts, she concluded, was a "national evil, as it promotes and encourages dissipation, mixes all characters promiscuously, and is the resort of the most unprincipled female characters."[45]

Great Public Service

When Mrs. Adams joined her husband, then minister to England, in 1784, she was a big help to him. Her patience and tact made up for his grumpiness over the fact that Europeans considered him something of a country bumpkin. Working behind the scenes, moreover, she also made a contribution of her own to the development of her country.

During the American Revolution, Holland had lent the United States several million dollars, and after the war Mrs. Adams thought another Dutch loan might help the newly independent nation put its finances on a firm basis. But her husband poohpoohed the idea. He didn't want to approach Holland again, hat in hand, and he thought England would eventually sign a treaty of commerce that would give his country's finances a big boost. But Mrs. Adams was not to be deterred. She wrote Thomas Jefferson, then minister to France, urging him to try his best to convince her husband that a new loan was a necessity. Jefferson thought Mrs. Adams was right, so he went to work on Adams and was so persistent he finally succeeded in persuading him to negotiate the loan. The upshot was that Holland made the loan and probably saved the young American republic from bankruptcy. Later on Adams heard about what his wife had done. "It is all your intrigue which has forced me to this loan," he grumbled. "I suppose you will boast of it as a great public service."[46]

Torrid and Frigid Zones

Once Abigail Adams, who was past fifty, wrote to express her disapproval of the "January and May" marriage of a young woman she knew to an older man. She said it was a union of "the Torrid and the Frigid Zones." Shot back Adams, who was entering his sixties: "But how dare you hint or lisp a Word about Sixty Years of Age? If I were near, I would soon convince you that I am not above forty."[47]

Splendid Misery

A few days after Adams became President, a friend of his wife called her situation as President's wife one of "splendid misery," and Mrs. Adams came soon to think she "was not far from the Truth." Still, she was proud of her husband and for the rest of his life referred to him as "the President."

As wife of the President, Mrs. Adams tried to use whatever influence she had to encourage American women to dress more decorously. When a Philadelphia clergyman attacked the new fashions as indecent she heartily agreed. Most women "wear their Cloaths too scant upon the body and too full upon the Bosom for my fancy," she said. "Not content with the *show which* nature bestows, they borrow from art, and litterally look like Nursing Mothers."[48]

For Ages to Come

The Adamses were the first to live in what eventually came to be called the "White House" after the capital was moved from Philadelphia to Washington in the fall of 1800. Mrs. Adams found the location of the President's new house magnificent but complained about the climate and the unfinished state of the Executive Mansion. Only six rooms were completed, she discovered, the stairs were not up, there were no call bells to summon servants, and the place was so cold and damp she had to keep thirteen fires going to make the place tolerable. "We have not the least fence, yard, or other conveniences without," she wrote her sister, "and the great unfinished audience room I make a drying-room of, to hang clothes in." Still, she was hopeful. "It is a beautiful spot," she acknowledged, "capable of improvement, and the more I view it, the more I am delighted with it." Besides, "this Home is built for ages to come."[49]

Adams and Jefferson

The Adamses had come to know Thomas Jefferson well during and after the American Revolution and esteemed him highly. Mrs. Adams thought Jefferson was "one of the choice ones of the earth" and Jefferson called her "one of the most estimable characters on earth." In June 1787, when Jefferson's younger daughter, Polly, stopped in London en route to Paris to join her father, Mrs. Adams took care of the little eight-year-old for a couple of weeks and formed a deep attachment for her. With the launching of the new American republic in 1789, however, political differences began eroding the Adams-Jefferson friendship, and the vituperative campaign of 1800, in which Jefferson wrested the Presidency from Adams, brought it to an end.

In April 1804 Mrs. Adams decided to write her erstwhile friend. The occasion was the death of Polly. When she heard the news, Mrs. Adams was deeply moved and wrote Jefferson a tender letter expressing her sympathy for him. Touched by her concern, Jefferson wrote back at once, regretting that "circumstances should have arisen which have seemed to draw a line of separation between us." He went on to recall his long friendship with John Adams, insisted he still admired him, and said that only one act of Adams—his appointment of Jefferson's political enemies to the Federal courts, just before leaving office—ever displeased him and that he had long since forgiven him. Abigail took the opportunity to write back to defend her husband's appointments and to remind Jefferson of the slanders the Jeffersonians had heaped on her husband during the 1800 campaign. Jefferson wrote again to recall the vilification Adams's followers had showered on him in 1800 and to denounce the Alien and Sedition Act of 1798 which Adams had sponsored. Mrs. Adams and Jefferson argued, too, about whether Jefferson had been personally responsible for the dismissal of John Quincy Adams from a position to which he had been appointed by a District Judge during John Adams's presidency.

After four letters Mrs. Adams brought the exchange to an end. A month later she turned the correspondence over to her husband. Somewhat astonished, Adams read his wife's four letters and Jefferson's three with care and then put them away after attaching a note: "Quincy, Nov. 19th, 1804. The whole of this correspondence was begun and conducted without my knowledge or suspicion. Last evening and this morning, at the desire of Mrs. Adams, I read the whole. I have no remarks to make upon it, at this time and in this place." A few years after Jefferson left office, however, Dr. Benjamin Rush brought Adams and Jefferson together again and they entered into a correspondence that lasted until the end of their lives. Mrs. Adams would have approved.[50]

Zeal

Not long after John Quincy Adams and his wife Louisa left Boston for St. Petersburg, where the former took up his duties as American minister to Russia, letters began arriving which made it clear that neither liked life in Russia very much. John Quincy found it hard to make ends meet on his small salary, and his wife felt lonely and neglected in St. Petersburg. Abigail Adams knew her son would never ask to be recalled so she decided on her own to help him. In the summer of 1810, without telling anyone about it, she wrote President Madison, pointing out how hard it was for her son to support his family in Russia on an inadequate salary and asking him to give John Quincy permission to return home. In a gracious response, Madison promised to give John Quincy permission to return if he requested it. But John Quincy never did. A few months later, when John Quincy learned how his mother had interceded for him and wrote her about it, Mrs. Adams sent him copies of her correspondence with Madison. She also apologized for her "zeal" for his welfare.[51]

CHAPTER 3

MARTHA JEFFERSON

1749–1782

Thomas Jefferson was a widower. When he became President in 1801 his wife Martha had been dead nineteen years. During his two administrations Dolley Madison, Secretary of State James Madison's vivacious wife, frequently acted as hostess at presidential dinners, and his eldest daughter Martha ("Patsy") also helped out on occasion. But Jefferson insisted on "republican simplicity" from the outset and did away with the formal levees which the first two Presidents and their wives had opened to the public every week. Once, to protest the new austerity, some Washington ladies called at the regular time, encountered the President in his muddy riding boots, and hastily retreated. For the British minister, Jefferson even appeared in bedroom slippers. Things might have been less casual had Martha lived to reside in the President's Mansion.

Martha Jefferson, six years younger than her husband, was by all accounts attractive and agreeable. "A little above middle height," Sarah Randolph, her great-granddaughter, described her, "with a lithe and exquisitely formed figure . . . a model of graceful and queenlike carriage . . . well educated for her day, and a constant reader."[1] An officer on Baron de Riedesel's staff who visited Monticello in January 1779 found her "in all respects a very agreeable, sensible & accomplished Lady"; and the Marquis de Chastellux, who met her in the spring of 1782, called her "a mild and amiable wife."[2] But she remains a shadowy figure. After her death in September 1782, Jefferson destroyed all her correspondence and rarely spoke of her thereafter. But there is every reason to believe they had a good marriage and that Jefferson never entirely recovered from his loss. In his autobiography he says simply that in 1782 he "lost the cherished companion of my life, in whose affections, unabated on both sides, I had lived the last ten years in unchequered happiness."[3]

Martha was a wealthy young widow when Jefferson began courting her in 1770 and she had many suitors. Born Martha Wayles in 1749, she had married Bathurst Skelton, a thriving young attorney, when she only only eighteen, and became a widow at twenty. Her love of music—she played the harpsichord—was one of her attractions for Jefferson, a violin-player, and he ordered an elegant "forté-piano" for her during the courtship (and hired a music teacher for her after they married).[4] One day, the story goes, two of Jefferson's rivals happened to show up at her place at the same time and overheard some music when the servants showed them in. It was Martha at the harpsichord and Jefferson on the violin and they were singing sweetly together. The two men listened for a moment, then shrugged their shoulders resignedly, took their hats, and left.[5] "In every scheme of happiness," Jefferson wrote his future brother-in-law, "she is placed in the foreground of the picture, as the principal figure. Take that away, and there is no picture for me."[6] Jefferson was now almost twenty-nine and Martha twenty-three.

The wedding came on January 1, 1772, at "The Forest," Martha's place, in Charles City County, not far from Williamsburg, and it was followed by several days of festivities. At length the newly-weds boarded a phaeton and started on the one-hundred-mile journey to Monticello. En route they encountered a light snow that turned into a heavy storm as they continued their way. "They were finally obliged to quit the carriage and proceed on horseback," according to Martha Randolph, their eldest daughter, writing many years later. "Having stopped for a short time at Blenheim, the residence of Colonel Carter, where an overseer only resided, they left it at sunset to pursue their way through a mountain tract rather than a road, in which the snow lay from eighteen inches to two feet deep, having eight miles to go before reaching Monticello. They arrived late at night, the fires all out and the servants retired to their own houses for the night. The horrible dreariness of such a house, at the end of such a journey, I have often heard them both relate." Fortunately, the young couple discovered a half-filled bottle of wine on a shelf behind some books and were able to celebrate their arrival at Jefferson's home after all.[7]

Martha, never robust, seems to have been worn out by child-bearing. In ten years with Jefferson she bore him six children and only two of them survived infancy: Martha ("Patsy") and Mary ("Polly"). She never regained her strength after the last child, a girl, was born in May 1782.

There is a story that as she lay dying in September, Jefferson promised never to remarry. He never did. When he was in Paris in 1786 he fell deeply in love with Maria Cosway, the beautiful young wife of a British artist, and wrote her one of the most charming love letters ever written by a President. But the love gradually cooled into friendship and Jefferson, so far as we know, was never in love again. Later on, Jefferson's enemies charged that he took up with Sally Hemings, an attractive young slave girl, and fathered several children by her. But historians today have pretty much disproven the Hemings story (to the disappointment of people who prefer titillation to truth) and have convincing evidence to show that it was Jefferson's nephews, Samuel and Peter Carr, not Jefferson himself, who got involved with Sally.[8]

If Martha had lived to become First Lady, she undoubtedly would have presided over presidential functions with ease and grace. She certainly would never have counseled her husband on politics, though, the way Abigail Adams did hers. Jefferson's views of women were far more restrictive than John Adams's. "The happiness of your life," he told his daughter Patsy when she got married, "depends now on the continuing to please a single person. To this all other objects must be secondary."[9]

Jefferson was interested in education and wrote extensively about it, but, as he acknowledged, "female education has never been a subject of systematic contemplation with me." Still, he gave his daughters Patsy and Polly a great deal of advice when they were growing up. Young women, he once wrote, should for the most part avoid novels and poetry and concentrate on dancing, drawing, French, and music ("where a person has an ear"). But learning to run a household was the mainstay of a woman's education. "The order and economy of a house," he declared, "are as honorable to the mistress as those of the farm to the master, and if either be neglected, ruin follows . . ." Needlework, though, was also highly desirable. "In the country life of America," he told Patsy, "there are many moments when a woman can have recourse to nothing but her needle for employment. In a dull company, and in dull weather, for instance, it is ill-manners to read, it is ill-manners to leave them; no card-playing there is among genteel people—that is abandoned to blackguards. The needle then is a valuable resource. Besides, without knowing how to use it herself, how can the mistress of a family direct the work of her servants?"[10]

But Jefferson wanted his women well dressed, too. To young Patsy he once wrote: "Nothing is so disgusting to our sex as a want of cleanliness or delicacy in yours. I hope therefore the moment you arise from bed, your first work will be to dress yourself in such a stile as that you may be seen by any gentleman without his being able to discover a pin amiss, or any other circumstances of neatness wanting."[11] There were limits, apparently, to republican simplicity when it came to women.

* * * *

Stupor of Mind

Because of his wife's poor health, Jefferson refused to accept a post abroad during the American Revolution and after the war he missed sessions of the Virginia General Assembly because she was "very dangerously ill." In her last illness he was by her side to the end. "As a nurse, no female ever had more tenderness or anxiety," his daughter Patsy wrote some years later. "He nursed my poor mother in turn with Aunt Carr and her own sister—sitting up with her and administering her medicines and drink to the last. For four months that she lingered, he was never out of calling; when not at her bedside, he was writing in a small room which opened immediately at the head of her bed."[12]

One day toward the end Martha opened Laurence Sterne's *Tristram Shandy* and started copying a passage from it: "Time wastes too fast: every letter I trace tells me with what rapidity life follows my pen. The days and hours of it are flying over our heads like clouds of windy day never to return—more every thing presses on—." At this point her hand faltered and Jefferson took the pen and finished what she was writing: "—and every time I kiss thy hand to bid adieu, every absence which follows it, are preludes to that eternal separation which we are shortly to make." Soon after she sank rapidly. "A moment before the closing scene," Patsy recalled, "he was led from the room in a state of insensibility by his sister, Mrs. Carr, who, with great difficulty, got him into his library, where he fainted, and remained so long insensible that they feared he never would revive. The scene that followed I did not witness, but the violence of his emotion, when almost by stealth, I entered his room at night, to this day I dare not trust myself to describe. He kept to his room three weeks, and I was never a moment from his side. He walked almost incessantly night and day, only lying down occasionally, when nature was completely exhausted, on a pallet that had been brought in during his long fainting-fit. My aunts remained constantly with him for some weeks, I do not remember how many. When at last he left his room, he rode out, and from that time he was incessantly on horseback, rambling about the mountains, in the least frequented roads, and just as often through the woods. In those melancholy rambles I was his constant companion, a solitary witness to many a violent burst of grief, the remembrance of which has consecrated particular scenes of that lost home beyond the power of time to obliterate."[13]

Edmund Randolph was struck by the magnitude of Jefferson's grief. "I ever thought him to rank domestic happiness in the first class of the chief goods," he wrote James Madison; "but scarcely supposed that his grief would be so violent as to justify the circulating report of his swooning away whenever he sees his children."[14] Madison thought stories about Jefferson's fainting "altogether incredible." But a few weeks after Martha's death Jefferson himself admitted to the Marquis de Chastellux that for some time he had been in a "stupor of mind" which "rendered me as dead to the world as she was whose loss occasioned it."[15]

CHAPTER 4

DOLLEY MADISON

1768–1849

When Dolley Madison died in 1849 at eighty-one, President Zachary Taylor declared: "She will never be forgotten because she was truly our First Lady for a half-century."[1] Few Americans would have disagreed. She was part-time hostess for the widower Thomas Jefferson for eight years and after that First Lady in her own right during James Madison's two terms as President. Even after leaving the White House in 1817, she continued to preside over lavish dinners at Montpelier in Virginia and receive a stream of visitors from here and abroad. And when Madison died in 1836 she returned to Washington and was again a major center of attraction there until almost the very end. John Tyler's first wife thought Mrs. Madison "added a new dimension to Washington society," and William Dunlap, the painter, called her "the leader of everything fashionable in Washington."[2] In Washington and elsewhere people called her "Queen Dolley," spoke of her *courteoisie de coeur*, praised her White House "Squeezes," and wanted to know "what Mrs. Madison wore" and what she served for dinner.[3] For years she was a kind of *grande dame* in the nation's capital. It was quite an achievement for a woman who began life as a Quaker.

Dolley's Virginian parents, John and Mary Payne, were members of the Society of Friends and reared their daughter (born in Guilford County, North Carolina, May 1768) in the strict discipline of their faith. They christened her Dorothea, but she was soon being called Dolley (sometimes spelled Dolly), possibly because her little brother William called her that. Dolley learned to read, write, and "do sums" from a tutor at the Scotchtown plantation in Virginia where the family lived for a time and at the Quaker school nearby which she attended with her brothers.[4] She also learned to cook, sew, and tend the garden, and, above all, to dress plainly, behave demurely, and shun ostentation like the plague. But Dolley didn't entirely conform. She ran races with her brothers; she also became interested in fine food and fabrics, largely because her maternal grandmother, an Anglican, taught her a great deal about those things. When she was a little girl her Grandmother Coles gave her a golden brooch which she wore proudly, but secretly, under her dress. She was heartbroken when she found it missing one day. She later said her grandmother was the single most important influence on her life.

In 1783, Dolley's father freed his slaves, took the family to Philadelphia, and began manufacturing laundry starch. But the business failed, the Society of Friends disowned him for getting into debt, and he went into seclusion. To support the family, Dolley's mother turned the home into a boarding house, and Dolley became chief cook. But at twenty-one she left home to marry John Todd, a young Quaker lawyer, settle down as a housewife, and begin building a family. Three years later a yellow fever epidemic hit Philadelphia and carried off her husband and one of her two babies. At twenty-five Dolley was a widow with a little boy, Payne, and, because she was extremely attractive, the center of attention whenever she went out shopping. "Really, Dolley," her friend Elizabeth Collins once scolded her, "thou must hide thy face, there are so many staring at thee."[5] Aaron Burr, New York Senator, who boarded with Mrs. Payne, was one of those who stared, but she rejected his advances; she knew he was married. It was Burr, though, who formally introduced James Madison, forty-two-year-old Congressman from Virginia, to her; and in May 1794 the young widow wrote Elizabeth Collins: "Friend—Thou must come to me,—Aaron Burr says that the great little Madison has asked to be brought to see me this evening."[6] Dolley saw Madison that evening, and many other evenings after that, for "little Jemmy" pressed his suit hard. Dolley had doubts at first but in the end she yielded, and the two were married on September 15, 1794. She never regretted her decision.

As Mrs. Madison, Dolley began to blossom socially both in Philadelphia, where Madison attended Congress, and at Montpelier, his estate in Virginia to which they periodically retired. Disowned by the Society of Friends for marrying outside the faith, she began doing things she had only dreamed of before: wearing colorful gowns and elegant shoes, presiding over fancy dinner parties and receptions, attending balls, and even playing cards. "Mrs. Madison," exclaimed President Washington after a party in the spring of 1795, "is the sprightliest partner I've ever had."[7] Abigail Adams observed that "an invitation to dine with Mrs.

Madison is prized by all who are asked to her home."[8] A foreign diplomat once told Dolley she was an excellent hostess because she was brought up to be one. Dolley was amused as she thought of her plain Quaker background. She eventually became an Episcopalian.

In the marriage James Madison flowered too. Once described as looking like a man on his way to a funeral, he began appearing in public a lot after the marriage, with Dolley beside him, obviously taking great pleasure in all the socializing. At assemblies, too, he became a graceful dancer and at dinners a lively conversationalist. When he went out on the stump, moreover, Dolley went with him and sat in the front row to hear him speak. The fact that people liked Dolley was an enormous asset to him politically. When he beat Charles Cotesworth Pinckney for the Presidency in 1808, the latter lamented: "I was beaten by Mr. and Mrs. Madison. I might have had a better chance had I faced Mr. Madison alone."[9] Dolley never really took much interest in politics and rarely tried to influence her husband's policies. In a letter to him during one of their brief separations she spoke of "her want of talents" when it came to political matters and of her "diffidence in expressing those opinions, always imperfectly understood by her sex."[10] But she was proud of her husband's achievements and anxious to back him up in every way she could. She undoubtedly helped him feel more comfortable in public than he had before the marriage; and, by her own easy, friendly, tactful, and sunny manner at official functions, created tremendous goodwill for him among both his political associates and the general public. British writer Harriet Martineau, who met her in the spring of 1835, surely got it right: "She is a strong-minded woman, fully capable of entering into her husband's occupations and cares; and there is little doubt that he owed much to her intellectual companionship, as well as her ability in sustaining the outward dignity of his office."[11] Madison was proud of his "beloved" from the beginning and she became deeply devoted to her "darling little husband."[12]

In 1801 Madison became Thomas Jefferson's Secretary of State and as a Cabinet wife Dolley quickly developed into the "Queen of Washington City."[13] Not only did she preside over diplomatic dinners for her husband; she also acted as official hostess for the widower President a great deal of the time. At her dinners she decided to feature American dishes; and she sought special recipes from friends and acquaintances in every part of the country. Soon people were talking about fine food she served at her dinner parties and were trying to get hold of her recipes. She became a leader of fashion too. She loved to try out new styles in dresses, hats, and shoes, and before long women everywhere were following her example. It got so that if she eschewed wigs, wore a train in public, put on emeralds, acquired a parrot, revealed a cleavage, or emphasized a certain color, other women began doing the same. The "Dolley Madison turban"—silk cloth coiled upward to look like a Turkish headdress—became the fashion for a time in Europe as well as in the United States. But Dolley designed her clothes to please herself, not to titillate the public. "I care not for newness for its own sake," she told her sister Anna. "I take and use only that which is pleasing to me."[14] And she never became a snob. She liked people of all kinds, had a remarkable memory for names and faces, and cheerfully followed President's Jefferson's rule of "republican simplicity" in official entertaining. She soon got into almost as much trouble as Jefferson did with rank-conscious foreigners.

In December 1803, Anthony Merry, the new British minister to the United States, and his imperious wife were invited to dine with President Jefferson. Jefferson had already received Merry, making an official call, in bathrobe and slippers, and the Merrys were determined to put up with no further insults from the easy-going President. They knew that at diplomatic dinners Jefferson's predecessor, John Adams, had always offered his arm to the British Minister's wife when dinner was announced and they expected Jefferson to do the same. But things didn't work out that way. On the evening of the party, when the time came for dinner, Jefferson offered his arm to Dolley Madison, his unofficial hostess, instead, and conducted her to the table. The next day Merry showed up at Madison's State Department office to make a formal protest. Madison listened quietly and said nothing. He was not about to apologize for the President's rule of democratic informality. And a few weeks later, when he and Dolley gave a dinner for the diplomatic corps and the Cabinet, they had no intention of departing from Jefferson's "Canons of Etiquette" themselves. Thus, when it came time for dinner, Madison offered his arm to Secretary of the Treasury Albert Gallatin's wife, while the French Minister escorted Dolley to the table, and Mrs. Merry found herself all alone until her husband rushed over and took her to the table. There was another angry protest to Madison the next day.

Dolley was eager to end the battle over dinner-table precedence. To mollify the Merrys she decided to throw a party in their honor at which her husband would escort Mrs. Merry to the table in all due solemnity. The Merrys accepted her invitation, but on the night of the party neither of them showed up at the appointed time. Finally Merry turned up an hour late; and when Dolley asked about his wife he simply said she was "indisposed." Dolley smiled sweetly and expressed hope for her speedy recovery. She was miffed, though, and so was her husband.[15]

But the Merrys were still not satisfied. They decided on a new offensive. Merry persuaded the Spanish Minister to go to Madison's office with him so they could lay down the law; the President, they said, was to give alternating preference to their wives at future diplomatic dinners and if he refused to do so they would boycott the President's House. Madison of course rejected the ultimatum. But when Dolley, acting for

Jefferson, threw another party for members of the diplomatic corps and their wives, neither Mrs. Merry nor the Spanish Minister's wife showed up. By this time the Madisons were both heartily sick of the "frivolous farce," as Madison called it, being enacted in Washington, and anxious to call it off.[16]

But the battle continued. About this time Napoleon's younger brother Jerome arrived in the United States, met Elizabeth Patterson, a Baltimore heiress, and soon won her as his wife. Since she was the niece of Secretary of the Navy Robert Smith and his brother Samuel, a New York Senator, Jefferson decided to throw a party for the couple. The night of the party, when it came time to dine, he accompanied Elizabeth Bonaparte to the table. The Merrys weren't there, but they chose to interpret Jefferson's action as another insult to Great Britain (which was at war with France). And a few nights later, when the French Minister gave a dinner for the diplomatic corps and offered his arm to Dolley, Mrs. Merry, it was said, stood nearby looking as though she were "suffering an apoplectic stroke."[17]

Mrs. Merry became increasingly obsessed with the necessity of outshining Dolley and taking first place on the Washington social scene. She finally decided her big moment had come when she and her husband received an invitation to attend a fancy ball which Robert and Samuel Smith sponsored in February 1804 for their niece and her husband. It was the biggest event of the social season, and President Jefferson himself was to be there. Mrs. Merry made careful preparations for the party. On that festive evening she showed up in elegant attire, bedecked with jewels, and for a few golden moments seemed to be the center of attention. Then things started going downhill for her. First, Elizabeth Bonaparte put in an appearance, wearing a gown of dampened muslin that clung to her body, and in an instant everyone forget about Mrs. Merry. Then Dolley arrived with her husband. She was wearing a gown of ivory satin with a square-cut neckline and on her head was a fancy turban from which rose two towering ostrich plumes. At the sight of Dolley all the women in the ballroom cried out in admiration and the men started applauding. Without half trying Dolley had upstaged the British Minister's wife again. The *National Intelligencer* called the ball "the most numerous and brilliant which has ever met in this district."[18]

A few days after the Smith ball, Merry called again at Madison's office to complain about his treatment in Washington. Madison told him no one wished to offend him and his wife and he pointed out that if Mrs. Merry had shown up for the informal party he and Dolley held for the Merrys a few weeks before she would have "received the first attention" that evening.[19] He also made it clear that at official functions he would never "deviate from the established course" of the Jefferson administration.[20] Social standards, he went on to say, varied from country to country, and in the United States the rule of *pêle-mêle,* that is social equality, with every guest for himself, reigned, and there was no desire to insult the Merrys in this policy. After Merry left, unappeased, Madison reported the conversation to James Monroe, American Minister in London, and added: "I blush having to put so much trash on paper."[21]

The Merrys absolutely refused to accept the rule of *pêle-mêle.* When Merry was invited to a bachelor's dinner at the President's House a little later, he announced he couldn't attend until he received instructions from his government and he persuaded the Spanish Minister to take the same line. By this time, though, Monroe had reported from London that the British government was ignoring all of Merry's protests about Jeffersonian etiquette, and the Madisons decided to offer the olive branch again. Dolley arranged another party for the Merrys and indicated that it was to be informal. The Merrys agreed to come and everything seemed to go well the evening of the party. When dinner was announced, Madison offered his arm to Mrs. Merry and both he and Dolley hoped this would put an end to all the friction. But Mrs. Merry was unhappy anyway. Dolley had invited a Washington haberdasher and his wife to the party, because they were interesting people, but the snobbish British Minister's wife interpreted this as another deliberate insult. After the party she sneered that Dolley's dinner had been "more like a harvest-home supper than the entertainment of a Secretary of State."[22] When Dolley heard this, she smiled and said that "she thought abundance was preferable to elegance; that circumstances formed customs, and customs formed taste; and as the profusion so repugnant to foreign customs arose from the happy circumstance of the superabandance and prosperity of our country, she did not hesitate to sacrifice the delicacy of European taste for the less elegant, but more liberal fashion of Virginia."[23] It was a nice retort courteous.

In March 1809, Dolley became First Lady (or "Lady Presidentess," as some people put it) and one of the first things she did was to invite some Senators and Congressmen to visit the President's House to see for themselves how run down it was.[24] Congress soon appropriated money for rebuilding and refurnishing it and Dolley worked closely with Benjamin H. Latrobe, the architect, in reconstructing the place. In a few months she was ready to start entertaining; she hired the best chef in Washington, took on a Master of Ceremonies, and expanded the guest list to include writers and artists as well as politicians and diplomats. The receptions, dinners, and "Wednesday drawing rooms" which she began holding quickly became the rage in Washington. But though her entertainments were lavish, she remained kind, friendly, and unassuming throughout and stressed wit, charm, and talent, as far as she could, not birth, wealth, or rank in her choice of guests. She was soon the nation's leading hostess—and, some people thought, America's most popular person—and at parties she was always the star attraction. Henry Clay thought she was "the most charming of ladies it has ever been my good fortune to encounter."[25]

During the War of 1812 Dolley cut down on her social activities. But she didn't suspend them entirely. Some entertaining, she thought, particularly to celebrate victories, was good for the nation's morale. But she probably gave Americans their biggest boost by her courageous behavior when the British invaded Washington. During that crisis she became something of a heroine.

Late in August 1814 news reached the President's House that the British had landed in Maryland and were headed for Washington. When Madison heard about it, he urged Dolley to leave at once for Virginia while he went out to army headquarters at the front to inspect preparations for the defense of the nation's capital. But Dolley insisted on staying in Washington until she was sure he was safe, so he rode off alone.

About 1:30 p.m. on August 24, Dolley sat down to eat. Suddenly she heard the roar of cannons in the distance. She rushed out, had her carriage brought to the door, and then commandeered a large wagon nearby and persuaded several men in the area to help her load it with papers, books, silver, and china from the Executive Mansion. While she was loading the wagon, Matilda Love, a good friend whose husband was at the front, came by to urge her to leave town at once. But Dolley wasn't quite ready to go. She wanted to save some paintings in the President's House first, and she got Matilda to help her carry them out to the wagon. About this time a young lieutenant came riding up, told her the British were approaching and urged her to get away at once. But Dolley had one more painting to rescue: the famous Gilbert Stuart portrait of George Washington. She did not herself climb a ladder and cut it from its frame, as legend later had it. But she did get her servants to mount the ladder and take it down. Years later, when Charles Carroll, son of signer of the Declaration of Independence Charles Carroll of Carrollton, claimed he had saved the Stuart painting, Dolley's friends were able to prove conclusively that it was Dolley, not Carroll, who deserved the credit.

After saving the Stuart portrait, Dolley drove off with Matilda, followed by the lieutenant and the wagon loaded with presidential valuables. Not long after, the British invaded Washington, set fire to the President's House and several other government buildings, and then, as a thunderstorm broke, left the city. That night Dolley sought refuge in the home of some Virginian friends, and the next night she and Madison met briefly in Falls Church, Virginia. Madison was soon off to the front again and he didn't get back to Washington until the night of August 27th. To his surprise, he found Dolley awaiting him there. Instead of staying in Falls Church, as he had urged, she had returned a few hours earlier when she heard the British were gone.

Dolley's return to Washington was a big event. Crowds of people gathered to cheer her arrival, and she smiled and waved at them as she passed and from time to time shook her fist in the direction of British forces. "We shall rebuild Washington City," she told a woman in Falls Church. "The enemy cannot frighten a free people."[26] Later on she visited the buildings the British had destroyed and launched a campaign for rebuilding the capital. Praised for saving valuable papers and paintings in time of danger, she said simply, "Anyone would have done what I did."[27]

In 1817 James Monroe succeeded Madison as President and the Madisons headed for Montpelier and retirement. But Dolley's old friend Elizabeth Lee told her: "Talents such as yours were never intended to remain inactive. . . . As you retire you will carry with you principles and manners not to be put off with the robe of state. . . ."[28] She was right. As mistress of Montpelier, Dolley went on entertaining on a grand scale and continued to impress people with her verve and vivacity. "Nowhere," said the Marquis de Lafayette after a visit in 1824, "have I encountered a lady who is lovelier or more steadfast."[29] As Madison's health declined she began reading to him and even taking dictation. His last message to the nation, "Advice to My Country," prepared shortly before his death in 1836, was in Dolley's handwriting. Dolley decided to return to Washington to live after her husband's death, and her arrival there in June 1837 created a great deal of excitement. There were hundreds of calling cards awaiting her there, and soon after she moved into her house there were formal calls from the Daniel Websters, the Winfield Scotts, John Quincy Adams, and President Martin Van Buren.

Dolley was almost seventy when she took up residence in Washington again, but she seemed as vigorous as ever, and she was just as charming. But she was hard-pressed financially. Her son Payne Todd had long since turned into a playboy and accumulated heavy gambling debts, and to pay them off and keep him out of prison, she was forced to sell the Montpelier estate. To help her out, Congress appropriated money to buy her husband's papers, including the notes he took at the Constitutional Convention of 1787, and later set up a trust fund providing her with a modest income. Dolley was forced to cut her entertaining down to one reception a month but she made each one a special event. "Mrs. Madison," exclaimed Van Buren after one of her receptions, "is the most brilliant hostess this country has ever known."[30] Daniel Webster told her she was the "only permanent power in Washington" and that all the others were transient.[31]

One day in 1846, after James K. Polk had become President, his wife

Sarah dropped by to see Dolley, then almost eighty, and joined the other visitors for a chat. Presently some of the women there asked Mrs. Polk whether she intended to *pay* calls in Washington the way Dolley had as First Lady or whether she intended to be "exclusive" like Mrs. Monroe and only *receive* calls. Mrs. Polk looked at Dolley and asked what she should do. Dolley thought for a moment and then shook her head. Times had changed, she told Mrs. Polk; new states had joined the Union, there were more Senators and Congressmen in town than in her day and the population of Washington had grown enormously. It would be hard, almost impossible, she said, to do things the way she had done them. Mrs. Polk, she concluded, would have to set her own pattern as First Lady.[32]

* * * *

Informal Introduction

One cold winter morning, it is said, Dolley was headed for the marketplace in Philadelphia and suddenly slipped on the ice. But Madison happened to be passing and quickly reached out and caught her in his arms. He then asked for a formal introduction. It's a charming story; unfortunately, it has no basis in fact.[33]

Good Husband

When President Washington and his wife heard rumors about James Madison and Dolley Payne Todd they decided to become matchmakers. One day Dolley was visiting Mrs. Washington and the President joined the ladies and began talking about his friend Madison. Madison, he said, was a man of honor and integrity, and at the same time gentle and considerate. He had a promising future, too, and would probably end up becoming President. He would also get married some day, Washington added, and no doubt his wife would consider herself the most fortunate woman on earth. Dolley stared at the floor during the President's discourse and later on joked about what she called "the President's Private Address to Mistress Todd."

Martha Washington (a distant cousin of Dolley by marriage) also made a "Private Address." She invited Dolley to tea a few days later and after a bit of small talk went straight to the point: "Dolly, is it true that you are engaged to James Madison?" "No," murmured Dolley, "I think not." "If it is so," persisted Mrs. Washington, "do not be ashamed to confess it; rather be proud; he will make thee a good husband, and all the better for being so much older. We both approve of it; the esteem and friendship existing between Mr. Madison and my husband is very great, and we would wish thee to be happy."

Madison's friend Jefferson also approved the match, but so far as we know made no "Private Addresses" to either James or Dolley. In the end Dolley decided to accept Madison's offer of marriage. In a letter to a friend just after making up her mind she wrote that "in the course of this day I give my hand to the man who of all others I most admire." But she misdated the letter and added a postscript: "Evening—Dolley Madison! Alas!" But despite her misgivings the marriage turned out to be a good one for both Dolley and James.[34]

Unfortunate Propensity

"She is still pretty," said Theodosia Burr of Dolley; "but, oh, the unfortunate propensity to snuff-taking!" It was true; Dolley did take out her snuffbox when conversation lagged and she soon had the Washington ladies doing the same. There is a story, probably apochryphal, that she once offered Henry Clay a pinch of snuff and after he accepted "put her hand into her pocket and pulling out a bandanna handkerchief, said, 'Mr. Clay, this is for rough work,' at the same time applying it in the proper place, 'and this,' producing a fine lace handkerchief from another pocket, 'is my polisher.' She suited the action to the words, removing from her nose the remaining grains of snuff."[35]

Oyster House

Late in 1802, the Oyster House, Washington's first public drinking and eating place, opened near the Capitol and because of its reasonable prices was an immediate success. Most of the customers were male and though women were admitted, the majority, it was said, were prostitutes from Baltimore. About three months after the place opened, Mrs. Madison expressed a wish to go there for some seafood. Somewhat startled, Madison sent a State Department clerk around to tell the proprietor that the Secretary of State and his wife planned to dine there the following day. The proprietor at once cleaned up the place, and when Dolley walked in the next day, the place was spotless, the customers, usually raucous, were conversing in low tones, and there were no women there from Baltimore. Dolley was delighted by the place, and after dining she announced that the oysters were delicious and promised to return. When word got out of her visit, the tavern's clientele changed overnight; soon fashionable ladies were dining there with their husbands. For the earthy—and for the Baltimore trollops—two new taverns soon made their appearance in the city.[36]

The Business of Men

"Politics is the business of men," Dolley once told her sister. "I don't care what offices they may hold, or who supports them. I care only about *people*." She loyally supported her husband's policies, bristled in private over attacks by his enemies, but remained largely uninvolved. "You know I am not much of a politician," she once told Madison, but went on to say she was "anxious to hear (as far as you think proper) what is going forward in the Cabinet." She added at once, though, that she knew he wouldn't want her "to be the active partisan that our neighbor [Mrs. L.] is, nor will there be the slightest danger, whilst she is conscious of her want of talents, and the diffidence in expressing those opinions, always imperfectly understood by her sex."

Yet Dolley got involved despite herself. While Jefferson was President, some of his Federalist enemies started spreading the story that she was his mistress. They also said she had borne Madison no children because she was oversexed. Madison was so angry at the charges that he wanted to challenge one particularly vituperative Federalist editor to a duel. But Jefferson persuaded him to ignore the man, and the stories soon died. In 1808, though, when Madison was running for President, his enemies revived the old stories and added some new ones. Not only did they accuse Dolley (and her sister Anna Cutts) of having affairs with foreign diplomats in Washington; they also said Dolley was playing around with various politicians in order to ensure an Electoral College majority that year for her husband. The Madisons ignored the smears, and in the end Madison won handily on election day.

In 1812, when Madison ran for a second term, some people thought Dolley helped bring about his re-election. "She saved the administration of her husband," wrote James G. Blaine many years later, "held him back from the extremes of Jeffersonism and enabled him to escape the terrible dilemma of the War of 1812. But for her DeWitt Clinton would have been chosen President in 1812." Blaine exaggerated Dolley's influence, but her enormous popularity as First Lady was unquestionably a major asset for the Madison administration.[37]

No Rouge

During the War of 1812, Dolley presided over a ball to celebrate the capture of the *Alert* and the *Guerriere* by American naval forces. At the height of festivities Lieutenant Paul Hamilton arrived with news of the capture of another British vessel, the *Macedonian,* and he was ushered into the ballroom with shouts of joy. Then he presented the flag of the captured ship to Dolley to great applause. One woman at the ball wrote afterward that Dolley didn't use rouge. "I am well assured," she said, "I saw her color come and go at the Naval Ball when the *Macedonian*'s flag was presented to her by young Hamilton."[38]

Gallant

In the fall of 1812, General William Henry Harrison defeated a British contingent near Detroit and shortly after his victory he visited Washington to consult with the President. After the conference Madison ordered him to return to his post. But later that evening Dolley announced that he would be at a party she was giving. Madison laughed, "General Harrison," he said, "should be thirty or forty miles on his way west by now." But Dolley told him: "I laid my command on him and he is too gallant a man to disobey me." "We shall soon see whose orders he obeys." That night Harrison was at her party.[39]

Evicted

When the British invaded Washington during the War of 1812 and Dolley fled the President's House, she sought refuge at one point in the house of a former acquaintance. But she was rudely evicted. "Mis' Madison," screamed the woman, "if that's you, come down and go out. Your husband has got mine out fighting and, damn you, you shan't stay in my house; so get out!" So Dolley moved on.[40]

To Tell My Gals About

One day two ladies from the West who were visiting Washington decided they wanted to get a glimpse of the First Lady. They asked an elderly gentleman they met in the street how to find the President's house and he offered to take them there. When they reached the Executive Mansion, the Madisons were having breakfast, but Dolley went out to greet them. She always dressed plainly in the morning and her visitors were surprised to see her in a dark-gray dress and white apron, with a linen handkerchief about her neck. Dolley chatted amiably with her visitors for a few minutes and they were delighted by her cordiality. Just before leaving, one of the ladies said timidly, "P'raps you wouldn't mind if I jest kissed you, to tell my gals about." Dolley kissed both of them.[41]

Outrun Me

Dolley had run races with her brothers as a little girl and at sixty she was still running. Once after the Madisons had retired to Montpelier, she took her sister Anna by the hand and said, "Come, let us run a race. I do not believe you can outrun me. Madison and I often run races here when the weather does not allow us to walk." And she began a brisk run.[42]

Message from Mrs. Madison

In 1844, the House of Representatives appropriated $25,000 for Samuel F. B. Morse's experiment demonstrating for the first time the electric telegraph connecting Baltimore and Washington. The Washington end had been set up in the Capitol and Dolley had been invited to be on hand with other dignitaries to witness the grand opening. First came the famous message from Baltimore, dictated by Annie Ellsworth: "What hath God wrought." Morse then turned to Dolley and asked her for a message. Dolley then dictated the first personal message sent by the Morse telegraph: "Message from Mrs. Madison. She sends her love to Mrs. Wethered." Her words went to a lady in the group at the Baltimore end.[43]

Seventy-two

William Wilson Corcoran, the banker who endowed the Corcoran Gallery of Art in Washington, was Mrs. Madison's creditor after her husband's death and came to know her well enough to talk freely with her. "Mrs. Madison," he once said, "may I ask, how old are you?" "I am seventy-two, Mr. Corcoran," she replied. The following year he repeated his question: "Mrs. Madison, how old are you?" "I am seventy-two, Mr. Corcoran," she repeated. The year after that he tried again. "Mrs. Madison, how old are you?" "I am seventy-two, Mr. Corcoran," she told him. He stopped inquiring.[44]

Bare Feet

One night a fire broke out at Montpelier and Dolley awoke to find smoke swirling around her. One of the servants rushed up to her room, broke down the door, rushed in, and cried: "Mistress, I have come to save you." But Mrs. Madison refused to leave until the two of them had gathered all of her late husband's letters and papers together to take out with them. Afterwards, when the fire had been extinguished, she went back into the house, clad in a black velvet gown and a nightcap, and in her bare feet.[45]

Nothing Worth Caring For

A few days before Dolley's death in July 1849, her niece went to her for sympathy over some little grievance and Dolley, who probably always remained a Quaker at heart, told her: "My dear, do not trouble about it; there is nothing in *this* world worth really caring for. Yes, believe me, I, who have lived so long, repeat to you, there is nothing in this world here below worth caring for. . . ."[46]

CHAPTER 5

ELIZABETH MONROE

1768–1830

When James Monroe became President in 1817, one newspaper said of his wife: "Mrs. Monroe is an elegant, accomplished woman. She possesses a charming mind and dignity of manners which peculiarly fit her for her elevated station."[1] But people who had liked Dolley Madison's style were soon disappointed by Elizabeth Monroe. From the outset she abandoned Dolley's practice of making calls on the wives of diplomats and Congressmen arriving in Washington as newcomers; she also terminated Mrs. Madison's continuous open houses and confined her official entertaining to specified occasions; and she stayed away from dinner parties at the White House, with the result that the wives stayed home too. Even Washington writer Margaret Bayard Smith, who liked her and thought her "charming and beautiful," was struck by the fact that although the Monroes had lived on and off in Washington for years, they remained "perfect strangers" there.[2] Yet Elizabeth Monroe's background and experience would have led one to expect a social regime under her guidance at least as brilliant as that of her predecessors.

Mrs. Monroe was born Elizabeth Kortright, daughter of a prosperous merchant, in New York in 1768, and though little is known about her early years, it is clear that by the time Monroe met her she was tall, stately, handsome, and graceful and, despite her family's Tory sympathies during the American Revolution, had developed the kind of self-confidence and even hauteur that goes with an assured social position.

When she married Monroe, a member of the Congress of the Confederation, in 1786, some of her friends, according to one observer, "twitted her with the amiable reflection that she was expected to have done better."[3] She did all right. Monroe, a young Virginia lawyer at the time, was ambitious and hard-working and had good connections, and his rise after the marriage was rapid: U.S. Senator under the new Constitution, Governor of Virginia, Minister to France and England, and Secretary of State in the Madison administration. By the time he entered the White House, his wife had moved in the highest political and social circles both here and in Europe.

Elizabeth Monroe liked France (where she was known as *la belle Américaine*) better than England, and it was there that she performed the one outstanding deed recorded of her: helping get Madame Lafayette out of prison. She seems to have enjoyed the artistic and literary life of Paris and felt at home in the cosmopolitan circles she encountered there. It was exhilarating, being cheered when she entered the Monroe box at the theater and hearing the orchestra play "Yankee Doodle" in honor of her and her husband. But in Paris she provoked some criticism for not doing all that was expected of a diplomat's wife. And in London both she and her husband encountered contumely and contempt. But that was largely because many Britishers still regarded the United States as a crude upstart in the world of nations.

Mrs. Monroe's familiarity with the Continental society—she spent almost ten years abroad—did not, as Washingtonians expected, lead her to become a social leader while she was in the White House. Her reluctance to shine socially may well have been due to ill health, though her critics charged her with snobbery. The nature of her illness remains obscure, but there is no doubt that for much of the time during Monroe's eight years as President she simply felt too indisposed to entertain the public the way Washingtonians expected her to. But their carping apparently bothered her. In January 1818 she asked Secretary of State John Quincy Adams's wife Louisa to come by to discuss the whole matter with her. To Louisa Adams she explained that Washington had become too crowded a city for her to pay calls on all the dignitaries there and that the best she could do with her strength was to receive visitors at the White House and perhaps have her daughter, Mrs. Eliza Hay, return calls. Mrs. Adams was sympathetic; she told Mrs. Monroe she must develop her own style as the wife of a President. She seems also to have suggested that Mrs. Monroe frankly cite her poor health when explaining why she was cutting down on social activities.[4] For a time the Washington's society women boycotted the Monroes. A drawing room reception on December 18, 1819, according to Esther Singleton, was opened to "a beggarly row of empty chairs."[5] There was sniping too, at Mrs. Hay, who, with Maria, the other Monroe daughter, tried to help their mother out as hostesses. Like her mother, Eliza Hay never did all that social Washington expected of her. She picked and chose the people she was willing to call on.

In the end, though, the Washington ladies were gradually reconciled and came to accept the pattern Mrs. Monroe established in the White House: weekly drawing rooms, open to the public, and fortnightly formal receptions at which she and her two daughters, Eliza and Maria, received guests. "Mrs. Monroe is certainly the Ninon of the day, and looks more beautiful than any woman of her age I ever saw . . . ," wrote Mrs. Edward Livingston approvingly. "She did the honors of the White House with perfect simplicity, nothing disturbed the composure of her manner."[6] Her manner, observed another guest, "is very gracious, and she is a regal-looking lady."[7] But the Monroes insisted on democratic informality in their entertaining, and some guests were startled by the motley crowd turning up at the weekly open houses. Of the Monroe receptions, one newspaper reported disgustedly: "The secretaries, senators, foreign ministers, consuls, auditors, accountants, officers of the navy and army of every grade, farmers, merchants, parsons, priests, lawyers, judges, auctioneers, and nothingarians—all with their wives and some with their gawky offspring—crowd to the President's house every Wednesday evening; some in shoes, most in boots, and many in spurs; some snuffing, others chewing, and many longing for their cigars and whisky punch left at home; some with powdered heads, some frizzled and oiled; some whose heads a comb has never touched, half hid by dirty collars, reaching far above their ears, as stiff as a pasteboard."[8]

By 1825, when the Monroes left the White House, the "good feeling" identified with President Monroe extended to his wife as well. But both were soon forgotten. In retirement at Oak Hill, Monroe's Virginia home, Mrs. Monroe sank deeper into invalidism and finally died, at sixty-three, in September 1830. Her husband passed away less than a year later, at seventy-four, on July 4, 1831. In a eulogy of the Monroes which John Quincy Adams gave some years later he could only think of conventionalities to utter about Elizabeth Monroe. "This lady," he declaimed, "of whose personal attractions and accomplishments it were impossible to speak in terms of exaggeration, was, for a period little short of half a century, the cherished affectionate partner of [Monroe's] life and fortunes. She accompanied him on all his journeyings thro' this world of care, from which, by the dispensation of Providence, she had been removed only a few months before himself. The companion of his youth was the solace of his declining years, and to the close of his life enjoyed the testimonial of his affection, that with the external beauty and elegance of deportment, she united the more previous and endearing qualities which mark the fulfillment of all the social duties, and adorn with grace and fill with enjoyment, the tender relations of domestic life."[9] It is clear that Adams knew as little about Elizabeth Monroe as anyone else in Washington.

* * * *

Happiest Effect

During the French Revolution the Marquis de Lafayette, America's old friend, ended up in a German prison. His wife and two children were also arrested and put in a prison near Paris. When Monroe was in France he heard that Lafayette's wife was going to be sent to the guillotine and decided to intervene and, in his capacity as U.S. Minister to France, give a "public demonstration of the interest which the United States took in her protection." His wife agreed to help. One morning in February 1795, she boarded a carriage her husband had procured and drove to the prison where Madame Lafayette was confined. A crowd gathered when she arrived, curious about the carriage (carriages had been banned in Revolutionary France) and its occupant. When the prison-keeper came out and wanted to know what she was doing there, she announced that she was the wife of James Monroe, American Minister to France, and had come to visit Madame Lafayette. To her surprise the keeper took her into the waiting room, asked her to be seated, and a few minutes later returned with Lafayette's wife. The latter rushed up in tears, threw herself at Elizabeth Monroe's feet, and told her she had been expecting a summons to prepare for her execution. Mrs. Monroe stayed with her a few minutes, then got up to leave and told her, in a voice loud enough for everyone to hear, that she would call again the following morning. It turned out that Madam Lafayette was, in fact, to have been executed that afternoon, and that Mrs. Monroe's visit that day had changed the minds of French officials. "The report of the interview," Monroe wrote later on, "immediately spread through Paris and had the happiest effect." French officials decided to release Madame Lafayette. They remembered how popular Lafayette was in the United States and they were eager to keep on good terms with the young republic. Monroe was proud of his wife's boldness on that occasion.[10]

CHAPTER 6

LOUISA CATHERINE ADAMS

1775–1852

Louisa Adams dreaded the idea of living in the White House. She was afraid it would "put me in a Prison."[1] It turned out to be worse than she had feared. After John Quincy Adams became President in 1825 and she became First Lady, everything seemed to go wrong. Congress indignantly rejected all of Adams's grandiose plans for unifying and developing the nation and, in the mid-term elections of 1826, his enemies won a majority there. For the rest of his term he became in effect a lame duck President and the venomous attacks on him, inside and outside of Congress, week after week, drove him increasingly into lonely and bitter isolation. For Louise, though, it was no better. Basically shy, she disliked the socializing expected of her, especially in the acrimonious atmosphere of the nation's capital, and her relations with her husband, never easy, were at times strained to the breaking point. She came to hate the White House. "There is something in the great unsocial house," she wrote, "which depresses my spirit beyond expression and makes it impossible for me to feel at home or to fancy that I have a home any where."[2]

Making a home for John Quincy Adams was hard for Louisa from the outset. Except for her short temper and stubborn spirit, she had little in common with her husband. He was stern, duty-ridden, dogmatic, and demanding; she, sensitive, outgoing, impulsive, and forgiving. Adams expected his wife to subordinate her wishes to his desires, for that is what many people after the Revolution expected of American women. Louisa conformed most of the time but she rebelled on occasion. She worked hard at being a loyal wife and a devoted mother, but was reluctant to submerge her own individuality completely in that of her husband. In 1825, she began *Record of a Life, or My Story,* and in 1840, *Adventures of a Nobody,* autobiographical accounts filled with wry and rueful regrets. But she was never in fact a mere nullity. At times she showed the kind of strength, courage, and determination that the Adamses admired in their women. And when her husband entered Congress after the White House years and became involved in the anti-slavery movement, she and John Quincy at long last succeeded in developing a great deal of sympathetic understanding for each other. Their last years together were probably their best.

As a young woman Louisa was far from being a nobody. Born in London in February 1775, the second of eight children of Joshua and Catherine Johnson, she had a happy and carefree childhood in which she was, as she later put it, "the first object of attention at home, every fault pardoned, every virtue loved."[3] Her father, a prosperous Maryland merchant who had come to England in 1771, left London for Nantes during the American Revolution and when he returned to England after the war, Louisa, who had attended a convent school in Nantes, was thoroughly French in manners, language, and dress and had to relearn her English. In London, she was surrounded by tutors, servants, and governesses and was encouraged to develop her taste for music and literature. She played the pianoforte and harp, wrote poetry, and danced at children's balls, and, though retiring and reserved, sang in public at her parents' insistence. She also read a lot and, at boarding school, fell under the influence of a Miss Young, who thought women were just as intelligent as men. By the time John Quincy Adams, the young American Minister to the Netherlands, met her, she was a prize catch: bright, charming, witty, attractive, and well-educated. She came, too, from a well-to-do family.

Louisa had her beaux before meeting John Quincy in 1795. Young Adams, too, had been in love before. He had fallen hard for the fourteen-year-old Mary Frazier when he was a young lawyer in Boston in 1788, but his parents, especially his mother, thought he was too young and unsettled to marry and forced him to break off with the girl. It was a wrenching experience for the lad and he seems to have drawn into his shell after that. At any rate his attraction for Louisa was far less intense than it had been for Mary. Louisa herself was drawn to the ambitious and promising young New Englander almost in spite of herself, and, as she once put it, she had to be "coaxed into an affection."[4] One thing that bothered her about Adams was his sloppy dress. Once, when she invited him to a family picnic, she teased him about his appearance and asked him to "dress himself handsomely and look as dashing as possible" for the get-together. John Quincy was irked by the request but, to her surprise and pleasure, when he arrived for the party he was handsomely dressed in an elegant blue suit and wearing a large Napoleon hat. But when she began complimenting him on his looks he lost his temper and told her that the woman he married must "never take the liberty" of telling him what to wear. Louisa then said he had better "choose a Lady who would be more discreet" and threatened to break up with him.[5] The two eventually made up and got married in London in March 1797, when she was twenty-two and he was thirty, but the angry exchange was a presage of things to come.

Shortly after the marriage Louisa's father went broke, and the newlyweds, who had been counting on his help, found themselves pressed for funds much of the time after that. Money problems—especially when John Quincy Adams was on foreign-service assignments abroad and was obliged to do a great deal of entertaining—were important, but they were minor, compared with disagreements over the role Louisa was to play in the marriage and how the children were to be brought up. Louisa had twelve pregnancies and seven miscarriages between her twenty-first and forty-second years, and partly as a result, her health was wretched a great deal of the time. She resented it when her husband was away from home during deliveries; she was also bothered by the harsh demands he made on their sons and by the way he made crucial decisions about their lives and hers without bothering to consult anybody. She once complained that "hanging and marriage were strongly assimilated."[6] JQA was fully aware of the differences of "sentiments, of tastes, and of opinions in regard to the domestic economy, and to the education of children" between him and his wife. There were "frailties of temper in both of us," he acknowledged; "both being quick and irascible, and mine being sometimes harsh."[7] But he still believed marriage was preferable to celibacy, and so did Louisa.

In 1797, JQA became the first U.S. minister to Berlin and the Adamses were to spend the next four years in the Prussian capital. JQA became totally absorbed in his work there and for a long time Louisa felt lonely and neglected. She also experienced the first of her many miscarriages, and the emotional and physical tensions that beset her sent her, as they did many middle-class American women in those days, to the spas for therapy. Things gradually improved; she began making friends, came to be regarded as a belle in Berlin society, and bore her first son, George, in 1801. But she and JQA clashed frequently. One day the Prussian Queen offered her some rouge because she thought she looked too pale, but when JQA told her to refuse the gift Louisa dutifully did as he commanded. When she was getting ready for the Carnival ball in the winter of 1800, however, she had second thoughts. All the other women would be wearing rouge, she knew, and she was afraid that if she didn't, she might "look a fright in the midst of Splendour." So she decided to accept the Queen's offer, and not long afterward, when dressing to attend a performance of *The Marriage of Figaro,* she put some rouge on her cheeks and decided it "relieved the dullness of her homemade dress and made me look quite beautiful." But when JQA saw what she had done, he lost his temper, grabbed a towel, and wiped it off. In December 1800, though, when she was getting ready for an appearance at court, she used some rouge again. This time, when JQA ordered her to take it off, she stoutly refused; at that, he abruptly turned away, ran out of the room and down the stairs to his carriage and went to court without her.[8] Louisa never regretted the defiance. It was necessary for her own self-respect, she felt, to remind her husband from time to time that she was not the conventionally submissive helpmeet that middle-class Americans seemed to admire in their wives in the early nineteenth century.

When the Adamses returned to the United States in 1801, Louisa dreaded meeting her Adams in-laws and decided to spend two months with her own parents in their home near Washington before joining her husband in Quincy. When she finally met Adams's family, she felt ill at ease and entirely out of her element in the austere Boston atmosphere. The Adamses, for their part, thought of her as "an *English* bride" at first, and wondered whether she could measure up to their own standards.[9] JQA's mother Abigail, in particular, found Louisa lacking in many ways and gave her a great deal of unsolicited advice on how to do the best she could by JQA. But John Adams was warm and kind and understanding from the outset and Louisa grew to love him. It wasn't until years later that she and Abigail came to appreciate each other.

In 1802, JQA was elected to the U.S. Senate and Louisa enjoyed settling in the milder clime in Washington. But when summer came her husband went to Quincy to visit his parents and left Louisa at home to manage things by herself for weeks on end. She resented having the family split up this way. "Oh this separation life," she wailed, "is not worth having on such terms."[10] But a far greater separation was in store for her. In 1809 President James Madison appointed Adams the first U.S. Minister to Russia and the latter decided to leave the two older boys, George and John, with relatives in Quincy and take only Charles, the youngest, with them to St. Petersburg. Louisa, who had not been consulted about the arrangements, was stunned. JQA's father, she was sure, would have supported her wish to keep the family together; but JQA, who left nothing to chance, had carefully planned things so as to keep his father out of the picture. In October 1809, JQA, Louisa, Charles, and Louisa's sister Catherine left for St. Petersburg. Louisa was not to see her two older sons again until 1815.

In St. Petersburg, JQA came into his own as a skillful diplomat. Because of Czar Alexander I's friendliness to the United States, with whom Russia was carrying on a flourishing trade, Adams occupied a preferred position among foreign diplomats despite his inability to match their expenditures when it came to entertaining. The Czar was kind to Louisa and her sister Catherine, too, but her years in St. Petersburg, which began and ended with the death of a child, were mostly sad and lonely. She missed her sons in America, found it hard to make ends meet on JQA's modest salary, and was plagued much of the time with headaches, eye trouble, and pain in her hands. She also found the constant round of balls, parties, dinners, suppers, and masquerades utterly exhausting. She did a lot of reading during her Russian years, but after going through Benjamin Rush's *Treatise on the Diseases of the Mind,* began questioning "the perfect sanity of my mind."[11] In 1814, JQA went to Ghent to help negotiate a treaty ending the War of 1812 with Britain and when it was concluded wrote Louisa telling her to dispose of their property in St. Petersburg and join him in Paris as soon as she could. Louisa was astonished at his request; he had never before given her so much responsibility. But she set to work at once winding up their affairs in Russia and making arrangements for the journey. In the end she impressed JQA and all the Adamses with the way she made the thousand-mile journey with little Charles all the way from St. Petersburg to Paris at a time when Napoleon, having escaped from Elba, was on the rampage again in Europe.

In 1815, JQA became U.S. Minister to England and Louisa's two years there were pleasant ones. The two older boys came to London, the family was united again, and JQA and Louisa shared their music (she played the pianoforte and he the flute) and poetry (they both wrote verse) with each other and with their three sons. There was a meeting of the minds with Abigail Adams, too, when the Adamses returned to the United States in 1817 so that JQA could take up his duties as James Monroe's Secretary of State. Better still: as wife of the Secretary of State Louisa soon found herself, for the first time, playing a significant part in her husband's career. Not only did she help him out with his paper work; she also helped him enormously in his quest for the Presidency.

Adams refused to campaign personally for the presidential nomination in 1824. He expected it to come as a reward for his many years of devoted public service. But there were other presidential aspirants pushing their cause during James Monroe's second term and it was absolutely necessary for Adams's backers to cultivate support for his candidacy among influential politicians and to keep his name before the public. As things turned out, it was Louisa who did the lion's share of the promotional work for JQA. She fully shared her husband's ambition to reach first place and went to work for him with zeal and energy; she visited wives of Congressmen, attended countless dinners, parties, balls, and receptions, and opened her home Tuesday evenings for carefully planned entertainments for people with political clout in

Washington. Socially, Adams himself was awkward and at times even rude; one visitor called him "doggedly and systematically repulsive."[12] Louisa, though, was charming, witty, and gracious (despite her shyness) and created a great deal of good will for her husband. After one of her parties one of the guests called her "the most accomplished American lady I have seen."[13] JQA was fully aware of what she was doing for him and he probably also realized that without her he had little chance of attaining his goal. The climax of Louisa's efforts came on January 8, 1824. On that day she was hostess for the biggest party she and her husband were ever to give: a lavish ball for Andrew Jackson (JQA's main competitor for the Presidency in 1824) to celebrate the ninth anniversary of Jackson's triumphant victory over the British in the Battle of New Orleans. Louisa sent out more than nine hundred invitations and carefully planned and supervised what was regarded as the biggest and most successful party of the decade. "In fact," cried *Harper's Bazar* afterward, "every body who was any body was there. . . ."[14] Louisa was no longer simply a home-maker. She had become in effect her husband's campaign manager and was working side by side with him for the first time in their marriage. It was an exhilarating experience.

Being First Lady, though, turned out to be a great letdown. Adams became President in 1825 but things turned sour for him and Louisa almost at once. For one thing, he had won neither a popular nor an electoral majority in the election of 1824; and when the House of Representatives chose him rather than Jackson (who had won more popular votes than JQA) as President, there was a general outburst of criticism which Adams felt keenly. For another thing, Adams's selection of Henry Clay (who had helped engineer Adams's victory in the House) as Secretary of State added to the fury, and he was charged with having made a "corrupt bargain" with the Kentucky Congressman in order to win the Presidency. Adams's grand plans for the country—using Federal funds to promote transportation, agriculture, education, and science—provided further criticism, and in the end Congress rejected all his proposals with scorn and contempt. Adams's Presidency, in short, was a failure from the outset and Louisa suffered mightily from her husband's disappointment and bitterness at the way things turned out. She and her husband did not mingle much socially during Adams's four years as President; they also kept their own entertaining to a minimum (some Congressmen even boycotted their receptions) and ended by spending long evenings, lonely, isolated, bored, and hostile to each other, in the White House. Louisa sought refuge from her unhappiness by reading, sketching, and cultivating silkworms; she also gorged herself on chocolate candy. She did a lot of writing, too, turning out plays, poems, and prose sketches that reflected her misery. And she went regularly to the spas to seek relief from all the ailments enveloping her.

JQA made regular visits to Quincy without her, and when they corresponded he addressed her as "Mrs. Louisa C. Adams" and she called him "The President." It wasn't until the death of their oldest son George, possibly by suicide, in April 1829 that in their sorrow they began to draw back together again.

The Adams's final years together were their best. JQA ran for Congress in 1830, won by a landslide, and began seventeen years of productive work in the lower house, work in which Louisa played an increasingly important part. Adams hadn't been in Congress very long before he, and Louisa with him, was drawn into the burgeoning anti-slavery movement, and, by extension, into the agitation for women's rights as well. Neither Adams nor his wife was an abolitionist at first; they disliked slavery, but stayed aloof from reformers who were demanding immediate emancipation. In 1836, however, the House of Representatives passed a "gag rule" against the reception of anti-slavery petitions and Adams, shocked by the action, protested bitterly and before he knew it found himself lined up with anti-slavery crusaders whom he had previously always avoided. Adams was primarily concerned with safeguarding the constitutional right of petition; but the fact that it was petitions against slavery he was championing made him increasingly sympathetic to the abolitionist cause itself. Louisa's sympathies broadened too; as she helped sort out, list, and summarize the petitions flooding into her husband's office, she too became increasingly abolitionist in her outlook. Adams fought hard for eight years, with every parliamentary device he could muster, to have the "gag rule" rescinded; and when he finally won a majority of votes for repeal in 1844 even his enemies were impressed by his achievement. By that time both he and Louisa had developed friendly relations with abolitionist leaders and revised their thinking about the slavery problem.

In their last years together John Quincy and Louisa did a lot of thinking about the place of women, as well as of blacks, in American society. In going over the letters and papers of John and Abigail Adams in preparation for a family history, JQA was impressed not only by his mother's strictures against slavery, but also by her insistent demands for a more creative role for women in the new republic struggling to be born in the Revolutionary War. Louisa was impressed, too, by Abigail's outspokenness on women's rights, and she particularly liked her mother-in-law's feisty letter to her husband in March 1776 demanding a declaration of independence for women as well as for men by the Continental Congress. Louisa was anxious for Abigail's letters to be given to the American public.

The abolitionist movement also shaped the thinking of John Quincy and Louisa about women's rights. Both of them were struck by the fact that women, formerly barred from public life, started taking an active role in the anti-slavery movement in the 1830s: drawing up petitions, preparing pamphlets and books, and even holding conventions and making speeches in public. They came to know Angelina and Sarah Grimké, the first female abolitionist agents in the United States, and by their association with them broadened their views on women's rights as well as on the slavery question. Adams presented anti-slavery petitions from women to Congress, told an assembly of women in Hingham, Massachusetts, that he considered them his constituents, too, and, when asked by Angelina Grimké "whether women can do anything in the abolition of slavery," assured her: "If it is abolished, *they* must do it."[15] Louisa read Sarah Grimké's *Letters on the Equality of the Sexes and the Condition of Women* with eagerness and excitement, entered into a correspondence with her on the subject, and, though she was never militant about women's rights, became convinced that in time women in America would come to be far more than mere appendages to their husbands. In a letter to her son Charles on February 21, 1838, she exclaimed: "When I see such Women as your Grandmother go through years of exertion, of suffering, and of privation, with all the activity, judgment, skill and fortitude, which any man could display; I cannot believe there is any inferiority in the Sexes, as far as mind and intellect are concerned, and man is aware of the fact. . . ."[16] By then, she thought, JQA was surely aware of that fact.

On February 1, 1848, JQA collapsed at his desk in the House of Representatives and was carried to the Speaker's room where Louisa was rushed to his side. But he never regained consciousness and died two days later at age seventy-seven. "You may conceive the dreadful shock which I sustained when sent for to the Capitol under the impression that he had only fainted," Louisa wrote her sister Harriet, "when I arrived there and found him speechless and dying, and without a moment of returning sense to show that he knew I was near him. . . . Dear Harriet, they tell me that it was the act of the Almighty, but oh, can anything compensate for the agony of this last parting on earth, after fifty years of union, without the pleasure of indulging the feelings which all hold sacred at such moments."[17]

In 1850, Louisa's son Charles Francis Adams took his twelve-year-old son Henry to visit his mother and the boy decided at once there was something exotic about "the Madame," as his grandmother was called in her old age. He "liked her refined figure," Henry Adams wrote many years later; "her gentle voice and manner; her vague effect of not belonging there, but to Washington or to Europe, like her furniture, and writing-desk with little glass doors above and little eighteenth-century volumes in old bindings labelled 'Peregrine Pickle' or 'Tom Jones' or 'Hannah More.' Try as she might the Madame could never be Bostonian, and it was her cross in life, but to the boy it was her charm. Even at that age, he felt drawn to it." Henry Adams was sure that he "did not wholly come from Boston himself" and liked to toy with the idea that there was something of Louisa in him. "The boy," he wrote of himself, "knew nothing of her interior life which had been . . . of severe stress and little pure satisfaction. He never dreamed that from her might come some of those doubts and self-questionings, those rebellions against law and discipline, which marked more than one of her descendants, but he might even then have felt some vague instinctive suspicion that he was to inherit from her the seeds of primal sin—that he was not of pure New England stock, but half exotic, that even as a child of Quincy, he inherited a quarter taint of Maryland blood."[18] Two years later, in May 1852, the Madame died in Washington at seventy-seven, and, in an unusual gesture, Congress adjourned in her honor and President Millard Fillmore and other high Washington officials attended her funeral.

* * * *

Perfect Beauty

On October 10, 1807, the Adamses headed for Washington, arrived by ship in New York, with their maid and two-month-old baby Charles, and Adams went off to get a carriage. He hadn't been gone long when a man ran up, snatched the baby from the maid's arms, and made off with him. Yelling and screaming, Mrs. Adams and the maid ran down the street after him, but when they turned the corner he had disappeared. Frantically they began pounding the doors of the houses, calling for help. Suddenly the man came out of one of the houses, placed the baby in Mrs. Adams's arms, apologized for what he had done, and explained that the child "was such a perfect beauty, he thought he would show him to his wife." Mrs. Adams was hardly mollified; but she was greatly relieved to have her baby back. After that she always kept her youngest son close to her.[19]

Horse's Tail

When John Quincy Adams was Secretary of State, with his eye on the Presidency, his political enemies attacked him so mercilessly that he was in a high rage much of the time. "Put a little wool in your ears," his wife advised him, "and don't read the papers." She tried hard to tease him out of his high dudgeon. Once, when he exploded in wrath because a newspaper said he neglected to wear a waistcoat or cravat and went to church barefooted, she told him that some guests she had entertained recently asked if he really went to church without shoes or stockings. "I replied," she said, "that I had once heard you rode to your office with your head to your Horse's Tail, and that the one fact was likely as the other."[20]

Long Journey

On January 20, 1815, Mrs. Adams received a letter from her husband that filled her with anxiety. The Treaty of Ghent, which he had helped negotiate, was signed, he told her; he had a new assignment in London and wouldn't be returning to St. Petersburg; and he wanted her to pack their things and come to Paris with their little boy Charles. "Conceive the astonishment your letter caused me if you can . . . ," she wrote back. "This is a heavy trial, but I must get through it at all risks, and if you receive me with the conviction that I have done my best, I shall be amply rewarded."

She deserved a reward. She set out for Paris on February 12, with eight-year-old Charles, an elderly French nurse, and two servants, and arrived there forty days later. While she was on her journey, Napoleon landed at Cannes and began the 100 Days Campaign that ended at Waterloo in July. The bitterly cold weather, the periodic breakdown of her carriage, and the hostility of strangers she encountered en route would have been difficult enough; venturing into the war zone, as she was forced to do toward the end, made her task even harder. Yet Louisa Adams, whose grit and granite the Adams family had long doubted, came through marvelously. Even her husband was impressed and encouraged her to write up her adventures afterward.

At first things went well. At Riga, the Governor entertained her royally. After that the difficulties began. It became so bitterly cold that "our provisions were all hard frozen . . . and even the madeira wine had become solid ice." In Courland, the carriage sank deep into the snow and they had to ask people there to dig them out with pick axes and shovels. At Mitau, the capital of Courland, the keeper of the inn where they stopped took her aside, told her about a terrible murder that had occurred the night before on the road they were on and added that Baptiste, one of her servants, was "a villain of the deepest dye." But Mrs. Adams remained undaunted. From "a proud and foolhardy spirit," she later wrote, she insisted on continuing the journey.

Shortly after leaving Mitau, night fell and the postilion announced he was lost. For the next few hours the little party was "jolted over hills, through swamps and holes, and into valleys into which no carriage had surely ever passed before, and my whole heart was filled with unspeakable terrors for the safety of the child." Then the horses broke down and she had to send Baptiste on ahead to find a place to spend the night. At length they crossed the Vistula, with the two men preceding the carriage with poles to be sure the ice would hold, and after a while they were in Prussia. But as they neared Berlin, Baptiste turned surly; there was "something threatening in his look that did not please me," Mrs. Adams recalled, "but I was afraid to notice it." But after an emergency stop to repair broken wheels, they finally made it safely to Berlin.

Berlin, where Mrs. Adams had been stationed with her husband fourteen years before, was "a haven of refuge, a joy, a resuscitation of pleasant recollections of the past," and she spent a week there, resting up, visiting old acquaintances, and having the carriage repaired. But when she and her party resumed their journey through Prussia, they began running into unruly soldiers everywhere, and, since she had learned that "anything that looked like military escaped from insult," she started wearing her little boy's toy soldier cap with its big plume and displaying his sword at the carriage window. Soon they learned that Napoleon had escaped from Elba and was back in France, and the two servants, fearing conscription, deserted, and she had to hire a fourteen-year-old Prussian boy as guide. The next thing she knew, they had driven right into the midst of Napoleon's Imperial Guard, on its way to rejoin the Emperor, and were being threatened with swords and bayonets. "Tear them out of the carriage!" yelled the soldiers, seeing the coach was Russian. "They are Russians! Take them out and kill them!" But Mrs. Adams finally convinced the soldiers she was an American, cried *"Vive l'Empereur!"* with enthusiasm, and ended by being escorted to safety by a general and his staff. "My poor boy," she recalled, "seemed to be absolutely petrified, and sat by my side like a marble statue" through it all. The following morning, Mrs. Adams's party learned that 40,000 of Napoleon's men were at the gate of Paris and that a battle was imminent. "This news startled me very much," she confessed, "but on cool reflection I thought it best to persevere. . . . I was sure that if there was any danger Mr. Adams would have come to meet me."

The last stage of the journey was easier. Because Mrs. Adams was the only private traveler on the road and because her coach had six horses, the rumor (which she encouraged) spread that she was Napoleon's niece, Stephanie, and she received respectful treatment from then on. As her party neared Paris, however, the rear wheels of the carriage became loose and then fell off and there was another emergency pause for repairs. But at eleven that night, March 23, Mrs. Adams and her party finally rolled into the courtyard of the Hotel du Nord where her husband was awaiting her. "My husband was perfectly astonished at my adventures," she recalled, "as everything was quiet in Paris, and he had never realized the consequences of the general panic in any other place." The nurse was taken off to the hospital with "brain fever," and Mrs. Adams went thankfully to bed. "I was carried through my journey and trials by the mercy of a kind Providence," she wrote at the end of her narrative of the journey, "and by the conviction that weakness, either of body or mind, would only render my difficulties greater and make matters worse."[21]

Almost as Glad

Like many—perhaps most—American whites, Louisa Adams for a long time took a condescending attitude toward blacks. When President Jackson held a "grand Negro Ball" to close Washington's social season in 1831, she was repelled by the thought of blacks eating and drinking in the White House and wrote some verses expressing her disgust.

> Says Coffee to Sambo, d'ya go to the ball?
> Yes! I was invited today
> And dey say we to dance in the great dining hall
> Dey be taking de Carpet away.
>
> * * *
>
> And dere's cake and dere's wine and all sorts of tings
> Jist what at the Draw Rooms we see;
> When the company comes and the waiter we brings
> With the punch and the coffee and tea.

Gradually, however, she developed active sympathies for the oppressed blacks. When she heard about a slave family that had been divided and sold, she contributed fifty dollars to a fund raised by some Washingtonians to free them. A few years later she bought the title to a slave woman who worked for her as a cook and set her free. She was "almost as glad," she said, "as if I was buying my own freedom."[22]

CHAPTER 7

RACHEL JACKSON

1767–1828

When Rachel Jackson learned that her husband had beaten John Quincy Adams in the election of 1828, she murmured: "Well, for Mr. Jackson's sake, I am glad; for my own part, I never wished it." For Mr. Jackson's sake she began making preparations for the move to Washington, but her heart wasn't in it. "I assure you," she told people, "I had rather be a doorkeeper in the house of God than to live in that palace at Washington."[1] At sixty-two Rachel was an extremely pious woman. Only her deep and abiding faith in the providential ordering of events seems to have carried her through the venomous presidential campaign of 1828. For Rachel herself was a major target of Jackson's enemies during the 1828 contest. Jackson tried hard to shield her from the slander, but she was aware of what was going on and it seems to have killed her spirit.

Jackson's campaign to wrest the Presidency from Adams began early and so did the attacks on his wife. In February 1827 Thomas D. Arnold, a candidate for Congress from East Tennessee, put out a handbill announcing that Jackson had "spent the prime of his life in gambling, in cock-fighting, in horse-racing . . . and to cap all tore from a husband the wife of his bosom." A vote for Jackson, the handbill declared, meant a vote for a man who thinks that if he takes a fancy to his neighbor's pretty wife he "has nothing to do but to take a pistol in one hand and a horsewhip in another and . . . possess her."[2] Shortly after Arnold's handbill appeared, Charles Hammond, editor of the *Cincinnati*

Gazette, took up the charge. "Gen. Jackson," he declared in his paper, "prevailed upon the wife of Lewis Roberts *[sic]* to desert her husband and live with himself."[3] The *National Journal,* an Adams paper, repeated the allegations. By this time Jackson's managers had had an emergency meeting, appointed a committee to handle the matter, and released the committee's lengthy statement defending the Jacksons for publication in Jacksonian papers throughout the country.

But Jackson's foes persisted. Arnold went on to announce that Rachel's first husband, Lewis Robards, once caught his wife and Old Hickory "exchanging most delicious kisses" and that not longer after that the guilty pair had "slept under the same blanket."[4] And in a pamphlet entitled *A View of Gen. Jackson's Domestic Relations,* Hammond gave more details about Rachel's misbehavior and asked the voters: "Ought a convicted adultress and her paramour husband to be placed in the highest offices of this free and Christian land?"[5] At this point Duff Green, editor of the Jacksonian *United States Telegraph,* invented a story about the Adamses: they had had premarital relations. "Let Mrs. Jackson rejoice," he wrote Jackson, "her vindication is complete." But Jackson didn't see it that way. "Female character should never be introduced or touched unless a continuation of attack should be made against Mrs. Jackson, and then only by way of *Just retaliation* on the known GUILTY. . . . I *never war against females* and it is only the base and cowardly that do."[6] Jackson thought that John Quincy Adams should have discouraged his supporters from warring against Jackson's beloved Rachel. He adored his wife and she him; and they were both shattered by the way her name was being dragged through the mud during the 1828 battle. Still, there is no question but that there were irregularities in their relationship back in their earliest days together, and Jackson's enemies were determined to make the most of them.

Rachel was a married woman when Jackson first met her in 1788, but her first marriage was a disaster. Born in 1767, daughter of a Virginian ironmaster who moved to Tennessee when she was only twelve, she had married Lewis Robards, a Kentucky landowner, when she was seventeen and soon regretted it. The two were badly mated. Rachel was lively and fun-loving and liked to socialize; Robards turned out to be petty, mean-spirited, cruel, and pathologically jealous. Whenever he saw his vivacious young wife engaging in friendly banter with another man, he suspected the worst, even though his own mother, with whom the young couple lived in Harrodsburgh, Kentucky, took Rachel's side whenever there was a quarrel. The fact that Robards was straying from the path himself may have added to his distrust of his wife.

One day Robards came across Rachel talking to a mutual acquaintance, Peyton Short, lost his temper, and sent her back to Tennessee, where her mother, the widow Donelson, ran a boardinghouse near Nashville. Not long after Rachel left Kentucky, though, he changed his mind, wrote to say he couldn't live without her, and begged for a reconciliation. Rachel agreed to it without much enthusiasm and he soon joined her in Nashville. About this time Andrew Jackson, an energetic young frontier lawyer, came to board at the widow Donelson's and he and Rachel clicked at once. Robards soon became suspicious and let it be known that he thought Jackson was "too intimate with his wife." When Jackson heard what Robards said, he took him aside and told him if he ever connected his name "in that way" with Rachel again, "he would cut his ears out of his head, and that he was tempted to do it any how."[7] Robards then had a peace warrant sworn out against Jackson. Followed then a comical contretemps. As the guards were taking Jackson to the magistrate, he asked for his butcher knife and when they let him have it he began examining it thoughtfully and looking Robards over carefully. Suddenly Robards took to his heels with Jackson in hot pursuit. And when Jackson returned a few minutes later, the magistrate decided to dismiss the warrant since the complainant had vanished.

Jackson deeply regretted the trouble he was causing Rachel—the two were probably in love by now—and he decided to leave the Donelson boarding house and find another place to live. First, though, he wanted to make it clear to Robards that the latter's suspicions were entirely unfounded. There was another confrontation and another quarrel. This time Robards threatened to whip Jackson and Jackson promised to "give him gentlemanly satisfaction" if he wanted a fight.[8] Robards ended by letting loose a flood of profanity and vowing he would never have anything to do with Rachel again. The upshot was that Jackson left the widow Donelson's boarding house for quarters elsewhere while Robards returned to Kentucky.

In the fall of 1790 Rachel heard that Robards was coming to Nashville to get her and that he planned to take her back to Kentucky with him, by force if necessary. She at once panicked; she was convinced that "after two fair trials" it would be impossible for her to live with him again.[9] So she decided to flee to Natchez, then in Spanish territory, and seek refuge among relatives and friends who were living there. Colonel Robert Stark, a family friend, agreed to accompany her on the perilous journey south and then Jackson came forward and volunteered to go along to help in case of Indian attacks en route. Jackson's decision to go to Natchez with Rachel of course played right into Robards's hands; it confirmed his worst suspicions. It is difficult to believe—Jackson's enemies certainly didn't—that he was not deeply involved with Rachel by this time. In any case, Rachel's party arrived safely in Natchez, and Jackson returned to Nashville to resume his law work there. Then, according to the report put out by his friends years later, Jackson heard that Robards had sued for and secured a divorce from Rachel, and he hurried at once back to Natchez. There, in the summer of 1791, he and Rachel were joined in holy and happy matrimony. That fall, the newlyweds returned to Nashville, settled down on a farm Jackson had acquired, and began winning the respect and affection of all their neighbors.

Two years later came the terrible news: Robards had not secured a divorce in 1791 after all and the Jacksons had been living in sin ever since the Natchez days. What had happened was this: the Virginia legislature (which still had jurisdiction over parts of Kentucky) had passed an enabling act on December 20, 1790, permitting Robards to bring suit against his wife in a Kentucky court and, after considerable delay, Robards had done so and, on September 27, 1793, received a decree of divorce from a court meeting in Harrodsburgh on the ground that "Rachel Robards, hath, and doth, Still live in adultery with another man" and that "it is therefore considered by the Court that the Marriage between the Plaintiff and the Defendant be disolved *[sic]*."[10] The Jacksons learned about the court's action in December and professed shock and horror. And when Jackson's friend John Overton suggested they remarry, Jackson at first recoiled from the suggestion, for it implied, he said, that they were not already married and that the charge of adultery was valid. In the end, though, he gave way. On January 14, 1794, he secured a wedding license and the following day he and Rachel were married by the justice of the peace in Nashville.

Despite the 1794 wedding, the Jacksons, unfortunately, were not destined to live happily ever after. There were too many discrepancies in the official story about their relationship put out by Jackson's friends when he was running for the Presidency, and the Adamsites harped on them continually during the 1828 presidential campaign. Questions persist today. There is no record of their marriage in Natchez, for one thing; there is evidence, for another, that they went to Natchez together in 1790, not in 1791, thus giving Robards cause to apply to the Virginia General Assembly for an enabling act later that year. That Jackson, an experienced lawyer, neglected to check up on the particulars of the divorce he thought Rachel had received in 1791 strains credulity. It is also hard to believe that he was unaware of the fact that in February and March 1792, the *Kentucky Gazette* carried eight notices summoning Rachel to appear before the court to answer charges of adultery, as required by the Virginia enabling act. It is possible of course that Jackson decided to go to Natchez with Rachel in 1790 in order to provoke Robards into seeking a divorce. It is also possible that Jackson and Rachel entered into a common-law marriage when they were in Natchez. One fact, though, is clear: when they returned to Nashville, their friends and neighbors accepted them as a respectable married couple, and Jackson's own civil and military career moved rapidly upward without any trace of scandal.[11] Years later, when the Jacksons finally came under attack, William B. Lewis, one of Jackson's neighbors, had this to say: "I would ask how it is possible that any man could have been held in such high estimation by a whole community if he had acted as has been alledged. Could any man, so destitute of moral virtue, and even setting at defiance the common dreams of life, no matter what his talents, and acquirements might be, maintain so high a standing? The thing is impossible and the mere supposition of its possibility is a vile slander upon the whole population of this State." Elsewhere Lewis wrote: "The Genl. and Mrs. Jackson both perhaps acted imprudently but no one believed they acted criminally—the whole course of their lives contradicts such an idea."[12]

The circumstances of their union in the early 1790s seems to have sobered the Jacksons. Rachel, once gay and carefree, became devoutly, almost obsessively, religious after Robards passed out of her life, almost as if she were atoning for her behavior as a young woman. And Jackson, somewhat wild, headstrong, and reckless as a young man, became a tender and thoughtful husband who tried to curb his temper for his wife's sake and came to share in part her great concern with religion. It was impossible, to be sure, for him to keep his temper reined in all the time. He had several fistfights and shooting matches with people who he thought had insulted him or his wife; and in 1806 he even killed a man in a duel after the latter had cast aspersions on Rachel's character. Rachel gloried in her husband's deep devotion to her; but she was never completely happy. As Old Hickory rose steadily in public life—as Congressman, Senator, Judge, and Major-General—he had to be away from home a great deal and the continual separations were hard for her to bear. "Do not, my beloved husband," she wrote him during the War of 1812, "let the love of country, fame and honor, make you forget you have [a wife]." "*My Love:* I have . . . your miniature," Jackson wrote back. "I shall wear it near my bosom; but this was useless for without your miniature my recollection never fails me of your likeness." Replied Rachel: "My thoughts are forever on thee. Where'er I go, where'er I turn, my thoughts, my fears, my doubts distress me. Then a little hope revives again and that keeps me alive. Were it not for that, I must sink; I should die in my present situation. But my blessed Redeemer is making intercession with the Father for us to meet again, to restore you to my bosom, where every vein, every pulse beats high for your health, your safety, and all your wishes crowned. . . . Think of me your dearest friend on earth."[13]

During Jackson's absences, Rachel had to run things at the Hermitage (as the Jacksons called the plantation near Nashville) by herself and she soon developed into an efficient household manager. She also gave loving care to one of the twin sons of her brother whom they adopted in 1809 and christened Andrew Jackson, Jr. Jackson was proud of the skill his wife developed in handling plantation business while he was away. He once advised his ward, Andrew J. Hutchins, to look for a wife "who will aid you in your exertions in making a competency and will take care of it when made" and then used his own wife as an example: "Recollect the industry of your dear aunt [Rachel], and with what economy she watched over what I made, and how we waded thro the vast expenses of the mass of company we had, nothing but her care and industry, with good economy could have saved me from ruin. If she had been extravagant the property would have vanished and poverty and want would have been our doom. Think of this before you attempt to select a wife."[14] For Rachel, though, running the "farm," as Jackson called his plantation, attending Sunday services, going to prayer meetings, reading the Bible, and singing hymns were not enough. To complete her happiness she wanted her husband always by her side, and she never concealed her wish that he abandon public life and settle down for good at the Hermitage. Jackson himself had mixed feelings. At times he yearned to retire to private life and sometimes even promised Rachel he intended to do just that. But his ambition to shine and his love for public life were overpowering and in the end he chose to remain a public figure. Rachel was well aware of how much her husband enjoyed operating on the national scene and she never tried to block his ambitions.

By the time Jackson was elected President, Rachel had long since ceased to be the sprightly woman who had caught his fancy more than three decades before. She was now stout, sober-minded, and a bit stodgy. Her health, too, was poor; she had aged rapidly during the scurrilous campaign of 1828 and at times suffered heart palpitations and shortness of breath. "Mrs. Jackson was once a form of rotund and rubiscund beauty, but [is] now very plethoric and obese . . . ," noted Henry Wise after visiting the Hermitage one day. "[She] talked low but quick, with a short and wheezing breath." Still, she continued to be something of a charmer. Wise found her "the very personation of affable kindness and of a welcome as sincere and truthful as it was simple and tender."[15] But the President-elect's wife was no woman of the world like Dolley Madison or Louisa Adams. She had traveled little, read little, knew little about fashion, and remained a provincial country woman to the end. While Jackson made preparations for his inauguration, Rachel stayed in her room reading the Bible, visiting with old friends, and smoking her beloved pipe. Some people wondered whether she would be presentable in Washington. Years later a Nashville woman recalled that it was "a matter of great anxiety how she would appear as mistress of the White House, especially as some of her warm but injudicious friends had selected and prepared an outfit for the occasion more suitable for a young and beautiful bride, than for a homely, withered-looking, old woman."[16]

Rachel wanted to skip the inauguration. She planned to move quietly and unobtrusively to Washington after the festivities were over. But

Jackson's friend John Eaton, Tennessee Senator, insisted she accompany her husband. "The storm has now abated, the angry tempest has ceased to blow," he told her. "A verdict by the American people has been pronounced . . . that for the honor of your husband you cannot . . . look back on the past. . . . No man has ever met such a triumph before. . . . If you shall be absent how great will be the disappointment. Your persecutors then may chuckle, and say that they have driven you from the field of your husband's honors."[17] Rachel, as usual, gave way. She began making half-hearted preparations for the inauguration herself.

Early in December, Rachel went into Nashville to shop for some clothes to take with her to Washington. She soon became tired and stopped to rest in the office of one of her relatives who was a newspaper editor. While waiting for her carriage she picked up a campaign pamphlet prepared by Jackson's friends defending her against Adamsite slanders. She was stupefied; until then she hadn't realized the depth to which her husband's enemies had sunk in the effort to besmirch her good name. When her friends arrived a little later, they found her crouched in the corner weeping hysterically. A couple of weeks later she became seriously ill, rallied briefly, and then died suddenly when Jackson left her bedside to get some rest. At her funeral on December 24, which ten thousand people, it was said, flocked to the Hermitage to attend, Jackson wept openly as the Rev. William Hume eulogized his wife. "I can forgive all who have wronged me," he told friends after the service, "but will have fervently to pray that I may have grace to enable me to forget or forgive any enemy who has ever maligned that blessed one who is now safe from all suffering and sorrow, whom they tried to put to shame for my sake!"[18]

* * * *

Soo-Blime

One day the Jacksons entertained a wedding party at the Hermitage and the conversation turned to religion. Someone mentioned Emmanuel Swedenborg and the next thing they knew Jackson was speaking with great animation about the "soo-blime" views of the Deity held by the Swedish theologian. When the local Presbyterian minister interposed an objection, Jackson, vehemently defending the "soo-blime" Swedenborg, exclaimed, "By God!" Then he looked sheepishly at his wife and quickly corrected himself: "By Jupiter!" Mrs. Jackson then turned to Henry Baldwin, one of the groomsmen, and said quietly: "Mr. Baldwin, dear, you are sleepy!" And though Baldwin denied he was sleepy, she rang for a servant to bring a candle "to light the dear child to bed," got up, and made it clear the argument was over.[19]

Babylon

In 1821, Jackson became Governor of Florida and headed south with his wife and "the two Andrews" (their adopted son Andrew Jackson, Jr., and their nephew Andrew Jackson Donelson). When they reached New Orleans, Mrs. Jackson was shocked by what she saw. "It reminds me," she wrote, "of those words in Revelation: 'Great Babylon is come up before me.' Oh the wickedness, the idolatry of the place! Unspeakable riches and splendor." Pensacola, where they settled soon after, was no better. There wasn't even a Protestant clergyman there; and though Mrs. Jackson and a few others attended prayer meetings regularly, she couldn't get used to the Continental Sabbath observed by the population there. "The Sabbath profanely kept," she wrote; "a great deal of noise and swearing in the streets; shops kept open; trade going on, I think, more than on any other day." Finally she got Jackson to issue a Sabbatarian order, and the following Sunday had the satisfaction of seeing its effect. "What, what has been done in one week!" she exulted. "Great order was observed; the doors kept shut; the gambling houses demolished; fiddling and dancing not heard any more on the Lord's day; cursing not to be heard." Still, she continued unhappy. She felt awkward and ill at ease being the wife of a high official and expected to dress elegantly and mingle with fashionable people in the theater, ballroom, concert hall, and salon. "Oh for Zion!" she cried. "I am not at rest, nor can I be, in this heathen land. . . ." She was overjoyed when her husband resigned his post later that year and she was able to return to the peace and quiet of the Hermitage and her little church there.[20]

Promise

One Sunday the Jacksons were walking to the little church at the Hermitage that Jackson had built for his wife in 1823 and she turned to him and asked him to join the church. "My dear," he said gently, "if I were to do that now, it would be said, all over the country, that I had done it for the sake of political effect. My enemies would all say so. I cannot do it *now,* but I promise you that when once more I am clear of politics I will join the church." He wasn't clear of politics until some years after his wife's death but he eventually joined the Presbyterian church.[21]

CHAPTER 8

HANNAH VAN BUREN

1783–1819

Like Jefferson and Jackson, Martin Van Buren was a widower when he entered the White House in March 1837. His wife Hannah had died almost eighteen years earlier and he never remarried. His daughter-in-law, Sarah Angelica, served as "Mistress of the White House" while he was President.

When Mrs. Van Buren (the first President's wife born a U.S. citizen) died in February 1819, the *Albany Argus* carried an obituary which was probably written by John Chester, minister of the First Presbyterian Church of Albany, New York, to which she belonged. Chester described her as "affectionate, tender and truly estimable" and then got to the main point about her: "Modest and unassuming, possessing the most engaging simplicity of manners, her heart was the residence of every kind affection, and glowed with sympathy for the wants and sufferings of others. Her temper was uncommonly mild and sweet, her bosom filled with benevolence and content—no love of show, no ambitious desires, no pride of ostentation ever disturbed its peace. . . . Humility was her crowning grace. . . . She was an ornament of the Christian faith."[1]

Chester's obituary is the major source of information about Mrs. Van Buren and essentially it tells us that in a day when wives were expected to be modest and unassuming Hanna Van Buren was precisely that. Van Buren doesn't even mention her in his autobiography. We have little more than the bare facts of her life to go on. She came from an old Dutch family, was born in Kinderhook, New York (the little Dutch town on the Hudson where Van Buren was also born), in 1783, attended the village school there with Van Buren, and was married in 1807, when she was twenty-four and Van Buren twenty-five. The following year the young couple (who spoke Dutch at home) moved to Hudson, a commercial center, where Van Buren served as surrogate for the county and practiced law. There was no Dutch Reformed Church in Hudson and in 1815 Jannetje (Van Buren called her by her Dutch name) joined the Presbyterian church there, apparently by "confession." In 1817 the Van Burens moved to Albany, the state capital, where Van Buren served as state's attorney and his wife became a member of the First Presbyterian Church there soon after. The first winter in albany was a severe one and she developed tuberculosis, became an invalid, and died in 1819 at the age of thirty-six. In all she had five sons, one of whom died in infancy.

What was Hannah really like? Contemporary descriptions, like Chester's obituary, emphasize her modesty and kindliness. One man who knew her says "there never was a woman of a purer and kinder heart" and one of her nieces speaks of her "modest, even timid manner—her shrinking from observation, and her loving, gentle disposition."[2] We know only two concrete things about her that make the generalizations meaningful. When John Chester organized a Sunday school in Albany "to teach the unlettered waifs of the street to read," there was considerable opposition among the women in the church, but Hannah warmly supported the project.[3] Shortly before her death, too, she again took thought of the poor. It was the custom then in Albany for pall-bearers to wear scarfs provided by the family of the deceased. But Hannah requested the money for scarfs go instead to the poor. Possibly the inscription on her gravestone says it all: "She was a sincere Christian, a dutiful child, tender mother, affectionate wife."[4] Probably it doesn't. We shall never know.

CHAPTER 9

ANNA HARRISON

1775–1864

When Anna Harrison heard that her husband had won the election of 1840, she signed: "I wish that my husband's friends had left him where he is, happy and contented in retirement."[1] She had been ailing and did not plan to be present at the inauguration on March 4, 1841. Her doctors advised her to wait until May, when the weather was better, before making the trip to Washington and she decided to let her widowed daughter-in-law, Jane Findlay Harrison, preside over White House receptions until she got there. But she never did make it to Washington and she was First Lady for only a month. A few days after the inauguration William Henry Harrison became ill and a month later died of pneumonia. Vice President John Tyler's wife Letitia replaced Anna Harrison as First Lady.

At sixty-eight, Harrison was the oldest man to become President (until Ronald Reagan, sixty-nine, in 1980) and his wife, at sixty-five, was the oldest First Lady. She had been a "remarkably beautiful girl," it was said, when she met and married Harrison in 1795, and she was still a fine-looking woman when her husband became President.[2] One man who visited the Harrisons in North Bend, Ohio, during the 1840 campaign exclaimed: "Mrs. Harrison is one of the handsomest old ladies I ever saw, a perfect beauty, and such a *good* person." Another visitor remarked: "She rules the General, apparently."[3] But there is no reason to think she was boss. Like Rachel Jackson, Anna Harrison wanted her husband to settle down as a gentleman farmer instead of seeking national fame, but he never did for long. He was away from home a great deal, like Andrew Jackson, on military and civil assignments, and Anna, like Rachel, learned to manage the household in his absence. Like Rachel she seems also to have found solace in religion for life's hardships and disappointments. Both Anna and Rachel were devout Presbyterians willing, apparently, to accept what they regarded as God's will without a murmur.

Anna was born in Flatbrook, New Jersey, in 1775, the daughter of John Cleves Symmes, a New Jersey judge. Her mother died not long after her birth and when she was four her father decided to entrust her to the care of her maternal grandparents on Long Island. He put on a British uniform, sneaked through British lines, delivered her safely to his in-laws in Southhold, and then rejoined the Continental Army in which he was serving as a Colonel. He did not see her again until the Revolutionary War was over. Anna's grandparents sent her to Clinton Academy in Easthampton for a time and then enrolled her in a boarding school in New York City run by Mrs. Isabella Graham, fresh from Edinburgh, who ran what was probably the best girls' school in the United States at the time. She was the first President's wife to have a formal education. But Anna's grandmother may have been the major influence; she inculcated thrift, diligence, and piety in the young woman. Years later Anna recalled that "from her earliest childhood, the frivolous amusements of youth had no charms for her."[4]

In 1794, when Anna was nineteen, Judge Symmes, who had remarried, headed for the Northwest Territory, where he had invested heavily in land, and took his daughter with him. There they settled down to frontier life in the tiny settlement of North Bend on the Ohio River, fifteen miles below Cincinnati. A few months later Anna met Captain William Henry Harrison, twenty-two, at the home of her older sister in Lexington, Kentucky, and promptly the two fell in love. Judge Symmes at first approved of the match and then changed his mind after hearing about a quarrel Harrison had when he was an ensign stationed at Fort St. Clair in southwestern Ohio a year or two earlier. But Harrison won the blessing of General Anthony Wayne, his commanding officer, for his suit, and one day in November 1795, when the Judge was away on business, he and Anna were married by a justice of the peace in the Symmes home in North Bend. For weeks afterward the Judge refused to speak to Harrison. Then one night, when they met at a farewell dinner for General Wayne, Symmes said testily: "I see you have married Anna. And how do you expect to support my daughter?" Harrison fingered his scabbard and said confidently: "My sword is my means of support, sir."[5]

The plough helped as well as the sword. Harrison was in and out of the army during the next few years and he also held a variety of civil posts during his forty-six years of marriage to Anna. With her help, too, he developed his farm at North Bend into a thriving enterprise. Anna had ten children (one of whom became the father of Benjamin Harrison, the twenty-third President) and took responsibility for their education as well as for that of many of the children in the neighborhood. She was a busy hostess as well as a conscientious schoolteacher; she welcomed streams of visitors to her home: relatives, travelers, itinerant preachers. She also got into the habit of inviting the whole congregation of her beloved church in North Bend to dinner on Sundays right after the morning service and serving food raised on the Harrison farm. Anna was proud of her husband's achievements: his rise from lieutenant to major-general in the U.S. Army, his creditable performance as a commander during the War of 1812, and his services as Governor of Indiana, Congressman, U.S. Senator, and, briefly, as Minister to Colombia. But she gave only lukewarm support to his quest for the Presidency in 1836 and 1840. The world of society and fashion in Washington held no attractions for her. She wished "Pah," as she called him, would stay home and cultivate his farm.

After Harrison's death Anna remained in North Bend until 1855—when a fire destroyed the home—and then went to live with her only surviving son about five miles away. Though she suffered many illnesses through the years, her mind continued to be alert until the end, and in her last years she followed Civil War battles with keen interest. "I never met a more entertaining person than your grandmother," her doctor told one of the grandchildren. "I could sit for hours and listen to her conversation."[6] Anna's offspring did not inherit her longevity. Only two of her ten children lived into maturity. But death held no fears for a hardy frontier woman like Anna Harrison. As she wrote her son William (in one of the few letters of hers to survive) in 1819 when he was a student in Transylvania College: "I hope my dear, you will always bear upon your mind that you are born to die and [as] we know not how soon death may overtake us, it will be of little consequence if we are rightly prepared for the event."[7] Presumably she was rightly prepared when death took her in February 1864, when she was eighty-nine.

CHAPTER 10

THE TYLER WIVES

Letitia Christian Tyler, 1790–1842
Julia Gardiner Tyler, 1820–1889

On April 6, 1841, right after William Henry Harrison's death, John Tyler, the first Vice President to become President, took his oath of office as Chief Magistrate and his wife Letitia became the "Lady of the White House." But Letitia Tyler was never able to perform the duties expected of a President's wife. She had suffered a stroke in 1838 and much of the time was confined to bed with paralysis after that. After moving into the Executive Mansion she received few visitors, returned no calls, and relied on her eldest son's wife Priscilla to serve as hostess in her stead. It was Priscilla, not Letitia, who presided over the big party held in honor of Washington Irving and Charles Dickens in March 1842. One of the few occasions when Mrs. Tyler left her bedroom was to be present when her third daughter, Elizabeth, was married in the White House in January 1842.

Even if Letitia Tyler had been in good health, she probably wouldn't have made much of a stir as First Lady. She seems to have been an exceedingly quiet presence throughout her thirty years of marriage to Tyler. The wife of a Virginia planter was expected to be modest and demure, but Letitia was more self-effacing than John Tyler wanted her to be. "She was perfectly content," it was said, "to be seen only as a part of the existence of her beloved husband; to entertain her neighbors; to sit gently by her child's cradle, reading, knitting, or sewing."[1] When Tyler was elected to the U.S. Senate he wanted her to join him in Washington, but she demurred; she said "she would feel better satisfied at home with her children."[2] She had seven children, two of whom died in infancy, and she supervised her large household with an eye to economy that her husband appreciated. She "maintained by her active economy the pecuniary independence of her husband under his continued public employment . . . ," according to an old account, "when in all probability, he would otherwise have been compelled to have relinquished the career of ambition in view of his family's necessities and requirements."[3] But she had no desire to shine outside the family; and she stayed quietly in the background when her husband became Senator, Governor of Virginia, Vice President, and then President.

Born in 1790 in Cedar Grove, twenty miles east of Richmond, Letitia came from a prominent Virginia family—her father, Robert Christian, was a wealthy planter—and she brought property and influence to John when she married him in 1813. The engagement was a long one: almost five years. Tyler wrote sonnets for her during the courtship; and he composed at least one love letter more notable for style than substance. "You express some degree of astonishment, my L.," he wrote on December 5, 1812, "at an observation I once made to you, 'that I would not have been willingly wealthy at the time that I addressed you.' Suffer me to repeat it. If I had been wealthy, the idea of your being actuated by prudential considerations in accepting my suit, would have internally tortured me. But I exposed to you frankly and unblushingly my situation in life—my hopes and fears, my prospects and my dependencies—and you nobly responded. To ensure your happiness is now my only object, and whether I float or sink in the stream of fortune, you may be assured of this, that I shall never cease to love you." He added that he was sending a copy of *The Forest of Montabano* for her to read and went on to discuss the main characters in the novel.[4]

Tyler's courtship was proper of course as well as prolonged. Not until three weeks before the wedding, he told friends, had he "ventured to kiss her hand on parting, so perfectly reserved and modest had she been."[5] Tyler was twenty-three and she twenty-two when they got married, but Tyler felt older. As he told a friend a few days before the wedding: "I had really calculated on experiencing a tremor on the near approach of the day; but I believe I am so much of the *old man* already as to feel less dismay at a change of *situation,* than the greatest part of those of my age."[6]

Virginia ladies were expected to be pious as well as chaste in Tyler's day and Letitia was a devout Episcopalian. One of her daughter Priscilla's earliest memories was that her mother "taught me my letters out of the family Bible."[7] Letitia may or may not have read the novel John sent her during the courtship, but later on her reading was exclusively religious in nature. By 1839, according to one observer, she was "always found seated in her large arm-chair, with a small stand by her side, which holds her Bible and prayer book—the only books she ever reads now."[8] One of the few times she seems to have asserted herself was when her husband contemplated sending their eldest daughter, Mary, to a school in Georgetown run by nuns. She put her foot down when it came to a Catholic education and saw to it that Mary went instead to a school in Williamsburg.

Tyler rarely mentioned his wife in letters to relatives and friends, but when he did so it was always approvingly. In March 1830 he told Mary: "I could not hold up to you a better pattern for your imitation than is constantly presented by your dear mother. You never see her course marked with precipitation; but on the contrary, everything is brought before the tribunal of her judgment, and her actions are all founded on prudence. Follow her example, my dear daughter, and you will be . . . a great source of comfort to me."[9] She passed quietly out of his life on September 10, 1842, at fifty-two, and the *National Intelligencer,* hostile to the Tyler administration, had the conventionally favorable things to say about her as "a Wife, a Mother, and a Christian."[10]

John Tyler's second wife, Julia, loved the White House and revelled in her role as White House hostess. "Under her short reign," observed a Virginia gentleman, "society was charmed by the splendor and propriety of affairs at the White House."[11] Julia Tyler found the Executive Mansion a "dirty establishment" when she first arrived there, but she cleaned it up while she was there and left it a sparkling place.[12] She taught the Marine Band to play "Hail to the Chief" whenever the President appeared in public, had the names of guests announced as they arrived at receptions, made calls on the proper people, held lively dinner parties, and mingled with eminent Americans and foreign notables with ease, grace, and wit. Every night she was so worn out by the day's activities that she fell asleep as soon as her head touched the pillow. "And yet," she told her mother, "not to have all the company and the very way I do would disappoint me very much."[13] While she was in the White House, newspapers called her "Lady Presidentess," a New York musician composed "The Julia Waltzes" in her honor, and the Washington correspondent for the *New York Herald* pushed her praises so much that he became in effect her press agent.

President Tyler was proud of Julia. He said he "never saw anyone in his life so fitted to a court."[14] But some observers thought she was too courtly, that she put on airs, and that she behaved at times like a Queen sitting on a throne. According to Mrs. Jessie Benton Frémont, "There was a little laughing at her for driving four horses (finer horses than those of the Russian Minister), and because she received seated—her large armchair on a slightly raised platform in front of the windows opening to the circular piazza looking on the river. Also three feathers in her hair, and a long-trained purple velvet dress were much commented on by the elders who had seen other Presidents' wives take their state more easily."[15] But James Buchanan thought the President was lucky: he had both "a belle and a fortune to crown and to close his Presidential career."[16]

Julia Tyler's social poise came easily. She was born to status. Her father, David Gardiner, was a man of wealth and leisure residing on Gardiners Island, at the eastern end of Long Island, and her mother Catherine also came from an old established family. Julia was born in 1820 on the family island, educated at home until she was sixteen, and then sent to a fashionable finishing school for young ladies in New York City before being introduced to the world of society and fashion at Saratoga Springs, New York. Her father, it was said, "never tired of imparting knowledge to his children, or of impressing upon them the necessity of the most elegant and high-toned accomplishments."[17] When she was still young, she was so bright, lively, attractive, and flirtatious that "venerable judges, grave senators, and dignified governors," one observer noted, sought her hand.[18] And when her father took her to Europe in 1840 on the Grand Tour, "English beaux on meeting her," said the popular writer, Lydia Sigourney, "seemed suddenly to become aware of the value of their lost colonies."[19] Julia was much sought after, too, in Washington, when her family settled there for a time after returning from Europe. "Sparkling and attractive, without affectation," according to writer Elizabeth Ellett, "she had a high and daring spirit" that made her the favorite Northern belle in the nation's capital.[20]

Then she met President Tyler. Official mourning for the death of Letitia, Tyler's first wife, in September 1842, ended early in 1843; and on February 7, when Julia attended a White House dinner with her family and played cards with the President, he fell hard at once. He flirted with her all evening, chased her around as she was about to leave so he could kiss her, and launched an assiduous courtship the following day. Two weeks later he proposed while they were dancing at the Washington's Birthday Ball in Washington and she shook her head and murmured, "No, no, no." But she continued seeing him regularly. His love letters were frequent and flowery—chock-full of "setting suns, stars peeping from behind their veils, the soul, music and memories, . . . raven tresses, brightest roses"—and she enjoyed reading them aloud to her family.[21] Still, she held off. He was over thirty years older than she.

In February 1844 came a tragedy that drew the two closely together. On the 22nd, at Tyler's invitation, Julia, her father, and her sister Margaret joined the President and several other distinguished guests for a pleasure trip down the Potomac on the warship *Princeton.* The guests boarded the *Princeton,* commanded by Captain R. P. Stockton, at Alexandria, proceeded downstream as far as Mount Vernon, and then sat down to a fancy "collation" in the cabin as the ship started back. About 4:30 p.m., at the request of Secretary of the Navy Thomas Gilmer, Captain Stockton agreed to fire off the "Peacemaker" (a big gun stationed in the bow of the ship). There was a terrible explosion and five people, including Julia's father and two members of Tyler's Cabinet, were killed. Julia collapsed when she learned what had happened to her father and Tyler carried her down the gangplank when the ship arrived at Alexandria. Not long after the funeral for her father, the two became engaged and Tyler wrote Julia's mother asking for her approval. In her reply Mrs. Gardiner expressed high regard for the President, but reminded him that her daughter was accustomed to certain luxuries. "After I lost my father," Julia said years later, "I felt differently toward the President. He seemed to fill the place and to be more agreeable in every way than any younger man ever was or could be."[22]

Late in June 1844, the President and his betrothed sneaked up to New York City and were married in the Church of the Ascension, an Episcopal church, had a wedding breakfast at the Gardiner mansion, and then headed back to Washington for a big reception in the White House. The bride, reported *Niles' National Register,* was a "beautiful and accomplished lady." Tyler was the first President to be married while in office, and the newspapers, which had been taken by surprise, made the most of the event. There were many jokes about "annexation," for Tyler had been pressing the Senate to act favorably on a treaty annexing Texas to the United States. "Treaty of Immediate Annexation," reported the *New York Herald* on June 28. "Ratified without the Consent of the Senate."[23] Julia was twenty-four when she married and Tyler fifty-seven. A few weeks before the wedding Tyler's old friend Virginia Congressman Henry A. Wise had argued against the January-June mating. "You are too far advanced in life to be imprudent in a love-scape," Wise had told him. "How imprudent?" Tyler wanted to know. "Easily," said Wise, "you are not only past middle age, but you are President of the United States, and that is a dazzling dignity which may charm a damsel more than the man she marries." "Pooh!" snorted Tyler. "Why, my dear sir, I am just full in my prime!" A few years later Wise encountered Tyler in Virginia seated beside a baby carriage. "Aha!" cried Wise, "it has come to that, has it?" "Yes," said Tyler proudly; "you see how right I was; it was no vain boast when I told you I was in my prime. I have a houseful of goodly babies budding around me. . . ."[24] In the end, Tyler, in two marriages, had more children than any other President: seven with Letitia and seven with Julia.

Whatever doubts Julia had about marrying an older man when Tyler was wooing her were stilled at once when she joined him in the White House. She came to adore him. She admired his writing; when he sent a message to Congress in December of 1844 she praised its style as "peculiar and beautiful, and of the truest simplicity." She was thrilled by his oratory; she called one of his addresses "supernatural." After one of his last speeches as President she told her mother that he had given expression to a "beautiful and poetic eloquence" that "could only be called *inspiration.* His voice was more beautiful than ever. It rose and fell, and trembled again. The effect was irresistible."[25] Tyler's causes—like Texas annexation—became her own and she excoriated his critics. After he signed the joint resolution of Congress taking Texas into the Union, she vowed she would always wear "suspended from my neck the immortal gold pen with which the President signed my annexation bill."[26] She also became as ardent a states-righter as her husband, and, if anything, more impassioned in her defense of the South and critical of the North before the Civil War than Tyler himself. In time she even learned to speak and write like a Southerner. Tyler reciprocated the adoration. "She is all I could wish her to be . . . ," he told one of his daughters, "the most beautiful woman of the age and at the same time the most accomplished."[27] Letitia's children were considerably less enthusiastic.

In November 1844, a *New York Herald* writer predicted that the social season, commencing with the return of Congress in early December, would be the most brilliant Washington had ever seen. He seems to have read the new First Lady's mind. "I intend to do something in the way of entertaining," Mrs. Tyler told her mother, "that shall be the admiration and talk of the Washington world."[28] And so it was. With the help of funds from the Gardiner family, she renovated the White House and purchased some fancy French furniture; she also acquired a fashionable Italian greyhound as a pet, enlarged her wardrobe, laid in a stock of good French wine, and made plans to introduce a new dance, the polka, at White House balls. Then, after grooming her sister Margaret and several of her cousins to be her "maids of honor," she gave her first big public reception at the White House on New Year's Day, 1845. "It was indeed a glorious assemblage," wrote her sister Margaret afterward, "and all acknowledged with *tongues and eyes* that such a court and such a crowd was never before seen within the walls of the White House. . . . After the shaking of hands was over, the President and Julia made two circuits around the East Room followed by her maids of honors, the crowd gaping and pushing to see the show. . . ."[29]

Mrs. Tyler's big valedictory party on February 18, two weeks before her husband left office, was even more dazzling. So jampacked was the White House with high muckamucks from the army, the government, and high society and so glowing the compliments afterward that the President, who had been disavowed by the Whigs for bucking their policies, permitted himself a rare quip: "Yes, they cannot say now that I am *a President without a party.*"[30] Nor was he a President without a partner. His wife was more than Washington's number one social butterfly during her short reign as First Lady. She gave her husband suggestions for patronage, pushed his policies, especially Texas annexation, with Congressmen and newsmen, and went over the numerous appeals for help she received with great care and passed on the ones she approved to her husband and the appropriate Cabinet officers for action. She enjoyed being a woman of consequence and regretted having to relinquish her role when Tyler's term ended.

After leaving the White House in March 1845, the Tylers retired to "Sherwood Forest," the Tyler plantation on the banks of the James River, and the former President, with the help of sixty or seventy slaves, took up the life of a gentleman farmer. The Tylers held numerous teas and dinner parties for friends and neighbors; they also spent part of each year at their summer home in Hampton, paid visits to Julia's mother in New York, and spent summers at fashionable resorts like Saratoga Springs, for, as Tyler put it, it was "but reasonable that Julia should like to look out on the great world once a year."[31] Sometimes of an evening Tyler got out his violin and Julia sang while he played or made a duet out of it with her guitar.

When the Mexican War came in 1846, Mrs. Tyler followed the war news eagerly and chided her brothers in New York for their lukewarmness. "Are you not interested in, and do you never think of the war?" she wanted to know. "It is full of thrilling interest in my opinion, but you do not seem even to think of it." Then she waxed patriotic. "What a glorious country is America! Who can recount such deeds of courage and valor as our countrymen? My opinion of them has never been half justice. I think that almost all are manly spirits. All nearly are capable of being heroes, and a coward constitutes the exception."[32] In 1853, when the Duchess of Sutherland and several other British ladies appealed to their counterparts in the South to take the lead in demanding an end to slavery, Mrs. Tyler spent a week composing a lengthy defense of the South's peculiar institution to send to the *New York Herald* and the *Richmond Inquirer*. She insisted slaveowners were kindly, that the slave "lives sumptously" compared with a British industrial worker, and that the Duchess and her friends had no business interfering in America's domestic concerns. "We are content to leave England in the enjoyment of her peculiar institutions," she declared, "and we must insist upon the right to regulate ours without her aid."[33]

Mrs. Tyler was a feminine "doughface"—a Northern woman with Southern principles. On the eve of the Civil War she was a passionate secessionist. She said the North had a "fiendish purpose" to dominate the South, told her New York mother she was "utterly ashamed of the State in which I was born, and its people," and when the war began declared: "The hand of Providence should assist this holy Southern Cause."[34]

Julia was with Tyler when he attended an abortive Peace Conference in Washington in the spring of 1861, and she also joined him in Richmond where he attended meetings of the Provisional Congress of the Confederacy. In Richmond, Tyler, who had been ailing for some time, died of a stroke at seventy-two in January 1862. Julia never really recovered from the loss of her husband. After the war she settled in Richmond, became a Catholic, and when she ran into financial difficulties applied for and received a pension from Congress. She died of a stroke in July 1889, at sixty-nine, in the same hotel where her husband had died twenty-seven years before. Tyler's first wife had been buried at Cedar Grove, the plantation where she was born, beside her parents. His second wife was buried with him in the Presidents' section of the Hollywood Cemetery in Richmond.

* * * *

Rose of Long Island

Late in 1839, the proprietors of Bogert and Macamly's, a New York dry goods and clothing establishment, distributed a handbill throughout Manhattan containing an endorsement of their wares by young Julia Gardiner. The advertisement, one of the first of its kind, pictured the nineteen-year-old society girl walking in front of the store and carrying a small sign announcing: "I'll purchase at Bogert and Macamly's, No. 86 Ninth Avenue. Their Goods are Beautiful and Astonishingly Cheap." In the picture, Julia was wearing a sunbonnet topped by large ostrich feathers and a heavy fur-hemmed winter coat; and at her side was an older man, also elegantly clad in B & M clothing. The caption: "Rose of Long Island."

The Bogert and Macamly advertisement, the first commercial ever made by a woman with Julia's standing, caused the Gardiners considerable embarrassment. They were even more dismayed a few months later when the *Brooklyn Daily News* made a front-page feature of some verses entitled "Julia—The Rose of Long Island." Julia had done the commercial for fun, without telling her parents, but soon realized her error. "Pa . . . talks of taking me to Europe in October—," she wrote a friend. "I think seriously." Soon she was off on the Grand Tour.[35]

Arduous Duties

Tyler's wedding in June 1844 was small and unpublicized, and the newspapers were taken by surprise when he announced his marriage to Julia Gardiner afterwards. Even John Jones, editor of *The Madisonian,* a pro-Tyler paper in Washington, had carried a routine story that day about the President's temporary departure from Washington to rest up from his "arduous duties" and seek a few days' "repose." After the wedding the *New York Herald* couldn't help teasing Jones. "John don't know what's going on," the paper crowed. "We rather think the President's 'arduous duties' are only beginning. 'Repose' indeed!"[36]

New Secretary of State

At Secretary of the Navy John Y. Mason's dinner party in November 1844, Mrs. Tyler was seated between Secretary of State John C. Calhoun and Attorney-General John Nelson. Both of them were "so exceedingly agreeable," she told her mother afterward, that "I cannot tell which was most so, but I like Mr. Calhoun the best. He actually *repeated verses to me.* We had together a pleasant flirtation." Later, when she told Tyler how the austere Calhoun whispered poetry of "infinite sweetness and taste" into her ear, he was vastly amused. "Well, upon my word," he said, "I must look out for a new Secretary of State, if Calhoun is to stop writing dispatches and go to repeating verses."[37]

Horticulture

Julia Tyler frequently visited Congress to listen to the debates and soon caught the eye of New York Congressman Richard D. Davis, a rather homely gentleman whom she called "an invincible old bachelor of 50." It got so that whenever she appeared, Davis would leave the floor, to the amusement of the other Congressmen, hurry up to the gallery, and begin showering her with compliments. One day she showed up wearing a lavishly flowered hat and Davis headed at once for the gallery. A few minutes later the vote on a minor bill came up and when the teller called out, "Mr. Davis," all eyes turned to the gallery. "Mr. Speaker," Representative James I. Roosevelt announced, "Mr. Davis has gone to the gallery to study horticulture."[38]

Julia's Dream

In January 1862, Tyler went to Richmond for the Virginia secession convention and his wife planned to join him a week later. But the night before she was to leave she had a bad dream. In it, her husband appeared before her, looking deathly pale, holding his collar and tie in his hand, and cried: "Are you awake, darling; come and hold my head." When she awoke she was so upset she decided to leave at once for Richmond and skip the stops she had planned en route. Upon reaching Richmond, she told Tyler about the dream and he was highly amused. But she continued to worry and in the middle of the night awoke with a headache. "Your forehead is so cold," said Tyler, putting his hand on her head; "shan't I send for the doctor? You see your dream is out; it is *your* head that *I* am holding, and not you mine."

The next morning, when Mrs. Tyler awoke, she saw her husband standing before the fire, partly dressed, looking tired. "Your dream is now out," he told her, "for I believe I have had a chill, and I have determined to go down to the breakfast table and take a cup of tea."

She wanted him to lie down while she sent out for tea, but he insisted on going himself. Shortly after he left she heard some noise in the dining room below, and a few minutes later, Tyler walked into the room, looking worn and haggard, holding his tie and collar in his hand. He had collapsed while having breakfast. Two days later he was dead of a stroke. Ever after the Tyler family took Julia's dream as precognitive, not coincidental.[39]

Flag

During the Civil War Mrs. Tyler's heart belonged to Dixie. She broke with members of her family who supported the Union, refused to take the Oath of Allegiance to the Union when she wanted to move to Staten Island, and had to get there by way of Bermuda. When she returned to New York to live in 1864 she continued to support the Confederate cause—bought Confederate bonds, distributed anti-Lincoln pamphlets, sent money and clothes to Confederate soldiers in Union prison camps, and worked for their exchange—until the bitter end.

On April 15, 1865, fifteen hours after Lincoln's assassination, three local toughs, armed with swords and clubs, burst into Julia's home in Castleton Hill and told her to hand over the Rebel flag she had hanging in the house. She denied she had such a flag, but when they saw a flag of sorts hanging over a picture in the parlor, they climbed up on chairs, ripped it down, knocked over some furniture, and left. Two days later the *New York Herald* printed an anonymous letter defending the action. "Secession, open or secret, will not be tolerated here," declared the writer. "You are aware that we are blest with having as a resident among us, Mrs. Tyler, widow of the deceased rebel ex-President John Tyler. She seems to be successful in passing the lines of our army, and of returning at her pleasure, and with her two eldest sons in the rebel army would seem to be a privileged person."[40]

CHAPTER 11

SARAH CHILDRESS POLK

1803–1891

Shortly after James K. Polk's inauguration as President in March 1845, his Vice President, George M. Dallas, said of Mrs. Polk: "She is certainly mistress of herself, and, I suspect, of somebody else also."[1] But Dallas got it wrong. Sarah did not rule her husband. Nor did she want to; she respected him too much. In some ways, however, she was a formidable woman. She was bright, lively, articulate, and well-read, and preferred discussing politics with the gentlemen attending the White House receptions to making small talk (as was expected) with the ladies there. One woman complained that Mrs. Polk never stayed in the sitting room with the ladies but was "always in the parlor with Mr. Polk" and the other gentlemen.[2] She was careful, though, not to overstep the bounds of feminine propriety when expressing her opinions in public. She always prefaced her remarks with the words, "Mr. Polk believes. . . ."[3]

But Sarah believed as Mr. Polk believed. She shared his ambitions as President to expand America's continental domain and did all she could to help him carry out his expansionist program. For years, too, she had been serving in effect as his private secretary and she continued to do so after he became President. She scanned newspapers for him, briefed him on current books, talked over the leading issues of the day with him, and kept him informed of political developments when he was out of town. Sarah was the only person Polk completely trusted when he was President, and he took her fully into his confidence. "None but

Sarah," he once remarked, "knew so intimately my private affairs." He often spoke of her "wisdom."[4]

Sarah Polk was unusually well-educated for a woman in those days. Her father, Joel Childress, a well-to-do merchant and planter in Murfreesboro, Tennessee, where she was born in 1803, saw to it that she received good schooling. First there was a tutor at home, then a school for girls in Nashville, and finally a Female Academy, run by the Moravians in Salem, North Carolina, where she learned something about history, literature, and geography as well as about music and drawing. At the Moravian Institute she became an avid reader and remained one all her life. She also got religion there and developed into a devout, even austere, believer for whom the Presbyterian church became an important part of her life. Sarah met Polk in 1821, shortly after leaving school, when he was serving as chief clerk for the Tennessee House of Representatives. When he proposed, she accepted, it is said, on condition that he run for the state legislature before they got married. He was surprised, expressed some hesitation at first, and then agreed to do so. Things worked out beautifully: he won the election and the two were married on January 1, 1824. When they moved into a two-room log house in Columbia after the wedding, she pronounced it perfect: small enough not to be so burdensome that she couldn't accompany her husband to the state capital. She was twenty at the time and he, twenty-nine, and they were considered a handsome couple. Someone said her eyes looked "as if she had a great deal of spice."[5] Polk later said she probably wouldn't have married him if he had remained a clerk. She was delighted when he ran for Congress in November 1824 and went to Washington with him the following year.

Sarah loved Washington. She enjoyed seeing her husband serve on one important committee after another in the Lower House and then become Speaker. She also liked mingling in Washington society and meeting people in the know. But she refused to go anywhere without her husband. "I wouldn't have a good time without Jim," she told people, "so there's no use going."[6] When Polk ran for Governor of Tennessee in 1839 she worked hard for his election; she mailed out campaign literature, arranged his schedule, and handled his correspondence. In the Governor's Mansion, after his victory, she impressed people with her charm and intelligence.

She was even more impressive as First Lady. Polk encountered much hostility while he was President but his wife received only praise. "Mrs. Polk is a very handsome woman," said an English lady after visiting the White House. "Her hair is very black, and her dark eyes and complexion remind one of the Spanish donnas. She is well-read, has much talent for conversation, and is highly popular."[7] Even Henry Clay, whom Polk had defeated in a bitter election in 1844, liked her. Once, as he left a White House party, he told her that although he had heard a lot of criticism of her husband's administration he had only "heard a general approbation expressed of her administration." Mrs. Polk replied that she was "happy to hear from him that her administration was approved" and then added that "if a political opponent of my husband is to succeed him I have always said I prefer you, Mr. Clay, and in that event I shall be most happy to surrender the White House to you." Clay left, Polk observed, "in an excellent humour."[8]

Sarah Polk's austerity as First Lady, though, was not universally liked. She banned cards and dancing in the White House and served no wine or refreshments at parties there. The night of Polk's inauguration, she went with him to the customary Inauguration Ball, but as soon as the Polks arrived the dancing stopped and it wasn't resumed until they left two hours later. There was never any dancing in the White House. "To dance in these rooms would be undignified," Mrs. Polk once explained, "and it would be respectful neither to the house nor to the office. How indecorous it would seem for dancing to be going on in one apartment, while in another we were conversing with dignitaries of the republic or ministers of the gospel. This unseemly juxtaposition would be likely to occur at any time, were such amusements permitted."[9]

Mrs. Polk was a Sabbatarian; she kept the Sabbath holy. When the Austrian minister called one Sunday morning to present his credentials, she instructed the servants to inform him that the President would not receive him on the Sabbath. Every Sunday morning, as church bells began ringing, she would enter the parlor with her husband's hat, gloves, and cane, smile sweetly, and say, "If we don't hurry, we'll all be late for church." To any visitors who made the mistake of calling on the President on Sunday she would say, "Do come with us." When the word got out, people began avoiding the White House on that day of the week.[10] But pious people heartily approved the First Lady's religious decorum. The *Nashville Union* was delighted by her "dignified and exemplary deportment since her occupancy of the Presidential Mansion," and declared: "As a proof of religion, doubtless Mrs. Polk deeply realized the responsibility of her position. . . . The example of Mrs. Polk can hardly fail of exerting . . . a salutary influence. . . . All will agree that by the exclusion of frivolities and her excellent deportment in other respects, she has conferred additional dignity upon the executive department of her government."[11]

Mrs. Polk took her work as well as her religion seriously. Like her husband, she was a prodigiously hard worker and disliked taking up too much of her time with what she regarded as wasteful activities. Entertaining was a necessity, of course, but if White House parties lasted longer than the Polks thought they should, they tried to make up for lost time by cutting down on their sleep so they could catch up on their work. Polk was one of the hardest working of all our Presidents, and his wife was usually toiling by his side. In the end he probably injured his health by overwork. After leaving office in March 1849 he planned a trip to Europe with his wife, but his health collapsed and he died within a few months at the age of fifty-five. For Mrs. Polk, "life was then a blank."[12] She was only forty-six at the time and was to live another forty years, but she continued to give her husband primacy in her life. She kept his study at Polk's Place in Nashville just as he had left it, had a marble tomb erected on the front lawn, and turned the mansion into a museum which exhibited his books, letters, diary, papers, and personal possessions. One of the many visitors through the years observed that "she loses herself in her admiration of all that Mr. Polk did that was noble."[13] Friends called on her frequently, but she never returned their calls. She stopped going anywhere except to church.

But Sarah Polk had too much energy to be happy dwelling entirely in the past. For a time she managed a plantation as well as a museum. Before leaving the White House her husband had acquired about 1,000 acres of cotton land on the Yalobusha River in Mississippi, and when he died she took over management of the enterprise. As an absentee owner she seems to have done a creditable job; she made the plantation profitable without abusing the slaves. After the Civil War she said that both she and her husband had always been emancipationists at heart. But though Polk's will recommended freeing the slaves upon his death, she never got around to doing it. It took the Civil War to free them. During the war she entertained Confederate generals at Polk's Place, but when Union forces took over Nashville she extended the same courtesies to Union officers. Her sympathies were basically Southern, of course, but she did try, as former First Lady, to rise above sectional loyalties as much as she could. Not long before her death in 1891, just before her eighty-eighth birthday, she told an interviewer that "when it came to actual conflict, and the lives of people with whom I always lived, and whose ways were my ways, my sympathies were with them; but my sympathies did not involve my principles. I have always belonged, and do now belong to the whole country."[14]

✲ ✲ ✲ ✲

All Superior

In 1823, James K. Polk, nearly thirty and discouraged about his future while serving as a clerk in the Tennessee legislature, asked Andrew Jackson what he must do to succeed in politics. "Stop this philandering," Jackson told him; "they tell me you are a gay Lothario; you must settle down as a sober married man." "Which lady shall I choose?" asked Polk. "The one who will never give you no trouble," replied Jackson. "Her wealth, family, education, health, and appearance are all superior. You know her well." "You mean Sarah Childress?" asked Polk thoughtfully. Then, after a brief pause, he started off. "I shall go at once and ask her."[15]

Hard Money

Polk was a "hard money" man; he was hostile to banks, opposed paper currency, and favored the use of gold and silver as a medium of exchange. During one long journey from Tennessee to Washington, he asked his wife to get some money out of the trunk. "I haven't enough in my pocket," he told her, "to pay expenses during the day." Mrs. Polk obligingly unlocked one of their trunks, dug into it, found nothing, so started going through another trunk. Halfway through it she finally came across several bags of coins. "Don't you see how troublesome it is to carry around gold and silver?" she said exasperatedly. "This is enough to show you how useful banks are." "Sarah," said Polk laughing, "you've turned your politics, I see." He and his wife agreed on most things, but not on the utility of banks. "Why," Mrs. Polk sometimes told her friends, "if we must use gold and silver all the time, a lady can scarcely carry enough money with her."[16]

Butter

During the 1844 campaign, an old lady who was supporting Henry Clay for the Presidency, told Mrs. Polk she thought Mrs. Clay would make a better First Lady because she was an economical housekeeper and made excellent butter. Mrs. Polk gave her a cold look and then assured her that if her husband won the election, she would live within the President's salary and wouldn't have to make her own butter to do it either.[17]

Marks of Respect

After Polk won the election of 1844 people began pouring into Columbia, Tennessee, to get a glimpse of the President-elect, and a large crowd came to the Polk house with a band. "Mrs. President," a gentleman on Polk's staff told Mrs. Polk, "some of your husband's friends wish to come into the house, but we will not let the crowd in, because the street is muddy and your carpets and furniture will be spoiled." "The house is thrown open to everybody," exclaimed Mrs. Polk. "Let them all come in; they won't hurt the carpets." Later, when the crowd had serenaded the Polks and left, Mrs. Polk told the man that, as she had expected, the people had "left no marks except marks of respect."[18]

Stop the Music

Late in February 1845, the Polks boarded a little steamer at Nashville to begin the journey to Washington for the inauguration. Shortly after the steamer started off, a band started playing in honor of the President-elect. It was Sunday, however, and Mrs. Polk thought it unseemly to desecrate the Sabbath with music. Polk acceded to her request to stop the music. "Sarah directs all domestic affairs," he told a friend with him, "and she thinks that is domestic."[19]

Gift

Once a White House guest found Mrs. Polk reading a novel and was surprised she had the time to do so. "I have many books presented to me by writers," Mrs. Polk explained, "and I try to read them all." Then she added: "At present that is impossible; but this evening the author of this book dines with the President, and I could not be so unkind as to appear wholly ignorant and unmindful of his gift."[20]

Many a Time

Whenever one of Polk's associates showed up at the White House on Sunday morning for consultations, Mrs. Polk insisted on taking him to church. But one politician decided to have some fun with her. When she told him she was particularly anxious to go to church that morning because there was a fine new preacher in the pulpit, he exclaimed: "Then I would like to go with you, Madame, for I have played cards with him many a time!"[21]

Woe unto You

One night at a White House reception there was a sudden lull in the conversation and suddenly a deep voice broke the silence: "Madame, I have long wished to see the lady upon whom the Bible pronounces a woe!" Mrs. Polk looked puzzled and several people gasped. Then the voice, which turned out to be that of a man from South Carolina, explained: "Does not the Bible say, 'Woe unto you when all men shall speak well of you?' " The unexpected compliment for the President's wife made everyone laugh.[22]

Lot in Life

One hot afternoon in mid-summer, Mrs. Polk put down her pen and walked over to the window, looked out, and saw some young blacks toiling away in the garden and perspiring profusely. "Mr. President," she said, "the writers of the Declaration of Independence were mistaken when they affirmed that all men are created equal." "Oh, Sarah," cried Polk, looking up from his desk, "that is one of your foolish fancies." "But, Mr. President," persisted Mrs. Polk, "let me illustrate what I mean." Then she pulled him over to the window. "There are those men toiling in the heat of the sun," she said, "while you are writing, and I am sitting here, fanning myself, in this house as airy and delightful as a palace, surrounded with every comfort. Those men did not choose such a lot in life, neither did we ask for ours; we are created for these places." Polk thought it over and decided it was an example of "Sarah's acumen." But for whites, at any rate, Mrs. Polk did believe that the equal-rights philosophy of the Great Declaration was a spur to personal growth and social progress, and so did her husband.[23]

Social Equality

Once a schoolteacher visited the White House with her pupils and got into a conversation with Mrs. Polk. "Mrs. President," she said, "this is the first time I have ever been invited to the White House." Mrs. Polk was not surprised. "Though a woman of culture and high character," she remarked afterward, "her occupation of schoolteacher barred her from social equality." Polk's wife took little interest in the woman's rights movement, but looked with the approval on the movement of women into the workaday world. "It is beautiful," she said toward the end of her life, "to see how women are supporting themselves, and how those who go forward independently in various callings are respected and admired for their energy and industry." Then she reflected, "It is now considered proper for young ladies, when they leave school, to teach or do something else for themselves. It was not so in my young days."[24]

Sagacity

When the Polks vacated the White House, Polk left his desk neat and clean and Mrs. Polk saw to it that the house was spotless. And although gas lighting had been installed while they were there, Mrs. Polk refused to remove the elegant candlelit chandelier in the reception hall because she thought it was so beautiful. The first night the gas lights were used at a reception after Zachary Taylor became President, the jets mysteriously went out and left the party in darkness. But Mrs. Polk's beloved chandelier with its candles saved the evening. The following day Washington newspapers spoke admiringly of her "sagacity."[25]

CHAPTER 12

MARGARET TAYLOR

1788–1854

When Margaret Taylor heard the Whigs were thinking of nominating her husband Zachary for President in 1848, she was thoroughly displeased. It was "a plot," she lamented, "to deprive her of his society, and shorten his life by unnecessary care and responsibility."[1] But when the nomination came his way, she went along with his decision to run, hoped for his sake he would win the election, and was pleased when he did. She had no intention, though, of assuming the duties of a President's wife; her health was too poor for that. She asked her twenty-two-year-old daughter, Betty Knox Bliss, to take over her duties as "Lady of the White House" and she herself remained quietly in the background during her husband's fifteen months in office. She attended the inauguration in March 1849, but after that stayed most of the time in her own rooms upstairs in the White House, seeing only relatives and friends and an occasional visitor like Daniel Webster. About the only time she left the White House was to attend church. She was a devout Episcopalian.

One aristocratic young lady from Natchez did get to see the new President's wife one afternoon in the White House and pronounced her "gentle" and "refined."[2] This came as a surprise to some Washingtonians. Taylor's enemies had been describing her as crude and illiterate. During the campaign they had been reluctant to attack Taylor himself, for he was a Mexican War hero, so they concentrated on belittling his wife. Mrs. Taylor, they said, came from a poor-white family; she was vulgar and ignorant. One cartoon pictured her puffing smoke from a long-stemmed corncob pipe in the manner of Rachel Jackson. The fact that tobacco made Mrs. Taylor "actively ill" and that Zachary refrained from smoking for that reason did not faze her critics.[3] They were too anxious to get at Taylor through his wife to worry about accuracy, and they set all sorts of mean-spirited stories about her in circulation in 1848. But when Mrs. Taylor arrived in Washington with her daughter for the inauguration, people were surprised to find that she was what John H. Bliss had found her to be years before: "a most kind and thorough-bred Southern lady" with simple tastes and gracious manners.[4] The legends, though, persisted. As late as 1937 the *New York Times* ran a story picturing her "peacefully smoking her corncob pipe in the White House, while Washington gasped."[5]

Both the Taylors came from prominent families: Zachary's in Virginia and Margaret's in Maryland. Margaret was born in Calvert County, Maryland, in September 1788, and her father, Walter Smith, was a well-to-do planter who had been a major in the Continental Army during the American Revolution. Her education, according to her critics, stressed "the practical rather than the intellectual," but that was unexceptionable for plantation ladies in those days.[6] Margaret met Zachary in the fall of 1809 when she was visiting her sister, Mrs. Samuel Chews, in Kentucky, and Zachary, a first lieutenant, was at the very beginning of what was to be a long tour of duty in the U.S. Army. She was nearly twenty-two and he was twenty-six when they married in her sister's double log house in June 1810. The setting was appropriate. Margaret was to spend much time in log cabins and crude army barracks during her forty years of marriage to a professional soldier. Life with Zachary took her all the way from Fort Snelling in the North to Tampa Bay in the South and sometimes meant log cabins in the winter and tents in the summer. It also involved continual hardship and danger in the wilderness and periodic separations from her husband.

Mrs. Taylor seems never to have complained about her hard life. She insisted on accompanying Zachary to the frontier posts where he was stationed and always tried hard to brighten the lives of the men under his command there and help out with the sick and wounded. Her favorite place was Baton Rouge, where Taylor was assigned in 1840, and where she stayed when he was off to command the Army of the Rio Grande during the Mexican War. At the Baton Rouge headquarters she let the officers take over the brick buildings while she moved into a picturesque old cottage that Spanish commanders had formerly occupied in pre-American days. With the help of her daughter Betty she transformed the run-down cottage into a comfortable home, started a garden, established a small dairy, and served the best food in the region. She also arranged to have a room set aside in the garrison building for religious services. During the Mexican War she was "always calm and cheerful, a constant source of comfort" to the wives of the other officers.[7]

Mrs. Taylor had six children (two died in infancy and one died very young) and she was seriously ill from time to time. At one point Zachary was so worried about her health that he applied for a leave. "At best her constitution is remarkably delicate," he informed the Quartermaster-General in Washington. "This information has nearly unmanned me, for my loss will be an irreparable one." Then he went on to say: "I am confident the feminine virtues never did concentrate in a higher degree in the bosom of any woman than in hers."[8] It is about the only record we have of his feelings for her.

Margaret Taylor undoubtedly enjoyed seeing her beloved husband rise from lieutenant to major-general, receive praise for his performance in the Black Hawk War along the Upper Mississippi in 1832 and the Seminole War in Florida in 1838, and, finally, achieve glory for his victories at Resaca de la Palma, Monterrey, and Buena Vista during the Mexican War. But she would have preferred living in retirement with him after the war in the Baton Rouge cottage to moving into the Executive Mansion in Washington. Neither she nor her husband went in much for pomp and circumstance; they were too accustomed to the simplicities of frontier life for that. Even if Mrs. Taylor had been up to it, she probably would have done her job as President's wife with a minimum of ceremony. Her lack of pretension impressed Mrs. Jefferson Davis who saw her several times. The pleasantest part of her White House visits, she said, were spent in "Mrs. Taylor's bright and pretty room where the invalid, full of interest in the passing show in which she had not the strength to take her part, talked most agreeably and kindly to the many friends who were admitted to her presence."[9] With young Betty in charge of entertaining there were "stately" dinners at the White House, but also lively dances for young people. Betty, someone observed, had the "artlessness of a rustic and the grace of a duchess."[10] She was probably much like her mother.

Margaret Taylor was not present when her husband presided at the laying of the cornerstone of the Washington Monument on July 4, 1850, and was felled by a stroke right after he finished the ceremony. She was horrified when she heard what had happened and recalled with some bitterness what she had said about her husband's health two years earlier when told he might become President. When he died a few days later she collapsed with grief. Betty was equally overcome. "We had thought of our mother's dying, for she is . . . seldom well," she later confessed; "but our father . . . we never expected to die!"[11] Right after the White House funeral Mrs. Taylor left Washington, never to return, and eventually settled in Louisiana to live with Betty and her husband in a little cottage in East Pasagoula. There she died in obscurity four years later.

* * * *

Dear Girl

One cold winter night, according to an old story, when the fourteen-year-old Margaret Smith and her father and brothers were sitting around the kitchen fire in their Maryland farmhouse, there was a knock at the door and two lads came in looking for shelter. One of them was Zachary Taylor, on his way to Washington to seek a commission in the U.S. Army. Margaret's father invited them in and Margaret, who had become housekeeper after her mother's death, served them supper, and then sat down shyly in the corner to listen to Zachary's stories about frontier life in the West. But when they got up to retire, she noticed Zach was limping; it was an old arrow wound, he told her, which had been swollen from so much hard riding during the past two weeks. She insisted on washing the wound, putting liniment on it, and bandaging it before he went to bed. The next morning, when Zach resumed his journey, he told his friend that Margaret was a dear girl, and his friend said she was likely to grow into an even dearer woman.

A few years later, the story goes, Zach, now in the Army, was in the Smith neighborhood again and came across Margaret, now a blooming young woman, riding an old farm horse to a quilting party. As Zach came galloping up, Margaret's horse dashed off in a panic, but Zach soon overtook her and stopped the horse by clutching the bridle. Apologizing for frightening her horse, Zach reminded her of their meeting a few years earlier; he also insisted on accompanying her to the quilting party. Zach ended by staying for something to eat; but he found it hard to get Margaret, still shy with him, to say much of anything. Finally he took a needle out of a sewing basket, threaded it with black thread, and then stitched the words, "Won't you talk with me?" In great embarrassment, Margaret begged him to rip the letters out, and he did so, but only after she agreed to let him take her home. On the way home she loosened up and by the time they got to her house he had her laughing with him. The next morning, when Zach left, they were engaged.[12]

The Old Woman and the Gal

Two young officers, fresh out of West Point, came across a casually dressed older man when they arrived for duty at Fort Smith. "Good morning, old fellow!" they greeted him. "Good morning," he returned. "How's crops?" they asked. "Purty good," he said. "Come on, take a drink with us." The old man joined them and though he didn't drink anything, he took their teasing in good humor. When he got up to leave, one of the West Pointers said, "Give our love to the old woman and the gals." Later they called on General Taylor, their commanding officer, in full dress, and to their chagrin found he was the "old fellow" they had been joshing earlier in the day. Taylor returned their salute and then introduced them to his wife and daughter Betty. "Here are the old woman and the gal," he told them.[13]

Sarah

The Taylors opposed the marriage of their daughter Sarah to Lieutenant Jefferson Davis. "I'll be damned," said Taylor, "if another daughter of mine shall marry into the army!" But Sarah eloped with Davis and when Taylor found out about it he was furious. "No honorable man would thus defy the wishes of parents," he exclaimed, "and no truly affectionate daughter be so regardless of her duty!" But three months after the marriage Sarah died of malarial fever and the Taylors were plunged into grief. Taylor made up with Davis when they met on the battlefield during the Mexican War; and, years later, when Mrs. Taylor came across him on a Mississippi riverboat, she, too, made her peace with him. She kept two letters from Sarah the rest of her life.[14]

CHAPTER 13

ABIGAIL FILLMORE

1799–1853

When Abigail Powers Fillmore died a few weeks after leaving the White House, the *Boston Journal* uttered the customary conventionalities: "She was a lady of great strength of mind, dignified manners, genteel deportment, and much energy of character. . . ."[1] The *Journal* overlooked the most interesting thing about her: she had once been a schoolteacher and was one of the greatest book-lovers of all our First Ladies.

When Millard Fillmore became President following Zachary Taylor's death in April 1850, Mrs. Fillmore was surprised at the dearth of books in the White House and decided to establish a library there. At her prompting, Millard Fillmore asked Congress to vote money for books and when the money was forthcoming he and his wife set aside a large room on the second floor of the Executive Mansion as a library and stocked it with books, current and old, and with maps and reference books as well. There Mrs. Fillmore also put her piano and harp (for she was a musician too). Though as the President's wife she hosted the usual entertainments—Tuesday morning receptions, Thursday evening dinner parties, Friday evening receptions—her preference was for quiet sessions with friends in the library, musical evenings there (she played, her daughter sang), and, as her health deteriorated, hours alone there with a good book before her. Visitors said it was the pleasantest room in the White House.

Abigail was a schoolteacher when Millard Fillmore first met her; in fact, he was her student for a few months. Her father, Lemuel Powers, a Baptist minister, died soon after her birth in Stillwater, Saratoga County, New York, in 1799, and her mother, Abigail, who moved to a frontier community in Cayuga County with relatives when Abigail was a little girl, seems to have taught school for a while and was primarily responsible for her daughter's education and that of her brother. The brother became a lawyer and a judge; Abigail became a country schoolteacher at sixteen. She was nineteen years old and teaching in Sempronius, New York, when, in the winter of 1818, Millard Fillmore, then eighteen, appeared one day in the classroom. Abigail wondered who the big farm boy was and motioned for him to come up to the desk. "I don't think I have your name," she said doubtfully. "It is Millard Fillmore," he replied and made it clear he was there to learn.[2] It turned out he was a clothmaker's apprentice who was eager to make his way in the world by becoming a lawyer. Abigail liked his energy and ambition, for she was full of enterprise herself, and the two were soon in love. They became engaged in 1819. In an autobiographical sketch he wrote years later, Fillmore summed it all up: "I pursued much of my study with, and perhaps was unconsciously stimulated by the companionship of a young lady whom I afterward married."[3]

Fillmore was Abigail's student for only a few months but her fiancé for years. As Fillmore later put it: Abigail was "eight years my sweetheart, twenty-seven my wife."[4] During the long engagement Fillmore served his apprenticeship as a lawyer and was admitted to the bar, while Abigail went on teaching school and also helped start a circulating library in Sempronius. For three years they courted by mail; they were separated by 150 miles and Fillmore couldn't afford to make trips to see her. And it was only a short time before the wedding that he felt bold enough to kiss her hand lightly upon parting.

At first Abigail's mother and brother objected to the match. Abigail's father, they reminded her, had come from a prominent old Massachusetts family, and the Fillmores were beneath her socially. But Abigail persisted and in the end they gave way. In February 1826 she and Millard were finally married by an Episcopalian minister in the home of her brother, Judge Powers, in Moravia, and then went to live in East Aurora, where Fillmore had built a house and set up a law office. After the marriage Abigail continued teaching until about the time her husband was elected to the State Assembly. In the spring of 1830 they moved to a larger home in Buffalo where Fillmore forged ahead as a lawyer. By then Abigail, who had two children, was no longer teaching; but she was still learning. She studied French, practiced the piano, and did a lot of reading. Whenever Fillmore went to New York City on business he always returned with an armful of books for her to devour.

As Fillmore gradually rose in the political world, Abigail began taking an interest in public affairs, listening to discussions by her husband

and friends, and sometimes joining in on the conversations herself. At times Fillmore even asked for her opinions on political matters. She accompanied him to Albany for sessions of the state legislature and after he was elected to Congress went to Washington with him. In 1840 a delegation of citizens called on her to ask that she speak at the dedication of a building in the nation's capital. It was an unusual request; for, except in abolitionist circles, women simply didn't appear in public to make speeches. Abigail refused the request and later told her Buffalo friends about it with great amusement. For all her intelligence and enterprise she took little interest in women's rights. After her husband became President, though, she did break precedent on a couple of occasions. Presidents' wives were never supposed to venture outside the White House without their husbands, but when Jenny Lind, the "Swedish nightingale," came to Washington, Abigail attended the concert by herself. She also took her daughter, Mary Abigail, to a public banquet held in honor of Louis Kossuth, the Hungarian liberator, who toured the United States in 1852. Some observers believe she persuaded her husband to abolish flogging in the U.S. Navy while he was President, but there is no evidence for this. She did, however, tell him that if he signed the Fugitive Slave Law, requiring the return of runaway slaves to the South, it would destroy him politically. She was right.

During Fillmore's last months in the White House, Abigail's health began to fail and it became necessary to turn many of her duties as White House hostess over to her daughter. She attended the inauguration of Franklin Pierce, her husband's successor, in March 1853, and then checked into the Willard Hotel with a severe chill. There she died a few weeks later, probably of bronchial pneumonia. It was not considered proper then for newspapers to feature the names of ladies in their columns, but the press, led by the *National Intelligencer,* broke the rule and gave considerable attention to her illness and sudden death. "For twenty-seven years, my entire married life," Fillmore later recalled, "I was always greeted with a happy smile."[5] Abigail's New York friends had presented her with a new silver-mounted harness for her carriage when she entered the White House. After her funeral Fillmore arranged to have the silver mountings removed, melted down, and made into pieces of plate, with her name inscribed on them, to give to the children. Five years later, when he married Carolyn C. McIntosh, the widow of a prominent Albany businessman, the nation was on the eve of the Civil War and most Americans had long since forgotten the days when the Fillmores were in the White House.

CHAPTER 14

JANE PIERCE

1806–1863

After visiting the White House during Franklin Pierce's administration, Charles Mason, Chief of the Patent Office, wrote in his diary: "Everything in that mansion seems cold and cheerless. I have seen hundreds of log cabins which seemed to contain more happiness."[1] The Pierce White House was bleak indeed much of the time. Shortly before Pierce became President his only son was killed in a train accident, and his wife Jane, never robust, collapsed in grief. She did not attend the inauguration on March 4, 1853, and when she was finally up to joining her husband in Washington, she came to be called "the shadow in the White House."[2] It wasn't until New Year's Day, 1855, that she began receiving White House guests at the side of her husband. But though she welcomed people with "a winning smile," according to one observer, "traces of bereavement" were still "legibly written on a countenance too ingenuous for concealment. . . ."[3]

Jane Pierce never wanted her husband to be President. She didn't really want him in politics at all. He was a member of the House of Representatives when she married him but she disliked Washington from the beginning. One of Pierce's friends was surprised to see how miserable she looked when she arrived there for the first time. And when Pierce was elevated to the U.S. Senate a few years later she took no pleasure in the honor. "Oh," she cried at one point, "how I wish he was out of political life! How much better it would be for him on every account!"[4] When his Senate term ended, Pierce settled down to law practice in Concord, New Hampshire, largely at her behest, and turned down President Polk's invitation to become Attorney-General in 1846. But when his friends began pushing him for the Democratic presidential nomination in 1852, Mrs. Pierce became alarmed. Pierce assured her he had no interest in running and would resist a draft, and she was reassured. But when he received the nomination after all and then went on to win the election, she decided it was God's, not Franklin's, will, and was reconciled. Later, when she learned that Franklin had wanted the nomination all along, she reproached him bitterly. This was a few days before the inauguration. It wasn't the only time they quarreled.

The Pierces were an unlikely pair. Franklin was bluff, hearty, and outgoing, and loved mingling with rowdy politicians; Jane was shy, sensitive, and retiring, and disliked the hurly-burly of democratic politics. Her background was Whiggish, too, with a touch of disdain for politicians, to boot, while Pierce was a good Jacksonian Democrat. She was a "consumptive," moreover, in frail health a great deal of the time and much given to melancholy. But Pierce had a problem too; he enjoyed the conviviality of the bar-room as well as the boisterousness of the political caucus, and had a tendency it was said, to tipple. But the two got along surprisingly well much of the time, for all their differences. Pierce tried to tailor his life to his wife's needs and desires as much as he could, and she was grateful for his solicitude.

But Jane could never bring herself to share her husband's love for public life. She couldn't help thinking there was something a bit childish about his fascination with the game of politics. Not long after the wedding she wrote her father-in-law from Washington a kind of progress report on the marriage. "We still continue to be pleased with our accommodations here," she said, "and are in fact as comfortably situated as we could be—have both generally been very well and not very unhappy. Frank does very well thus far, sir, and is as you say a *pretty good boy*—it is to be sure rather *soon* to judge but I hope I shall have no reason to alter my opinion—*in such a case,* I shall appeal to you, who I am sure will lend me your *countenance.*" She went on to say that it was "so excessively windy that I am disappointed in my wish of going to church which is a deprivation to which I hope I shall not often be subjected. . . . I have been out very frequently and intend to take your advice my dear sir, to exercise as much as possible in the open air. . . . We have an invitation to dinner to Gov. Cass' on Wednesday which is accepted notwithstanding my predilections for a quiet dinner at home."[5] About this time her husband was also telling a friend how things were going. "Jane's health," he said, "has prevented her from mingling as much in gay society as we might otherwise have done, but we have been to several large parties in the evening and she is now enjoying better health. I think she has hardly been better since I have known her."[6]

An auspicious beginning: Frank, on good behavior, and Jane, in pretty good health. It was never to be much better than that in thirty years of marriage.

Jane was devoutly, even, at times, fiercely religious, and insisted on daily services, prayers, and Bible-readings in the family. Her father, Jesse Appleton, was a Congregational minister who became president of Bowdoin College in Maine a year before she was born in Hampton, New Hampshire, in 1806. He had, it was said, "a morbid sense of responsibility for the religious and intellectual welfare of his students," consumed himself in his work, and died when Jane was thirteen.[7] After her father's death, her mother, who came from a well-to-do family with connections to the Lawrences of Lowell, returned to her home in Amherst, Massachusetts, and saw to it that Jane received "a careful and thorough education" involving a great deal of concern for divine wisdom and human mortality.[8] In 1826, Franklin Pierce, a Bowdoin graduate, arrived in Northampton to study law with Judge Samuel Howe, met "dearest Jeanie" across the river soon after, and entered into a long engagement with her. Jane's relatives seem to have had doubts about Franklin, even though his father was Governor of New Hampshire, for they took a dim view of politics as a profession and knew that Jane's beau was politically ambitious. They were bothered, too, apparently, by rumors that he spent too much time in hotel taverns and that drink ran in his family. But Jane had her way; and in 1834 she and Franklin, now a member of Congress, were married in Amherst by the Reverend Silas Aiken, her brother-in-law, and took up residence in Hillsborough, New Hampshire.

From the outset Jane's health was "delicate." Since she was too frail to do the household chores, Pierce arranged for a married couple to live in and take care of the house and cook the meals. When he went to Washington to attend sessions of the House, and, later, the Senate, she usually stayed home because she wasn't strong enough to travel. Once, when she did go to the capital with Pierce, one of his associates noticed that she was "in very delicate health and wanting in cheerfulness."[9] Jane seems to have liked Hillsborough as little as she did Washington, and Pierce finally decided to move. In 1842, he took her to Concord, where she was much happier and, when his Senate term expired, he settled down to a lucrative law practice there. In Concord, as in Hillsborough, Jane was spared housework because of her frail condition, and she devoted what energies she had to her religion and to her children. She had three sons, one of whom died in infancy. The second boy died at four, and she came to focus all her attention on the third, Benjamin ("Bennie"). She joined the Congregational church shortly after arriving in Concord, and saw to it that Bennie received rigorous religious training. He attended family worship every morning, said prayers every evening, and went to church with his parents every Sunday. He also listened to his mother read Bible stories and learned to sing hymns. Soon he was having religious experiences.

In 1846, when James K. Polk invited Pierce to become his Attorney-General, the latter told the President that his wife's health was worse than ever and that he would be forced to avoid public service, "except at the call of my country in time of war."[10] The Mexican War provided the exception. Soon after the war began, Pierce volunteered for service, arranged for friends to look after his wife and son while he was away, and left for Mexico in May 1847. "I cannot trust myself to talk of this departure," he wrote Jane shortly after leaving Concord. "My heart is with my own dear wife and boy, and will be wherever duty may lead my step."[11] He ended by spending nine months in Mexico and becoming a Brigadier-General, but injuries following a fall from a horse prevented his active participation in the fighting after that. When the war ended, he resigned his commission and returned to Concord. Things improved: Jane's health took a turn for the better, Pierce's law practice flourished; and, with Bennie as the main object of their affections, he and Jane entered what was probably the best period of their marriage.

Pierce's return to politics in 1852 ended his domestic happiness. Jane fainted when she heard the news of her husband's nomination for President that year, and little Bennie thoroughly absorbed her anxieties. "I hope he won't be elected," he told his mother, "for I should not like to be at Washington and I know you would not either."[12] After the election Jane reconciled herself to the inevitable and began making preparations for the move to Washington. Then came the tragic accident and death of Bennie. Both she and her husband were consumed with guilt as well as grief at the loss of their son and both sought solace in religion. Jane finally decided that God had taken Bennie from them so he wouldn't be a distraction to his father when he assumed his duties as President. And Pierce himself seems to have looked on his son's death as punishment for his own shortcomings. At his inauguration, almost as if in penance, he refrained from taking the usual oath of office; instead, he simply affirmed his loyalty to the Constitution.

After the inauguration, when Jane finally joined her husband in the White House, there was little for her to do there. Pierce had engaged a New Hampshire couple to run the place—supervise the servants, hire caterers for dinners and receptions, and handle the accounts—and also provided Jane with a companion, Mrs. Abby Means, who could act as hostess until Jane felt strong enough to do so herself. Jane remained upstairs most of the time at first, writing pathetic little notes to Bennie and reproaching herself for not having shown him more affection while he was living. She did, though, receive relatives and friends (including her husband's old Bowdoin College classmate Nathaniel Hawthorne) and she got to church every Sunday. She also urged the servants and members of the White House staff to attend services "for my sake."

She was as strict a Sabbatarian as Sarah Polk had been. "Many a time," recalled Pierce's private secretary, "have I gone [to church] from respect to her, when, if left to my own choice, I should have remained in the house."[13] At length she began making public appearances—presiding at state dinners, holding Friday receptions, attending the President's levees—but she did so spiritlessly. To Washington society she was little more than the "invalid" in the White House.

In March 1857, James Buchanan, a bachelor, succeeded Franklin Pierce as President and his niece, Harriet Lane, succeeded Mrs. Pierce as White House hostess. That fall the Pierces began three years of travel—to the West Indies and to Europe—hoping that a change of scene and climate might restore Jane's health. Everywhere they went she carried Bennie's Bible with her as well as a little box containing locks of hair from her "precious dead," her sons, her mother, and her sister. When the Civil War came, Pierce became a bitter foe of the Lincoln administration and alienated his friends in Concord for criticizing the Union war effort. There is some reason to believe that Jane did not fully share his views. In a letter Pierce wrote her during the war, at any rate, he seemed to be trying hard to explain his position to her. "My purpose, dearest, is immovably taken," he told her. "I will never justify, sustain, or in any way or to any extent uphold this cruel, heartless, aimless, unnecessary war. Madness and imbecility are in the ascendant. I shall not succumb to them, come what may. I have no opinions to retract, no line of action to change."[14]

On December 2, 1863, Mrs. Pierce died in Andover, Mass.; her last words were: "Other refuge have I none." Nathaniel Hawthorne came at once for the funeral, and when he looked at her shrunken little figure in its elegant coffin, he couldn't help feeling that she had never had anything to do with "things present." In her will she provided for bequests to the American Bible Society, the American Society of Foreign Missions, and to the American Colonization Society. Then came gifts to relatives and to two servants "long hired in her family." The rest went to her husband.[15]

* * * *

Thunderstorm

One day, when Jane Appleton was doing some reading in the Bowdoin College library, she glanced out of the window, saw a storm coming up, and rushed out of the library to return home. No sooner did she get outside when the storm broke in full force and a loud crash of thunder sent her reeling against an old oak tree. While she was crouching there in terror as the lightning flashed and the thunder pealed, another student, Franklin Pierce, who had seen her leave the library and followed her, came running up, told her she was in the worst possible place for an electric storm, caught her up in his arms, and saw that she got home safely. It was the beginning of a romance.[16]

Accident

On January 5, 1853, two months before Pierce's inauguration, he and his wife visited Boston with their thirteen-year-old son Bennie, and then boarded the Boston & Maine Railroad for the return trip to Concord. Somewhere between Andover and Lawrence the axle of one of the passenger trains broke and all the cars tumbled down a steep embankment. Pierce, bruised and badly shaken, rescued his wife from the wreckage and then went to find his son. He soon found the boy, his head crushed and pinned under a beam. The death of Bennie shattered Mrs. Pierce's already precarious health. Her "agony," reported a clergyman who had been with them, "passes beyond any description. She could shed no tears, but, overcome with grief, uttered such affecting words as I can never forget. . . . She was conveyed to a house nearby, and there she gave vent to the grief that rent her heart. . . ." She never recovered from her loss.[17]

The Comfort of It

One day, while visiting some friends in the country, President Pierce began to wax eloquent about his friend Nathaniel Hawthorne, and, as he talked, he thrust his hands deep into his pockets while pacing up and down the veranda. Suddenly he saw his wife, the soul of propriety, looking disapprovingly at him. "No," he said stubbornly, "I won't take them out of my pocket. I am in the country and I like to feel the comfort of it."[18]

CHAPTER 15

MARY TODD LINCOLN

1818–1882

Mary Lincoln was the first wife of a President to become a storm center while she was in the White House. From the outset she was almost as much a target of abuse as her husband was, and until Eleanor Roosevelt she was the most vilified of all our First Ladies. The slander began almost as soon as she entered the White House in March 1861. She was called crude and vulgar, vain and pretentious, frivolous and flighty, stupid and ignorant, wasteful and extravagant, meddlesome and conniving, greedy and corrupt. By August 1861 the attacks on her had become so vicious that the *Chicago Tribune* exclaimed, "HOLD ENOUGH!" and came to her defense. If Mrs. Lincoln "were a prize-fighter, a foreign danseuse or a condemned convict on the way to execution," observed the editor, "she could not be treated more indecently. . . . No lady of the White House has ever been so maltreated by the popular press. . . . The sighs and sneers of sensible people all over the land and the mockery of the comic papers are the natural consequence."[1]

Mrs. Lincoln had expected better. When Mr. Lincoln, as she called him, received the Republican nomination for President in May 1860, she was pleased and proud; and when he won the election in November it seemed as though her long-held ambitions for him had at last reached fulfillment. Though the nation was poised on the brink of civil war by the time of the inauguration, she continued to hope for the best. She even looked forward to the great service her husband would render the American people, with her by his side, in reducing sectional tensions and restoring harmony to the nation. She was eager, too, to take up her duties as First Lady. She had been admired for her social skills in Lexington, Kentucky, where she was born Mary Ann Todd in 1818, the daughter of a merchant and banker, and in Springfield, Illinois, where she had met and married Abraham Lincoln in 1842, and she could hardly wait to start displaying her talents on the national scene and winning friends for her husband. In Springfield she had been the local belle. From a proud and prosperous Kentucky family ("One 'd' is good enough for God," Lincoln is said to have said, "but not for the Todds"),[2] she was smart, witty, articulate, and well-read, and had even learned French at Madame Mentelle's finishing school for girls in Lexington. She expected to shine in the nation's capital.

Like Lincoln himself, however, Mrs. Lincoln found heartbreak, not glory, in the White House. Washington socialites looked askance at the takeover of the White House by outsiders from the Middle West and boycotted the inaugural ball and the first few White House receptions. And although many of those who came were impressed by the new First Lady's charm, grace, wit, and vivacity, there was a great deal of grousing about her by people who counted in Washington. "All manner of stories about her were flying around," noted Charles Francis Adams; "she wanted to do the right thing but, not knowing how, was too weak and proud to ask. . . ."[3] If Mrs. Lincoln had been quiet and retiring, like most of her predecessors in the White House, perhaps all would have been well, even in a time of national crisis. But there was nothing meek and mild about her; she was bright, high-spirited, energetic, and full of good talk; she was also short-tempered, impulsive, and outspoken. She had no intention of staying in the background; she was anxious to do everything she could to help make her husband's administration a success. As a result she became almost as controversial a figure in Washington as Lincoln himself. Social commentator Laura C. Holloway, for one, sympathized with the First Lady from Illinois. "She found herself," observed Holloway, "surrounded on every side by people who were ready to exaggerate her shortcomings, find fault with her deportment on all occasions, and criticize her performance of all her official duties."[4]

For some people Mrs. Lincoln could do nothing right. They called her parsimonious for cutting down on White House dinners because of the war crisis; but they said she was callous when she arranged ambitious entertainments to boost the morale of Unionists and demonstrate that the Federal Government was still a going concern. They deplored her taste, but criticized her fine gowns as extravagant; they also tore into her for spending the taxpayers' money renovating the somewhat seedy White House. They even expressed shock at her low décolletage. The "weak-minded Mrs. Lincoln had her *bosom* on exhibition," exclaimed one hostile visitor after attending a White House reception. Sneered another: "She stuns me with her low-necked dresses and the flower beds which she carries on top of her head."[5] Soon she was receiving so many letters filled with hate that she asked W. O. Stoddard, White House mail clerk, to screen her mail for her.

Hardest of all to bear were the charges of disloyalty. Mrs. Lincoln came from Kentucky, a border state, and had relatives who supported the Confederacy; but her own devotion to the Union cause was wholehearted. Her opposition to slavery was equally firm, and, like Lincoln, she became an emancipationist as well as a Unionist. One of her best friends in Washington, in fact, was the aristocratic Charles Sumner, Radical Republican from Massachusetts who was continually urging Lincoln to take stronger action to end slavery. But there were rumors all the same that the First Lady was at heart pro-Confederate in her sympathies and that she was even secretly aiding the Southern cause. "The President's wife," noted Stoddard, "is venomously accused of being at heart a traitor and of being in communication with the Confederate authorities."[6]

So clamorous did the charges of treason become, according to a dramatic tale first appearing in print in 1905, that a Congressional committee finally decided to look into the matter. But one morning, the story goes, just as the hearing opened, Abraham Lincoln suddenly appeared, unannounced, "at the foot of the table, standing solitary, his hat in hand, his tall form towering above the committee members. . . ." With his face filled with "almost unhuman sadness," it was said, Lincoln started speaking, "slowly with infinite sorrow," as the room grew still. "I, Abraham Lincoln, President of the United States," he declared, "appear of my own volition before this Committee . . . to say that I, of my own knowledge, know that it is untrue that any of my family hold treasonable communication with the enemy." Then, the story ends, he left "as silently and solitary as he came," and the committee adjourned without taking any further action. Sad to say, the touching little tale of 1905 is entirely fanciful. But it does recall vividly the widespread distrust of Lincoln's wife during the Civil War.[7]

Mrs. Lincoln was not of course without her defenders. Ben: Perley Poore, Washington journalist, praised her hospitality, charity, graceful deportment, and goodness of heart, and thought she was worthy to stand beside the glamorous Dolley Madison as a First Lady. In the *Sacramento Daily Union,* moreover, California reporter Noah Brooks blasted the "he-gossipers and envious retailers of small slanders" for their sniping at Mrs. Lincoln and told his readers: "Mrs. Lincoln, I am glad to be able to say from personal knowledge, is a true American woman, and when we have said that we have said enough in praise of the best and truest lady in our beloved land." When Mrs. H. C. Ingersoll called on Mrs. Lincoln in the spring of 1864 to seek support for a plan to aid the Union cause, she was astonished to discover how different the First

Lady was from reports about her in the press. But when she suggested that Mrs. Lincoln publicly refute some of the most outrageous charges against her, the latter exclaimed: "Oh, it is no use to make any defense; all such efforts would only make me a target for new attacks. I do not belong to the public; my character is wholly domestic, and the public have nothing to do with it. I know it seems hard that I should be maligned, and I used to shed many bitter tears about it. . . ." Then she added: "but since I have known real sorrow—since little Willie died—all these shafts have no power to wound me."[9]

Little Willie's death of typhoid fever in February 1862 shattered the Lincolns. He was only eleven at the time and seemed more like his father than either Robert, the eldest son, or Thomas ("Tad"), the youngest. In Mrs. Lincoln's case, the loss of Willie came close to producing mental collapse. She had convulsions, stayed in bed for days, had visions of Willie visiting her at night, and began attending spiritualistic séances in hope of communicating with her lost boy. At one point, it is said, her grief was so great that Lincoln took her to a window, put his arm around her, and, pointing to a building for mental patients in the distance, said gently: "Mother, do you see that large white building on the hill yonder? Try and control your grief, or it will drive you mad, and we will have to send you there."[10] To her half-sister, Emilie Todd Helm, who came to the White House to help out, Lincoln confessed: "I feel worried about Mary, her nerves have gone to pieces; she cannot hide from me that the strain she has been under has been too much for her mental as well as her physical health."[11]

But the "stabs given Mary," as Lincoln called them, continued unabated.[12] Her enemies pointed out that many parents had lost their sons in battle and that at least the First Lady had the privilege of seeing her son die rather than having to send him out to be shot on a distant battlefield. There was criticism, too, because Robert, who had just graduated from Harvard, was not in uniform. Mrs. Lincoln had a morbid fear of losing Robert too and kept him out of the army as long as she could despite the public clamor. In the end, though, Lincoln, embarrassed by it all, found a place for him on General Grant's staff, and the criticism subsided. By this time Mrs. Lincoln seems to have recovered some of her equilibrium. She resumed her White House entertaining, spent much time and energy visiting hospitals with presents for the wounded soldiers, and tried to forget her troubles by indulging in an orgy of spending on elegant clothes and jewelry in New York's smartest stores. But her "nervous spells," Lincoln noticed, persisted: migraine headaches, extravagant suspicions of the President's associates, periodic outbursts of temper, and brooding fears about losing her husband too.[13] She also began worrying about the huge debts she was incurring in New York and shrank from telling her husband about them.

Lincoln's bid for a second term in 1864 produced one of the vilest campaigns in American history. Not only did Lincoln's opponents shower him with abuse; they also took after his wife with almost fiendish glee. An editorial in the *Illinois State Register* on October 30 was typical. There was an attack on her as a person: "a sallow, fleshy, uninteresting woman." Then came an assault on her taste: "She introduced sensationalism with the White House economy; courted low company in her innocence of what was superior." Followed then charges of callousness: "She shamed the country with her famous ball on the battle night, which George H. Boker commemorated with the scathing poem, 'The Queen Must Dance.' " Then came a spurious story: that Mrs. Lincoln called a dentist out of bed one night to pull one of her son's teeth and was then too cheap to pay him anything. Finally a miscellany of aspersions: she refuses to pay hotel and department store bills, omits gratuities when given the best seats in theaters, is vastly overweight, wears "absurd costumes," is "a coarse, vain, unamiable woman," with "no conception of dignity" and with "the peevish assurance of a baseless parvenue." Upshot: "We only write this to show the ladies of the land that the reelection of Old Abe must involve the installation of his wife in the capitol, to which we are certain they will object."[14] The *Register*'s attack was not the most vehement of the 1864 campaign.

Lincoln's victory in November boosted Mrs. Lincoln's spirits, and the collapse of the Confederacy in April 1865 produced great joy. To celebrate the end of the war she sent big bouquets to all of her friends—and to some of her enemies as well. "Dear husband," she exclaimed, as they prepared to go out for a drive Friday afternoon, April 14, "you almost startle me by your great cheerfulness." When she asked whether to invite someone to go with them on their little drive, he said, "No, I prefer to ride by ourselves today." On what turned out to be their last ride together, the Lincolns talked of many things: of trips to California and to Europe some day and of where to settle down after leaving the White House. When she again remarked on Lincoln's great happiness, he told her: "We must both be more cheerful in the future. Between the war & the loss of our darling Willie—we have both been very miserable."[15] They both felt tired, however, and Mrs. Lincoln was developing another headache, and they talked of canceling plans to go to the theater that evening to see the comedy, *Our American Cousin,* with General Grant and his wife as company. But Lincoln thought they ought to go; the diversion might be good for them and, besides, people expected them to make an appearance. When the Grants begged off the last minute, they asked Major Henry R. Rathbone and his fiancée, Clara Harris, to take their place.

It was about 8:30 when the Lincolns arrived in their box at Ford's Theater that night and, after receiving a big ovation, settled down to enjoy the play. During the third act Mrs. Lincoln took her husband's hand, nestled close, and whispered, "What will Miss Harris think of my hanging on to you so?" "She won't think anything about it," Lincoln smiled.[16] Moments later John Wilkes Booth burst into the box, shot the President, wrestled with Major Rathbone, shouted, *"Sic semper tyrannis,"* and leaped to the floor below to make his escape. When the stunned First Lady realized what had happened she gave a piercing scream and then collapsed. In the confusion that followed while the President's body was being taken to a bedroom in a house across the street, she lost sight of him in the crowd outside and wailed, "Where is my husband? Where is my husband?"[17] Hours later, as his life ebbed away, she begged him to take her with him.

Mrs. Lincoln was too distraught to attend the President's funeral, and it was weeks before she was able to pull herself together and make arrangements to move out of the White House. "God, Elizabeth, what a change!" she told Elizabeth Keckley, the black seamstress who became her close friend during the White House years. "Did ever woman have to suffer so much and experience so great a change? I had an ambition to be Mrs. President; that ambition has been gratified, and now I must step down from the pedestal. My poor husband! Had he never been President, he might be living today. Alas! all is over with me."[18]

But Mrs. Lincoln's ordeal was far from over. Not long after leaving the White House she settled in Chicago with Robert and Tad and began wondering how to pay off the huge debts she had contracted during her clothes-buying sprees after Willie's death. Meanwhile William ("Billy") Herndon launched his inquiries into her husband's early years and began gathering information that placed her in a bad light. Herndon had been Lincoln's law partner in Springfield, Illinois, for many years, but from the beginning he and Mrs. Lincoln had clashed. He thought she was highty-tighty and supercilious; she regarded him as somewhat disreputable because of his slovenly ways, disapproved of his drinking habits, and never deigned to invite him for dinner in the Lincoln home. After Lincoln's tragic death, when he decided to prepare a biography of his former partner, he was delighted when he came upon tales about Ann Rutledge, supposedly Lincoln's first love, and jumped to the conclusion that they were a major key to understanding the sixteenth President. On November 16, 1866, Herndon presented the Rutledge story to the public for the first time in a lecture, widely reported, in Springfield. Young Ann, Herndon announced, was the only woman Lincoln had ever loved; he had loved her "with all his soul, mind, and strength." And when she died of a raging fever in New Salem, Illinois, at nineteen, he "slept not . . . ate not, joyed not." Never again, averred Herndon, did Lincoln love any woman the way he had loved Ann. She was his "grand passion."[19]

When Herndon's lecture was reported around the nation, there was considerable indignation at first. Critics did not dispute the facts so much as their revelation, which they thought in poor taste. But Herndon stood firm; in the years that followed, in fact, he expanded on his theme. Not only was Lincoln's heart given only to Ann, he insisted, his marriage to Mary Todd, who stood above him in the social scale, in 1842, was a mere "policy marriage" to advance his own fortunes.[20] Even so, Herndon said, Lincoln was plagued with doubts after becoming engaged to Mary, and he failed to show up on their wedding day, set for January 1, 1841. And when they were finally reconciled and married the following year, the stage was set for a unhappy union. Mary never forgave Lincoln for jilting her in 1841, insisted Herndon, and her violent temper made his home life a veritable hell much of the time. To prove his point Herndon put together miscellaneous bits of gossip and snippets of hearsay, extrapolated mightily, and ignored all the counter-evidence. "My friend," he quoted Lincoln as telling a man who had trouble with Mrs. Lincoln, "I regret to hear this, but let me ask you in all candor, can't you endure for a few moments what I have had as my daily portion for the last fifteen years?"[21]

The Ann Rutledge story is one of the most enduring of all the Lincoln myths and it is enshrined in fiction, film, poetry, drama, and popular biography. But it is without foundation in fact. There was an Ann Rutledge, to be sure, and Lincoln knew her and was saddened by her untimely death, but there is not a shred of evidence that historians can take seriously for a romantic attachment between the two. The story of the wedding for which Lincoln failed to make an appearance is equally fictitious. As for the Lincoln marriage, which Herndon thought so unhappy, there seems little reason to doubt that, despite the inevitable ups and downs of the Lincolns' early years together, their affection and esteem for each other grew steadily as they came to know each other better. They had much in common: a fascination with politics, a love of poetry and good writing, a hatred of slavery, an eagerness to rise in the world. Their sons, too, meant a great deal to both of them and most observers thought they were loving, even over-indulgent parents.

The evidence is overwhelming: the Lincoln marriage was a good one and did much for both partners. They met in Springfield in 1839, became engaged soon after, separated temporarily because of the opposition of the Todds to Mary's marrying beneath her, and then came back together and were married in 1842. Mrs. Lincoln idolized her husband, though she was upset at times by his rather causal ways and recurrent spells of abstraction. Lincoln himself was proud of his wife; he enjoyed her wit and intelligence and learned to handle her periodic flare-ups with kindness and compassion. She was his "child-wife," he liked to say, and she didn't mind it a bit.[22] "There never existed a more loving and devoted husband," Mrs. Lincoln told friends. Once, when a woman who was chatting with the President in the East Room teased him for looking over at Mary, he laughed and exclaimed: "My wife is as handsome as when she was a girl, and I, a poor nobody then, fell in love with her; and what is more, I have never fallen out."[23] Mrs. Lincoln's older sister Frances, who knew the Lincolns well, probably got it right: "He was devoted to his home, and Mrs. Lincoln . . . almost worshiped him. . . . And they certainly did live happily together—as much so as any man and woman I have ever known."[24]

But there was no happiness for Lincoln's widow. The Herndon assault was only one of the burdens that made her last years almost insupportable. The pressure of her creditors, the delay in settling her husband's estate, and her own increasingly pathological fear of poverty led her to seek financial aid from Congress and to make public appeals for assistance, too, and she came again under heavy attack for greed and extravagance. The *Springfield Journal* observed that she had been "deranged for years, and should be *pitied* for all her *strange acts,*" but not everybody was so charitable.[25] When she tried to arrange a sale in New York of her gowns and jewelry that ended in failure, there were new attacks on her character and behavior. "If I had committed murder in every state in this blessed Union," she told a relative, "I could not be more traduced. An ungrateful country, this."[26]

In the end Lincoln's estate was settled and the creditors paid off; but Mrs. Lincoln's obsession with the poorhouse persisted and so did her compulsion for reckless spending. Her eldest son Robert (now a lawyer in Chicago), fearful that she might dissipate her estate, began questioning her sanity, and the newspapers were quick to publicize his doubts. In 1868 she fled to Europe with Tad to escape "persecution from the vampyre press" and began wandering from country to country there, suffering violent headaches, brooding about her sad lot, and continuing her pressure on Congress for financial support.[27] In July 1870, her old friend Charles Sumner arose in the Senate to take up her cause. Not only did he warmly support a pension for the former First Lady; he also made an impassioned defense of her against the charges that had been hurled at her since Lincoln's assassination. "Surely the honorable members of the Senate must be weary of casting mud on the garments of the wife of Lincoln . . . ," he cried. "She loved him. I speak of that which I know. He had all her love."[28] Not long after Sumner's speech Congress voted her a lifetime pension of $3,000 (raised to $5,000 in 1882) and she decided to come home.

In April 1871, Mrs. Lincoln arrived in the United States, with Tad, now a fine-looking lad of eighteen, eager to see her grandchild, whom Robert and his wife Mary had named after her: Mary Todd Lincoln. But Tad had been suffering on and off from a cold for several weeks, and not long after they reached Chicago he took a turn for the worse and died soon afterward. There was no real recovery from this latest blow. "I feel that there is no life for me, without my idolized Taddie," she wrote. "One by one I have consigned to their resting place my idolized ones, & now, in this world, there is nothing left me, but the deepest anguish & desolation."[29] She continued her spending sprees; she also developed the delusion that someone was trying to hurt her and it got so she was afraid to eat or even sleep. One night, when she headed for a hotel lobby only half-dressed and Robert tried to stop her, she began screaming hysterically at him.[30]

In May 1875, after a sanity hearing which Robert requested after consultation with her doctors, Mrs. Lincoln was declared legally incompetent by a hand-picked judge and jury in Chicago and sent to a sanitarium specializing in nervous diseases in Batavia, Illinois. "Oh, Robert," she cried after the hearing, "to think that my son should ever have done this!"[31] In September, however, she succeeded in obtaining a release and in June 1876 was pronounced "restored to reason." Shortly afterwards she fled to Europe again, settled in France, and spent the next four years traveling around in an effort to forget her troubles. In the fall of 1880, after a bad fall which injured her back, she returned to the United States and settled in Springfield, Illinois, with her older sister, Elizabeth Todd Edwards.[32] Eventually she reached a reconciliation of sorts with her son and developed a close friendship with young Edward Lewis Baker, Jr., her grandnephew, that meant much to her.

But fears and anxieties continued to plague the President's widow and she kept increasingly to her bedroom on the second floor of her sister's house, paralyzed and half-blind. In July 1882 she suffered a stroke, went into a coma, and died soon after at sixty-four. In the eulogy at her funeral in the First Presbyterian Church in Springfield, the Reverend James A. Reed urged Mrs. Lincoln's relatives and friends to think of her as the woman she was before her husband's assassination. The "bullet that sped its way and took her husband from earth," he told them, "took her too."[33]

✲ ✲ ✲ ✲

One of My First Guests

Mary Todd grew up in Kentucky and like other Kentuckians had a great admiration for Henry Clay. One day, when she was thirteen, she took her little pony out to Ashland, Clay's estate about a mile from Lexington, to show the Kentucky Senator. "He can dance—look!" she cried, when Clay came out to greet her; and she put her pony through his paces. Clay praised the performance and then invited the little girl to dinner. "Mr. Clay," said Mary at dinner, "my father says you will be the next President of the United States. I wish I could go to Washington and live in the White House. I begged my father to be President but he only laughed and said he would rather see you there than be President himself. He must like you more than he does himself. I don't think he really wants to be President." "Well," laughed Clay, "if I am ever President I shall expect Mary Todd to be one of my first guests.

Will you come?" "If you were not already married," said the little girl, "I would wait for you."[34]

Handsomest Man in Town

On September 29, 1832, President Andrew Jackson visited Lexington and there was great excitement when he passed through town. Mary Todd, a loyal little Whig, watched the parade with a friend who was a Democrat but refused to applaud. "I wouldn't think of cheering General Jackson, for he is not our candidate," she said, "but he is not as ugly as I heard he was." "Ugly!" cried her friend. "If you call General Jackson ugly, what do you think of Mr. Clay?" "Mr. Henry Clay is the handsomest man in town," insisted Mary, "and has the best manners of anybody—except my father. We are going to snow General Jackson under and freeze his long face so that he will never smile again." But Jackson beat Clay overwhelmingly in the 1832 election.[35]

Well Enough

Stephen A. Douglas courted Mary Todd for a while. One day she was sitting on the porch, weaving a wreath of roses for a party, when Douglas came by and said, "Mary, will you come down and have some ice cream?" "I will come," she said impishly, "if you will wear this wreath of roses upon your head." To her surprise he put the wreath on his head and off they went. By the end of the year, however, she had lost interest in Douglas and became engaged to Lincoln. "Mr. Lincoln may not be a handsome figure," she told people who questioned her choice, "but people are perhaps not aware that his heart is as large as his arms are long." Years later a relative told her, "I used to think Mr. Douglas would be your choice." "No," said Mary, "I liked him well enough, but that was all."[36]

Worst Way

"Miss Todd," said Abraham Lincoln going up to Mary Todd at a cotillion, "I want to dance with you the worst way." "And he certainly did," Mary is said to have told her cousin Elizabeth afterwards.[37]

Rebecca

In September 1842, the *Sangamo Journal,* a Whig paper, published a letter written by Lincoln but signed "Rebecca," which criticized James Shields, state auditor and a Democrat, for his views on taxes and banking with which Lincoln heartily disagreed. Lincoln's was the second of four "Rebecca letters" appearing in the *Journal,* and the last one, written by Mary Todd and her friend Julia Jayne, contained a lot of personal ridicule. A few days later the *Journal* also published a little jingle Mary had written teasing Shields. At this point Shields blew up, sought out the *Journal* editor, who told Lincoln, and Lincoln at once accepted responsibility for all the Rebecca materials. Two weeks later Shields challenged him to a duel.

Lincoln was surprised, but he had no choice but to accept the challenge. As the challenged party, he had choice of weapons, decided on broadswords, and began practicing with them. Because of his height he thought he could disarm Shields before any damage was done. "I didn't want the d——d fellow to kill me," he told a friend, "which I rather think he would have done if we had selected pistols." A few days later, Lincoln and Shields met on Bloody Island, opposite St. Louis, with their seconds and swords, and then, to the delight of everyone concerned, agreed to call the whole thing off. Mary deeply regretted the way she had unintentionally put Lincoln's life on the line. "I doubtless trespassed, many times & oft, upon his great tenderness & amiability of character," she wrote a friend long afterward. She and Lincoln agreed never to speak of it again. Years later, an Army officer started teasing Lincoln about it. "Mr. President," he said, "is it true, as I have heard, that you once went out to fight a duel and all for the sake of the lady by your side?" "I do not deny it," said Lincoln, his face flushing, "but if you desire my friendship, you will never mention it again."[38]

Statute

At the Lincoln wedding, November 4, 1842, the minister, following the Episcopal prayer book, started repeating the words, "With this ring I thee endow with all my goods and chattels, lands and tenements," and an Illinois judge who had never witnessed this kind of service before thought the proceedings absurd. "Lord Jesus Christ, God Almighty, Lincoln," he burst out, "the statute fixes all that!"[39]

Not Pretty

In 1846, when Ward Hill Lamon was in Springfield, Lincoln invited him to a party at his house. When he introduced him to his wife, Lamon told her that Lincoln was a big favorite in eastern Illinois. "Yes, he is a great favorite everywhere," said Mary. "He is to be President of the United States some day; if I had not thought so I never would have married him, for you can see he is not pretty. But look at him! Doesn't he look as if he would make a magnificent President?"[40]

Very Small

Sundays the neighbors would see Mary go off to church in Springfield and Lincoln pull a little wagon with the baby in it up and down the street. He always held an open book in one hand and read absorbedly as he walked. One Sunday, however, the baby fell out of the wagon and when Mrs. Lincoln returned from church she found the little fellow bawling and squalling on the ground while Lincoln went on with his reading. She gave a loud shriek and then, when Lincoln saw what had happened and picked up the baby, the two of them began laughing.

Sometimes, though, Mrs. Lincoln chided her husband for his absentmindedness. "He is of no account when he is at home," she once sighed; "he never does nothing except to warm himself and read. He never went to market in his life; I have to look after that. He just does nothing. He is the most useless, good-for-nothing man on earth." But she added that if anyone else ever said that about him she would scratch the person's eyes out. Once, when she was criticizing Lincoln for his carelessness, a relative remonstrated, "If I had a husband with a mind such as yours," she said, "I wouldn't care what he did." Mary was pleased by the compliment to her husband. "It is very foolish of me," she agreed; "the things of which I complain are really very small."[41]

Little Woman on Eighth

When the Republican convention was meeting in Chicago in July 1860, Lincoln was waiting for the news in the telegraph office in Springfield with some of his friends. The afternoon of the third day a messenger handed him a telegram and cried, "Mr. Lincoln—you've been nominated!" As his friends cheered, Lincoln got up to leave. "There's a lady over yonder on Eighth Street," he said, "who is deeply interested in this news; I will carry it to her."[42]

Elected

On election day, November 1860, Lincoln spent most of the day at the State House, voted in the afternoon, and in the evening followed the returns in the telegraph office with some of his friends. When the news flashed that New York, the key state, had gone for Lincoln, the crowds gathered outside cheered loudly and called for a speech. But Lincoln ignored their cries and hurried home. When he got there, he found his wife sound asleep, touched her gently on the shoulder, and said, "Mary." When she did not respond, he tried again, this time more loudly, "Mary, Mary! *We are* elected." Then he took her to a victory supper given by some Republican women.

Years later, Herndon concocted a story about how Lincoln had a big fight with his wife election night, which Robert Sherwood made use of in his broadway play, *Abe Lincoln in Illinois,* in 1938. Lincoln did, it seems, tease his wife the next day about having locked him out the night before. She had told him that if he didn't get home by ten she would lock the door. But when she "heard the music coming to her house," Lincoln was supposed to have said, "she turned the key in a hurry." Mrs. Lincoln was annoyed when he told the story, but there was no quarrel. She was in high spirits about the victory.[43]

Diamond Brooch

One of the most damaging stories told about Mrs. Lincoln by her enemies concerned her last hours in Springfield. Hermann Kreismann, a German-American politician who disliked her intensely, went to the Lincolns' hotel room just before they left to board the train for Washington and found Lincoln in a state of turmoil and his wife lying on the floor, "quite beside herself." "Kreismann," Lincoln is supposed to have said, "she will not let me go until I promise her an office for one of her friends." The friend had promised her a diamond brooch, he said, and Mrs. Lincoln, who loved jewelry, didn't want to be deprived of her gift. The story is implausible; it is unlikely that Lincoln would invite Kreismann into the hotel room with his wife on the floor. But the tale appeared after Lincoln became President and was used by critics as evidence of his wife's greed.[44]

Long and Short of It

When the Lincolns stopped in Ashtabula, Ohio, en route to Washington early in 1861, the crowd gathered there cheered the President-elect lustily and then called for his wife. Lincoln said he "didn't believe he could induce her to come. In fact," he added, "he could say that he never succeeded very well in getting her to do anything she didn't want to do." This brought a big laugh. But at Pennsylvania Station, where he did introduce his five-foot-three wife, he told the people there that he had concluded to give them "the long and the short of it."[45]

First Reception

The Lincolns held their first reception in the White House on March 6, 1861. Mrs. Lincoln was so nervous about it that she was in tears when Elizabeth Keckley, the black seamstress on her staff, arrived with her new gown shortly before the party. "I won't be dressed," said Mrs. Lincoln. "I will stay in my room. Mr. Lincoln can go down with the other ladies." But the sight of the new dress tempted her to try it on. Meanwhile Lincoln, pulling on his white kid gloves, came in to sit on the sofa and watch her get dressed. His praise and her good reflection in the mirror gave her self-confidence and she finally went downstairs on Lincoln's arm. After the party the guests remarked on how delightful a hostess she was.[46]

At This Juncture

In July 1861, when the Confederate guns could be heard at Bull Run and Washington seemed to be in imminent danger of being invaded, General Winfield Scott suggested that Mrs. Lincoln leave with the children. "Will you go with us?" Mrs. Lincoln asked her husband. "Most assuredly I will not leave at this juncture," said Lincoln. "Then I will not leave you at this juncture," she said firmly.[47]

A Few Ladies

At one point during the Civil War, Confederate troops got so close to Washington that the sound of cannot fire could be heard in the city. Lincoln visited the forts and outworks several times during the crisis, and on one occasion, when Mrs. Lincoln was with him at Fort Stevens, some rifle balls came alarmingly close to the presidential couple. When the rebel forces finally returned, almost unharmed, to Richmond, there was a great deal of criticism of Union forces. Mrs. Lincoln was inclined to blame Secretary of War Edwin Stanton for the almost defenseless condition of Washington. Two or three weeks later, Stanton called on the Lincolns and during the course of the conversation told Mrs. Lincoln playfully, "Mrs. Lincoln, I intend to have a full-length portrait of you painted, standing on the ramparts at Fort Stevens overlooking the fight." "That is very well," said Mrs. Lincoln sharply, "and I can assure you of one thing, Mr. Secretary. If I had had a *few ladies* with me, the Rebels would not have been permitted to get away when they did!"[48]

Flirtations

Mary Lincoln was a bit of a coquette herself, but she was always upset by the sight of attractive women flocking around her husband. Once, just before Lincoln went down to a reception, he came into her room as she finished dressing and, "with a merry twinkle in his eyes," asked her which of the ladies he would be permitted to talk to that evening. He mentioned several names, but his wife called them all deceitful or detestable, and finally said she didn't approve of his "flirtations with silly women, just as if you were a beardless boy, fresh from school."

"But, mother," he sighed, after she had said something disparaging about every woman he mentioned, "I insist that I must talk with somebody. I can't stand around like a simpleton and say nothing. If you will not tell me who I may talk with, please tell me who I may not talk with." After she had named three women whom she particularly detested, he gave her his arm and they went down.[49]

Strong Opinions

Mrs. Lincoln had strong opinions about the men her husband worked with. She thought Secretary of the Treasury Salmon P. Chase untrustworthy and called Secretary of State William H. Seward a hypocrite. "If I listened to you," Lincoln once told her, "I should soon be without a Cabinet." "Better be without it than to confide in some of the men that you do," she returned. She was just as hard on Lincoln's generals. She called General McClellan a "humbug" and thought General Grant was a "butcher" who had no regard for human life. "Well, mother, supposing that we give you command of the army," Lincoln said, "no doubt you would do much better than any general that has been tried." Though Mrs. Lincoln insisted her husband "placed great confidence in my knowledge of human nature," her attempts to influence his appointments had little success."[50]

Dinner

When the Count de Paris and the Count de Chambord visited Washington in 1862, Secretary of State William Seward urged Mrs. Lincoln to entertain them in the White House. But Mrs. Lincoln was trying to economize at the time and refused to make arrangements for dinner unless the Government paid for it. When Seward insisted she pay for it herself, she decided to sell the mountain of manure that had just been delivered for the White House lawns and use the proceeds to pay for the dinner.[51]

Rebel in the House

One day, shortly after Lincoln invited his wife's half-sister Emilie, a Confederate sympathizer, to stay with Mrs. Lincoln after the death of Willie, General Sickles and Senator Harris came for a call, and at one point Harris asked Mrs. Lincoln: "Why isn't Robert in the Army?" Then as Mrs. Lincoln bristled, he went on: "He is old enough and strong enough to serve his country. He should have gone to the front some time ago." "Robert is making is preparations now to enter the Army, Senator Harris," said Mrs. Lincoln coldly; "he is not a shirker as you seem to imply for he has been anxious to go for a long time." She went on to defend her oldest son. "If fault there be, it is mine," she said, "I have insisted that he should stay in college a little longer as I think an educated man can serve his country with more intelligent purpose than an ignoramus." Harris interposed at this point. "I have only one son," he said, "and he is fighting for his country." Then he turned to Emilie. "And, Madam," he said, "if I had twenty sons they should all be fighting the rebels." "And if I had twenty sons," cried Emilie angrily, "they should all be opposing yours." And she left the room in tears with Mary following her.

A little later General Sickles went upstairs to report what had happened to Lincoln. Lincoln's eyes twinkled when he heard Emilie's retort, but Sickles saw nothing amusing about it. "You should not let that rebel in your house," he cried, pounding the table angrily. "Excuse me, General Sickles," said Lincoln quietly, "my wife and I are in the habit of choosing our own guests. We do not need from our friends either advice or assistance in the matter." Then he added: "Besides, the little 'rebel' came because I ordered her to come [to help Mary], it was not of her own volition."[52]

Stanton

A man arrived in Secretary of War Edwin Stanton's office with a card from Mrs. Lincoln requesting his appointment as a commissary. "There is no place for you," snapped Stanton, "and if there were, the fact that you bring me such a card would prevent my giving it to you." With this, he tore up the card. The following day the man appeared in Stanton's office again; this time he had a formal letter from Mrs. Lincoln making the same request. Stanton again threw him out of the office. He also sought out the First Lady to complain about the imposition on his time. "Yes, Mr. Secretary," she said, "I thought that as wife of the President I was entitled to ask for so small a favor." "Madam," he said, "we are in the midst of a great war for national existence. Our success depends upon the people. My first duty is to the people of the United States; my next duty is to protect your husband's honor, and your own. If I should make such appointments, I should strike at the very root of all confidence of the people in the government, in your husband, and you and me." Mrs. Lincoln realized at once that he was right about it. "Mr. Stanton," she said, "you are right, and I will never ask you for anything again." And she never did. She and Stanton, moreover, came to think highly of each other. Years later, when the irascible Radical Republican died, she wrote a friend: "I, too, dearly loved Mr. Stanton & greatly appreciated the services he rendered his country, our loved, bleeding land, during the trying rebellion."[53]

Not Well

On March 23, 1865, the Lincolns boarded the *River Queen* and headed for City Point, Virginia, to visit General Grant's headquarters, review the troops, and arrange conferences between the President and Grant. For the next few days Mrs. Lincoln, plagued with migraine headaches ever since the death of little Willie, was so short-tempered and irritable that she lost the respect of many of Lincoln's associates. During an excursion to the front (to review the Army of the Potomac) on March 26, when General Adam Badeau, one of Grant's aides, happened to mention that one of the officer's wives had obtained a special permit from the President to be at the front, Mrs. Lincoln exploded in wrath. "What do you mean by that, sir?" she cried. "Do you mean to say that she saw the President alone? Do you know that I never allow the President to see any woman alone?" Badeau hastily left.

The following day Mrs. Lincoln was in an even blacker mood. The Lincolns went to the front again, this time to review the Army of the James, commanded by General Edward Ord, but Lincoln went on horseback with the officers and Mrs. Lincoln went in a carriage with Mrs. Grant. But General Ord's young and extremely attractive wife went with the men, since there was no room for her in the carriage, and when Mrs. Lincoln arrived at the front and heard about it she was furious. "What does that woman mean by riding by the side of the President and ahead of me?" she stormed. "Does she suppose that *he* wants *her* by the side of *him?*" Mrs. Grant tried to pacify her and Mrs. Lincoln got angry with her too. "I suppose you think you'll get to the White House yourself, don't you?" she cried. Mrs. Grant murmured something about being satisfied with her present position which was, she said, far greater than she had ever expected to attain. "Oh! You had better take it if you can get it," cried Mrs. Lincoln. " 'Tis very nice." At this point Major Seward, nephew of Secretary of State Seward, rode up, totally unaware of the contretemps. "The President's horse is very gallant, Mrs. Lincoln," he said brightly. "He insists on riding by the side of Mrs. Ord." "What do you mean by that, sir?" yelled Mrs. Lincoln Seward disappeared at once. Moments later the carriage caught up with the Lincoln-Grant party and Mrs. Ord came riding over to greet the President's wife. But Mrs. Lincoln gave her such a severe dressing-down that she retreated in tears. At dinner that night Mrs. Lincoln engaged in a tirade against General Ord that astonished Grant, while Lincoln, with what Badeau called "an expression of pain and sadness that cut to the heart," tried to quiet her. During the remaining days at City Point, Mrs. Lincoln kept to her room with splitting headaches, and Lincoln explained her absence by saying "she is not well." She was far from well. Unfortunately, her outbursts at City Point, not long before Lincoln's assassination, though unusual for her, won her the unjust reputation among Lincoln's admirers of being a lifelong termagant.[54]

Woman-like

In October 1867 Mrs. Lincoln unexpectedly ran into her old friend Massachusetts Senator Charles Sumner and was bitterly reminded of better days. She was sitting in a railroad diner, she wrote a friend, with "my black veil . . . doubled over my face," and "immediately felt a pair of eyes gazing at me. I looked him full in the face, and the glance was earnestly returned." "Mr. Sumner," she murmured, conscious of her abject appearance, "is this indeed you?" His face, she observed, was "as pale as the table-cloth," as he said gently, "How strange you should be on the train and I not know it." Mrs. Lincoln wanted to escape and sought an excuse. "I must secure a cup of tea for a lady friend with me who had a headache," she told him and left her seat. When she returned, Sumner entered unexpectedly with a cup of tea "in his own aristocratic hands." But he was so deeply moved by Mrs. Lincoln's misery that his hands shook and he spilled tea all over her *"elegantly gloved"* hands. "His heart," Mrs. Lincoln wrote, "was in his eyes. . . . I never saw his manner so gentle and sad. . . ." After he left, Mrs. Lincoln was so overwhelmed by the turn her life had taken since her husband's death that *"woman-like* I tossed the cup of tea out of the window, and tucked my head down and shed *bitter tears.* . . ."[55]

Mary and Billy

Billy Herndon, Lincoln's law partner, got off to a bad start with Mary Lincoln. At the end of a dance with her, shortly after she married Lincoln, he told her she seemed "to glide through the waltz with the ease of a serpent." "Mr. Herndon," said Mrs. Lincoln coldly, "comparison to a serpent is rather severe irony." She didn't like him anyway; she thought he drank too much, was offended by the way he called himself an agnostic, and considered him in general rather sleazy.

Years later, Ward Hill Lamon published the *Life of Abraham Lincoln* (1872), based on material he had purchased from Herndon, containing the Ann Rutledge story, Herndon's belief that the Lincoln marriage had been unhappy, and the statement that Lincoln was not a Christian. Mrs. Lincoln at once denounced the book as filled with "sensational falsehoods" and "base calumnies," but the book soon touched off a controversy about her husband's religion.

On December 12, 1873, Herndon gave a lecture in Springfield in which he announced that Lincoln, like himself, was an infidel. To back up his statement he quoted Mrs. Lincoln as having told him in an interview in 1866 that her husband was not a "technical Christian." The

Herndon lecture produced a public quarrel between Mrs. Lincoln and her husband's former law partner. On December 19 the *Illinois State Journal* reported that Mrs. Lincoln "denies unequivocally that she had the conversation with Mr. Herndon, as stated by him." Herndon then called her a liar and gave full details of an interview he had had with her in 1866. Mrs. Lincoln then made it clear that she hadn't denied the conversation with Herndon; she had denied the way he had interpreted her remarks. "I told him in positive words," she wrote, "that my husband's heart was naturally religious," and that he often turned to the Bible for comfort and frequently attended church with her, though not a member. "What more can I say in answer to this man," she ended, "who when my heart was broken with anguish, issued falsehoods against me & mine, which were enough to make the Heavens blush?"[56]

Not God's Will

When Mrs. Lincoln returned from Europe on *L'Amerique* in October 1880, Sarah Bernhardt, the noted actress, was a fellow passenger. Early one morning Bernhardt went out on deck for some fresh air and met a woman in black "with a sad, resigned face." She was about to engage her in conversation when a wave hit the ship, the ship lurched, and the woman in black was almost thrown down the stairway. Bernhardt grabbed the legs of a bench to keep from falling and then caught the woman by the skirt to prevent her from tumbling down the stairs. "She thanked me in such a gentle, dreamy voice," Bernhardt recalled, "that my heart began to beat with emotion." "You might have been killed, madame," Bernhardt told her, "down that horrible staircase." "Yes," said the woman, "but it was not God's will." Then she revealed that she was President Lincoln's widow. Bernhardt couldn't help thinking: "I had just done this unhappy woman the only service that I ought not to have done her—I had saved her from death."[57]

CHAPTER 16

ELIZA JOHNSON

1810–1876

In May 1868, when Colonel William H. Crook, Andrew Johnson's bodyguard, learned that the President had been acquitted in his impeachment trial, he rushed out of the Capitol, dashed down the steps, ran along Pennsylvania Avenue as fast as he could, and burst into the White House library to tell Johnson and his friends the good news. Then he hurried up to Mrs. Johnson's room on the second floor and yelled: "He's acquitted! The President is acquitted!" The President's frail little wife at once got up from her chair, grasped Crook's right hand warmly with both hands, and, with tears in her eyes, exclaimed: "Crook, I knew he'd be acquitted; I knew it. . . . Thank you for coming to tell me."[1]

Eliza Johnson's faith in her controversial husband was unbounded. She knew he was short-tempered and stubborn, and at times had a chip on his shoulder, but she also knew he was a man of integrity. "Tell it as it is or not at all!" was the slogan of the Johnson household, and it had gotten the President into trouble more than once.[2] On the eve of the Civil War, when he was a U.S. Senator from Tennessee, Johnson had insisted on touring his state to speak against secession and had his life threatened. And when he became President upon Abraham Lincoln's assassination and fought against the Reconstruction policies of Radical Republicans in Congress, he touched off a movement to remove him from office. During the three-month impeachment crisis he visited his ailing wife every morning to seek aid and counsel. If he followed his conscience, she told him, he had nothing to fear. Her support was always moral, not political, for she never involved herself in policy.

Johnson's policies as President produced hatred and contempt in and outside of Congress; but his practices as a resident of the White House evoked nothing but praise. Even his worst enemies could think up nothing unkind to say about his family while Johnson was President. When the Johnsons left Washington after General Grant's inauguration as President in March 1869, everyone agreed that their behavior had been impeccable. "The occupants of the White House leave it with the most spotless social reputation," wrote a Washington journalist. "There is no insinuation, no charge against them. They have received no expensive presents, no carriages, no costly plate. They will be remembered in Washington as high-minded and honorable people. The House has been kept in order, elegance, and liberal hospitality. No old friends cut, no new ones toasted; but an even tenor of sociability has made all feel welcome."[3]

It was Andrew Johnson's daughter, Martha Patterson (whose husband represented Tennessee in the Senate), who was primarily responsible for running the White House during his tenure of office. His wife Eliza had tuberculosis, and in those days the treatment prescribed was to stay indoors and rest as much as possible. When the Johnsons moved into the White House following Mary Lincoln's departure in August 1865, Mrs. Johnson selected a small room on the second floor (across from the room where her husband planned to work) as her bailiwick and from then on lived most of her White House life there, sewing, reading, crocheting, and playing with her grandchildren. On rare occasions she appeared at the breakfast table; most of the time, though, she ate meals in her room. She did, to be sure, oversee her husband's wardrobe and see that he got the food he liked; she also gave him sensible advice when he took his problems to her. "Now, Andrew," she said reprovingly, if he got excited; sometimes when he was sounding off, she put her hands on his shoulders and cried sternly: "Andy!" But she left all the White House entertaining to her daughter.[4]

Mrs. Patterson was an unpretentious but gracious hostess. "All parties agreed," according to one Senator, "that the White House was never more gracefully kept and presided over than by Mrs. Patterson." But because flocks of curious sightseers had invaded the White House after Lincoln's assassination and helped themselves to souvenirs, the place was in shambles when the Johnsons moved in: curtains torn, carpets filthy, chairs broken. Congress appropriated $30,000 for renovation, not nearly enough, but Mrs. Patterson managed rehabilitation so efficiently that people were astonished at what she accomplished. "When it was opened for the winter season," reported New York writer Mary Clemmer Ames, "the change was apparent and marvellous, even to the dullest eyes, but very few knew that the fresh, bright face of the historic house was all due to the energy, industry, taste and tact of one woman, the President's daughter."[5] Mrs. Patterson also acquired a couple of Jersey cows to graze on the White House grounds, supervised the dairy, and saw to it that there was fresh milk and butter for the White House occupants every day.

Simplicity was the watchword of the Johnsons. "We are a plain people from the mountains of Tennessee," Mrs. Patterson told guests at an early White House reception, "called here for a short time by a national calamity. I trust too much will not be expected of us."[6] Both her parents had humble beginnings. Her father, Andy, was a tailor's apprentice, just starting out on his own, when he first met her mother, Eliza McCardle, in Greeneville, Tennessee, in 1826. Eliza was the daughter of a shoemaker who died when she was young, and her widowed mother supported the two of them by making patchwork quilts and cloth-topped sandals. After their marriage in 1827, the Johnsons moved into a two-room house—the front room was Johnson's tailor shop—and Eliza helped Andy with reading, writing, and arithmetic, for he was utterly without schooling, and she had attended Rhea Academy in Greeneville for a time. Johnson's rise was rapid—in politics (mayor, state legislator, Congressman) as well as in business—and by the time he entered the U.S. Senate in 1858 he was well off.

In March 1862, when Lincoln made Johnson, a War Democrat, military governor of Tennessee, with headquarters at Nashville, Mrs. Johnson remained in Greeneville, as always when her husband was away, to manage the family business. In April, however, Jefferson Davis imposed martial law on all of East Tennessee and Unionists like Mrs. Johnson were given thirty-six hours to leave. Mrs. Johnson pleaded illness and was allowed to stay; but the Johnson property was taken over by the Confederates and she found refuge along with her eight-year-old son at the home of a near-by relative. In September she applied for and received permission to cross Confederate lines to join her husband; and after a grueling journey, involving cold, hunger, and much intimidation, she finally arrived in Nashville, the following month, worn-out and seriously ill. In March 1865, when Johnson went to Washington to take his oath of office as Vice President, she remained in Nashville. Lincoln's assassination the following month made her fearful for her husband's safety. "The *sad, sad* news has just reached us, announcing the death of President Lincoln," Martha wrote her father from Nashville. "Are you safe and do you feel SECURE? . . . Poor Mother, she is almost deranged fearing you will be assassinated."[7]

Mrs. Johnson brought a carload of people with her when she moved into the White House: two sons, daughter Martha, Martha's husband, and the three Patterson children; and daughter Mary Stover, a widow, and her two children. They were her main company—the young ones adored "grandma"—for she saw few people outside of the family and refused all requests for interviews. Only twice did she go downstairs to appear in public: in August 1866, when Hawaii's Queen Emma, on a world tour, visited the White House; and in December 1868, when Martha arranged a children's ball (the first ever held in the White House) on the occasion of her father's sixtieth birthday. Both times Mrs. Johnson received guests while seated. For the "Juvenile Soirée," as it was somewhat pretentiously (for the Johnsons) called, she greeted the scores of youngsters as they arrived and then watched them dance in the East Room: the Esmeralda, the Varisoivienne, the Basket Quadrille, the Quadrille Sociable, and even a Waltz. Mrs. Ames, who covered the event, described her as "a lady of benign countenance and sweet and winning manners" who charmed everyone. "At that time," wrote Mrs. Ames, "she was seated in one of the republican court-chairs of satin and ebony. She did not rise when the children or guests were presented, but simply said, 'My dears, I am an invalid,' and her sad, pale face and sunken eyes proved the expression."[8]

From almost the day she arrived in the White House, Mrs. Johnson looked forward to leaving. "Crook," she told the President's bodyguard, "it's all very well for those who like it—but I do not like this public life at all. I often wish the time would come when we could return to where I feel we best belong."[9] Her daughter Martha felt the same way about it. At the last official White House dinner she said frankly: "I am glad this is the last of entertainments—it suits me better to be quiet and in my own home. Mother is not able to enjoy these things. Belle [Martha's daughter] is too young and I am indifferent to them—so it is well it is almost over."[10]

After leaving Washington, the Johnsons returned to Greeneville to take up residence in their restored and enlarged house there. But Johnson did not abandon public life; anxious to vindicate himself with the American public, he ran for Congress and was defeated twice. In 1874, however, the Tennessee legislature elected him to the Senate and the following year he appeared triumphantly on the Senate floor again and was warmly greeted by his colleagues. Shortly afterwards he died of a stroke. His wife survived him by only six months.

* * * *

Beau Ideal

One morning in the fall of 1826, young Andy Johnson arrived in the little town of Greeneville, Tennessee, looking for work as a tailor. As he trudged down the street, with his mother and stepfather and a wagon loaded with chairs, tables, pots, and pans, a little group of girls by the roadside were amused by the sight. Whispered one of the girls: "What a handsome beau he will make for some Greeneville girl—when he gets his face washed." "Don't laugh, Sarah," said Eliza McCardle, "I like him." Andy liked her too. When he asked where he could spend the night and Eliza helped him find a place, her girlfriends teased her about her strange "beau" and she is supposed to have said that he might really be her beau someday. Johnson eventually found work in Greeneville, began courting Eliza, and they were married on May 17, 1827. Years later a White House official called their union the "nearest approach to ideal marriage I've ever seen or known."[11]

CHAPTER 17

JULIA DENT GRANT

1826–1902

Right after Ulysses S. Grant finished reading his inaugural address on March 4, 1869, he turned to his wife Julia, took her hand, and said quietly: "And now, my dear, I hope you're satisfied."[1] Mrs. Grant was more than satisfied; she was elated. Her beloved husband—the Hero of Appomattox—had once again lived up to her belief when she first met him years before that he was destined for great things. The inaugural ball that evening may have been a bit disorganized, but she was pleased and happy through it all. She was eager to renovate and refurbish the White House and then make her debut there as the nation's primary hostess.

With the Grants in the White House the Gilded Age (as Mark Twain called it) had an appropriate First Family. The President and his wife both admired the Captains of Industry and Lords of Finance who were taking over the country after the Civil War, and they also liked the splendor and elegance with which the newly rich surrounded themselves and entertained their friends. President and Mrs. Grant were, to be sure, modest and unassuming people when they appeared in public and they strove to be loyal to Jeffersonian canons of republican simplicity in the Executive Mansion; at the same time they enjoyed hobnobbing with the high and mighty: rich industrialists, powerful bankers, high-ranking military and naval officers, influential newspaper publishers, distinguished diplomats, and fashionable people in general. They also took great pleasure in entertaining royal personages from abroad in the White House. After the Civil War, democratic America's passion for Kings and Queens and for the fancy titles forbidden by the U.S. Constitution reached astonishing proportions.

In Julia Dent Grant the American people had a President's wife for a change who thoroughly enjoyed herself in Washington. Eight years there, in fact, were not enough for her; she wanted another four years after her husband completed two terms in office, and was sorry she didn't get them. Mrs. Grant was generally liked by the American public; even her husband's enemies left her alone for the most part. She was, according to the *New York Tribune,* writing a year before her husband's accession to the Presidency, "a sunny, sweet woman; too unassuming to be a mark of criticism, too simple and kindly to make the mistakes which invite it."[2]

Simple and kindly in many ways she was; but she was also a woman of fashion. From the beginning Washington society looked with approval on her demeanor as White House hostess. "This has been a roystering season," announced a society writer after Grant had been President a year. "There seems to be a social rebound after the anxieties and estrangements of the war. One week of our gala life will show any sojourner or spectator that we are in truth upon an era of good feeling. I do not think we ever had so brilliant a season. . . . We have had more good dinners and pleasant reunions than in any Presidency that I can remember." Mrs. Grant's receptions were indeed elegant, he wrote a year later, but they were also truly "republican" in nature: "There the poorest working woman can go, whether *en train* or in her plain work-day dress, touch the hand and gaze into the plain but kindly sympathetic face of the President's wife. Not a few avail themselves of this privilege who neither wear trains, low corsages or high colors."[3] But social observer Benjamin Perley Poore (who signed his first name "Ben:") was amused by the mingling of the low and the lofty in the Grant White House. "There were ladies from Paris in elegant attire and ladies from the interior in calico," he noted; "ladies whose cheeks were tinged with rouge, and others whose faces were weather-bronzed by outdoor work; ladies as lovely as Eve, and others as naughty as Mary Magdalene; ladies in diamonds, and others in dollar jewelry; chambermaids elbowed countesses, and all enjoyed themselves."[4]

Mrs. Grant held weekly receptions, open to all, at first on Tuesday and then on Saturday afternoon. Departing from custom, she invited the Cabinet ladies to join her on the receiving line; and when Secretary of State Hamilton Fish toyed with the idea of leaving the Cabinet, she anxiously urged him to stay on so Mrs. Fish could continue helping out at public functions. The Grants held state dinners and New Year's receptions as well as weekly levees; they also had special receptions for special people like Prince Arthur (third son of Queen Victoria), the Grand Duke Alexis (third son of the Russian Czar), and, last but probably least, King Kalakua of Hawaii (who may or may not have been a third son, for all that Washington cared). Sometimes the best-laid plans of Mrs. Grant and the White House staff went ruinously agley. Washington was so bitterly cold at the time of the second inaugural ball in March 1873 that many people left early, those who stayed tried dancing all bundled up, and a woman with bronchial trouble suddenly met her Maker on the dance floor. But the resplendent wedding in May 1874 of charming young Nellie Grant to Britisher Algernon Sartoris, nephew of actress Fanny Kemble, had all Washington a-twitter; there were two hundred distinguished guests, eight bridesmaids, lavish floral decorations, expensive wedding presents (including a 500-dollar handkerchief), and a many-coursed banquet culminating in *Epigraphe la fleur, de Nelly Grant,* which the White House chef had dreamed up for the occasion.[5]

The world of fashion was not new to Mrs. Grant when she entered the White House. Born in St. Louis in 1826, the daughter of a Missouri planter, she grew up in comfortable surroundings, with slaves in attendance, and, though plain-looking and a bit cross-eyed, she moved comfortably in St. Louis society as a young woman and was popular with young men because of her warm heart and merry ways. Her father, Frederick Dent, liked young Lieutenant Grant well enough when he first met him in the spring of 1844, but he was upset when Grant started courting his daughter. He wasn't sure that the quiet lad from Illinois was good enough for his favorite child, Julia, and he was afraid that the lass would find life as an army wife lonely and oppressive. But Julia and "Ulys," as she called him, took to each other at once. They both loved horses and were expert riders; they also shared a common interest in growing things, especially flowers. To Colonel Dent's chagrin they fell deeply in love.

For Grant it was love at first sight; Julia was the "lady of his dreams," the only woman he ever loved. She was, as a friend later suggested, "his salvation."[6] She gave him the kind of deep affection and warm companionship he had never known as a boy growing up; and after their marriage in St. Louis in 1848 he always felt confused and aimless and took to drinking when separated long from her. As for Julia, the shy, handsome Ulys was the first young man she ever took seriously; she not only loved him deeply but was also sure he had a bright future ahead of him. Grant's career had its frustrations, disappointments, and setbacks; but his marriage never faltered. After their wedding day, he and Julia were always happy when together and unhappy only when apart. "They were a perfect Darby and Joan," observed Horace Porter, Grant's aide-de-camp, who saw much of them at Grant's headquarters during the Civil War. "They would seek a quiet corner of his quarters of an evening, and sit with her hand in his, manifesting the most ardent devotion; and if a staff-officer came accidentally upon them they would look as bashful as two young lovers spied upon in the scenes of their courtship."[7] The Grants were affectionate, even indulgent, as parents, too, and their daughter and three sons adored them. "I have never seen a case of greater domestic happiness than existed in the Grant family," declared George W. Childs, editor of the *Philadelphia Ledger,* one of their wealthy friends.[8] When they toured the world after leaving the White House, one of their friends saw them in Switzerland and was amused to catch them flirting with each other in public.

Julia Grant was proud to be the wife of a Great Man—a General and a President—and she took her duties as hostess seriously because she wanted to win as much good will as she could for her husband. But she was no clinging vine. She was a strong woman, physically and emotionally, and she did not hesitate to assert herself when she felt strongly about her husband's associates and thought it important to warn him against people she considered untrustworthy. At times, too, she had vigorous opinions about policy as well as people and enjoyed passing them on to her husband. Sometimes Grant took her advice and sometimes he didn't; and frequently he went ahead and did what he wanted to do without consulting her at all. Adam Badeau, Grant's secretary in the White House, thought that her judgment of people was occasionally better than Grant's, and that she was a real help to him. In the main, though, he observed, she advised him "simply and strongly to do what he thought right, and perhaps induced him to do it; although he, as little as any man, I believe, required such inducement."[9]

But there were clashes. Mrs. Grant admitted as much in the sunny memoirs she wrote after her husband's death in 1877. Her memoirs (not published until long after her own death) cast a rosy light on their years together, even the discouraging ones, but they do include wry recountings of the times she and Ulys tested their strength. At one point during the Civil War, when she urged him to relieve General William Rosecrans of his command, she recalls that he became impatient. "Do not trouble yourself about me, my dear little wife," he told her. "I can take care of myself." On another occasion she reveals that it took bitter tears and two hours of sullen silence to persuade him to release an ailing Confederate officer (whose friends had interceded with her) from harsh confinement. And once, after he finally yielded to her impassioned plea to pardon a young man about to be shot as a deserter (and whose wife had sought her out), he told her: "I'm sure I did wrong. I've no doubt pardoned a bounty jumper who ought to have been hanged."[10]

In April 1863, Mrs. Grant even tried her hand at running the war. She had hardly arrived at Grant's headquarters for a visit when she asked him, "Why do you not move on Vicksburg at once? Do stop digging at this old canal. You know you will never use it. Besides, that is what McClellan has been doing. Move upon Vicksburg and you will take it." Apparently amused, Grant asked her if she had a plan of action to propose. She did of course. "Mass your troops in a solid phalanx at a point north of the fortress," she advised, "rush upon it, and they will be obliged to surrender." "I am afraid," said Grant with a smile, "your plan would involve great loss of life and not insure success. Therefore I cannot adopt it. But it is true I will not use the canal. I never expected to, but started it to give the army occupation and to amuse the country until the waters should subside sufficiently to give me a foothold and then, Mrs. Grant, I will move upon Vicksburg and will take it, too. You need give yourself no further trouble. This will all come out right in good time, and you must not forget that each and every one of my soldiers has a mother, wife, or sweetheart, whose lives are as dear to them as mine is to you. But Vicksburg I will take in good time."[11] It was her turn to yield. After he took Vicksburg, she began calling him "Victor."

When Grant became President, he teased his wife about having to hide his list of Cabinet appointments to keep her from interfering with his choices. Later on, to be sure, he did talk things over with her from time to time and he also sought her opinions on various issues. Sometimes, though, he thought her advice injudicious and other times quite gratuitous. "Mrs. Grant," he once told her good-naturedly, "I think you are a—what shall I call you—mischief!" Another time he exclaimed: "It is lucky you are not the President. I am afraid you would give trouble."[12] It's a pity she didn't warn him against some of the sleazy men he appointed to high office who by their shenanigans succeeded in making his administration one of the most corrupt in American history. But then she, too, was naive in many ways.

Mrs. Grant loved being First Lady and hoped for a third term. But Grant was ready to retire after two terms and she never forgot how he tricked her into accepting his decision. Early in the summer of 1875, she recalled, her suspicions were aroused when members of the Cabinet began assembling unexpectedly one Sunday afternoon in her husband's office. "Why is it you have happened to call today?" she wanted to know. "I am sure there is something unusual." There was much hemming and hawing, and, when Grant arrived, she left the room without having her curiosity satisfied. Later she learned that after she left Grant told his Cabinet that he planned to send a letter to the Republican State Convention in Pennsylvania (which had just formally proposed his renomination) announcing that he had no intention of running again. And when the Cabinet meeting broke up, he did just that; he wrote the letter, showed it to his secretary, Adam Badeau, and told Badeau he was sending it to the press without telling his wife, for he wanted to present her with a *fait accompli.* He even went out to mail the letter himself. As soon as he returned, Mrs. Grant cornered him. "I want to know what is happening," she exclaimed. "I feel sure there is something and I must know." With a smile on his face Grant told her about the letter he had just sent out rejecting proposals for a third term. "And why did you not read it to me?" she cried. "Oh, I know you too well," said Grant, "it would never have gone if [you] had read it." "Bring it and read it to me now," she persisted. "No," said Grant, "it is already posted. . . ." "Oh, Ulys!" she wailed, "was that kind to me? Was it just to me?" In her memoirs, Mrs. Grant wrote that she was "deeply injured" by what her husband had done; later, she reproached Badeau for not having kept Grant from mailing his letter.[13] Her disappointment was profound and took time to get over. She wept upon leaving the White House, and told her husband: "Oh, Ulys, I feel like a waif, like a waif on the world's wide common."[14]

But Mrs. Grant was never a waif on the world's common. Not long after ceasing to be First Lady, she became a triumphant Penelope (as she playfully called herself) accompanying her beloved Ulysses on an ambitious Odyssey around the world. In May 1877 the Grants set sail for England; and for the next two years and a half they made the rounds: in Europe, the Middle East, and the Orient. Everywhere they went they were greeted warmly and entertained lavishly; and they had no trouble holding their own with the great and the near-great (and the not-so-great) whom they encountered in one country after another around the globe. In Britain they met Benjamin Disraeli, Anthony Trollope, Robert Browning, and Arthur Sullivan, and slept at Windsor Castle.[15] In Belgium, their host was King Leopold; in Austria, Emperor Francis Joseph; in Russia, Czar Alexander II; and in Germany, Prince Bismarck. (They also met Richard Wagner in Germany, though, in her memoirs, Mrs. Grant got him confused with Franz Liszt.) In the Middle East, they took a trip up the Nile, visited the Pyramids, and to Mrs. Grant's immense pleasure (for she was a good Methodist), they spent some time in the Holy Land. In China, General Li Hung-chang, "the Bismarck of the East," took a fancy to the Grants, and in Japan, Emperor Mutsuhito gave Mrs. Grant a set of drawing-room furniture after she had lavished praise on it.

Everywhere the Grants went they made headlines; and in the United States the press chronicled their adventures with tender-loving care. When they finally got back to the United States in the fall of 1879, Old Guard Republicans were pushing the Hero of Appomattox for the presidential nomination the following year. Grant said he was "supremely indifferent" to the idea, but, to his wife's joy, he made it clear he was available. But by the time the Republicans met to pick their candidate, an anti-third term movement had developed within the party and it was clear that Grant was no shoo-in for the nomination.

As the Republicans began balloting in Chicago in June 1880, Grant approved a letter withdrawing his candidacy in case his chances for the nomination became hopeless. But when Mrs. Grant heard about it, she protested vigorously; Grant's managers were "*not* to use that letter," she ordered. "If General Grant were not nominated," she said, "then let it be so, but he must not withdraw his name—no, never." Since Grant was going to Milwaukee for a Union Army veterans' meeting about this time, she suggested stopping off in Chicago en route and appearing on the floor of the Convention to rally support for the nomination. But Grant adamantly refused to do so; he said he would rather cut off his right hand. "Do you not desire success?" she asked him. "Well, yes, of course," he said, "since my name is up, I would rather be nominated, but I will do nothing to further that end." "Oh, Ulys," she sighed, "how unwise, what mistaken chivalry. For heaven's sake, go—and go tonight. I know they are already making their cabals against you. Go, go tonight. I beseech you." But Grant refused to go hat in hand to the Convention. "Julia," he exclaimed, "I am amazed at you."[16] When the balloting began, Grant's name led on the first trial; and his faithful supporters stuck loyally by him in ballot after ballot for the rest of the week. In the end, however, his opponents came together to unite on James A. Garfield and succeeded in putting Garfield over on the thirty-sixth ballot. Mrs. Grant always regretted the way things turned out, but she and her husband loyally supported Garfield in the campaign that followed.

The Grants's last years together were pleasant, if placid, for a time, and then turned sorrowful. In May 1884, a brokerage firm with which Grant was associated collapsed and the Grants found themselves strapped for funds. To earn money to support Julia comfortably and to pay off his debts, Grant began writing articles on Civil War battles for *Century Magazine;* then, with a generous contract arranged for him by Mark Twain, he began working on his *Personal Memoirs.* About this time he learned he had cancer of the throat and that it was too late to operate. Mrs. Grant couldn't bring herself to believe at first that he was critically ill, but she stayed close by, supervised his medication, and tried to cheer him up with light-hearted talk and in-house jokes. At length, though, she realized, as he had long since realized, that there was no hope. Toward the end his pain was almost insupportable and he wrote: "A verb is anything that signifies to be; to do; or to suffer. I signify all three." Mrs. Grant suffered with him. "My tears blind me," she wrote a friend in February 1885. "Genl. Grant is very ill. I cannot write how ill."[17]

When Grant died in July 1885, with his *Personal Memoirs* completed and ready to be published, his wife's grief for a time was almost unbearable. "My beloved, my all, passed away," she wrote later, "and I was now alone."[18] She was not of course alone; she had her sons and daughter and grandchildren, and she also had countless friends and admirers. She was able to travel, too, for the royalties from her husband's book made her a rich woman. Eventually she settled down to a comfortable life in Washington and worked on her own memoirs. There she died in December 1902, at seventy-seven, and was buried beside her husband in the huge mausoleum his friends had established for him years before in New York City's Riverside Park overlooking the Hudson River.

* * * *

Crossing a Bridge

One day Grant arrived at the "White House," the Dent home near St. Louis, and found Julia's family all dressed up and ready to go to a wedding. They invited him to go along, so he drove Julia in a buggy to the church. On the way they came to a flimsy bridge nearly submerged in the waters of a creek swollen by heavy rains, and Julia became timorous. "Are you sure it's all right?" she asked. "Oh, yes," said Grant; "it's all right." "Well, now Ulysses," said Julia, "I'm going to cling to you if we do go down." "We won't go down," Grant assured her, and drove across the bridge with the frightened Julia holding on to his arm. When they reached the other side, she released her hold, and he drove on for some time in thoughtful silence. Then, clearing his throat, he said, "Julia, you spoke just now of clinging to me, no matter what happened. I wonder if you would cling to me all my life." Her answer was favorable.[19]

Frankness

One morning in May, Lieutenant Grant called on Julia's father in uniform and announced: "Mr. Dent, I want to marry your daughter, Miss Julia." "Mr. Grant," said Dent, after thinking things over, "if it were Nelly [Julia's sister] you wanted, now, I'd say Yes." "But I don't want Nelly," said the young soldier firmly. "I want Julia." "Oh, you do, do you?" cried Dent. "Well, than I s'pose it'll have to be Julia." Emma, Julia's other sister, who overheard the exchange, thought Grant's "frankness had pleased him, and had, I think, won him over in spite of himself."[20]

Not Too Young

During the first few months of the Civil War, Grant had his eleven-year-old son Fred with him a great deal of the time, but before moving into Missouri in July 1861 he sent him home. "We may have some fighting to do," he wrote Mrs. Grant, "and he is too young to have the exposure of camp life." Mrs. Grant disagreed. She wrote urging Grant to keep the lad with him. Knowing her military history, she reminded him that Alexander the Great had been just as young when he accompanied his father, Philip of Macedon, on his military ventures. Grant was amused by her argument, but her letter reached him too late.[21]

Sorrowful

One afternoon, not long before the successful movement on Richmond, General Grant and his wife were discussing the war in his cabin office. "Ulys," said Mrs. Grant, "why do you not tell me something of your plans?" "Would you like to know of my plans?" cried Grant. "Of course I would," said Mrs. Grant eagerly. "I will show you with pleasure," said Grant, turning to a map on his office table. Then he began locating the various Union and Confederate forces on the map and when he had made the situation clear he said, "Now you have the position of all the armies. You observe it in a perfect cordon from sea to sea again." "Yes," said Mrs. Grant, "what next?" "Do you want to know more?" said Grant. "Yes, of course," said Mrs. Grant; "everything I ought to." "Well," said Grant, "I am going to tighten that cordon until the rebellion is crushed or strangled." The thought of this made Mrs. Grant feel sorrowful and she told her husband how she felt. "Yes," said Grant, "war is always sorrowful, but only think how dreadful it would be if a cordon like that I have just pointed out to you encircled our Union."[22]

Agreement

When Lincoln returned from a conference with Confederate officials at Hampton Roads in February 1865, Mrs. Grant waylaid him on his way to confer with General Grant, and asked if he had reached an agreement on ending the war. "Well, no," said Lincoln sadly. "No!" cried Mrs. Grant. "Why, Mr. President—are you not going to make terms with them? They are our own people, you know." "Yes," said Lincoln. "I do not forget that." Then he took a large piece of paper from his pocket, unfolded it, and read aloud the terms he had proposed. Mrs. Grant thought them extremely generous and cried, "Did they not accept those?" "No," said Lincoln wearily. "Why, what do they want?" exclaimed Mrs. Grant, indignantly. "That paper is most liberal." Said Lincoln with a little smile: "I thought when you understood the matter you would agree with us."[23]

Modest and Not Aggressive

One day, shortly after she entered the White House, Mrs. Grant sat in the library, dressed for a reception, and one of the ushers appeared at the door and, bowing low, said, "Madam, if any colored people call, are they to be admitted?" "This is my reception day," Mrs. Grant said, after thinking about it for a minute or two. "Admit all who call." But no blacks called that afternoon; "nor did they at any time during General Grant's two terms of office," she wrote in her *Memoirs*, "thus showing themselves modest and not aggressive, and I am sure they, as a race, loved him and fully appreciated all that he had done for them."[24]

Installments

When Mrs. Grant was in the White House, she encouraged all the servants in the White House to purchase homes of their own while the city was still small and prices modest. When one of the blacks who waited on table ignored her advice, she took him aside one day to lecture him. "Harris," she said, "if you do not buy a home at once, and commence paying for it while houses are cheap, your opportunity will soon be gone. The time is coming when there will be a great change in real estate values all over the city. Washington will grow into a big place so suddenly that you will never again have the chance that you now possess. If you do not go out and select a home and commence to pay for it, I will buy one for you myself; and I will take out of your wages each month enough to pay for the installments." She finally convinced him.[25]

The President, Immediate

Whenever Mrs. Grant needed to interrupt her husband when he was busy in his White House office, she would take a sheet of note paper, fold it in three to make it into an envelope, and write, "The President, immediate" on it. On May 22, 1875, her note read: "Dear Ulys: How many years ago to day is [it] that we were engaged: Just such a day as this too was it not? Julia." Soon she got her answer: "Thirty-one years ago. I was so frightened however that I do not remember whether it was warm or snowing."[26]

All the Talking

Grant's taciturnity was well-known. Once when the train he was traveling on stopped at a station and was soon surrounded by crowds of people, a woman in the crowd suddenly cried: "I want to see the man that lets the women do all the talking!"[27]

A Little Strategy

Mrs. Grant was annoyed when newspapers reported that her husband was "no conversationalist; in fact, a sphinx." Having lived with him for years, she knew he could talk interestingly and at length on many matters. "But I must confess," she wrote in her memoirs, "that I had to resort to a little strategy sometimes to start him off when I wished him to be especially interesting to others, and this plan I confided to several of the General's friends, who often used it with great success. I would begin to tell something with which I knew he was perfectly familiar and would purposely tell it all wrong. Then the General would say, 'Julia, you are telling that all wrong,' and seemed quite troubled at my incompetency. I would innocently ask, 'Well, how was it then?' He would begin, tell it all so well, and for the remainder of the evening would be so brilliant and so deeply interesting."[28]

Same Eyes

Mrs. Grant was cross-eyed and while her husband was President she toyed with the idea of having corrective surgery. But just as she was about to make final arrangements for an operation, Grant decided to talk her out of it. ". . . I don't want to have your eyes fooled with," he told her. "They are all right as they are. They look just as they did the very first time I ever saw them—the same eyes I looked into when I fell in love with you—the same eyes that looked up into mine and told me that my love was returned. . . . This operation might make you look better to other people; but to me you are prettier as you are—as you were when I first saw you. . . ." "Why," cried Mrs. Grant, "it was only for your sake that I was even thinking of having anything done." She immediately canceled surgery.[29]

Conundrum

One morning General Grant came into his wife's room and said, "Julia, I have a new conundrum for you." "Yes?" said Mrs. Grant curiously. "Why is victory like a kiss?" Grant exclaimed. Mrs. Grant repeated the question and then confessed she had no answer. "Why," exclaimed Grant, "it is easy to Grant." "Ah, that is charming," laughed Mrs. Grant. "Is it original with you?" "No," said Grant, and then told her that he learned it from a young woman at a little party the day before and "when she gave the flattering answer, I accepted the challenge with promptness and took the kiss. Yes, two of them. Do you not think I was right?" "Why, Ulys," wailed Mrs. Grant, "how could you? I really think some girls very bold—dreadful." But Grant seemed to enjoy her annoyance more than she did his conundrum.[30]

Jumper

One afternoon, when the Grants were resting on the porch of the summer White House at Long Branch after lunch, their son Buck suddenly got up, vaulted over the porch railing, and disappeared. "Did you see that, Julia?" said Grant playfully. "Now if you were sitting here alone, and the railing extended across the steps, and the cottage caught fire, you must be rescued or burned." "You think so!" cried Mrs. Grant defiantly. Then she jumped up and vaulted over the railing the way Buck had done and then stood below laughing at her husband.[31]

Wife of Great Ruler

When the Grants dined with Queen Victoria and she said something to Mrs. Grant about her laborious duties, Mrs. Grant replied, "Yes, I can imagine them: I too have been the wife of a great ruler."[32]

Enchanted

The Grants were royally entertained by Prince Bismarck and his wife when they were in Germany, and Mrs. Grant was particularly taken by the Prince. When she gave her hand to Bismarck to say farewell, he bent low over her hand and kissed it. "If that were known in America, Prince Bismarck," said Mrs. Grant, laughing, "every German there would want to be kissing *my* hand." "I would not wonder at all at them," said the Prince, looking admiringly at her hand. "I was, of course," Mrs. Grant said years later, "enchanted with Prince Bismarck."[33]

A Little Joke

When the Grants were visiting some glacier fields in Switzerland, Grant stopped for a smoke with son Jesse, and his wife decided to walk on ahead and join her husband when the carriage caught up with her. After walking a while, she sent her maid Bella to see if her husband and son were coming and when she was told the carriage was on its way, she decided to have a little fun. "I am going to hide," she told Bella. "You tell them I am lost, that you do not know where I am." "Oh, Madam," said Bella, "I couldn't say that, for it would not be true." "Well, then," cried Mrs. Grant, "say 'When I went to see if you were coming, Mrs. Grant was on that bank. When I returned, she was not there.' That is literally true." Then she hid behind a big tree, just as the carriage drove up. "Why, Bella," cried Grant, "where is Mrs. Grant?" Holding her veil up to conceal her laughter, Bella answered as her mistress had instructed. Grant at once leaped out of the carriage and began looking over the precipice. "Why, of course she has gone on up the mountain," he finally decided. "Drive on at once," he told the driver. At this point Mrs. Grant came out of her hiding place and cried, "Oh General, here I am!" Grant was not amused and he made his displeasure clear. But his wife continued with her teasing. "What did you think had become of me?" she wanted to know. "I did not know what to think," confessed Grant. "But," he added severely, "the next time, I will drive on and leave you."[34]

Enchantments

One evening, when the Grants were cruising the Mediterranean on the *Vandalia,* one of the officers came running up and cried: "Mrs. Grant, hasten and get out your cotton wool. We are nearing the island of Calypso. Remember the fate of Ulysses of old. Take warning and bring out the cotton wool and fill the General's ears at once." "Oh," said Mrs. Grant, getting into the spirit of things, "my Ulysses has no music in his soul. And am I not with him? Do you suppose that I, after such a lesson from dear old Penelope, would remain at home? No, indeed! I defy all outside enchantments." And so the Grants passed the fabled island in safety. "Not a single strand of music did we hear," Mrs. Grant reported, "save that of the waves as they dashed against and washed the sides of our gallant ship, which, on the afternoon of December 28, dropped anchor at Malta."[35]

Xantippe

When the Grants visited the Acropolis, General Read pointed out where Socrates had died in prison and after reflecting on "poor old Socrates" for a few minutes Mrs. Grant spied a building nearby and asked: "What temple is this?" "Why, did you not know?" exclaimed General Read. "It is the Temple of Victory. The eagle has alighted there and has never left his perch." Mrs. Grant invited her husband to come up and stand by her in the Temple of Victory. As the party moved on, she repeated her invitation, but Grant continued to decline and pretended he was going to leave her. "Come up," insisted Mrs. Grant again. "It is a little thing to yield, and I so much wish it." Suddenly Grant turned, rapidly mounted the steps, came close to her side, and, taking her hand, exclaimed: "I am here, my Xantippe." Mrs. Grant enjoyed the quip about Socrates' sharp-tongued wife almost as much as her husband did.[36]

Cup of Tea

Musical-comedy (and later movie) star Marie Dressler took the train to Lake George for a much-needed vacation, but because of the low esteem in which actresses were held by respectable people in those days she was snubbed by everyone in the hotel where she stayed. One afternoon, feeling lonely and depressed, she found a piano in one of the hotel rooms, sat down, and started playing and singing to console herself. "Delightful!" murmured a friendly voice at the window. "I wonder if you sing, 'Believe Me, If All Those Endearing Young Charms'?" Dressler looked up to see a sweet old face framed in graying hair. "It's one of my favorites too," she said. "Won't you come in?" "I live at the cottage above here," said the little old lady, entering the room. "I was passing and heard your voice." For the next hour or so Dressler played and sang to her audience of one and when she finished the old lady got up, invited her home for a cup of tea and to meet her daughter. Dressler readily accepted, but told her she was a musical-comedy actress, and, to her surprise and joy, it didn't seem to bother her at all. Dressler had tea, met the lady's daughter, and spent the first happy hour she had had on her vacation. That evening, when she entered the hotel dining room, people began rising and offering her a place at their tables. "Do come over here, Miss Dressler," said one formerly haughty dowager. "There's such a nice view of the lake from our window." "We've been saving a place for you," exclaimed a plump, bewhiskered gentleman, whose wife was making pleasant little clucking sounds as he drew out a chair for her. For a moment, Dressler thought she was losing her mind, but she accepted the sudden hospitality and remained dazed all evening. The next morning things became clearer. The gracious old lady who had taken her under her wing, she learned, was the widow of General Ulysses Grant, and she had introduced herself to Dressler simply as Mrs. Grant. And it had never occurred to Dressler that anyone so forthright and friendly could be socially prominent, much less a former First Lady.[37]

CHAPTER 18

LUCY WEBB HAYES

1831–1889

While Rutherford B. Hayes was making a plea for sectional harmony in his inaugural address in March 1877, reporter Mary Clemmer Ames was watching Lucy Hayes closely. "She has a singularly gentle and winning face," she told readers of the *Independent,* a Protestant weekly, a few days later. "It looks out from the bands of smooth dark hair with that tender light in the eyes which we have come to associate with the Madonna. I have never seen such a face reign in the White House." But Mrs. Ames had some questions to raise. "I wonder what the world of Vanity Fair will do with it?" she mused. "Will it frizz that hair?—powder that face?—draw those sweet, fine lines away with pride?—hide John Wesley's discipline out of sight, as it poses and minces before 'the first lady of the land'?"[1]

Mrs. Ames need not have worried; Vanity Fair never got to Mrs. Hayes. Being the "first lady" (Ames was supposedly the first to use the expression) didn't turn her head a bit. She neither frizzed her hair nor powdered her face; nor, as a good Methodist, did she ever neglect John Wesley's discipline. Not long after the inauguration, *Philadelphia Times* editor Alexander K. McClure, a scathing critic of the pretensions and extravagances of Washington society, was hailing the new First Lady with joy. "Mrs. Hayes deserves the thanks of every true woman for the stand which she has taken against extravagance in dress," he exclaimed. "She has carried to the White House the same quiet dignity and lady-like simplicity for which she was distinguished at home; and her dress on public occasions, while invariably handsome and becoming the wife of the President, has also been invariably unostentatious."[2] During the ensuing years Mrs. Hayes's demeanor as First Lady was so quietly tasteful that in 1880 Mrs. Ames composed a celebratory sonnet for her which, in the hyper-sentimental style of the period, was filled with stupendous similes and mighty metaphors. And when the Hayeses departed Washington in 1881, Unitarian clergyman Thomas W. Higginson declared that the White House was "whiter and purer because Mrs. Hayes has been its mistress."[3]

Not everyone was thrilled by the new Wesleyan White House: the daily prayers, the Bible readings, the Sunday night hymn-singings. The food, critics conceded, was excellent; but the temperance policy, they wailed, made the customarily dull state dinners duller than ever. The Hayeses served wine but once. When they were making plans to entertain two Russian Grand Dukes—Alexis and Constantine—in August 1877, Secretary of State William Evarts pleaded with them to serve wine with dinner; Russia, after all, was "our perpetual ally," an Evarts intermediary told them, and "a dinner without wine would be an annoyance, if not an affront" to the young men.[4] The Hayeses yielded; six wine glasses appeared on the table. They were the last to appear. Shortly after the Grand Duke dinner, Hayes announced a no-liquor policy for the White House.

Some people thought Mrs. Hayes ("Lemonade Lucy") was responsible for the liquor ban, but it was in fact the President's own idea.[5] Like his wife, he favored abstinence in principle; he also thought it would help the Republican party with temperance people. Protestant groups, especially Methodists, were delighted with the new policy, and were inclined to give Mrs. Hayes major credit for it. But there was grousing among many of the people, native and foreign, who had to attend White House dinners. In May 1877, *Puck,* humorous weekly, ran a cartoon showing Mrs. Hayes's smiling face peering out of a jug of water but glaring angrily out of a bottle of wine. Under the bottles came the lines:

> How wine her tender spirit riles
> While water wreathes her face with smiles.[6]

A compassionate steward, it was rumored, sneaked rum into the refreshments for some of the guests at White House dinners, but the President denied that this was so.

The Women's Christian Temperance Union (WCTU), founded in 1874 and rapidly gaining strength in the late 1870s, adored the President's wife. But Mrs. Hayes never joined the organization; nor did she become affiliated with any other temperance society, not even the one in her hometown, Fremont, Ohio. She resisted all efforts, too, by WCTU president Frances E. Willard to have her become publicly associated with the temperance movement. On one occasion, moreover, she even let slip a remark which offended some of her temperance admirers. Cruising along the Philadelphia waterfront with her husband in 1878, she told people on the steamer who were helping themselves to claret punch that although she was an abstainer she didn't want to dictate to other people. "I want people to enjoy themselves in the manner that is the most pleasing to them," she assured them. When her statement leaked out, members of the Lucy Hayes Temperance Society in Washington promptly dropped her name from their title.[7] Still, she committed no further gaffes while she was First Lady and she remained in good standing with most temperance workers to the end. When she was visiting San Jose, California, with her husband in the last year of his Presidency, she was given a big reception by the WCTU and presented with a silk banner inscribed with her name and the words, "She hath done what she could."[8]

Some people thought she could have done more. Temperance advocates sought her help in vain; so did women's rights crusaders. But Mrs. Hayes stayed firmly aloof from their activities; she was surprisingly inactive for a woman with her background. Born the daughter of a physician in Chillicothe, Ohio, in 1831, she had a good education—attended classes at Ohio Wesleyan College in Delaware, Ohio, and earned a degree at Wesleyan Female College in Cincinnati—and took some interest in the feminist movement as a young woman. "It is acknowledged by most persons," she once wrote, that a woman's mind "is as strong as a man's. . . . Instead of being considered the slave of man, she is considered his equal in all things, and his superior in some."[9] But as she grew older she came to accept most of the womanly conventionalities of her day without protest.

Hayes first met Lucy in the summer of 1847 when she was only sixteen, "a bright sunny hearted *little* girl not quite old enough to fall in love with," he wrote; "so I didn't." But his mother came to think the girl would make a good wife for her son and Hayes himself was not uninterested. "Youth," he wrote, "is a defect that she is fast getting over and may be entirely out before I shall want her."[10] But he teased his mother about it. "I wish I had a wife to take charge of my correspondence with friends and relatives," he wrote her. "Women of education and sense can always write good letters, but men are generally unable to fish up enough entertaining matters to fill half a sheet. I hope you . . . will see to it that Lucy Webb is properly instructed in this particular."[11] But he didn't see much of her until late in 1849, when he moved to Cincinnati, where she was attending Wesleyan Female College, opened his law practice in the city, and began attending Friday receptions at the college. Soon he was writing, "I *guess* I am a great deal in love" and making entries about Lucy's "low sweet voice" and "soft rich eyes" in his diary. But he was not neglectful of her mind. "Intellect, she has too, a quick spritely one, rather than a reflective profound one," he wrote somewhat condescendingly in May 1851. "She sees at a glance what others study upon, but will not, perhaps, study out what she is unable to see at a flash. She is a genuine woman, right from instinct and impulse rather than just judgment and reflection." Then he let himself go. "It is no use doubting or rolling it over in my thoughts," he added. "By George! I am in love with her!"[12]

On a June night soon after Hayes proposed. He was with her at the college, got up suddenly, seized her hand, and exclaimed: "I love you!" When she remained silent, he repeated his declaration. "I must confess," said Lucy finally, "I like you very much." And they plighted their troth. "I don't know but I am dreaming," she told him afterward. "I thought I was too light and trifling for you."[13] They were married in the Webb home in Cincinnati on December 30, 1852, and lived happily thirty-six years after. Cried Hayes two months after the wedding: "Blessings on his head who invented marriage!"[14]

For a time after her marriage Lucy Hayes continued her girlhood interest in women's rights. After hearing Lucy Stone lecture on the woman question in 1854, she told her husband she favored better wages for women and thought "violent measures" were sometimes justified to dramatize the need for reform.[15] When it came to woman suffrage, however, she drew back, even in those days, although two of her aunts strongly supported it. She knew what her husband thought. "My point on this issue," he declared, "is that the proper discharge of the functions of maternity is inconsistent with the like discharge of [the political duties of] citizenship."[16] In this matter, as in all others, Mrs. Hayes followed her husband. Hayes entered Congress, became Governor of Ohio, and, then, after the bitterly disputed election of 1876, took up his duties as President with the utmost seriousness; but Mrs. Hayes's life during all these years remained centered almost exclusively on home and hearth (eight children, three of whom died in infancy), with time out of course for entertaining and charity work.

When Mrs. Hayes became First Lady, some feminists thought she would represent the "new woman" in America, that is, the woman who was trying to find her place in the sun on her own in a man's world. Reporter Laura Holloway hailed her as a symbol of "the new woman era" and predicted: "Her strong healthful influence gives the world assurance of what the next century woman will be."[17] But she could not have been more wrong, for the new First Lady turned out to be firmly traditional. Woman, for Mrs. Hayes, was "an elevating influence—man's inspiration"; man should "go forth to duty," she once wrote, while woman stays at home and "weaves the spell which makes home a paradise to which he may return, ever welcome, whether he is victor or vanquished."[18] Though she was the first President's wife with a college degree, she refrained from taking up the cause of education for women and seems in fact to have been opposed to professional education for women. In 1878, her former schoolmate Rachel L. Bodley invited her to attend commencement at the Women's Medical College of Philadelphia, where she was Dean, to show her support for the "Higher Education of woman as developed in professional study." But Mrs. Hayes turned down the invitation, though she later visited the college with her husband and let him do all the talking.[19] She also turned down the request of Emily Edson Briggs, Washington reporter, to ask the President to appoint Briggs to represent "American womanhood" at the International Exhibition in Paris in 1878.[20]

Mrs. Hayes's refusal to get involved as First Lady irritated some of her initial admirers. "As the *First Lady* in the United States," Missouri businesswoman Mary Nolan wrote her, "you from your official position must take a greater interest in the development of Woman's Industries, than any lady in the land. . . ."[21] Emily Briggs was more critical; she wrote to say that women in the United States had the right to know whether the President's wife approved of "the progress of women in the high road of civilization or whether you are content because destiny lifted you to an exalted position, so high and far away that you cannot hear the groans of the countless of our sex. . . ."[22] But Mrs. Hayes kept her peace; she remained as aloof from the "new woman" movement as she did from the temperance campaign with which it became associated. President Hayes eventually supported the National Woman's Suffrage Association request to send a women's delegation to the Paris Exposition to represent the industries of women in the United States, but there is no evidence that his wife played any part in his action. Nor did she shape her husband's decision to approve a bill in February 1879 allowing women to practice law before the Supreme Court, although Susan B. Anthony and Elizabeth Cady Stanton had asked her to do so.

Despite Mrs. Hayes's refusal to get involved in politics, some people thought she had considerable influence over her husband; they even called her the head of the nation as well as mistress of the White House. Once, when she made a trip to Ohio without her husband, some newspapers said that "in the absence of his wife, Mr. Hayes is acting President."[23] But Mrs. Hayes never interfered with her husband's work and she rarely made suggestions to him even on minor matters. When a minister's wife called at the White House right after the 1877 inauguration to ask her not to serve liquor in the White House, she exclaimed: "Madam, it is my husband, not myself who is President. I think a man who is capable of filling so important a position as I believe my husband to be is quite competent of establishing such rules as will occur in his house without calling on members of other households."[24] She did, it is true, help two or three women get minor government posts by mentioning their names to her husband, but most of the time she stayed out of the picture. On one occasion, when she thought Colonel Thomas

L. Casey deserved to be made head of the U.S. Army Signal Corps, she asked her son Webb to push Casey's case with his father. "I do not dare these days," she told him, "to express my opinions and desires about people."[25] Yet despite her diffidence, she won acclaim as "the new woman." "In her successful career as the first lady of the land," wrote one admirer, "was outlined the future possibilities of her sex in all other positions and conditions."[26] Statements of this kind kept popping up while she was in the White House and even afterward, but they rested largely on wishful thinking.

After the Hayeses turned the Executive Mansion over to the Garfields in March 1877 and returned to Ohio, Frances Willard presented President Garfield with a portrait of Mrs. Hayes, commissioned by the WCTU, to be hung in the White House. She had words of high praise for the former First Lady at the presentation ceremony. Some Illinois temperance women also did her high honor: they presented her with six big morocco-bound, gilt-edged volumes containing autographs and tributes from prominent people through the land: writers, painters, musicians, government officials, army and navy officers, scientists, lawyers, clergymen, teachers, and temperance workers. Mrs. Sarah Polk (who had also banned wine from White House functions) signed her name; and there were verses lauding Mrs. Hayes penned by Oliver Wendell Holmes, Henry Wadsworth Longfellow, and John Greenleaf Whittier. Mark Twain also good-naturedly joined the chorus of praise. "Total abstinence is so excellent a thing," he wrote, "that it cannot be carried to too great an extreme. In my passion for it I even carry it so far as to totally abstain from Total Abstinence itself."[27]

After settling down to comfortable retirement with her husband at their fine estate in Spiegel Grove, Mrs. Hayes busied herself with Sunday School teaching, church suppers and fairs, reunions with her children and grandchildren, and with occasional trips with her husband. She continued to resist Frances Willard's efforts to get her to work for the WCTU; about all she did for the cause was to lead the Fremont temperance league in prayer now and then. She did, though, reluctantly agree to serve as president of the Woman's Home Missionary Society of the Methodist Episcopal Church, formed in 1880 to work among the poor and destitute, especially immigrant women, in the United States. The position was largely honorary, though once a year it required her to speak briefly (and much against her will) at annual meetings. One afternoon in June 1889 while sewing and watching a tennis match she suffered a stroke and died a few days later. "She was," wrote Hayes sorrowfully in his diary, "the embodiment of the Golden Rule."[28] "She was a good woman . . . ," declared the *Detroit Free Press,* "with an abundant stock of old-fashioned virtues."[29] In January 1893, Hayes himself died of a heart attack and was buried next to her in Fremont's Oakland Cemetery.

* * * *

Stopped To Do Baltimore and Washington

In September 1862, Hayes was wounded at the battle of South Mountain and arranged to send a telegram to his wife asking her to come at once. But the telegram was botched; it told her neither where she could find him nor how severe his wound was. But she headed east anyway, accompanied by her brother-in-law, and spent six days hunting for him in hospitals in Baltimore and Washington. At last, worn out and discouraged, she visited the Patent Office, then being used as a hospital, and suddenly came across some wounded soldiers with the badge of the 23rd Ohio, Hayes's unit, on their caps. "Why," cried the soldiers, when they saw her, "it is Mrs. Hayes." They told her Hayes was in a hospital in the little village of Middletown, Maryland, and she resumed her journey at once. Twelve hours later she was at her husband's side, relieved to find he was making a good recovery. After she told him about her frantic search, he tried to tease her. "And so, dear," he said, "you stopped to do Baltimore and Washington before coming to me, did you?" She didn't find it funny.[30]

Wise Woman

When it came time to retire after the inaugural dinner, Mrs. Hayes told her overnight guests that all the rooms in the White House were spacious and elegant but that one of them, called the State bedroom, was reserved for especially distinguished guests, including royalty. "Now," she said, "I am too wise a woman to make a choice among you who may each be supposed to have a right to this distinction, and thus make you my enemies. I have thought of an expedient; you shall draw cuts." So one of the guests ran to the steward for broom-straws, broke them into the required number of pieces, and held them for each of the guests to draw. But the person who drew the State bedroom was more impressed with the First Lady's tact than with the fancy furnishings of the room.[31]

One Level

When the Hayeses attended church in Washington for the first time after the inauguration, people in the congregation flocked around to see them at the end of the service. One of their friends suggested there be an unwritten law keeping everyone in place until the Presidential party had left, to save crowding and crushing by the curious. "No, dear," Mrs. Hayes told her. "Here we are all on one level."[32]

The Sergeant's Stripes

One day Sir Edward Thornton, British minister to the United States, called at the White House with some of his friends, and, as they entered the Blue Room, they were surprised to see Mrs. Hayes sitting on the floor, needle and thread in hand, repairing the uniform of an old ordnance-sergeant. The callers were about to make a hasty retreat when Mrs. Hayes got up, shook hands with the veteran, assured him his uniform was now perfect, and turned to greet the Thornton party. It turned out that the veteran had fought in the War of 1812 and the Mexican War and the Hayeses had sent him a full-dress uniform and arranged to have him photographed in the White House. But the stripes had been left off, and when the veteran arrived at the White House, he was so upset by the thought of being photographed without them that Mrs. Hayes sent for needle and thread so she could stitch them on. She was just finishing the task when the British Minister arrived. When the Britishers heard about the old veteran they were filled with admiration for Mrs. Hayes.[33]

Stout Old Woman

Harriet Blaine, wife of Maine Senator James G. Blaine, resented the way President Hayes ignored the advice of her husband, and she took out her resentment on Mrs. Hayes. She complained that the White House was dirty; she also told people Mrs. Hayes traveled with the President to keep people from insulting him. Once Mrs. Hayes and her friend Bettie Evarts (daughter of the Secretary of State) inadvertently sat next to Mrs. Blaine at a concert, who, as soon as she saw them, ostentatiously moved to another seat. "Bettie," said Mrs. Hayes in a loud whisper, as people gaped, "who was that stout old woman in purple?"[34]

Autographs

When President and Mrs. Hayes took a trip to the West in the fall of 1880, Mrs. Hayes tried to avoid the autograph-hunters who flocked to their train at every stop. At one town she kept out of sight until it was time for the train to start and then sat down by a window. But as soon as the people outside spied her, they began handing autograph books, cards, and newspapers through the window for her to sign and soon she was on the point of collapse from signing her name so many times. As soon as the train started, her mischievous son Rutherford came running into the President's car and handed his mother an album he had been passing in through the window to her. He had gotten her autograph fifty-six times.[35]

CHAPTER 19

LUCRETIA GARFIELD

1832–1918

One day Lucretia Garfield was kneading a batch of bread for her family, and feeling suddenly overwhelmed by the drudgery of it all, when an idea flashed into her mind: if she tried to make the best bread in the world, she might overcome her dislike of baking. At once her attitude changed. "It seemed like an inspiration, and the whole of my life grew brighter," she recalled years later. "The very sunshine seemed flowing down through my spirit into the white loaves, and now I believe my table is furnished with better bread than ever before; and this truth, old as creation, seems just now to have become fully mine—that I need not be the shrinking slave of toil, but its regal master, making whatever I do yield its best fruits."[1]

But getting on top of things wasn't really all that easy for James Garfield's wife. Her marriage took a long time to become more than toilsome. Both she and her husband agreed in retrospect that their early years together had been highly unsatisfactory. They did, to be sure, see eye to eye on most matters from the outset. They were both devout members of the Disciples of Christ and took their religion seriously; they accepted the Protestant work ethic—diligence, thrift, temperance, discipline—without question and at times put the Puritan virtues above good health; and they were devoted to education, prized literature and the fine arts, and enjoyed reading books together and attending plays, concerts, and lectures.

But Mrs. Garfield wasn't as sure as her husband, at least in the beginning, that a "Woman's province is her home." In an essay with the title "Women's Station" which she wrote as a college student, she wondered why women had contributed so little to the arts and sciences and decided it was mainly because of lack of educational opportunities. She also expressed some exasperation over the fact that women were paid less than men, as tailors and schoolteachers, even when they did equally competent work. "Is it reasonable," she asked, "that the same number of stitches equally good should be worth less because taken by woman's weaker hand? Is it equitable that the woman who teaches school equally well should receive a smaller compensation than man, who is so much more able to support himself in other ways?" In the end, though, she came around to the conventional (and Garfieldian) wisdom on the subject. "How dare we ask for increased privileges while we so poorly perform our present responsibilities?" she exclaimed in concluding her ruminations. "We should remember greatness is not in station, but as is often said, '*Act well* your part, *there* all honor lies.' "[2] Mrs. Garfield was anxious to act well her part as wife and mother. She did everything she could to advance her husband's career and to rear the children she bore him (there were eight, of whom five survived) the best way she knew how. She came to share his hostility, too, to the woman's rights movement, especially to the demand for the vote. Garfield thought woman suffrage was "atheistic" and would result in "the utter annihilation of marriage and the family," and his wife accepted his views without dissent.[3] The Garfields agreed on fundamentals.

The clashes were primarily temperamental. Lucretia was shy and reserved and kept her feelings in; James was outgoing and gregarious and needed a wider circle for his energies than home and family. The courtship itself was long and troubled. The two first met in 1849 at Geauga Seminary in Chester, Ohio, where they were classmates, but they didn't strike up an acquaintance until they were together again in the Western Reserve Eclectic Institute, a Disciples' school in Hiram, three years later. In December 1853, Garfield began courting "Crete," as he called young Lucretia Rudolph (born on a farm near Hiram in 1832), but he worried about it a great deal. "There is one question I have not yet settled in regard to her social nature," he wrote in his diary, "viz., whether she has that warmth of feeling, that adhesive nature which I need to make me happy. . . . For myself, I feel that under the proper circumstances, I could *love* her, and unite my destiny to hers. With regard to her feelings upon it, I am not certain."[4] But her feelings soon became clear: she loved him deeply. Before he went off to Williams College in July 1854 to complete his education, he reached an informal understanding with her that was close to being an engagement.

Crete treasured the letters Garfield wrote her from college; she feared at times that she loved him more than she did God. But Garfield himself was having second thoughts. ". . . There is no delirium of passion," he lamented, "nor overwhelming power of feeling that draws me to her irresistably *[sic]*."[5] When he saw her in Hiram again, at the end of the school year, the meeting was awkward and strained. Crete was bewildered; "at times," she wrote in her diary, he "almost overwhelms with affection and then . . . turns and in his own intellectual strength and greatness seems unapproachable."[6] Just before he returned to college, she decided to let him take a look at her diary. He was deeply moved by what he read there. "Never before," he wrote in his own diary, "did I see such depths of suffering and such entire devotion of heart. . . . From that journal I read depths of affection that I had never before known that she possessed."[7] He returned to Williams with his love for her reaffirmed. "My soul . . . ," he wrote jubilantly, "is full to overflowing now. From foundation to topstone its halls are choral with joy."[8]

But the doubts returned. During his last year at Williams, Garfield struck up an acquaintance with an attractive and vivacious young Disciple named Rebecca Selleck who, he told Crete, somewhat defensively, reminded him of her. Both Crete and Rebecca attended Garfield's graduation exercises in August 1856 and applauded his Metaphysical Oration. And when he returned to Hiram with Crete afterward to begin teaching at Eclectic, he corresponded with Rebecca for a time. But his engagement to Crete continued; he begged her to be patient with him and she was more than willing to do so. When a friend reproved him for "unpardonable neglect and coldness" toward Crete and blamed it on Rebecca, Garfield told him: "If I ever marry, I expect to marry Lucretia Rudolph."[9] In September 1856, Crete finally offered to release him from the engagement, but he rejected her offer. "No, I can't lose you," he exclaimed, "but give me time."[10] She did. But it was not until the spring of 1858 that he at long last decided to "try life in union" with her.[11] There was something cheerless about his decision. "I don't want much parade about our marriage," he told her. "Arrange that as you think best."[12] On November 11, 1858, they were married at her father's house in Hiram, with relatives and a few friends present, and then went to live in a near-by boarding house. There was no honeymoon.

The "years of darkness," as the Garfields later called their early life together, continued.[13] Garfield was away from home a great deal of the time, first in Columbus, attending sessions of the Ohio Senate, to which he was elected in 1859, and then, when the Civil War came, in the Union Army, where he became commanding officer of the 42nd Ohio Volunteer Infantry. At one point before going into the service, he told his wife frankly he considered their marriage "a great mistake," and in great despair she had thrown herself on his mercy. "If you are so constituted," she told him, "that the society of others is more desirable than a wife's or if the strange untoward circumstances of your life in its relations to me have driven you for sympathy to others . . . then you are not to blame. . . . But you must not, O, No, you *must not* blame me if I feel it. . . . I try to hide it. I crush back the tears just as long as I can for I know they make you unhappy."[14] In July 1862, however, there was a sudden coming together; when Garfield returned home on sick leave after several months' absence, he and his wife went off to a secluded farmhouse nearby at Howland Springs for rest and recuperation, had what amounted to a honeymoon at long last, and found they cared deeply for one another. "It is indeed a 'baptism into a new life' which our souls have received," exulted Garfield, "and which, after so many years of hoping and despairing has at last appeared in the fulness of its glory."[15]

But the "new life" took time to develop. There were more crises. In September 1862 Garfield was in Washington on army business, met Secretary of the Treasury Salmon P. Chase's attractive young daughter Kate, and began squiring her around town. He was not seriously interested in her (nor she in him), and he told Crete all about it in his letters home, but she couldn't help being a bit jealous. She was even more upset when he decided to look up his old friend Rebecca when he had to go up to New York for a few days. The reunion with Rebecca turned out to be innocent enough, as Crete soon recognized, but the interlude with Lucia Calhoun, a young widow he met while he was in New York, was something else again. When Crete somehow learned that her husband had been burned by "the fire of lawless passion," as she put it, she was beside herself with grief. *"James,"* she wrote him sorrowfully, *"I should not blame my own heart if it lost all faith in you.* I hope it may not . . . but I shall not be forever telling you I love you *when there is evidently no more desire for it on your part than present manifestations indicate."*[16] Crete's reproach irritated Garfield at first. If she felt so estranged, he told her, "I should consider it wrong for us to continue any other than a business correspondence." Then he came down off his high horse, appealed to the "new life" they had discovered at Howland Springs, and asked for forgiveness. Crete, as always, forgave him; they agreed on a "truce to sadness," and from then on things began getting better between them.[17] Years later, when Crete had a dream about Rebecca Selleck and wrote her husband about it, he told her: "Don't let any dreams trouble you . . . You never had so whole and complete a power over all my heart as you have now."[18]

Garfield left the army at the end of 1863 and after a brief stay in Hiram went to Washington to take up his duties in Congress, to which he had been elected while still in the service. When Congress adjourned and he was back home again, his wife suddenly handed him a piece of paper. On it she had summed up their life together thus far: they had been married for four and three-quarters years and had lived together for only twenty weeks. "I then resolved," Garfield wrote afterwards, "that I would never again go to Washington to a session of Congress without taking her."[19] He rented rooms and then a house and thereafter always took his family to Washington when Congress was meeting. Gradually he and his wife became great companions there; they read Dickens and Scott to each other, attended meetings of the Literary Society of Washington, made social calls, dined out, and took trips together. On Crete's forty-second birthday Garfield thanked her for "being born" and for being his wife.[20] "We no longer love because we ought to," he once exclaimed, "but because we do. The tyranny of our love is sweet. We waited long for his coming, but he has come to stay."[21] For her part, Crete was surer than ever that her affections had been rightly bestowed all along. "Very many men may be *loved* devotedly by wives who know them to be worthless," she told him. "But I think when a man has a wife who holds him in large esteem, knows that in him there is no pretense, nothing but the *genuine*—then he has reason to believe in his own worth." Years after Garfield's death, she confessed to her children that her parents had never been openly affectionate and she had had to learn from her husband how to let herself go.[22]

Garfield served sixteen years in Congress and became Speaker of the House in 1874. But while making a name for himself in Washington, he was also coming to focus his attention increasingly on his wife and family. His letters and journal in the 1870s were filled with tributes to Crete: "the best woman I've ever known"; "the light of my life"; "the solid land on which I build all my happiness and hope."[23] On November 11, 1875, he reflected with satisfaction on the way things had worked out for them. "This is the anniversary of our wedding which took place 17 years ago," he wrote. "If I could find the time and had the ability to write out the story of Crete's life and mine, the long and anxious questionings that preceded and attended the adjustment of our lives to each other, and the beautiful results we long ago reached and are now enjoying, it would be a more wonderful record than any I know in the realm of romance."[24] Somewhat to his own surprise he became quite domestic. He still enjoyed socializing, to be sure, but he found much of Washington's social life "uncomfortable and meaningless" and preferred evenings at home reading or playing casino and bezique with Crete.[25] Washington sophisticates smiled at the Ohio couple's "splendid rusticity" and found evenings with them too homespun for their own tastes. "Very good people I am sure they are," wrote one guest after dinner with the Garfields and their friends, "but a plainer, stiffer set of village people I never met."[26]

Crete followed her husband's rise to eminence in Congress with adoration but not surprise, for even during the "dark years" of their marriage she had thought he was destined for great things. As early as 1875 she was thinking of him as a presidential possibility and telling him that a "far-seeing spirit" was saving him for the time when he would become (in Tennyson's words),

> The pillar of a people's hope
> The centre of a world's desire.[27]

She was delighted when he received the Republican nomination for President (albeit not until the thirty-sixth ballot) in June 1880. She helped him handle the torrent of guests during his "front porch campaign" in Mentor, Ohio, after the nomination, and rejoiced when he defeated General Winfield Hancock (though by a plurality of only about 10,000 popular votes) in November. After the election she urged him to take on the Stalwarts (who had backed Grant for the nomination), "fight them *dead,*" and "put every one of them in his political grave."[28] For her, Garfield's inauguration in March 1881 was "the greatest spectacle" she had ever seen and her husband "almost superhuman" when he gave his inaugural address.[29] But she was timid about being First Lady and sought advice from Mrs. James Blaine, the Secretary of State's wife, on how to behave as White House hostess.

Mrs. Garfield was First Lady for only a few months and was seriously ill with malaria much of that time. On June 18, 1881, when her husband took her to the railroad station for a trip to the Jersey coast, where he had rented a cottage for her to recuperate in after her illness, a demented office-seeker named Charles Guiteau was waiting there for him. But when he saw the President's wife, pale and sickly, he abandoned his plan to kill the President. "Mrs. Garfield looked so thin," he said later, "and she clung so tenderly to the President's arm, that I did not have the heart to fire on him."[30] On July 2, however, he was ready to act. When Garfield returned to the station for a trip to New York, where his wife was to join him for a visit to New England, Guiteau felled him with two shots. Rushed back to Washington to be at his side in the White House, Crete became his constant bedside companion, tried to bolster his spirits, and hoped against hope for his eventual recovery. When he began talking about how to handle the children after he was gone, she refused to let him go on. "Well, my dear," she exclaimed, "you are not going to die as I am here to nurse you back to health; so please do not speak again of death."[31] But after lingering eighty days, Garfield died on September 19, and his wife was shattered. "Oh!" she wailed, "why am I made to suffer this cruel wrong?"[32]

Mrs. Garfield survived her husband by thirty-seven years. A Congressional pension of $5,000 a year and a subscription launched by Cyrus Field that eventually raised $300,000 enabled her to live comfortably at Lawnsfield, the splendid estate her husband had developed in Mentor, and even expand it into an elegant mansion. There she collected memorabilia about her husband, corresponded with relatives and friends on black-bordered stationery, and followed the careers of her children and grandchildren with loving care. She also read widely, took notes on her reading, translated Victor Hugo, wrote poetry, gave talks on literary topics to women's groups, and began (but never got very far with) a biography of her husband. In 1882, when rumors reached the press that she was planning to remarry, she expressed "humiliation that anyone could believe me capable of ever forgetting that I am the wife of General Garfield."[33] And in 1913, when a reporter sought her out for an interview, she complained that none of her husband's photographs "suggest at all the way his face lit up in conversation. His face was very responsive, not at all settled or fixed as the photographs invariably suggest."[34] She died of pneumonia at her winter home in Pasadena, California, in March 1918 at the age of eighty-six.

CHAPTER 20

ELLEN HERNDON ARTHUR

1837–1880

Chester Arthur's wife Ellen ("Nell") never made it to the White House. She died of pneumonia in January 1880, some twenty months before Arthur took his oath of office as President following the death of James Garfield. She remains an elusive figure, partly because she never attracted the attention of the press during her lifetime and partly because her husband destroyed all of his personal papers just before his own death in 1886. She was, it appears, high-spirited but decorous, deeply devoted to her family, ambitious for her husband, and skillful as a hostess. If she had become First Lady, Washington society would probably have approved. She was, according to a New York politico, "one of the best specimens of the southern woman."[1]

Nell Arthur was indubitably Southern. Born Ellen Lewis Herndon in Fredericksburg, Virginia, in August 1837, she came from an old Virginia family, proud of its roots in the region's history. Her mother, a Hansbrough, moved with ease in Washington society when Nell was growing up; and her father, William Lewis Herndon, made a name for himself as a young naval officer when he headed an expedition to explore the Amazon River in 1851. Nell first met Arthur in 1856, when she went up to New York to visit her cousin, Dabney Herndon, one of Arthur's roommates. Despite the differences in background—Arthur, son of a Baptist minister, was seven years older and just beginning to make his way upward in the world of law and politics—the two seem to have taken to each other at once and were soon deeply in love. "I know you are thinking of me," Arthur wrote her in one of the few letters between them that survives. "I feel the pulses of your love answering to mine. If I were with you now, you would go & sing for me 'Robin Adair' [;] then you would come & sit by me—you would put your arms around my neck and press your soft sweet lips over my eyes. I can feel them now."[2]

In 1857, Nell's father went down with his ship in a storm off Cape Hatteras after seeing that the passengers were rescued, and his valor won him a memorial column at the Naval Academy in Annapolis. Arthur was in the West at the time, purchasing some land in Kansas, but he returned at once when Nell wired for help, and ended up taking over the management of Mrs. Herndon's financial affairs. In February 1858 he visited Nell's relatives in Fredericksburg and passed muster with them. "He is a fine looking man," observed Nell's uncle, Dr. Brodie Herndon, after the encounter, "and we all like him very much."[3] Shortly afterward, Arthur proposed to Nell on the porch of the United States Hotel in Saratoga Springs, New York, and in October 1859 they were married in the Calvary Episcopal Church in New York City. Eventually they had three children, two boys and a girl, but their first-born died when he was about two and a half years old.

The Civil War posed problems for the Arthurs. Arthur's father was an abolitionist; and though Arthur himself was more cautious in his commitments, he had, as a young lawyer before the war, handled one case that won freedom for some Virginia slaves brought to New York and another that led to racial integration on New York City's streetcars. Nell came from a slaveholding family; and her sympathies, like those of her mother and all her relatives (some of whom fought in the Confederate Army and Navy), were with the South. During the war Arthur referred good-naturedly to his "little rebel wife," but there appears to have been considerable tension between the two, especially after Arthur became Inspector-General and then Quartermaster-General of the State of New York, responsible for supervising supplies for New York volunteers in the Union army.[4] But Arthur helped his wife's relatives whenever he could during the war. On two occasions he saw to it that Nell's cousin Dabney Herndon (his onetime roommate) a Confederate officer, was released and returned to Virginia after being taken as prisoner of war. And when he was in Fredericksburg on an inspection tour in 1862, he did everything he could to provide for the comfort of the Herndons in that war-devastated area. Arthur, according to Nell's uncle, "was very affectionate & kind. He thought we might be suffering and deliberately proffered aid. We told him we forgot the General in the man."[5]

Arthur was a conservative Republican. Insisting that the preservation of the Union, not the eradication of slavery, was the North's main war aim, he became increasingly critical of the Lincoln administration after the Emancipation Proclamation. In May 1864, Arthur's sister observed that he "was so down upon the administration that he was almost a copperhead," and that he was not planning to vote for Lincoln's re-election in 1864.[6] By this time he had returned to law practice, no doubt to his wife's relief, and was concentrating on amassing enough money to support the good life to which both he and Nell aspired.

The Arthurs enjoyed fine living. As Arthur's law practice prospered during and after the war, they became upwardly mobile; they moved into a fine two-story brownstone house on Lexington Avenue, filled it with fine furniture, acquired servants, hired tutors for their children, took expensive vacations, entertained lavishly, and moved with delight in upper middle-class circles. They both dressed elegantly, too, and Arthur, who imported some of his clothes from London, came to look like a "well-fed Briton."[7] Mrs. Arthur did more than act as hostess for her husband's friends and associates. She had a nice contralto voice and enjoyed singing at fashionable charity fund-raisers and giving little recitals for friends at home.

After the war, Arthur became active in the Republican party, associated himself with Senator Roscoe Conkling's political machine, and soon won the praise of party leaders as "one of the most rigid of organization and machine men" in the state.[8] In 1871, President Grant rewarded him for his diligence by making him Collector of the Port of New York; and in that capacity he soon became such an expert at managing the spoils of office for New York Republicans that he was known as a "spoilsman's spoilsman." His wife was upset by the attacks on his character and by his removal from office by President Hayes, a civil-service reformer, in July 1878. She was also distressed by the fact that he spent so much time smoking, drinking, and talking politics with his cronies that she began feeling lonely and neglected. There was talk of separation.

But Arthur was deeply devoted to his wife and her death in 1880 came as a terrible blow. He was in Albany when she caught cold and developed pneumonia, and he entrained at once for New York when he heard she was ill. To his sorrow, by the time the Sunday milk train deposited him in New York late that night she was no longer conscious. He remained twenty-four hours by her bedside, hoping for the best, but the next day she was gone. She was only forty-two. Arthur was filled with remorse for having neglected her, and after her death he talked about how different things would have been had she survived. "Honors to me," he told her uncle, "are not what they once were."[9] When he was President, he gave a stained glass window in her memory to St. John's Episcopal Church in Washington, where she had sung in the choir as a young woman. And he had it placed in the south side of the building so he could see it from the White House.

After Arthur became President in September 1881, he persuaded his youngest sister, Mary McElroy, to spend several months each year in the White House acting as "Mistress of the White House" for him. And despite his bereavement, he tried to make the Executive Mansion a cheerier place than it had been under the Hayeses and the Garfields. When some temperance people called to urge a continuation of the no-liquor policy, he told the head of the delegation: "Madam, I may be President of the United States, but my private life is nobody's damn business."[10] According to one social observer, White House entertainments under Arthur "are elaborate and elegant. The dinners, so some say who have survived dinners with a series of Administrations, were never so good; and not only Diplomats, but other people, receive the full allowance of wine and the entire variety prescribed by social law."[11]

From time to time there were rumors that the widower in the White House planned to remarry. Didn't he have a picture of a woman in his bedroom and didn't he place some fresh roses in front of it every day? Some gossipers mentioned the daughter of a State Department official as a possibility; others quite soberly suggested Frances Willard ("Stranger things have happened") as the lucky woman.[12] But it turned out that the picture was of Nell and the rumors abruptly ceased. Arthur never remarried; nor did he return to law practice in New York, as he had planned, after leaving the White House in 1885. His health deteriorated rapidly while he was President and he was forced to retire after leaving office. In the fall of 1886 he died of a cerebral hemorrhage at fifty-six.

CHAPTER 21

FRANCES CLEVELAND

1864–1947

When the forty-eight-year-old Grover Cleveland was elected President in 1884, "mammas with marriageable daughters," according to Illinois Senator John A. Logan's wife, "began to plan for [the] opportunity to meet the President-elect."[1] But it all seemed a waste of time; Cleveland acted like a confirmed bachelor. Once, when his sisters asked him whether he had ever thought of getting married, he replied: "A good many times; and the more I think of it the more I think I'll not do it."[2] Ten days after his inauguration in March 1885, his sister Rosa Elizabeth ("Libbie") held a reception at the White House which charmed everybody. It looked as if Cleveland's sister was to be mistress of the White House for the next four years.

Libbie Cleveland was probably the most learned lady ever to preside over White House social functions. Before agreeing to help her brother out in the Executive Mansion, she taught at exclusive girls' schools, gave lectures in "racy rhyme" to advanced classes at these schools, and published essays on George Eliot and other writers. When receiving lines at the White House became long and tiresome, she silently conjugated Greek verbs to keep from getting bored. She was, surprisingly for a Cleveland, a bit of a feminist. In a book of essays dedicated "to my fellow countrywomen," she gave Mohammed's wife Kadijah major credit for his achievements and Thomas Carlyle's wife Jane high praise for recognizing and encouraging his talents. "We can do no better or braver thing," she advised American women, "than to bring our best thoughts to the everyday market; they will yield us usurious interest."[3] Not, though, with her brother. He was thoroughly traditional when it came to the place of women in the scheme of things. A "good wife," he insisted, is "a woman who loves her husband and her country with no desire to run either."[4]

In April 1885, Libbie and her brother Grover entertained Mrs. Oscar Folsom and her twenty-one-year-old daughter Frances at the White House for a week or two, and there was an outburst of speculation among Washingtonians about Cleveland's single state. Was it possible, after all, that the new President was about to abandon his bachelorhood? And that the widow of his old Buffalo friend and law partner, Oscar Folsom, was the object of his affections? Washington gossipers got it only half right; Cleveland was planning to get married all right, but it was the daughter—the lovely young Frances, now in her senior year at Wells College—in whom he was interested, not the mother. He had been corresponding with her ever since she went off to college in Aurora, New York, with the approval of the Widow Folsom. But he thought marriage ought to wait until she graduated.

Cleveland had known Frances Folsom, who was born in Buffalo in 1864, ever since she was an infant. When her father died in a horse-and-buggy accident in 1875, he became administrator of his estate as well as a kind of guardian for the eleven-year-old "Frankie." By the time he entered the Governor's Mansion in Albany, New York, she had grown into an attractive young woman with charm and poise, and he was taking a serious interest in her. He sent her flowers as well as letters when she was at college; and, in a remark no one understood at the time, he told one of his sisters he was "waiting for his wife to grow up."[5] Frances reciprocated his affection; and the two became secretly engaged in August 1885, after she graduated from Wells. Then she went off to Europe with her mother in the spring of 1886 to gather her trousseau and do some sight-seeing. But the betrothed kept their plans a strict secret from everybody except their immediate families. "Poor girl," Cleveland told her later on, "you never had any courting like other girls."[6]

The press continued its probing; newsmen had the feeling President Cleveland was up to something. "WASHINGTON GOSSIP—," screamed the *New York Herald* on April 19, 1886. "Society Incredulous About the President's Marriage—What if it Prove True?—Official Precedence and Other Matters to be Affected."[7] But gossipers were still focusing on Frances's mother. When an old friend showed Cleveland a newspaper clipping which suggested that the President was going to marry Mrs. Folsom, the latter growled, "I don't see why the papers keep marrying me to old ladies all the while—I wonder why they don't say I am engaged to marry her daughter."[8]

On May 27, 1886, Frances and her mother returned from Europe.

To help them escape the mob of reporters awaiting them in New York, Daniel Lamont, Cleveland's White House secretary, took a boat out to their ship in New York harbor and then whisked them off to Gilsey House in Manhattan before newsmen knew what was happening. On Memorial Day, Cleveland went up to New York to review the parades there and sneak in a visit to Gilsey House; and as he stood on the reviewing stand in Madison Square, people cheered as the bands played Mendelssohn's Wedding March and then segued into "He's Going to Marry Yum-Yum" and "Come Where My Love Lies Dreaming." There were also cries of "Long Live President Cleveland and his bride!" The next morning, newsboys fanned out in the city yelling, "Here's your morning *Sun;* all about the President's wife!"[9] The announcement from the White House later that day that Cleveland was going to marry the daughter, not the mother, astonished the know-it-alls, but it delighted most people. It would be the first presidential marriage in the White House.

Cleveland insisted that the wedding, set for June 2, be plain and simple. "I want my marriage to be a quiet one," he told his fiancée, "and am determined that the American Sovereigns," by which he meant the press, "shall not interfere with a thing so purely personal to me."[10] He revised and shortened the marriage ceremony, substituting the word, keep, for obey, in the bride's vows, and arranged for only about thirty guests to be present for the service. His own minister in Washington, the Reverend Byron Sutherland, a Presbyterian, officiated; John Philip Sousa conducted the Marine Band in the Mendelssohn March, and church bells rang and the Navy Yard fired a twenty-one-gun salute after the ceremony. But when the newly-weds went off to Deer Park, Maryland, to be "far from the madding crowd" for their honeymoon, the "American Sovereigns" were hot at their heels. Scores of newspapermen set up observation posts near the honeymoon cottage, watched the couple day and night with field glasses, and reported their every move; they even waylaid waiters taking food to the cottage so they could report the cuisine to the newspapers.

Cleveland was outraged by the intrusion into his privacy. When he returned to Washington with his bride, he wrote a blistering letter to the *New York Evening Post* excoriating the busybody reporters. "They have used the enormous power of the modern newspaper to perpetuate and disseminate a colossal impertinence," he exclaimed, "and have done it, not as professional gossips and tattlers, but as the guides and instructors of the public in conduct and morals. And they have done it, not to a private citizen, but to the President of the United States, thereby lifting their offence into the gaze of the whole world, and doing their utmost to make American journalism contemptible in the estimation of people of good breeding everywhere." At a dinner in his honor at Harvard University a few months later, when he saw reporters staring at his wife, he interrupted his speech to exclaim: "O, those ghouls of the press!"[11] Mrs. Cleveland took it all more calmly than he did. But, then, the press liked her better than it did the President.

Mrs. Cleveland, at twenty-two, the youngest of all our First Ladies, got off to a good start in the White House. As Cleveland watched her in action at her first reception, he nudged her mother and chortled, "She'll do! She'll do!"[12] Her "vim and dash" pleased all the guests, and the friendly luncheons, teas, and Saturday afternoon receptions she presided over received accolades in the press. She had to shake hands with so many people at first—some people even sneaked back for a second welcome—that it took massage treatments afterwards to reduce the pain. "No more brilliant and affable lady than Mrs. Cleveland has ever graced the portals of this old mansion," observed "Ike" Hoover, Chief Usher at the White House. "Her very presence threw an air of beauty on the entire surroundings, whatever the occasion or the company."[13] When the presidential couple visited the West in 1887, the *Ohio State Journal* announced that Cleveland had seen at least 10,000 men there as well fitted to be President as he was, but that almost nobody was so well fitted to be a President's wife as Mrs. Cleveland. "My only regret about it," Republican leader Chauncey Depew told Cleveland's secretary, Dan Lamont, "is that it will be so much harder for us to win against both Mr. and Mrs. Cleveland."[14] One of Cleveland's political enemies was reduced to putting it this way: "I detest him so much that I don't even think his wife is beautiful!"[15]

Still, there was slander, especially at election time. In 1884, when Cleveland, still a bachelor, ran for President the first time, the Republicans discovered he had fathered an illegitimate child as a young man in Buffalo and went around chanting: "Ma, ma, where's my pa?" In 1888, when he ran for a second term, against Republican nominee Benjamin Harrison, they circulated stories about how miserable his marriage had turned out to be and how unhappy his young wife was. Cleveland got drunk, they charged, and beat his wife; he even threw her out of the White House once in the middle of the night. So insistent did rumors of this kind become that Mrs. Cleveland, in an unusual step for a First Lady, was forced to issue a statement (in the form of a letter to a woman in Worcester, Massachusetts) putting the gossip to rest. "I can wish the women of our Country," she declared, "no greater blessing than that their homes and lives may be as happy, and their husbands may be as kind, attentive, considerate and affectionate as mine."[16] Cleveland lost to Harrison in 1888, but not because of his wife. When he ran again (and successfully) in 1892, in fact, his supporters put his wife's picture on campaign posters slightly above and between him and his running mate. No other First Lady was ever so honored.

During Cleveland's 1892 campaign, a New York woman proposed

organizing a "Frances Cleveland Influence Club" to help him out, but he firmly refused her offer. "We trust you will not undervalue our objection," he told her, "because it rests upon the sentiment that the name now sacred in the home circle as wife and mother may well be spared in the organization and operation of clubs created to exert political influence."[17] Cleveland distrusted women's clubs as a matter of principle, for he thought they encouraged wives and mothers to neglect their families, but he thought clubs involving women in politics were the worst of all. He certainly didn't intend for "Frank," as he called his wife, to meddle in politics. "I have my heart set upon making Frank a sensible, domestic American wife," he told one of his sisters a few months before the marriage, "and I would be pleased not to hear her spoken of as 'The First Lady of the land' or 'The Mistress of the White House.' I want her to be happy and to possess all she can reasonably desire, but I should feel very much afflicted if she lets many notions in her head. But I think she is pretty level-headed."[18] Once, when Cleveland asked White House aide William H. Crook to notorize a deed for him and his wife, it was necessary for Crook to take Mrs. Cleveland aside and query her as to her willingness to sign the paper. When she said she was signing without any mental reservations, Cleveland turned around and snorted: "I think that such a requirement of the law is silly—I mean the clause that required a notary privately to examine a woman before she signs a deed like this." Then, after thinking it over, he added: "Still, I suppose the requirement was caused by reason of impositions practiced upon some poor women, who felt compelled to sign papers under their husband's insistence."[19]

There is no evidence that Mrs. Cleveland ever developed "notions," as Cleveland called them, or ever tried to intrude herself into the domain the President considered strictly his own. It is doubtful, indeed, that Cleveland ever bothered to discuss public issues with her, even casually, or that it ever occurred to him to ask her advice on policies or appointments. Mrs. Cleveland was young and inexperienced when she married him, and seems never to have developed much interest in politics. She was, so far as we know, content to become the "sensible, domestic wife" that Cleveland wanted her to be. For his part, he "idolized" her, according to "Ike" Hoover, "thought of her as a child, was tender and considerate with her always."[20] In domestic matters, though, Mrs. Cleveland had her say, and her husband usually listened. When it came to dress, food, and deportment, she "would watch over him as though he were one of the children," Hoover observed, and she succeeded in toning down his irascibility, polishing his manners, smoothing his relations with people, and (since he was a "workaholic") getting him to relax and take vacations from time to time.[21] Once he bought an orange-tawny suit she disliked and she did everything she could to keep him from wearing it. Finally she told him he would lose the Irish vote if he appeared in public with it and won her point.[22]

Mrs. Cleveland wept when she left the White House after William McKinley's inauguration in March 1897; so did some of the White House employees, for she had always taken a personal interest in them and given them birthday and Christmas presents every year. For retirement, the Clevelands first thought of New York City; but then they decided a smaller town would be better for their children. One morning, after a trip to Westchester to look at houses, Mrs. Cleveland came down to breakfast and said, "I have an inspiration about our future home." "So have I," said Cleveland promptly. "What is it?" they both asked at the same time. Then they both exclaimed: "Princeton!" And so Princeton it was.[23]

In Princeton, New Jersey, the Clevelands acquired a place they called "Westland" (after their friend Andrew F. West, Dean of the Princeton Graduate School) and settled down to a quiet life in a college town. Cleveland eventually became a trustee at Princeton University and his wife took pleasure in entertaining the faculty and students, and, according to one newspaperman, soon became "a sort of patron saint and goddess in human form" to the college crowd.[24] Cleveland did some writing—on politics and on (or, rather against) women's clubs—enjoyed the children (five in all, three girls and two boys), hustled his wife off to football games, and took her for trips on his friend Commodore E. C. Benedict's yacht, *Oneida* (on which he had had his secret operation for cancer of the jaw in 1893).[25] For summers they acquired a place in Tamworth, New Hampshire, in 1905.

Not long after he retired, Cleveland's health began to go downhill. Long overweight and accustomed to a sedentary life, he was soon suffering from a variety of ailments, including heart trouble, and eventually became a semi-invalid, requiring all of his wife's good spirits as well as nursing skills to keep him comfortable. After he died in June 1908, at seventy-two, his wife, who was only forty-four, struck up a friendship with Thomas J. Preston, Jr., professor of archaeology at Wells College, and they were married in February 1913. Preston joined the Princeton faculty soon after the marriage, and his wife went on entertaining college people at their home, as she had before. She also engaged in charity work, was active in the Women's University Club, and helped raise money for Wells College. She lived on until she was eighty-three. When she died in her sleep in October 1947, she was visiting her elder son, a lawyer, in Baltimore. Five presidential widows were still living at the time of her death: Caroline Harrison, eighty-nine; Edith Roosevelt, eighty-six; Edith Wilson, seventy-five; Grace Coolidge, sixty-eight, and Eleanor Roosevelt, sixty-three.[26]

* * * *

New Life

Cleveland was one of the hardest working of all our Presidents. "It was work, work, work," according to "Ike" Hoover, "all the time." On the day of his wedding, he worked as hard as ever, though he did take time out to chat with Frances and her mother and go on a short drive in the afternoon. Late in the afternoon, however, when the Postmaster-General sent a messenger over to inquire whether he had time to sign some documents, he cried: "Yes, I will sign—but tell him to get those documents here as quick as the good Lord will let him!" After the wedding, a Cabinet officer's wife declared: "You may depend that the President will not work himself to death now. He has begun a new life. . . . You may look for a decided change in favor of leisure. . . ." She was partly right.[27]

Marrie

A deluge of letters flooded the White House after Cleveland's marriage, but one of his favorites was sent by Joe James, a Chinese living in Philadelphia. "I am glad you marrie to bear plenty good fruits to the nation," wrote James, "and I congratulate to your marriage all enjoy yourself. I read your letter you so kind to our Chinese living here, and instruct the Government to protect the Chinese and be please to live everywhere. So I thank your kindness ever so much. I heard you on June 2d to be marrie. So I send a little present to you and the bride. I hope you enjoy yourself to receive it. One china ivory fan, with sandalwood box, for the bride; one ivory card-case for the President, all sent by mail. I hope God bless you in prosperity in all things." The letter, to which the Clevelands sent a hearty thanks, was signed in Chinese as well as English.[28]

Lesson

One afternoon shortly after the marriage, the Clevelands were to go out driving and at the appointed time Cleveland put on his overcoat and stood waiting for his wife. Minutes passed and he became impatient. Finally he decided to teach his wife a lesson in punctuality, took off his coat and gloves, and resolved to cancel the drive. Presently he heard her voice at the foot of the stairs. "Come along," she cried. "I am ready now." Cleveland wavered for a moment. "And what do you suppose I did?" he said afterward. "Why, I got up, put on my coat and gloves again, and went driving."[29]

Rabble of Shop-Girls

Soon after coming to the White House, Mrs. Cleveland began holding informal receptions on Saturday afternoon for women in Washington. But a Washington official, concerned for the dignity of the First Lady, urged she give them up. "For what reason?" asked Mrs. Cleveland. "Well, you see," said the official, "about half of all the women who come Saturday afternoon are clerks from the department stores and others—a great rabble of shop-girls. And of course a White House afternoon is not intended for them." "Indeed!" exclaimed Mrs. Cleveland. "And if I should hold the little receptions some afternoon other than Saturday they couldn't attend, because they have to work all the other afternoons. Is that it?" "Certainly," said the official. "That's it exactly." To his chagrin, Mrs. Cleveland at once issued orders that nothing should ever interfere with her Saturday afternoon receptions.[30]

Arsenic

While Frances Cleveland was in the White House, some advertisers—for pills, patent medicines, and soap—began using her name without her permission. One advertisement, taking up a full page in a magazine, attributed her beauty to the use of arsenic. "Mrs. Cleveland's Remarkable and Beautiful Complexion," went the heading. "How Such Lovely Complexions Are Best Retained." Came then the explanation. "Many have been the remarks . . . about the complexion of the First Lady. In fact some of her close personal friends have even gone so far as to insinuate that some forces other than those supplied by Nature were brought into use by Mrs. Cleveland. But of course this was only conjecture. The secret of her beautiful complexion . . . is simply the use of arsenic, which can safely be taken and which can be procured from the New York doctor whose name is signed to this advertisement." Letters reached the White House criticizing Mrs. Cleveland for lending her name to advertisers, and the National Woman's Temperance Convention in Washington passed a resolution protesting "the immoral exhibition of the faces and forms of noted women as trade-marks and advertisements." Mrs. Cleveland had never, to be sure, sanctioned the use of her name by advertisers but she felt powerless to stop it.[31]

Exit the Bustle

When Congress was in recess, reporters found it hard to dig up stories worth sending their papers, and some of them met every day to see if they couldn't come up with something worth writing about. But one day they couldn't think of anything. "Can't we send a society item?" one reporter finally suggested. "Yes, if you've got one," said another;

"there isn't a line in sight now." "Then let's manufacture one," said the first man. And he went into deep thought. "I've got it" he suddenly exclaimed. "Let's say that Mrs. Cleveland has decided to abolish the bustle." "Brilliant!" cried his colleague. So they went to work and scribbled out a story that went out an hour later and revolutionized the fashion industry: that the young and beautiful First Lady, a leader in fashion, had stopped wearing a bustle. When Mrs. Cleveland saw the story, she didn't quite know how to react. It didn't seem worth denying; but if she appeared in a bustle again it would require a great deal of explaining. So she did the simple thing; she immediately ordered a gown without a bustle. Upshot: exit the bustle![32]

Coming Back

On March 4, 1889, when the Clevelands left the White House after Benjamin Harrison's inauguration, Jerry Smith, White House retainer, escorted Mrs. Cleveland to her carriage and bid her a fond farewell. "Now, Jerry," she said, "I want you to take good care of all the furniture and ornaments in the house, and not let any of them get lost or broken, for I want to find everything just as it is now, when we come back again." "Excuse meh, Mis' Cleveland," said Jerry, astonished, "but jus' when does you-all expec' to come back, please,—so I can have everything ready, I mean?" "We are coming back just four years from to-day," she smiled. And they did.[33]

Every Damn Sliver

When Kansas newspaperman William Allen White interviewed Cleveland in Princeton, he noticed a lovely old colonial piece and stopped to admire it. "Like that stuff?" asked Cleveland. "Know anything about it, or just like the looks of it?" White said he had a lot of affection for old furniture. "Well," said Cleveland, "I suppose if you like it, you like it." Then he told how he had acquired that piece—and many others—without intending to.

The preceding summer, according to Cleveland's tale, his wife had visited an island off Maine and soon after her return "a great, damn vanload of junk came rolling down the avenue toward the house," when she was in New York. "What in the hell have you got there?" he asked the mover. "Old furniture for Mrs. Cleveland," said the mover. "It came from some place up in Maine." Cleveland eyed the old chairs, bureaus, beds, and desks with disgust and cried: "You turn around and take that damn kindling pile back downtown." After the mover left with the furniture, Cleveland forgot about it.

Two weeks later Cleveland went to New York for the day and when he got back he saw the piece that White was admiring. "Frank," he said to his wife, "where'd you get that? Where'd it come from?" "That?" cried Mrs. Cleveland airily. "Why, we've always had that. We've had that for years." Cleveland grunted and dismissed it from his mind. But the next time he went away for a few days a big chair showed up downstairs and he cried: "Where'd that come from, Frank?" "Oh, that old chair," said his wife nonchalantly, "we got it out of the attic; we've had that for a long time." And then Cleveland went into their spare bedroom one day and saw a new bed, shiny and polished, there. But he knew he had seen the bed before. Then he suddenly remembered: the bed had been on the kindling pile he had sent away weeks before. But it was a good-looking bed, so he decided not to say anything about it. "And in six or seven months, what do you think?" he cried to White. "Frank had slipped every damn sliver of that furniture into the house, one piece at a time, and she thought I didn't know it. And one day I told her, and we had a good laugh. I suppose it is all right," he told White; "if that's the sort of thing you like, why, you like it!" White left impressed with Cleveland's pride in his young wife.[34]

In the White House

At a White House luncheon given by President Harry Truman, his daughter Margaret had General Dwight D. Eisenhower as her luncheon partner. When Ike came up to join her, she was chatting with Mrs. Preston, Grover Cleveland's widow who had remarried, and though Margaret introduced him to the former First Lady, he didn't seem to realize who she was. "And where did you live in Washington, ma'am?" he asked, joining the conversation. She looked at him demurely. "In the White House," she said, as the General's face fell.[35]

CHAPTER 22

CAROLINE HARRISON

1832–1892

When the Harrisons—Benjamin and Caroline—took over the White House from the Clevelands in March 1889, the venerable old structure, long in need of major repairs, was in such a dilapidated condition that there was serious talk of replacing it with an entirely new Executive Mansion elsewhere in Washington. The architect even got around to constructing a model for the proposed new building and, because of the new First Lady's interest in the project, called it "Mrs. Harrison's place." But the plan never got off the ground. Traditionalists objected to abandoning a historic landmark, and Congress failed to come up with the necessary funds. When the legislators finally got around to voting $35,000 for mere renovations, Mrs. Harrison swallowed her disappointment and threw herself with zest and energy into the work of refurbishing the old place.

"We are here for four years," Carrie Harrison told a reporter as she went about her task; "I do not look beyond that, as many things may occur in that time, but I am anxious to see the family of the President provided for properly, and while I am here I hope to get the present building put into good condition." She was especially bothered by the plethora of offices and dearth of living space in the White House. "Very few people understand to what straits the President's family has been put at times for lack of accommodations," she added. "Really there are only five sleeping apartments and there is no feeling of privacy."[1] The First Lady knew whereof she spoke; the Harrison White House was chockablock with people, including John Scott (her 90-year-old father), Mrs. Mary Dimmick (her young widowed niece), daughter Mary McKee and her husband James, and three lively little grandchildren. Son Russell and his family were also around a great deal of the time. "President Harrison," observed one newspaper, "is the only living ruler who can gather at his table four generations in the direct line from Great-grandfather Scott."[2] Benjamin Harrison McKee—the two-year-old "Baby McKee"—was especially popular with the public. He was, said Kansas editor William Allen White, "forever crawling over the first page of the newspapers."[3]

Mrs. Harrison herself made few headlines. Washington society thought her somewhat provincial and excessively domestic, though conceding her grace and poise at White House functions. But the socialites got it wrong; the new President's wife was not a simple homebody. She painted watercolors, was a fine pianist, cultivated orchids, and, unlike her reserved husband, liked to socialize. As a young woman, in fact, she had been bored with housework and was even a bit careless in her dress. But she was peppy and outgoing, and the somewhat strait-laced young Ben, who first met her in 1848, when she was only seventeen, was charmed by her company. He got to know her when he was a freshman at Farmers' College (in the Walnut Hill section of Cincinnati), where her father, an educator as well as a Presbyterian minister, taught mathematics and science. Ben studied physics and chemistry with Dr. Scott; he also visited the Scott home on occasion and was much taken with the "charming and loveable" little Carrie with the brown hair and eyes, who was enrolled in a girls' school run by her father.[4]

In 1849, Dr. Scott took his school to Oxford, Ohio (where Carrie had been born in 1832), expanded its facilities, and renamed it the Oxford Female Institute. Soon after, Ben transferred from Farmers' College to Miami University, which was close to the Institute, and began courting Carrie. Before long he was spending so many evenings at Carrie's house—as well as taking her on buggy and sleigh rides—that his classmates began calling him a "pious moonlight dude."[5] Carrie was pious, too, but she bent the rules a bit; and although the Presbyterianism both she and Ben espoused frowned on dancing, she got Ben to take her to dances. Ben always sat it out, for he didn't know how to dance, but Carrie found other partners and enjoyed herself immensely on the dance floor. In 1852 the two were secretly engaged; and upon his graduation from Miami that year Ben went off to read law with a firm of attorneys in Cincinnati. The following year Carrie graduated from the Oxford Female Institute and began teaching music there and elsewhere.

Marriage posed problems for the young couple. Ben was living on a shoestring in Cincinnati and it seemed the better part of wisdom to postpone marriage until he was established in his profession. But the young man's mind boggled at the thought. He threw himself furiously into his law study, but he couldn't keep Carrie long out of his thoughts. Post office clerks thought him something of a joke because he came around so often looking for letters from Carrie; "she sometimes neglected *me* for a full week," he once complained.[6] Then Carrie fell ill from overwork and Ben was plunged into despair. Knowing how separation was making his own life miserable, he was sure the long engagement was primarily responsible for his fiancée's delicate health.

In desperation Ben finally turned to his friend John Anderson for advice but was dismayed by what he was told. Anderson, a practical sort, thought Ben was crazy to think of marriage before he had a steady income. "Love is powerful as an incentive," he acknowledged, "but will it pass current for potatoes and beef? Coffee and muffins for two are not paid for by affection between the 'two.' Hard cash buys! *Where will it come from?*"[7] But Ben saw things differently. Marriage, he knew, meant a hard struggle to make ends meet; but more delay meant further deterioration in Carrie's health. "The question, then, John," he said, "is narrowed down to this: Shall I marry Carrie now and thus relieve her of those harassing doubts and fears which wear away her life, or shall I agree to stand aside and let her hasten to an early grave?"[8] Put this way, there was no further hesitation. In August 1853, he and Carrie decided to get married that fall and then settle down on his father's farm at The Point, at the mouth of the Big Miami, while he completed his studies under the direction of Cincinnati attorney Bellamy Storer. On October 20, the two were married by Carrie's father in a simple ceremony in Oxford and then headed for The Point.

The Harrisons' early married years were hard ones. In the summer of 1854, when Ben finished his law studies, they moved to Indianapolis and Ben began looking for clients in the rapidly developing "Heart of Hoosierdom." But there were not many of them at first and Ben was forced to borrow money from family and friends to pay for food and lodgings. "Indeed I would almost be willing to work for nothing," he told Carrie, "just for the sake of being busy."[9] After a few years he did become busy—with Republican politics as well as law work—and he was eventually able to provide Carrie and their two children, Russell and Mary, with a fine home in a nice part of town. But he spent so much time at work making money that Carrie began to feel neglected. Ben didn't realize this for a long time. Then, when he went into the Union Army after the outbreak of the Civil War and had plenty of time to think about his marriage during the long, lonely days between engagements, he began to regret the way he had put work above wife.

Harrison entered the service in the summer of 1862 as commanding officer of the 70th Indiana Volunteers which he had raised at the request of the Governor Oliver P. Morton. Six months of separation from Carrie produced self-reproach as well as loneliness. "I now see *so* many faults in my domestic life that I long for an opportunity to correct," he wrote her. "I know I could make your life so much happier than ever before. . . ."[10] Carrie visited him twice, when he was stationed at Bowling Green, Kentucky, and each reunion increased his determination to do better by her after the war. "I hope to be a better husband and father," he told her, "a better citizen and a better Christian in the future than I have been in the past."[11] He was filled with anxiety when he heard she was ill, and he was deeply depressed whenever her letters took a long time to reach him. "Your daily domestic life I feel to be a part of my life," he wrote her towards the end of the war, "and I love to know every little event of it to feed my love of home upon, and direct my imagination when I go in fancy to my present home."[12]

One afternoon, when Harrison was on garrison duty in South Carolina, he went for a walk, came across a garden, and picked two rosebuds to send his wife. "Imagine," he wrote her, "my whispering in your ear with this simple gift all that could be delicate and affectionate in a lover in his first declaration."[13] In May 1865, when the war was over and he was preparing to return to civilian life, he wrote her a lengthy letter renewing his pledge never to put her in second place again. He also asked her to remind him of his promise if he ever started backsliding. He even suggested that she give him "some little article of apparel of ornament" when he got home to remind him of the pledges he had made. Better still, he would select a keepsake for her to wear at all times. "And when I deliver it to you," he said, "we will weave a spell about it [as] a constant reminder of the resolutions and vows I have made in the army." He hoped she approved his suggestion about the keepsake. "Will you wear it," he asked her, "and promise always to hold it up before me when you see a cloud on my brow or hear hasty words from my lips?"[14] Carrie treasured the keepsake letter; she put it in her pocketbook and had it with her when she met him on his return to Indianapolis with his regiment in June 1865.

Home again after the war, Harrison seems to have worked about as hard as ever, in his increasingly lucrative law practice and then in the U.S. Senate to which he was elected in 1880. But he also gave more time to Carrie; he took her on buggy rides, went to the opera and theater with her, and joined her for strawberry festivals and church suppers. He also encouraged her in her own efforts at self-expression: doing watercolors in her home studio and teaching china painting to young ladies. He even looked the other way when she and several other mothers organized dancing lessons for their children. For her part, she took no great interest in politics, but followed with interest the proceedings of the Republican National Convention in Chicago in the summer of 1888. "Well, Mamie," she told her daughter when crowds began gathering in front of the Harrison home after her husband's nomination for President on the eighth ballot, "your father's got it!"[15]

Mrs. Harrison's health had been "delicate," as people put it primly in those days, ever since a bad fall on the ice in 1881. Still, she entered on her duties as First Lady in 1889 with enthusiasm and energy and was soon receiving accolades from Washingtonians who had been lamenting the departure of Frances Cleveland from the White House. Her main activity at first was overseeing the rehabilitation of the Executive Mansion: modernizing the kitchen, installing a new heating system, replacing moldy old floors with new ones, painting and repapering the bedrooms, cleaning the chandeliers, acquiring new curtains, upholstery, and furniture, and putting private bathrooms in each of the bedrooms. Her husband was especially pleased by the new bathrooms. "They would tempt a duck to wash himself every day," he exclaimed.[16] But neither he nor Carrie took to the new electric lights and call buttons that White House aide "Ike" Hoover arranged to have installed. To Hoover's amusement, they were so afraid of getting shocks that they continued using the old gas lights; they also left the electric lights on all night if he forgot to turn them off before leaving. But they did like the new central switchboard which permitted several telephones, instead of just one, in the big building.

While carpenters, plumbers, bricklayers, and electricians were swarming about the place, Mrs. Harrison went carefully through closets, drawers, and cabinets, weeded out broken dishes and worn-out utensils, and arranged for the purchase of a new set of china—of her own design, the four-leafed clover—for the White House. From the broken and incomplete sets she came across in her explorations, she also assembled a collection of china acquired by former First Families that soon became a major tourist attraction. She found time, too, to develop a huge assortment of plants and flowers (especially orchids, her favorite flower) in the White House greenhouses and filled the White House with them. On one visit, a reporter estimated that in the East Room alone there were about 5,000 decorative plants and "about a mile of smilax" for chandelier festoons; and on mantels and window seats elsewhere, he figured there were "5,000 azalea blossoms, 800 carnations, 300 roses, 300 tulips, 900 hyacinths, 400 lilies of the valley, 200 bouvardias, 100 sprays of asparagus ferns, 40 heads of poinsettia, and 200 small ferns."[17] The Clevelands were astonished at the way the White House looked when they returned for a second term in 1893.

While giving the White House a facelift, Mrs. Harrison continued to pursue one of her favorite hobbies, needlework, and donated most of her handiwork to church bazaars and charities. She did a lot of painting, too, decorating White House candlesticks, milk pitchers, flower-pot saucers, and crackerboxes with her favorite designs: flowers, leaves, shepherdesses, milkmaids, and, of course, four-leafed clovers. She even painted her favorite decorations onto porcelain dishes for Washington ladies looking for White House souvenirs. "Many a baby whose parents have named him for the president," noted one reporter, "has received a milk set painted by Mrs. Harrison with cunning Kate Greenaway children."[18] In 1890, the First Lady helped organize the Daughters of the American Revolution, served as first President-General, and, in an address to the D.A.R.'s First Continental Congress, declared: "We have within ourselves the only element of destruction; our foes are from within, not from without. Our hope is in unity and self-sacrifice."[19]

Until Mrs. Harrison became ill in 1891 she thoroughly enjoyed the social life of the White House and more than made up for her husband's iciness in personal relations. She presided graciously at official functions and saw to it that dancing again became a regular part of public receptions. She organized painting classes for the wives and daughters of government officials and engaged a tutor to give French lessons too. When Christmas came, she helped decorate the first Christmas tree to be set up in the Executive Mansion and on Easter Monday she stood with her husband on a wooden stand to watch thousands of Washington children roll their eggs on the White House lawn. There was an "air of genteel gaiety" in the White House, it was said, when Harrison was President, and it was primarily the work of the President's wife.[20]

But there was sadness as well as gladness in the Harrison White House. Mrs. Harrison was upset by the political attacks on her husband and surprised and deeply hurt when she, too, became a target of her husband's enemies. In July 1890, when Postmaster General John Wanamaker and several other Philadelphians presented her with a cottage at Cape May Point in New Jersey, there was an outburst of criticism in the opposition press. "Who are those generous individuals," cried the *New York Sun,* "that have bestowed upon MRS. BENJAMIN HARRISON a cottage at Cape May Point, clear of encumbrance, and with floors swept clean for BABY McKEE to creep over this summer?" The *Sun* demanded that the names of the donors be made public for, the editor noted, "the President who takes a bribe is a lost President." Other newspapers quickly joined the attack; and although it turned out that Harrison had insisted on sending Wanamaker a check for $10,000 when the gift was first made, his critics charged that he had made his payment only after he had been exposed as a "gift grabber."[21] Mrs. Harrison was shattered by the slurs on her husband's integrity. "Oh, Colonel Crook," she exclaimed to the White House Chief Usher one day, "what have we done?" Shocked by her distraught appearance, Crook said, "I do not understand, Madam. What do you mean?" "What have we ever done," she wailed, "that we should be held up to ridicule by newspapers, and the President be so cruelly attacked, and even his little helpless grandchildren be made fun of, for the country to laugh at!" Then, as Crook was trying to think of some way to cheer her up, she continued: "If *this* is the penalty for being President of the United States,

I hope the Good Lord will deliver my husband from any future experience."[22]

Harrison lost to Cleveland when he tried for a second term in 1892, but by then Mrs. Harrison was no more. She developed a bad cough in the summer of 1891, became weak and listless during the winter of 1892, and was soon bedridden. Though her illness was at first diagnosed as nervous prostration, it turned out that she had developed tuberculosis and that there was little hope of recovery. During the 1892 campaign, Harrison reduced his personal appearances so he could be with his wife, and Cleveland also refused to appear in public out of respect for the President's wife. On October 25, Mrs. Harrison died, at sixty-one, with the President by her bedside; and after a funeral in the East Room on the 27th, she was buried in Indianapolis. She was the second First Lady—Letitia Tyler was the first—to die in the White House. Her daughter, Mary Harrison McKee, became the White House hostess for the remainder of Harrison's term.

After leaving office, Harrison returned to Indianapolis, converted part of his home into an office, and resumed his law practice. In April 1896, he married Mary Scott Dimmick, his wife's niece, a Pennsylvania woman who had been widowed early, lived with the Harrisons in the White House, helped Carrie out with secretarial work, and provided the President with much comfort after Carrie's death. Neither Russell Harrison nor Mary McKee attended the wedding; they strongly opposed their father's remarrying. Harrison tried hard to win their approval. "A home is life's essential to me and it must be the old home," he told his son. "Neither of my children live here—nor are they likely to do so, and I am sure they will not wish me to live the years that remain to me in solitude."[23] But his children remained unreconciled.

In his last years, Harrison attended dinner parties, church suppers, and flower shows with his new wife; he even took her to concerts. "I am not devoted to music," he said, "but Mrs. Harrison is, and I am devoted to her."[24] For vacations he took his wife and little girl (born in 1897) to the Adirondack Mountains, the Jersey seashore, and Europe. In 1901, he died of pneumonia, and his wife, who eventually published a biography of him, survived him by forty-seven years. When she died in New York in January 1948, she was almost ninety.

CHAPTER 23

IDA SAXTON McKINLEY

1847–1907

When William McKinley first met Ida Saxton a couple of years after taking up his law practice in Canton, Ohio, she was the belle of the town: beautiful, witty, cultured, vivacious, and full of energy. She was also headstrong. There "is no let up," said a friend, "when she does set her head."[1] Ida (born in Canton in 1847), who was twenty-two when William met her, had taken a job in her father's bank after attending a girls' school in Pennsylvania and spending six months in Europe. With her handsome features, languorous blue eyes, and lively disposition, she had many suitors. But she soon set her head on the handsome young attorney she met at a church picnic. William, then twenty-seven, also fell deeply in love and made frequent visits to the bank where she worked to hand her little deposits and large bouquets. On Sundays, the two of them stopped to talk while she was on her way to teach Sunday School at the Presbyterian church and he was headed for the Methodist Sunday School. There is a story that one Sunday morning, when they came to the spot where they usually separated, he exclaimed: "I don't like these partings. I think we ought not to part after this." "So do I," Ida is supposed to have rejoined.[2]

The marriage, which was solemnized on January 25, 1871, by William's as well as Ida's minister, began auspiciously. Ida's father, who was "rather particular" when it came to his daughter, had great faith in the future of his hard-working son-in-law and gave the young couple a fine house on North Market Street in which to begin their family.[3]

From the outset, William was extremely devoted to his wife; he catered to her every wish and spent evenings at home with her when he was not working. She, for her part, idolized him; she liked to call him "the Major" in public, for she was proud of his Civil War service, but in private she always addressed him as "Dearest." The birth of a little girl, whom they named Katherine after Ida's mother, on Christmas Day, added to their happiness, for they both adored children. Mrs. McKinley saw that little "Katie" was much photographed and even arranged to have an oil painting done of her.

In 1873 fortune turned sour for the McKinleys. A series of deaths in the family almost unhinged Ida McKinley: first, Grandfather Saxton, to whom she had been close, passed away, and then her mother, to whom, as the eldest daughter, she had been almost like a younger sister. Her mother's death came just before her second child, whom they named Ida, was born. The birth was difficult, and the baby was frail and sickly and lived only five months. After that the once sprightly young homemaker was never well again. She developed phlebitis, which left her partly crippled, and she always walked with difficulty after that; she seems also to have suffered brain damage about that time, for the splitting headaches and epileptic seizures, mild and major, that plagued her for the rest of her life date from this period. Mrs. McKinley was convinced that the good Lord was displeased with her for His own good reasons and she spent a lot of time brooding about her troubles and reproaching herself. She also became obsessed with Katie's well-being, fearful that an angry God might take her too. Once, when McKinley's brother Abner asked Katie to take a walk with him, the little girl told her uncle: "No, I mustn't go out of the yard, or God'll punish Mama some more." In 1876, Katie died of typhoid fever when she was only four. The once-attractive young banker's daughter with the "languorous eyes" was by now a sickly, self-centered, and at times querulous semi-invalid who was totally dependent on her husband for getting around in the world.[4]

Historians don't rate McKinley highly as a public figure; except for his break with isolationism while President and his venture in "internationalism," though brief, in the Spanish-American War of 1898, they have not been particularly enthusiastic about his administration. As a private person, though, the twenty-fourth President stands tall; in any Kantian kingdom of ends, he would undoubtedly be a prominent citizen. His solicitude for his ailing wife after the sorrows of their early married years was wonderful to behold, and some of his friends considered him a near-saint. "There is nothing more beautiful than his long devotion to his invalid wife, whose invalidism was not of the body alone," observed journalist Charles Willis Thompson. "The terrible illness which ruined her health permanently impaired her spirit, but this only made her man her knight; it made him more unselfishly devoted, more tender."[5] For McKinley there were no lamentations, no recriminations, no resentments for the turn his life had taken; there was only a deep and tender concern at all times for his wife's comfort and happiness. McKinley thought she would eventually get well again, and each good period she experienced—fewer headaches, milder and less frequent seizures—filled him with hope. He continually sought new medical advice in a quest for proper therapy, and tried out various medicines for alleviating the pain she suffered. In the end, though, only massages for the headaches and bromides for the convulsions were of much help; and, above all, constant companionship, acute sensitivity to her moods and needs, and a complete absence of tension.

In Canton, people whispered that Mrs. McKinley had epilepsy, but the word was never uttered by the McKinleys, their doctors, or their friends. Until well into the twentieth century, indeed, when dilantin was developed for controlling the ailment, epilepsy was a frightening word. As late as 1917, a medical book contained this grim definition: "Epilepsy is a terrible disease to look upon, not painful in itself, but productive of great distress and misery. It is not attended with immediate peril to life, but is liable to terminate in worse than death—in insanity, or fatuity—and carries with it perpetual anxiety and dismay."[6] McKinley's policy was to minimize his wife's illness as much as he could. She had a headache, he would tell people, as if it were a minor problem; or, he would say, she wasn't feeling good today. At dinner parties, he sat next to her and if he saw a minor seizure coming, he promptly tossed a napkin or handkerchief over her face while continuing to talk with the guests; and his wife, once the spell was over, would resume her part in the conversation as if nothing had happened. Always on the alert, though, McKinley unobtrusively got her out of a room full of people if he saw that she was on the brink of a major attack. Most of the time, in short, he seemed to take a casual view of his wife's illness, and so, in the end, did most people who came into contact with her. McKinley's strategy worked perfectly. None of the people who voted for him (or against him) in the 1896 and 1900 presidential contests were aware of the nature of his wife's illness. All they knew was that her health was "delicate." Once, when one of Mrs. McKinley's nieces heard the word, epilepsy, applied to her aunt, she was outraged that the Democrats would stoop so low as to invent lies about the President's wife.[7]

In 1876, the year of Katie's death, McKinley was elected to Congress, and the following year, when he went to Washington to enter on his duties, his wife was well enough to go with him. In Washington he rented two rooms in the Ebbitt House to live in; he also hired a maid to help take care of his wife. When he was at the Capitol, his wife stayed at home with the maid, working on her embroidery or crocheting bedroom slippers for friends and charities. But McKinley spent most of his free time with her; he took her for carriage rides when she was up to it and on occasion even to the theater, which she loved. From time to time, too, he took her to dinner at the White House, for President and Mrs. Hayes, their Ohio friends, were extremely hospitable to the young Congressman and his wife. Mrs. McKinley even had lunch by herself a few times with Lucy Hayes and was proud of being in and out of the White House without her husband. McKinley had become a heavy smoker by this time; it was his only indulgence. But since tobacco bothered his wife, he always smoked his cigars after supper while walking up and down the sidewalk in front of the hotel; it was his only exercise.

Whenever McKinley had to be out of town, he invariably wrote his wife at least once a day. A friend asked Mrs. McKinley one day what her husband could possibly have to say in all those letters, and she said simply: "He can say he loves me."[8] But one separation produced a major crisis. In 1898, when Congress was in session and Mrs. McKinley went to Canton without her husband to visit her sister Pina, she suddenly began having such violent convulsions that the family doctor feared for her life. Notified of the emergency by telegram, McKinley left for Canton at once. When he reached his hometown, the doctor told him his wife was unconscious and there was no hope for her. But McKinley dismissed the doctor, went to his wife's bedside, and began rubbing her hands, stroking her forehead, and caressing her face. "Ida, it is I," he said gently, and begged her to speak. For hours he tried to rouse her without getting any response. Finally, early the next morning, she began moving slightly; a little later she opened her eyes, grasped his hand, and whispered: "I knew you would come."[9]

In 1892 McKinley became Governor of Ohio, and when he moved to Columbus, the state capital, his wife's health was much improved; there were fewer attacks and she seemed at times on the road to recovery. She was certainly happier. While her husband was at work in his office, she kept busy at her hobbies; turning out bedroom slippers, hundreds of them; making black satin neckties for her husband and his friends; and cleaning jewelry. She did a fair amount of socializing too; she played hostess for McKinley's friends and associates at the hotel where they lived and thoroughly enjoyed the weekends at the elegant homes of wealthy friends like Mark Hanna of Cleveland and Bellamy Storer of Cincinnati. Every morning, when McKinley left the hotel, crossed the street, and entered the Capitol grounds on the way to his office, he always paused for a moment, turned around, and raised his hat to his wife, who sat watching him from the hotel window. At exactly three every afternoon, moreover, he stopped all work in his office, excused himself, and went to the window to wave his handkerchief and wait for his wife to wave back.

Some of McKinley's associates resented the way his wife dominated his life, but McKinley himself never felt hemmed in. He told the Storers that his wife was once "the most beautiful girl you ever saw," and added: "She is beautiful to me now."[10] Ida McKinley, for her part, worshipped her husband. "In her estimation, he was perfect," observed Mrs. John A. Logan, "and she discoursed upon his good qualities with the fervor of a girl in her teens over her lover."[11] Once a friend of the McKinleys referred to McKinley as a politician, and Mrs. McKinley was indignant; he was a statesman, she announced, not a politician. In 1893, when McKinley was having financial difficulties, she offered to give him all the money she had inherited from her father. Her attorneys advised against it, but she told them: "My husband had done everything for me all my life. Do you mean to deny me the privilege of doing as I please with my own property to help him now?"[12]

When Mark Hanna and other Republican bigwigs began pushing McKinley for President, some people wondered whether his wife would be up to the job of First Lady. As if to prove that she would be, Mrs. McKinley used the occasion of her silver wedding anniversary in February 1896 to preside over a big party in Canton for Ohio Republicans. So many people showed for the occasion—close to six hundred—that the McKinleys had to entertain them in two shifts; but for six hours Mrs. McKinley received all the guests in her wedding gown by the side of her husband, and after the party was over, seemed little the worse for wear. She would do splendidly in the White House, averred a *Philadelphia Press* reporter who interviewed her some weeks later, for she possessed "the steel badge of courage."[13] To Mrs. McKinley's delight, the Republicans nominated her "Dearest" for President in June 1896 and launched his "front porch" campaign in Canton soon after. During the campaign she occasionally joined her husband to greet the countless delegations that turned up in Canton to pay their respects and listen to his little words of welcome and wisdom. Once, when a little boy who came by to see the fun asked her why there were so many pictures of her husband all around town, she explained: "Because he's a dear good man and I love him!"[14]

When McKinley became President in March 1897, his wife was happy to be in the White House and resolved to do the best she could as First Lady. She presided at the public receptions, occasionally standing, but usually seated; she also appeared with her husband at State dinners. But her determination to play the part of hostess despite her illness meant revising the time-honored procedures. It was unusual for the mistress of the White House to receive guests while seated and even more unusual for her not to offer her hand in greeting. Jennie Hobart, the Vice President's wife and a good friend of the McKinleys, suggested that Mrs. McKinley hold a bouquet of flowers in her hands at receptions so that people filing past her chair would know that there was to be no handshake. This usually worked; but at one reception a woman who didn't grasp the situation kept her right hand extended toward the First Lady for such a long time that Mrs. Hobart finally leaned over, put out her hand, and said, "Won't you shake hands with me instead?" McKinley at once flashed a look of heartfelt gratitude to his friend.[15]

The State dinners posed an even bigger problem than the receptions. White House etiquette called for the President to escort the wife of the guest of honor to the dinner table and place her on his right, while his wife took her place at the table opposite him. But McKinley simply couldn't afford to let his wife sit that far away from him; and from the beginning he escorted her himself into the dining room and placed her next to him. Protocol-conscious officials, domestic and foreign, were aghast at the horrendous breach of decorum, but McKinley knew exactly what he was doing. One evening his wife somehow ended up at the table across from him rather than by his side and he spent the next two hours "anxious to the point of distraction," Mrs. Hobart observed, lest she have a spell that he couldn't handle right away.[16] "Could it possibly offend anyone for me to have my wife sit beside me?" he asked Mrs. Hobart afterwards. "Mr. President, you are Chief Executive," she told him. "This is your home. It is your power to do as you choose."[17] So while the fastidious grumbled, McKinley continued to seat his wife next to him at formal White House dinners.

While she was in the White House, Mrs. McKinley continued to have seizures, usually mild ones, and her husband continued to deal with them in the seemingly offhanded way he had long since perfected. One day, when William Howard Taft was conferring with the President, he asked for a pencil with which to jot something down; and as McKinley reached into his pocket for one, his wife suddenly made "a peculiar hissing sound." To Taft's bewilderment, McKinley calmly put a handkerchief over his wife's face with one hand while handing Taft a pencil with the other, and a minute or two later he removed it and Mrs. McKinley rejoined the conversation where she had left off.[18] For the more violent seizures the President had the White House staff arrange things so that his wife could be quickly whisked away from the crowds at receptions and taken upstairs by elevator.

Mrs. McKinley's plucky performance as First Lady despite her poor health received high praise from the public. She was "an inspiration to all women," declared *Harper's Bazar,* "who for one reason or another are hindered from playing a brilliant individual role in life."[19] Her insistence on receiving guests at receptions by the side of her husband, though over-exertion might produce "fainting spells," seemed courageous, even heroic, to many people. "I believe her determination to be with you," Captain H. O. S. Heistand told the President, "is the determining factor in her strength."[20] But the President's tender devotion to his ailing wife received even higher praise. "President McKinley," said Mark Hanna, "has made it pretty hard for the rest of us husbands here in Washington!"[21] Opposition newspapers never dared raise the question of Mrs. McKinley's health; they knew it would produce an outpouring of sympathy for the President.

For a time Mrs. McKinley seemed to be getting better. In June 1898, George Cortelyou, McKinley's secretary, found her "vastly improved" and noted: "Now she can almost walk alone."[22] By the end of the year, however, she seems to have had a relapse. The strain of White House entertaining took its toll; so did the murder of her brother George by his mistress in October 1898, though she acted as though she knew nothing about it. Soon she was behaving peevishly in public, unusual for her, and McKinley was increasing his watchfulness over her. In June 1899, while visiting friends with her husband, she had a violent attack and fell into a deep depression. Back in the White House again, she was more dependent on McKinley than she had been in some years. One evening Cortelyou dropped by to tell McKinley that some papers they were working on could wait until morning. McKinley expressed great relief, for, he said, his wife was waiting for him. That afternoon, he went on to say, he had left to do an errand and was gone longer than he had intended; and when he got back he found her sobbing like a child and overwhelmed with fear that "something might have happened to him."[23] The little trips McKinley took with her helped restore her spirits; so did his purchase of their old Canton house, which he had sold in 1877 because of financial need. The two spent many happy hours planning improvements to the house and talking of the day when they would retire from public life and live quietly together there again. But Mrs. McKinley had no objections to her husband's bid for a second term in 1900, rejoiced in his victory, and appeared proudly at his second inauguration in March 1901.

In April 1901, to get his second term off to a good start, McKinley planned a grand tour of the South and the West, culminating in an appearance at the Pan-American Exposition in Buffalo, New York, in June, for which he scheduled a major address on the tariff. Mrs. McKinley loved to travel; and with her maid, the White House physician, and her niece along to help out, she embarked on the six-week journey with excitement and enthusiasm. The tour went swimmingly at first; McKinley was warmly received wherever he went and his wife took enormous pleasure in the ovations for her "Dearest" at every stop. Just before reaching Los Angeles, however, she developed an infected felon on one of her fingers, and, despite surgery, was seriously ill by the time the train arrived in San Francisco. For several days she was on the point of death; and at one point McKinley gave up hope and began planning funeral arrangements. Then, suddenly, to everyone's astonishment and relief, she rallied, and was soon well enough to board the train for Washington. McKinley cancelled the rest of the tour and postponed his Buffalo appearance until September. In July he took his wife to Canton for the summer. There, in the town where they had first met and married and where they expected to spend their retirement, the McKinleys had two happy months together, and by September Mrs. McKinley was well enough to accompany her husband to Buffalo.

On September 5, 1901, McKinley spoke at the Pan-American Exposition in Buffalo and was gratified by the favorable response to his address. The following morning he and his wife visited Niagara Falls with John G. Milburn, president of the Exposition, and his wife; but Mrs. McKinley was too worn out to attend the reception scheduled for that afternoon and decided to rest at the Milburns' house instead. Before leaving for the reception, McKinley took his wife to her carriage, handed her smelling salts, and said with mock-formality: "Good afternoon, Mrs. McKinley. I hope you will enjoy your ride; good-bye."[24] Then, as she drove off, he began waving and she turned around to smile back. A few hours later, anarchist Leon Czolgosz, anxious to produce a better world, shot the President at the reception in the Temple of Music; and, as a dozen men moved in on the assassin, McKinley gasped: "Don't let them hurt him." Then, as his friends helped him to a chair, he told his secretary: "My wife—be careful, Cortelyou, how you tell her—oh, be careful!"[25]

Mrs. McKinley's reaction, when she finally heard what had happened, was surprisingly stoic. She was determined, apparently, to put on a brave front for the sake of her husband, who, despite a serious wound, seemed to have a chance for recovery. The doctors permitted her to see him for only a brief period every day, but her appearance always bolstered his spirits. For several days the President lingered, and for a time the doctors thought he would recover. By September 13, however, he was failing rapidly and that evening he asked to see his wife again. When the doctors took her into his bedroom, his face lit up; she went to his bedside, took his hands in hers, smoothed the hair on his brow, put her arms around him, and kissed him on his lips. Then, as she stood by his bed and watched his suffering, she murmured, "I want to go too; I want to go too." "We are all going," he whispered; "we are all going; we are all going."[26] Early the following morning he was dead at fifty-eight. "He is gone," his wife told her friends, "and life to me is dark now."[27]

After the funeral Mrs. McKinley returned to Canton to live in the house on North Market Street, with her sister Pina keeping watch over her. "I am more lonely every day I live," she told one of her friends.[28] At first she thought only of joining her husband and prayed every day that she might do so. Later on she said she wanted to live until the McKinley Mausoleum, erected with Federal funds in Canton, was completed. She died on May 26, 1907, just before her sixtieth birthday.

The great memorial tomb where she was laid to rest beside her husband, was dedicated four days later.

* * * *

Carried the Party

Mrs. McKinley worshipped her husband; she thought he was the greatest man in the world. Whenever anyone compared him to Daniel Webster or even Napoleon, she waved the compliment aside as if it was honoring the other man, not McKinley. Her husband, said Julia Foraker, was the "Hero of all time—why 'drag in' anybody else?" Once, in an effort to make small talk, Senator Joseph Foraker told her: "Your husband is lucky. He doesn't have to go home to vote. Now I have to drop everything and go to Ohio [where there was a spring election] to vote." To his surprise, Mrs. McKinley took his remark with the utmost seriousness. "Well, I am glad to hear that," she said. "I think it's about time you men did something. My husband has carried the Republican Party for twenty years. Now I'd like to see somebody else do something."[29]

Red Eyes

When H. H. Kohlsaat arrived for dinner at the White House one evening in April 1898, there was a piano recital in the Blue Room and Mrs. McKinley was seated near the pianist, looking frail and tired. As soon as Kohlsaat entered the room, McKinley motioned for him to sit beside him and then whispered: "As soon as she is through this piece, go and speak to Mrs. McKinley and then go to the Red Room door. I will join you." Kohlsaat did as the President requested and then went into the Red Room with him. McKinley sat down and, in great distress, rested his head on his hands with his elbows on his knees. "I have been through a trying period," he told Kohlsaat. "Mrs. McKinley has been in poorer health than usual. It seems to me I have not slept over three hours a night for over two weeks. . . ." Then, after talking about how Congress was trying to drive the country into war with Spain, he broke down and cried like a boy of thirteen. Kohlsaat put his hand on the President's shoulder and waited until he calmed down. "Are my eyes very red?" McKinley finally asked. "Do they look as if I had been crying?" "Yes," said Kohlsaat frankly. "But I must return to Mrs. McKinley at once," said the President. "She is among strangers." "When you open the door to enter the room, blow your nose very hard and loud," advised Kohlsaat. "It will force tears into your eyes and they will think that is what makes your eyes red." McKinley did as Kohlsaat suggested and the guests seemed not to have guessed that their President had been weeping.[30]

Surprise

When the McKinleys visited the Berkshires in the summer of 1897, John Sloan and his wife arranged a large dinner party for them at their fashionable place in Lenox, Massachusetts, and told the butler to arrange a nice surprise for them. That evening, when the McKinleys and the other guests entered the great oak-beamed dining-room, they saw a long table gleaming with silver and brilliant with flowers and with a large object in the center covered by a silken American flag. After a bishop said grace, the butler whisked the flag off and there was a big stuffed American eagle, with outstretched arms, staring at them. Then, to everyone's surprise, there was the sound of a lock clock whirring and the bird began to wink and nod toward Mrs. McKinley and flap its wings up and down in a jerky fashion as if it were about to soar to the ceiling. Mrs. McKinley jumped up in a panic and her husband came rushing over to take her by the hand and help her out of the room. The Sloans quickly disconnected the butler's masterpiece, carried it to the lawn to run down, and resumed their dinner party when the President was able to bring his wife back, quieted and reassured.[31]

Ruled by My Husband

When a young English woman admitted at a White House dinner party that much as she liked America, she still loved England best and preferred living there, Mrs. McKinley looked at her severely and exclaimed: "Do you mean to say that you would prefer England to a country ruled over by *my husband?*"[32]

CHAPTER 24

THE ROOSEVELT WIVES

Alice Lee Roosevelt; 1861–1884
Edith Kermit Roosevelt; 1861–1948

When Republican bigwigs proposed Theodore Roosevelt for second place on the McKinley ticket in 1900, the Rough Rider, then serving as Governor of New York, had mixed feelings. He was reluctant to turn down the chance for a place on the national ticket, but he looked on the Vice-Presidency without enthusiasm. His wife Edith, however, had no doubts: she thought T.R. would be bored to death spending four years merely presiding over the U.S. Senate and she was convinced it would mean his political demise. And she was probably right.

Despite his wife's misgivings, T.R. bowed to the wishes of party leaders. At the Republican convention in June 1900, indeed, he seemed to revel in the big ovation he received, as his wife looked on, when the delegates picked him as McKinley's running mate. By this time Mrs. Roosevelt was reconciled. She gave "a little gasp of regret" at first, according to the *New York World,* and then "accepted the situation with a grace worthy of a true patriot." But she was not particularly elated. "I had hoped to the last moment that some other candidate would be settled upon," she told her sister after the convention. "I am trying to look at the bright side."[1]

Then came one of those startling events that is continually upsetting the logic of history: President McKinley was assassinated a few months after he began his second term, and T.R. became President in September 1901. In no time at all Roosevelt was presiding over one of the liveliest administrations in American presidential history and his wife was becoming enormously popular. The two, everyone agreed, made a great team.

Theodore and Edith Roosevelt (born in Norwich, Connecticut, in 1861) had known each other since childhood. They grew up together in New York City, moved in the same privileged circles as youngsters, corresponded when T.R. went abroad with his family in 1869, and perhaps even talked about marriage before he went off to Harvard in the fall of 1876. Edith visited T.R. at Harvard when he was a freshman and he proudly showed her off to his classmates. "I don't think I ever saw Edith looking prettier," he exclaimed afterward; "everyone . . . admired her little Ladyship intensely, and she behaved as sweetly as she looked."[2] But the high-spirited Edith didn't always behave sweetly; she had "bad days," T.R. discovered, as well as "good days." The two seem to have quarreled just before he began his second year at Harvard and her name disappeared from his diary. Then, in October 1878, when he was a junior, he met the beautiful young Alice Hathaway Lee at a party in Boston and fell head over heels in love with her. "As long as I live," he recalled, "I shall never forget how sweetly she looked and how prettily she greeted me."[3]

Like Edith Kermit Carow, Alice Lee (born in a Boston suburb in 1861) had impeccable social credentials; her father was a partner in one of Boston's leading banking firms and her mother came from a prominent old Massachusetts family. Alice, who was only seventeen when the twenty-year-old T.R. began courting her, was warm, gentle, and full of fun—her friends called her "Sunshine"—and she was a bit intimidated by T.R.'s vigor, talkativeness, and assertive ways. She was in no hurry to get married, moreover, and for months kept him on tenterhooks. But though T.R. spent many anxious days and sleepless nights during the long courtship, he never gave up hope. "See that girl?" he cried at one Harvard party. "I am going to marry her. She won't have me but I am going to have her."[4] He was in ecstasy when she finally gave her consent, but continued in a state of anxiety. At one point he was so afraid someone was going to run off with her before the marriage that he sent abroad for some French duelling pistols so he could be ready for the emergency. "How I love her!" he exalted. "She seems like a star of heaven, she is so far above other girls; my pearl, my pure flower. When I hold her in my arms there is nothing on earth left to wish for; and how infinitely blessed is my lot. . . . Oh, my darling, my own bestloved little Queen!"[5]

On October 27, 1880, T.R. and Alice were married in the First Parish Church (Unitarian) in Brookline, Massachusetts, and after a short honeymoon took up residence in New York where T.R. began studying law at Columbia University. But T.R. was bored by law, got into politics, and in 1882 became a member of the New York legislature. Alice went up to Albany with him whenever the legislature was in session, but after she became pregnant in 1883 they decided she should remain in New York City at his mother's place to have the baby. "How I did hate to leave my bright, sunny little love yesterday afternoon," T.R. wrote her after one short visit. "I love you and long for you all the time, and oh so tenderly; doubly tenderly now, my sweetest little wife."[6]

On February 13, 1884, while T.R. was doing committee work in the state capital, he received a telegram announcing the birth of a baby, whom they named Alice Lee; but as he began making preparations to return to New York, he received another telegram informing him that his wife was seriously ill. T.R. boarded the next train for New York, but it moved so slowly through the dense fog that night that he didn't reach his mother's house until close to midnight. To his horror he found both his wife and his mother critically ill. "There is a curse on this House!" exclaimed his brother Elliott. "Mother is dying and Alice is dying too."[7] T.R.'s mother died of typhoid fever the following morning and his wife, stricken with Bright's disease, and barely able to recognize T.R. when he arrived, died the next afternoon. She was only twenty-two. T.R. drew a large cross in his diary that night and wrote under it: "The light has gone out of my life."[8]

For days afterward T.R. was "in a dazed, stunned state," according to a friend. "He does not know what he does or says."[9] But he returned to Albany after the funeral and plunged into work there; and when the legislature adjourned he went out to the Dakota Badlands to forget his sorrow in the strenuous life of a cowboy. In Dakota he wrote a moving little memorial to his wife for circulation in the family and among friends, and then tried to erase her from his mind. He never mentioned her again, not even to their daughter, who later became famous as Alice Roosevelt Longworth. "It was pathetic, yet very tough at the same time," mused Mrs. Longworth years later. "I think my father tried to forget he had ever been married to my mother. To blot the whole episode out of his mind. He didn't just never mention her to me, he never mentioned her name to *anyone*. Never referred to her ever again. It was most curious."[10]

But T.R. was a romanticist; he sincerely wanted to be true to Alice forever. When a friend said time would assuage his grief, he exclaimed: "Don't talk to me about time will make a difference—time can never change me in that respect."[11] There was "that awful sentimentality," Mrs. Longworth observed, "about the concept that you loved only once and you never loved again."[12] And yet T.R. had once had a childhood sweetheart, Edith Carow, and she was still a close friend of his sisters Corinne and Anna. T.R. hadn't forgotten her. After Alice's death, in fact, he went out of his way to avoid seeing her when he was in New York; he even instructed his sister Anna to let him know in advance when Edith was visiting her, so he could stay away. And Edith, who had once been much drawn to T.R., tried just as hard to keep from seeing him. But what now seems inevitable finally occurred: one afternoon in the fall of 1885, the two met by accident in Anna's house, talked briefly, and then began seeing each other again. Soon T.R. found he was deeply in love with her and that Edith felt the same way about him. For a time his feelings bothered him. "I have no constancy!" one of his friends overheard him muttering one evening, "I have no constancy!"[13] Even after he and Edith became secretly engaged in November 1885, he was still deeply troubled. "I utterly disbelieve in and disapprove of second marriages," he told Anna. "I have always considered that they argued weakness in a man's character. You could not reproach me one half as bitterly for my inconstancy and unfaithfulness as I reproach myself."[14] Then, by a sheer act of will, he seems at long last to have succeeded in wiping the memory of Alice from his mind and he was finally able to act. On December 2, 1886, he and Edith were married in London, where she had gone to live with her mother and sister after the death of her father, an improvident New York businessman who had left the family somewhat short of money.

T.R.'s second marriage was just as happy as his first. Perhaps it was even happier. Edith Roosevelt once remarked to her son Ted that his father would eventually have become bored by his first wife. This may be unfair; Alice would surely have matured, as T.R. himself did, as she grew older. But one thing is clear: Edith was no "Sunshine," even as a young woman. At Miss Comstock's School in Manhattan, where she received her education, one of her classmates, puzzled by her aloofness, exclaimed: "Girls, I believe you could live in the same house with Edith for fifty years and never really know her."[15] T.R. did come to know and adore her, but he was aware of her reticences. Personable, witty, and warm-hearted she surely was; but she was also forthright, unsentimental, a bit austere, and, as T.R. told Anna shortly before the marriage, "naturally reserved."[16] T.R. called his first wife "Baby." He would never have dreamed of addressing his second wife that way. She was "Edie" and he was "Theodore." (The nickname, Teddy, was intolerable to him after the death of Alice.)

T.R.'s life was a frenetic one, particularly as the family grew (eventually four boys and two girls, including Alice), and as T.R. moved steadily up in the political world. Edith managed the family finances, for T.R. was no good at this; she also handled the children sensibly (Alice gave her stepmother high praise in this regard), enjoyed the family fun and games almost (but not quite) as much as her husband did, entertained his political cronies good-naturedly, and, above all, provided T.R. with the kind of sympathetic understanding and warm companionship that was essential to his well-being. His impetuousness and propensity for getting in harm's way bothered her at first, but she learned to accept his restless ways with stoic resignation. She didn't really want him to get into the Spanish-American War fighting in 1898 or to go big-game hunting in Africa in 1910, but she kept her misgivings to herself, for she knew these things were important to him.

One day Mrs. Roosevelt was packing to leave Washington with the children for Sagamore Hill, the Roosevelt home at Oyster Bay on Long Island, and her friend Mrs. Bellamy Storer asked her whether T.R. was going too. "For Heaven's sake!" exclaimed Mrs. Roosevelt. "Don't put it into Theodore's head to go too; I should have another child to think of."[17] Mrs. Roosevelt didn't really mind the adolescent streak in her husband's personality, for she knew it was part of his charm, but she never hesitated to tease him gently when she thought his boyish exuberance was getting out of hand. After a dinner they gave in Washington for their British friend Cecil Spring-Rice, T.R. confessed to Henry Cabot Lodge that his "tendency to orate" that night was "only held in check by memory of my wife's jeers."[18]

But T.R. accepted his wife's little sallies in good humor. One day, after cutting his head trying to repair a windmill at Sagamore Hill, he appeared in the house with a big cut on his scalp and his wife said calmly: "Theodore, I do wish you'd do your bleeding in the bathroom. You're spoiling every rug in the house."[19] On another occasion, T.R. went off on a picnic with the children and they all went swimming with their clothes on. When they got home, Mrs. Roosevelt, afraid that the children might catch cold, gave them doses of ginger syrup and ordered them to bed. But when they complained to their father about it, he told them: "I am afraid there is nothing I can do. I'm lucky that she did not give me a dose of ginger too."[20] Toward the end of his Presidency, when T.R. was planning a trip abroad after retirement, he told his secretary, Archie Butt, that he didn't want to wear knickerbockers and silk stocking at court functions in Europe, and Butt said that as a veteran of the Spanish-American War he was entitled to wear the uniform of a cavalry colonel. T.R. perked up at this, for he loved elegant uniforms, and announced he was going to wear a fancy colonel's uniform with patent-leather boots. "Theodore," interposed Edith, "I would never wear a uniform that I had not worn in the service, and if you insist upon doing this I will have a *vivandière's* costume made and follow you throughout Europe."[21] He at once decided to do without the fuss and feathers. Sometimes Edith merely cautioned him when he got carried away about something: "Theodore! Theodore!" "Why, Edie," he would say meekly. "I was only—."[22]

T.R. seems to have appreciated his wife's restraining influence. When Theodore, Jr., his first born son, got engaged in 1910, he wrote to congratulate him and took the occasion to reflect on his life with Edith during the past quarter of a century. "Greatly though I loved Mother," he told Ted, "I was at times thoughtless and selfish, and if Mother had been a mere unhealthy Patient Griselda I might have grown set in selfish and inconsiderate ways." His wife was "always tender, gentle, and considerate, and always loving," he went on to say; "yet when necessary pointed out where I was thoughtless and therefore inconsiderate and selfish, instead of submitting to it. Had she not done this it would in the end have made her life very much harder, and mine very much less happy."[23] But T.R.'s wife made a similar point about herself in her own congratulatory letter to Ted. "One should not live to oneself," she told her son. "It was a temptation to me, only Father would not allow it. Since I have grown older and realize that it is a great opportunity when one has a house that one can make pleasant for younger—and also older—people to come to, I have done better."[24] Once, though, when T.R. began praising her in flowery language to a friend of theirs, she became impatient. "Theodore," she cried, "you're talking just like an obituary!" "Why, *Ee*-die," he wailed.[25]

The Roosevelts flourished in the White House. "I don't think any family has ever enjoyed the White House more than we have," T.R. exclaimed a few months before he made his successful bid for a second term in 1904.[26] With six children (ranging from teen-ager Alice to baby Quentin) and innumerable pets (dogs, cats, guinea pigs, snakes, lizards) overflowing the place, the Roosevelts presided over what Chief Usher "Ike" Hoover called "the wildest scramble in the history of the White House."[27] Mrs. Roosevelt managed it all with imperturbable effortlessness, imposing discipline on the youngsters but permitting them a great deal of freedom. She also supervised the household with quiet efficiency and won the devotion of the White House staff by her thoughtfulness. As hostess, too, for countless dinners, parties, teas, receptions, and "at homes," she was always "dignified and wise," according to her husband, and took her place beside Dolley Madison in the opinion of Washington social observers with a historical bent.[28] "The list of entertainments for which Mrs. Roosevelt sent out invitations would appal any American woman," noted White House aide W. H. Crook. "Yet so great was her capacity for carrying through her share of her husband's life, in addition to her own particular duties as wife, mother, homemaker, that she was able to live those seven busy years without losing health, strength, or the youthful, vivacious, charming presence that made her personality as remarkable as that of her husband."[29]

T.R. thought his wife was the "ideal great lady and mistress of the White House."[30] And Anne O'Hagan, writing in *Harper's Bazar* for May 1905, was inclined to agree. Mrs. Roosevelt, she observed, "is not only the President's wife; she is Theodore Roosevelt's wife—vocation enough for an ordinary woman. . . . Mrs. Roosevelt must know public affairs keenly, sympathetically; she must ride horseback; she must love the out-of-doors in which her husband delights; she must be prepared at luncheon on Monday to talk understandingly to the latest exploiter of the blazed trail, on Tuesday to listen with ready intelligence to police stories from Jacob Riis, on Wednesday to question Booker Washington about Tuskegee, and on Thursday to quote *The Single Life* to Pastor Wagner. She must share or she must moderate the manifold enthusiasms of Theodore Roosevelt; to do either she must, of course, understand them. . . ."[31]

Sometimes Mrs. Roosevelt helped her husband in his work as President. She sorted his mail, went over papers prepared by his secretaries for him, and read the New York and Washington newspapers to mark items for his attention. She gave advice, too, when asked, and some people thought she kept her husband from many a gaffe. "Never, when he had his wife's judgment," according to reporter Mark Sullivan, "did he go wrong or suffer disappointment."[32] Above all, though, she saw to it that the two of them had some time together all to themselves, either in daily rides, reading to each other, or in walks around the White House. And when his work was pressing him hard, she insisted that he got proper rest and relaxation. When he worked evenings in his study, she had a habit, it was said, of tapping her foot on the floor of her living room promptly at 10:30 and calling, "Theodore!" "Yes, Edie," he would say and put his work away.[33]

There was an aristocratic flavor to the Roosevelt White House. Mrs. Roosevelt acquired a social secretary, held weekly meetings with the wives of Cabinet members to discuss White House protocol and proprieties, arranged a regal wedding for twenty-two year old Alice in 1906, and a fancy coming-out party for the seventeen-year-old Ethel in 1908. Yet Mrs. Roosevelt was no more of a snob than her husband was. She did entertain the "Very Smart," as Archie Butt called them, at the White House; but, like her husband, she much preferred evenings with people from varied backgrounds who were bright, lively, informed, and entertaining.[34] At one White House reception she suddenly noticed that the guests were snubbing a young woman who had once been well-to-do but was now working as a saleswoman in a New York department store she patronized. As the woman was about to leave, Mrs. Roosevelt went up to her, held out her hand, and said, "I think we need hardly be introduced, since we are old friends. I am so glad to see you here." Then she put her arm around the woman's waist, led her to a sofa, and sat and chatted with her for fifteen or twenty minutes.[35]

In February 1909, just before T.R. went out of office, Archie Butt asked Mrs. Roosevelt whether she regretted leaving the White House and she said she felt somewhat as Mrs. Cleveland did in a similar situation: relief at not having the family subject to public criticism any longer. She then showed him her husband's new calling cards, bearing only his name, Theodore Roosevelt, instead of President, on them. She had wanted the word, Mr., added, she said, but T.R. had opposed it. When Butt said he agreed with the President, she cried: "Why should he not have *Mr.* Theodore Roosevelt, as any other gentleman would have on his card?" "Because he is not like any other gentleman," said Butt. "Then why not simply have Roosevelt, or Theodore, or even The ex-President?" she cried. "I want him to be the simplest American alive after he leaves the White House, and the funniest thing to me is that he wants to be also and says he is going to be, but the trouble is he has really forgotten how to be." She went on to say she hoped his year hunting big game in Africa and her year in her sister's little cottage on the Mediterranean would make them both glad to settle down at long last to a quiet life in retirement at Oyster Bay.[36]

But there was to be no real retirement. Only fifty-two when he left the White House, T.R. did not settle down for long. As Mrs. Roosevelt half-expected, he got involved in politics again after the African trip and ended up running for President, with her lukewarm support, on the Progressive ticket in 1912. After his defeat, he took a trip into the Brazilian wilderness with his son Kermit that almost cost him his life. When the United States went to war with Germany in 1917, he was proud that all four of his sons were in uniform and tried hard to get into the fighting himself. In March 1918, when news came that Archie had been wounded in France and received the *Croix de Guerre*, Mrs. Roosevelt sent for a bottle of Madeira and, as T.R. wrote his son, "all four of us filled the glasses and drank them off to you; then Mother, her eyes shining, her cheeks flushed, as pretty as a picture, and as spirited as any heroine of romance, dashed her glass on the floor, shivering it in pieces, saying, 'That glass shall never be drunk out of again'; and the rest of us followed suit and broke our glasses too."[37] But the news in July that Quentin had been shot down and killed behind German lines shattered the family. Mrs. Roosevelt nearly collapsed and T.R. seemed suddenly to become old and ailing; the little boy in him, according to Hermann Hagedorn, finally died at that point.[38] On November 11, the day of the Armistice, T.R. was taken to the hospital with inflammatory rheumatism and told he might be in a wheel-chair the rest of his life. "All right!" he said gamely, "I can work that way too."[39] But he did not long survive his return home. Sunday afternoon, January 5, 1919, he and his wife read aloud to each other for a while, as of old, and then, as she left the bedroom, he said: "I wonder if you will ever know how I love Sagamore Hill."[40] A few hours later he passed away in his sleep.

After T.R.'s death, Mrs. Roosevelt did a lot of traveling, in Europe and the Far East, but she was also active in charity and church work in Oyster Bay and in the local Women's Republican Club. She kept in close touch with her children and grandchildren, of course, and presided over annual family reunions at Sagamore Hill on the occasion of her birthday. She also entertained visiting dignitaries from time to time, but for the most part kept out of the news. When Prohibition came, she served cocktails at her parties, not because she cared much for drinking, but to protest what she regarded as governmental instrusion on individual freedom. She remained a true-blue Republican to the end; she moved far away, in fact, from the progressive principles her husband had championed in 1912 and became a bitter foe of Franklin D. Roosevelt (a "distant cousin of my husband") and the New Deal. "We expected Franklin Roosevelt to take us out of the mud when he went into office," she declared after F.D.R. had been President a few months, "but he has led us into the mire!"[41]

In 1936, Mrs. Roosevelt warmly supported Alfred Landon, the Republican candidate for President, pointing out that he had backed T.R. in 1912. She also went public that year to help the Republican cause. When Democratic Senator Vic Donahey of Ohio compared F.D.R. to T.R., she issued a vigorous public protest. T.R.'s Progressive party, she announced, "was built on American principles. Its liberalism was in accord with our theories of democracy and personal liberty, and in no way resembled the policies of the present administration in Washington." Her husband, she added, "believed in shielding the individual from oppression and giving him freedom to develop himself," while the sponsors of the New Deal "believe in ordering the actions of the individual by mandate of the Federal Government. The schemes which they would have us consider as progressive embody the theories of government developed in various European countries, and are incompatible with our American democracy and liberty."[42] Herbert Hoover couldn't have said it better. In December 1948, a few weeks after her death at eighty-seven, *Life,* in an editorial on First Ladies, called her "one of the strongest-minded and strongest-willed presidential wives who ever lived in the White House."[43]

* * * *

Little Ships

Shortly after marrying Alice Lee and settling in New York to study law at Columbia University, T.R. began working on his first book, *Naval History of the War of 1812*. He and Alice also did a lot of socializing, taking in numerous teas, dinners, balls, receptions, and concerts. "Every moment of my time," T.R. wrote in his diary, "is occupied." One evening he was at home standing at a table, surrounded by books on navigation and making sketches for the ships he was writing about in his book, and Alice came rushing in. "We're dining out in twenty minutes," she wailed, "and Teedy's drawing little ships!"[44]

Love, Honor, and Obey

When the Roosevelts were in Washington, the children usually attended the Episcopal Church with their mother, while T.R. went to the Dutch Reformed Church, in which he had been reared. When the children misbehaved, however, they had to accompany their father to church the following Sunday on the theory that he represented discipline more clearly than Mrs. Roosevelt did. On one occasion, though, the conduct of one of their little boys fell just short of such a measure, and Mrs. Roosevelt said sternly, as they walked to church: "No, little boy, if this conduct continues, I shall think that you neither love, honor, nor obey me!" T.R. enjoyed her use of the marriage service.[45]

Biggest of Her Children

T.R. admitted that his wife thought he was "just the biggest of her children." Once she absent-mindedly kissed him good-bye after having kissed the children, and added, "Now remember to be good while I'm away!"[46]

Many Knots

During the 1904 presidential campaign, when T.R. told a reporter that "nothing would make me a candidate again," his wife thought he was making a mistake. "Theodore," she said, "has tied so many knots with his tongue that he can't undo with his teeth." T.R. finally realized her wisdom when he decided to run for President in 1912; he once said that every time he ignored her advice he soon came to regret it.[47]

Not Then Have Known

At a dinner party at Sagamore Hill one evening, T.R. began talking excitedly about his life as a hunter and ranchman in the Dakota Bad Lands. "With your love of that free existence, Mr. President," interposed one of his guests, "I wonder how you ever settled down in the humdrum East. Honestly, now, don't you wish you had been born and reared on a ranch?" T.R. was about to say yes, when he paused, glanced at Mrs. Roosevelt, who was seated at the farther end of the table, and began, "No, because—." "I know why," interjected one of the ladies. "Why?" returned T.R. "Because," said the lady, "you would not then have known Mrs. Roosevelt." Chortled T.R.: "That was what I was going to say!"[48]

Wolf Movies

One evening T.R. invited some friends to the White House to see a kinetoscope run off some moving pictures of Oklahoma Marshal Abernathy catching wolves alive by forcing his hands into their throats. One scene was particularly thrilling; it showed Abernathy in the water fighting with a wolf weighing more than 125 pounds and finally getting control of the animal by seizing him by the upper jaws as he made a spring for the Marshal's face. Afterwards T.R. sent White House aide Archie Butt upstairs to tell Mrs. Roosevelt what she had missed.

Mrs. Roosevelt made it clear she had no regrets. Catching wolves, she said, had no interest for her; her main concern was that her husband might insist on showing the wolf movies some evening when there were dinner guests at the White House. But T.R. had, in fact, already asked Abernathy to bring his movies to the White House again the following week to show at a dinner party. Mrs. Roosevelt was upset when she heard about it. "How would you like to be asked to sit through a kinetoscope performance of a lady showing antique fans?" "Why, Edith," protested Roosevelt, "I cannot understand being so tortured." "It would not be worse," she said, "than your wolf pictures to the average dinner guest in Washington."

A little while later, when Archie Butt was planning the guest list for T.R.'s wolf-movie party, he asked Mrs. Roosevelt for advice. "Don't ask me," she cried. "It is yours and the President's party: I won't have anything to do with it." "But the President told me to see you," persisted Butt. "Then you and the President must fix it up," chuckled Mrs. Roosevelt. "I do not want dogs and wolves introduced to my dinner guests. These pictures may make my guests sick, especially after eating a big dinner. You can't tell what will happen. [My husband] thought the young members of the diplomatic corps would enjoy it, but I positively refuse to invite foreigners to witness this strangling of wolves." "But, Mrs. Roosevelt," wailed Butt, "I did not have—" "Oh, yes, you did," said Mrs. Roosevelt. "You encouraged him in this idea and even told me it was a most enjoyable affair." "Only by his order," said Butt. "Well, then," laughed Mrs. Roosevelt, "you arrange the guests by his orders." In the end Butt eliminated foreign attachés from the guest list and substituted some U.S. Army couples.[49]

Fall Guy

Edith Roosevelt enjoyed teasing her husband about his precipitousness, but she didn't like other people to be critical. Once T.R. took Secretary of War Elihu Root on a long hike and made him climb so many cliffs that Root vowed he would never endanger his life again that way. Later on, when Mrs. Roosevelt told Root that T.R. had had a bad fall while on a hike and injured one of his legs, Root burst into laughter and said he was glad. Mrs. Roosevelt was angry; she thought his reaction was brutal and it was a long time before she forgave him.[50]

Stepdaughter

Edith Roosevelt got along well with her stepdaughter Alice (later, Alice Roosevelt Longworth), but wasn't sure she had done right by her.

"Mother," cried Alice gaily right after her White House wedding in 1906, "this has been quite the nicest wedding I'll ever have. I've never had so much fun." "I want you to know that I'm glad to see you leave," sighed Mrs. Roosevelt, worn out by the festivities. "You have never been anything but trouble!" "That's all right, Mother," shot back Alice. "I'll be back in a few weeks and you won't feel the same way." "And I was and she didn't," added Alice, when she related the playful exchange years later. "Well, I don't *think* she did. We were certainly able to laugh and jeer about a good many things together." Edith Roosevelt once told Alice Longworth she thought she had been unkind to her, but Mrs. Longworth heartily disagreed; she thought Mrs. Roosevelt had in many respects treated her better than she did her own children.[51]

Of Greater Importance

When T.R. returned from Africa in 1910 and Mrs. Roosevelt went to New York to meet him at the dock with the children, she told Archie Butt: "Think! For the first time in two years I have them all within reach!" Then she ran to the cabin where T.R. was talking politics to some friends and cried: "Come here, Theodore, and see your children. They are of far greater importance than politics or anything else!"[52]

Tyrannical Humor

At Sagamore Hill, T.R. liked to chop wood, but one day a tree he felled accidentally landed on the telephone wires. When Mrs. Roosevelt smiled wryly at what her impetuous husband had done, T.R. told her playfully it was her fault because in her role as "forester of this establishment," she hadn't marked which trees he was to cut. But, he added, "I will say nothing more about it and will not hold you up to scorn before your children if you will let the subject drop once and for all." Later, when he told his friend, French Ambassador Jules Jusserand, about the incident, the latter assured him Mrs. Roosevelt hadn't said a word about it to anyone. "Ah! But you don't know my wife!" wailed Roosevelt. "She has a language all her own. That telephone will never ring now that my wife will not begin to chuckle to herself, and if the cursed thing ever gets out of order, which it most frequently does, she will tell the servant to see if the wires are still up or if the trees are down." Then he added with a grin: "No, my dear Mr. Ambassador, people think I have a good-natured wife, but she has a humor which is more tyrannical than half the tempestuous women of Shakespeare."[53]

In Love

On November 17, 1917, T.R. told his youngest son Quentin: "This is the 32nd anniversary of Mother's and my engagement, and I really think I am just as much in love with her as I was then—she is so wise and good and pretty and charming."[54]

Managing the Hospital

In October 1912, when a crank named John Shrank shot and wounded T.R. in Milwaukee, where he had gone to make a campaign speech, Mrs. Roosevelt was watching a musical comedy in a New York theater. But when the campaign worker who brought her the news assured her T.R. hadn't been seriously hurt and was able to deliver his speech, she remained for the rest of the play and then stayed overnight in the city awaiting more details. The next day she learned that T.R. was scheduled for surgery in a Chicago hospital to remove a bullet lodged in his chest; she also learned he was ignoring his surgeon's orders to remain quiet and was receiving a stream of visitors in his hospital room. "I am the only one who can manage him," she cried, "and make him obey the doctor's orders." So she boarded the train for Chicago, and when she reached the hospital, installed herself in a room next to his, and soon "was practically managing the hospital," according to the *Chicago Tribune,* "as far as the colonel was concerned."

With Mrs. Roosevelt in charge, there were no more visitors; there was meek compliance, too, with whatever the doctors ordered. "That sedate and determined woman, from the moment of her arrival in Chicago, took charge of affairs and reduced the Colonel to pitiable subjection," wrote reporter Charles Willis Thompson. Before her arrival, he told *New York Times* readers, T.R. had been "throwing bombshells" at his doctors; but once his wife showed up, "a hush fell upon T.R. [and] he became meek as Moses." T.R. may or may not believe in the women's-rights plank in the Progressive party platform, Thompson went on to say, but it was clear that he accepted women's rights in his own home. "This thing about ours being a campaign against boss rule is a fake," Thompson quoted T.R. as telling him. "I never was so boss-ruled in my life as I am at this moment." But he smiled fondly at his wife as he said this, according to Thompson, and "it was evident that the Great Unbossed likes being bossed for once."[55]

CHAPTER 25

HELEN HERRON TAFT

1861–1943

On March 4, 1909, right after William Howard Taft's inaugural ceremony ended, Mrs. Taft broke a precedent: she rode from the Capitol to the White House with her husband. No First Lady had ever done that before. It was customary for the outgoing President to accompany the new President to the White House; but when Mrs. Taft learned that Theodore Roosevelt planned to head for the railroad station instead, she decided on the innovation. "For me," she wrote later, "that drive was the proudest and happiest event of Inauguration Day." She was thrilled, too, when she and her husband descended from the carriage and walked into the White House together. "I stood for a moment," she recalled, "over the great brass seal, bearing the national coat of arms, which is sunk in the floor in the middle of the entrance hall. 'The Seal of the President of the United States,' I read around the border, and now—that meant my husband!"[1]

Taft might never have become President without the prodding of his wife. His bent, from the beginning of his career, was for the bench, not political office, and his biggest dream as a lawyer and judge was for an appointment to the U.S. Supreme Court. But Helen Taft had her eye on the Presidency; and when President Roosevelt offered to appoint her husband to the highest court, as he did several times, she always succeeded in persuading him to turn down the offer. One evening early in 1908, when the Tafts were dining in the White House with the Roosevelts, T.R. closed his eyes at one point and pretended to be a clairvoyant. "I see a man standing before me weighing about 350 pounds," he intoned. "There is something hanging over his head. I cannot make out what it is; it is hanging by a slender thread. At one time it looks like the Presidency—then again it looks like the Chief Justiceship." The response of the Tafts was predictable. "Make it the Presidency!" exclaimed Mrs. Taft. "Make it the Chief Justiceship!" cried Taft.[2] Taft was, his wife lamented, "a most difficult candidate for his loyal and eager supporters to manage." But manage they did, and they finally got him into the White House.[3]

The White House was old hat to Helen ("Nellie") Taft, for the Roosevelts had entertained her and her husband there frequently when T.R. was President. But she first visited the place when she was only seventeen, and Rutherford B. Hayes (her father's friend and law partner) was President. Afterward, when she returned to Cincinnati, where she had been born Helen Herron in June 1861, she announced that she liked the Executive Mansion so much that she had decided to marry a man who was destined to be President of the United States. As a young woman, though, she wondered at times whether she would ever get married at all. She wanted intellectual companionship as well as love, she confided to her diary, and the young men who sought her company gave her little of the former. "Why is it so very rare in a man and woman to be simply intimate friends?" she wrote in the fall of 1880. "Such a friendship is infinitely higher than what is usually called love, for in it there is a realization of each other's defects, and a proper realization of their good points without that fatal idealization which is so blind and, to me, so contemptible. . . . From my point of view a love which is worthy of the name should always have a beginning in the other. . . . To have a man love you in any other way is no compliment."[4]

Young Nellie Herron met Will Taft for the first time at a bobsled party one winter evening in 1879, but the two didn't become seriously interested in each other until he began attending a "salon" she organized at her house to discuss "topics intellectual and economic."[5] Nellie and Will had a lot in common: they both grew up in Cincinnati, had fathers who were attorneys and public servants, and they both moved comfortably in upper-crust circles in their hometown. But the contrasts were striking: Will was tall, handsome, sweet-tempered, and easy-going, while little Nellie, for all her charm and good looks, was somewhat intense, outspoken, and critical. She also had a touch of unconventionality; she liked to smoke cigarettes and drink beer with her friends now and then. She became a schoolteacher, though she didn't need to, after graduating from Miss Nourse's school in Cincinnati and taking courses at Miami University; and she studied music assiduously and became a competent pianist. Some of her suitors were intimidated by her intelligence, energy, and forthrightness, but Will, who shared his father's liking for brains as well as beauty in women, was much taken by her. "She wanted to do something in life and not be a burden," he told his father, when discussing her school-teaching. "Her eagerness for knowledge of all kinds puts me to shame. Her capacity for work is wonderful."[6] Will began courting her soon after joining her salon, but for a long time he despaired of getting anywhere with her. "Oh, Nellie, do say that you will try to love me," he beseeched her in one love letter. "Oh, how I will work and strive to be better and do better, how I will labor for our joint advancement if you will only let me."[7] When Nellie finally agreed to a secret engagement in May 1885 he walked on clouds and showered her with ecstatic letters. They were married in June 1886 at the Herron home in Cincinnati, and after a honeymoon in Europe took up residence in a new house on Walnut Hills that Taft had borrowed money to build for her.

Taft was in a law partnership when he married Nellie and was also serving as assistant to the prosecuting attorney of Hamilton County. One day, less than a year after the wedding, he came home and exclaimed: "Nellie, what would you think if I should be appointed a Judge of the Superior Court?" "Oh, don't be funny," sniffed his wife. "That's perfectly impossible."[8] But he was serious; he received the appointment to the Ohio court in 1887, when he was only twenty-nine, and was exceedingly proud of the honor. His wife was pleased, too, but she deplored the association with men so much older than he was, and "dreaded to see him settled for good in the judiciary and missing all the youthful enthusiasms and exhilarating difficulties which a more general contact with the world would have given him."[9] His appointment by President Harrison in 1890 as Solicitor-General of the United States was more to her liking; the position took them to Washington, which she soon came to love, and it gave him "an opportunity," she thought, "for exactly the kind of work I wished him to do; work in which his own initiative and originality would be exercised and developed."[10] His return to Cincinnati in 1892 as a member of the Sixth Federal Circuit Court of Appeals dismayed her; it meant associating with men twice his age once more and she feared he would be "fixed in a groove for the rest of his life."[11]

Despite Nellie's reservations, Taft loved his work as Circuit Judge; he began eyeing the Supreme Court, too, as a distant but no longer impossible achievement. While he made the rounds on his circuit, which covered six states, Nellie busied herself with the children (eventually two boys and a girl) and plunged into cultural activities; she attended art classes, joined a book club, and played a leading role in the formation of the Cincinnati Symphony Orchestra. When her husband was out of town, she exchanged affectionate letters with him; it was clear that their marriage contained the intellectual companionship, as well as love, that both prized so highly. She tended to treat him like a child at times—he was her "dear darling, lovely, beautiful, sweet precious boy"—but he seemed to need and want her occasional mothering.[12]

There were altercations, of course, but the reconciliations were always swift. "I know that I am very cross with you," she told Taft after one little spat, "but I love you just the same."[13] She did have a sharp tongue and couldn't help lecturing him from time to time, but he didn't really mind a bit. He wanted her, in fact, to keep him in line: nudge him when he dozed off in company, take over the carving at dinners when he absent-mindedly put most of the meat on the first plate, and, above all, tell him exactly what her opinions were about things. "Thought of you has so much intellectual flavor and sweet sentiment, too," he once told her. "I am so glad that you don't flatter me and sit at my feet with honey. You are my dearest and best critic and are worth much to me in stirring me up to best endeavor."[14] In one respect, however, her endeavors were in vain. As he began gaining weight in early middle age, he wanted her to force countermeasures on him. "Nothing will do for me . . . but regular and hearty exercise," he told her. "This is my fate and is essential to my living to a good old age. You must, therefore, make yourself a thorn in my side to that end, my darling."[15] She did her best, and so did he, but it wasn't nearly enough, and he continued to balloon.

One day in January 1900 a telegram arrived at the Taft home. "What do you suppose that means?" mused Taft as he opened it. It turned out to be a request from President McKinley to see him on "important business" as soon as he could, and Taft was soon on his way to Washington. "In three days," Mrs. Taft recalled, "he came home with an expression so grave that I thought he must be facing impeachment." But the news filled her with joy. "The President wants me to go to the Philippine Islands," he told her. "Want to go?" "Yes, of course," she cried. "The President and Mr. Root [Secretary of War] want to establish a civil government in the Philippines," explained Taft, "and they want me to go out at the head of a commission to do it." Mrs. Taft could hardly wait to leave. "I wasn't sure what it meant," she said later, "but I knew instantly that I didn't want to miss a big and novel experience. I have never shrunk before any obstacles when I had an opportunity to see a new country and I must say I have never regretted any adventure." Before the family started for Manila in April, she read everything she could lay her hands on about the Philippines.[16]

The years in the Philippines were mostly happy ones for the Tafts. After the Philippines Commission headed by Taft ended American military rule there and established civil government for the islands the United States had acquired in the Spanish-American War, Taft became the first governor, displayed great competence as administrator, and soon won the respect and devotion of the Filipinos in Manila and elsewhere. Mrs. Taft toured the eighteen provinces with her husband, bearing the hardships cheerfully, ended the color line observed by the army under General Arthur MacArthur, and presided with verve over receptions and dinners at the elegant Malacanan Palace on the Pasig River. "I hadn't been brought up with any such destiny in view," she wrote later, "and I confess it appealed to my imagination."[17]

In January 1903, Theodore Roosevelt, who became President after McKinley's assassination, offered Taft an appointment to the Supreme Court. To Nellie's delight her husband turned down the offer on the ground that he still had much to do in the Philippines. When T.R. persisted and Taft finally yielded, there was "a buzz of astonishment and incredulity" at the news, Mrs. Taft noted with glee, and soon the Filipinos were blanketing Manila with placards proclaiming, "*Queremos Taft*" (We Want Taft).[18] There were parades, speeches, and cables to Washington on Taft's behalf as well, and in the end T.R. capitulated. "Taft, Manila," he cabled, "All right stay where you are. I shall appoint someone else to the Court. ROOSEVELT."[19] A few months later, however, he asked Taft to join his Cabinet as Secretary of War and, since the post permitted him to continue dealing with the Philippines, Taft decided to accept. "This was much more pleasing to me than the offer of the Supreme Court appointment," Mrs. Taft noted with satisfaction, "because it was in line with the kind of career I wanted for him and expected him to have. . . ."[20] Just before leaving Manila she threw a big farewell party: a Venetian masked ball at which her husband and she, dressed as the Doge and His Lady, stood for hours receiving guests at the Pasig River landing, "with as much mock stateliness as we could command in the midst of such a merry throng. It will linger in my memory always as one of the most entrancing evenings of my life."[21]

Not long after the Tafts returned to the United States from the Philippines and settled in Washington, a woman approached Mrs. Taft at a tea party and exclaimed: "Why, out there you were really a queen, and you come back here and are *just nobody!*"[22] But Mrs. Taft wasn't a nobody; she was the wife of a prominent Federal official who was beginning to be talked about as a possible successor to his friend Theodore Roosevelt. Mrs. Taft planned to do all she could, in the way of meeting and charming Republican bigwigs, to help her husband on the road to the White House. But Taft wondered whether they could afford much entertaining, for his salary as a Cabinet officer was modest. To economize, T.R. advised giving up champagne for dinner parties and substituting Sunday evening high teas whenever possible for the dinners. "You should see Nellie's lip curl," Taft told a friend, "at the suggestion of Sunday high teas and dinner parties without champagne."[23] In the end Taft's brother Charles helped him out with funds.

In 1906 came another Supreme Court vacancy and another offer from T.R. Taft was tempted to accept the appointment this time, but Mrs. Taft argued vigorously against it. She even went into a huddle with T.R. about her husband's future, and T.R. told Taft afterward that he had a real chance at the Republican nomination for President in 1908. But Mrs. Taft didn't trust him. Though the President had publicly stated he would not seek a third term, she was convinced he was quietly promoting himself anyway and counted on a stampede on his behalf at the nominating convention.

Mrs. Taft found it hard getting her husband to work for the nomination. "Politics makes me sick," he kept telling her; and even after a Taft boom got under way, he remained lackadaisical.[24] To keep his name before the public, though, he did agree to make a trip around the world, at T.R.'s suggestion; he also made some speeches when he got back, putting forth his views on the issues. Just before speaking in Columbus, Ohio, he wrote his wife disconsolately that he expected his somewhat wordy address to be a dud. "I shall be made fun of because of its length," he told her. "But I am made this way and 'I can do no other.' That is the kind of old slow coach you married. . . . Keep me in your thoughts about this speech. By some mental telepathy you may help out." His wife wrote back reassuringly. "Never mind if you cannot get off fireworks," she said. "It must be known by this time that is not your style. . . . If people don't want you as you are, they can leave you, and we shall be able to survive it."[25] But the Columbus speech was a success, and the movement for Taft continued to gather steam. And despite Mrs. Taft's suspicions, T.R. pushed his friend's candidacy with great vigor.

When the Republican Convention was meeting in Chicago in June 1908, the Tafts were with friends in the War Department office, following the proceedings by means of telegraph bulletins and telephone calls. On the second day of the Convention there was a lengthy demonstration for T.R. that aroused all of Mrs. Taft's old suspicions. The following day, when her husband's name was placed in nomination and the cheering began, Taft sat quietly as bulletins arrived describing the demonstrations on his behalf, but Mrs. Taft was on the edge of her chair in great anxiety. "I only want it to last more than forty-nine minutes," she announced. "I want to get even for the scare that Roosevelt cheer of forty-nine minutes gave me yesterday." "Oh, my dear, my dear," protested Taft mildly. The ovation for Taft fell twenty minutes short of the one for T.R. the day before, but when the nominating speeches ended, Mrs. Taft began to relax. Then her son Charlie came in with a new bulletin and her face went white as she read it: "A large portrait of Roosevelt has been displayed on the platform and the convention has exploded." There was a deathly silence in Taft's office. Then the balloting began and soon bulletins arrived reporting that Taft had received the nomination. "It is needless to add that Mrs. Taft's face had more than regained its normal color," wrote one of their friends afterwards. "She was the personification of a proud and happy wife.'"[26] The election in November had a happy ending too: Taft won nicely in his contest against William Jennings Bryan.

Once in the White House, Mrs. Taft turned her attention to managing the household over which she had been so anxious to preside. She converted some of the public rooms into private ones, for the use of her family, and replaced the White House steward with a housekeeper, Elizabeth Jaffray, whom she expected to relieve her of "the supervision of such details as no man, expert steward though he be, would ever recognize," i.e., keeping the rooms spotlessly clean.[27] She also dismissed most of the White House ushers and replaced them with black footmen in blue livery who were to stand at the White House entrance to receive visitors and give instructions to sightseers. She even contemplated having the houseboys wear full-dress uniforms, but dropped the idea when she overheard one of them smirking, "Won't we look fine cleaning windows in tails?"[28] Her innovations produced considerable criticism; T.R.'s fans thought her changes were an insult to Mrs. Roosevelt, and some people thought she was substituting Malacanan regality for democratic simplicity. "She seems to have forgotten entirely that the White House is, in a measure, a public institution," grumbled Chief Usher "Ike" Hoover, "and made an effort to conduct it along the lines of a private household."[29] Mrs. Taft was surprised by the criticism. "Perhaps I did make the process of adjusting the White House routine to my own conceptions a shade too strenuous," she acknowledged; but, she pointed out, "I could not feel that I was mistress of any house if I did not take an active interest in all the details of running it. . . . I made very few changes, really."[30]

Some of Mrs. Taft's changes were well received: the musicales she sponsored in the White House and the Shakespeare performances on the White House lawn. Her efforts to beautify Washington also came in for considerable praise. With a park in Manila called the "Luneta" in mind, she arranged to have Potomac Drive converted into a promenade, with bandstands at each end, where people could walk or ride and listen to band music in the late afternoon two or three times a week. To beautify Potomac Drive, moreover, she decided to reproduce something she had admired when she was in Tokyo: the cherry-blossom festival. When a search of nurseries in the United States came up with only about a hundred saplings, the mayor of Tokyo, a friend of the Tafts, offered to present the city of Washington with three thousand cherry trees. Unfortunately, the trees he sent over turned out to be infected and had to be destroyed. But when he heard the news he merely smiled and remarked: "Oh, I believe your first President set the example of destroying cherry trees, didn't he?"[31] Then he sent over more trees and within a few years the blossoming of cherry trees in the spring had become a popular tourist attraction in the nation's capital. Not so popular, though, was the cow Mrs. Taft had grazing on the

White House lawn; there was fresh milk for the Tafts every day, but plenty of fresh remarks about it too. Taft had the last presidential cow and the first presidential automobile.

Mrs. Taft didn't entirely abandon politics when she became First Lady. She kept track of official appointments for her husband, continued to discuss issues with him, and sat quietly in on important White House conferences, and talked over the agenda with her husband afterward. "Ike" Hoover thought she overdid things; many was the time, he reported, when he saw a guest at a party get the President off in the corner for a private talk only to have Mrs. Taft spot them, rush over, and insert herself into the conversation. Hoover—and other members of the White House staff—found Mrs. Taft too stern and demanding; they longed for a less assertive First Lady.

But Mrs. Taft's activistic reign was was short-lived. One afternoon in May 1909, she and her husband boarded the *Sylph* for a trip to Mount Vernon; the vessel had barely left the dock when she suffered a stroke that left her partly paralyzed and barely able to speak. "The President looked like a great stricken animal," reported White House aide Archie Butt. "I have never seen greater suffering or pain on a man's face."[32] Mrs. Taft's recovery was slow; it was a year before she was able to walk and talk normally again. During her convalescence, Taft sat by her bedside several hours every day, reading to her, telling her funny stories, and helping her regain her speech. He also brought her flowers from the White House garden. The first time he picked them, the gardener mistook him for a thief and started yelling. Thereafter Taft always handed his daily bouquet to his wife with the remark, "I stole these for you!"[33]

Mrs. Taft eventually recovered from the stroke, but she never resumed the vigorous regimen of her first weeks in the White House. But Taft was overjoyed when she became her old self after about a year or so. He finally reported joyfully to his brother Horace: "She is quite disposed to sit as a pope and direct me as of yore which is an indication of the restoration of normal conditions."[34] In June 1911 she gave a big party (reminiscent of the Venetian ball in Manila) to celebrate the Tafts's silver wedding anniversary, which Archie Butt thought was "the most brilliant function ever held" in the Executive Mansion.[35] When a stream of silver gifts unexpectedly poured into the White House from all over the world, Mrs. Taft thriftily decided she had enough wedding gifts for young brides to last her the rest of her life.

In 1912 Mrs. Taft supported her husband's decision to run for a second term, even though the split in the Republican party between Taft and Roosevelt supporters made the victory of Woodrow Wilson, the Democratic candidate, almost a certainty. When T.R. contested Taft's nomination (and later ran on an independent ticket of his own), Taft seemed surprised, but his wife wasn't. "I told you so four years ago," she exclaimed, "but you would not believe me." "I know you did, my dear," said Taft good-naturedly. "I think you are perfectly happy now. You would have preferred the Colonel to come out against me to have been wrong yourself."[36] Leaving the White House, after Wilson's inauguration in 1913, was easy for Taft. "I'm glad to be going—," he told the Wilsons. "This is the lonesomest place in the world."[37] But Mrs. Taft was miserable on Inauguration Day. She stayed upstairs packing while Taft lunched with the Wilsons after the inaugural ceremony, and left without saying good-bye to the White House staff. Later, when a reporter asked her daughter Helen whether her mother was relieved to be free of White House responsibilities, the latter said frankly: "Mother was never much for relief. She always wanted something to be happening."[38]

The happenings were quieter after the White House. It was law now, not politics, for William Howard Taft: first, a professorship in the Yale Law School in 1913 and then, at long last, an appointment as Chief Justice of the Supreme Court in 1921. Taft was sublimely happy doing what he had wanted to do all along and Mrs. Taft discovered that the life of the law wasn't so bad after all. Like her husband, she was glad to be out of the public eye and she didn't miss politics the way she thought she would. In New Haven, and then back in Washington again, she played golf, read widely, attended plays and concerts, and entertained her husband's friends and colleagues two or three nights a week. She also discussed the issues with her husband as of old. They didn't always agree: Taft supported prohibition, while his wife firmly (though privately) opposed it; and, while Taft came finally to favor woman suffrage, his wife was dead set against it until it became a reality in 1919. She apparently thought most women intellectually inferior; in Washington she always preferred discussing politics with Senators, Congressmen, and Cabinet members to making small talk with their wives.

Taft worked hard as Chief Justice; he even lost weight and was finally able to get dressed without the help of a valet. But his health deteriorated steadily—there were digestive ailments and heart attacks—and by early 1930 he was confined to his bed and experiencing periods of unconsciousness. Just before his death in March 1930, at seventy-three, he could say only, "darling," when his wife came to his bedside.[39] Mrs. Taft survived her husband by thirteen years. She continued living in Washington, took a dim view of the New Deal, as her husband undoubtedly would have, and followed the careers of her children with pride: Robert in the U.S. Senate, Charles as civic leader in Cincinnati, and Helen as Dean of Bryn Mawr College. She was almost eighty-two when she died in May 1943 and was buried beside her husband in Arlington National Cemetery. The *New York Times* credited her with "having played an important part in planning the career of her distinguished husband."[40]

* * * *

Excellent Manners

After Will and Nellie got married, they lived for a time with Taft's parents in Cincinnati, awaiting completion of their own home. Mrs. Taft liked her in-laws; she found they had "strong minds, intellectual tastes, wide culture and catholic sympathies." Not long after moving in with them, however, Taft came to her in great agitation. "Nellie," he said, "Father has got himself into rather a difficulty and I hope I can rely on you to help him out—not make it too hard for him, you know—make him feel as comfortable about it as you can." Then he told her his father had befriended a black messenger when he was head of the War Department in Washington and that the man, now a porter, was in town, had looked up his old boss, and had received an invitation to lunch. "He's downstairs now," Taft went on to say. "Father came to me and said he had just realized it was something of a difficulty and that he was sorry. He said that he could take care of Mother if I could take care of you. So I hope you won't mind." Mrs. Taft was amused by her husband's concern. "I went down to luncheon," she wrote later. "I didn't mind and Will's mother didn't mind, but the expression on the face of Jackson, the negro butler, was almost too much for my gravity. I will say that the porter had excellent manners and the luncheon passed off without excitement."[41]

Delinquency

When Taft went to Washington as Solicitor-General in 1890, he asked his wife to call on Supreme Court Chief Justice Horace Gray's wife as soon as possible, but she was so busy getting settled in her new home that it was months before she got around to making the courtesy call. When she finally made the call, she apologized profusely for the delay. Mrs. Gray was amused. "I should have waived ceremony and come myself to welcome you to Washington," she said, "except for the one thing which I could not very well overlook, and that is—that Mr. Taft has not yet called on Mr. Justice Gray!"[42]

Expenses

When Taft became Federal judge for the 6th Circuit in 1892, Mrs. Taft was "dreadfully upset" because he stinted on his traveling expenses and even returned some of the money he hadn't spent at the end of the year. She told him the travel allowance was really intended to supplement his modest salary and added that the other Federal judges, including Supreme Court justices, kept it all. "You would not be very popular with the Supreme Court any more than your own bench if your views were known," she told him, "and I advise you to keep them to yourself." Taft began taking a more liberal view of his travel money.[43]

Callers

Soon after the Tafts arrived in Manila some Filipino neighbors made a courtesy call on them: el señor, la señora, and four señoritas. The Filipinos knew no English and the Tafts very little Spanish, so everyone sat around for some time simply nodding and smiling and emitting "ohs!" and "ahs!" After a half hour or so the six callers got up, much to Mrs. Taft's relief, and started to leave. "Why, no!" cried Taft in a burst of hospitality. Then he decided to try out a little Spanish. *"Porque?"* he exclaimed. *"Tenemos bastante tiempo.* Why hurry?" Somewhat embarrassed, the visitors politely sat down. Mrs. Taft regretted the fact that her husband had taken time out to study Spanish at all; she also regretted the fact that he knew so little about Filipino manners at this point, and that he didn't realize his visitors expected some sign from him when it was time to leave and would consider it discourteous to leave when he urged them to stay. So the visit dragged on with more nodding and smiling and gestures of goodwill. As the dinner hour approached, Mrs. Taft darted warning glances at her genial husband, but even he was beginning to look hard-pressed. After another hour the Filipinos again indicated an intention to leave and this time Taft uttered no words of protest in either English or Spanish. Afterwards Mrs. Taft instructed him in Filipino manners and customs and "never again," she reported, "did we urge [callers] to reconsider their sometimes tardy decision to depart."[44]

Could Handle It

Mrs. Taft was astonished when a typhoon hit the Philippines one day, breaking windows and rocking the foundations of buildings, and her husband slept soundly through it all. "The chair was shaking—," she told him after the storm had passed; "how could you sleep?" "Now, Nellie," said Taft defensively. "You know it is just my way. I knew you could handle it." No wonder she liked to call him "Sleeping Beauty."[45]

A Stitch in Time

Once, when Taft was Governor-General of the Philippines, he had occasion to visit the court of Czar Nicholas. As he and Mrs. Taft alighted from their carriage at the imperial palace, there was a loud ripping sound and they discovered that the seam of his trousers had given way. Since there was no time to return to the hotel to change suits, Mrs.

Taft borrowed needle and thread from a lady-in-waiting and hastily sewed up the rent. But Taft was afraid that his wife's stitches might not hold up during the audience, so he was careful to move crabwise before the Czar and, upon leaving, back out of the room.[46]

Celebrity

One summer Mrs. Taft visited England with her children while her husband, then Secretary of War, took a Congressional party that included T.R.'s celebrated daughter, Alice, on a tour of the Philippines. When Mrs. Taft was ready to leave London she found her trunks had been mislaid. She was frantic; the train taking her to the steamer on which she had booked passage to the United States was due to leave in five minutes. "I am Mrs. William Howard Taft of Washington," she told the station-master. "I must get my trunks on that boat-train. They'll be here in a few minutes. Can't you hold it for me?" The station-master looked blankly at her. "My husband is the Secretary of War of the United States," she went on desperately. "I am very sorry, Madam," he said. Finally she had an inspiration. "You must have heard of him," she cried. "He's traveling now with Miss Alice Roosevelt." The station-master's face lighted up at once, and he ordered the steamer train held. Then he sent a flock of his subordinates out to look for the trunks, and when they located them, saw her off on the boat-train. For days afterwards Mrs. Taft's children—and her friends—referred to her as "*The* Mrs. Taft whose husband was traveling with Miss Alice Roosevelt."[47]

Mimic

When Mrs. Taft invited Alice Roosevelt Longworth to attend her husband's inaugural luncheon in March 1909, the latter said she might not be able to make it in time because she was seeing her father off at the train station. Mrs. Taft then offered to send her a ticket to admit her to the White House in case she was able to make the luncheon. Alice was furious; the White House had been her home for almost eight years. "Instead of taking it as obvious routine," she wrote in her memoirs, "I flew shouting to friends and relatives with the news that I was going to be allowed to have a ticket to permit me to enter the White House—I—a very large capital I—who had wandered in and out for eight happy winters! Indeed, I gave myself over to a pretty fair imitation of mischief-making." Thereafter Mrs. Longworth did everything she could to belittle Mrs. Taft as First Lady. She criticized her changes in the White House staff; she also accused her of being a stuffy and pretentious hostess. For years she went around imitating Mrs. Taft's supposed stiff and awkward ways. "It was a really good piece of mimicry," she recalled. Sometimes she sat on the front seat of the surrey with her family, pretending to look like Mrs. Taft, and then tell members of the family: "This, my darlings, is what is coming after you!" But Mrs. Taft got even; she persuaded her husband to turn down the suggestion that Alice's husband, Nicholas Longworth, be appointed Minister to China.[48]

Letter

When Taft bought the first stamp advocating votes for women, someone asked him what he would do with it. "I'll put it on a letter to Mrs. Taft," he replied.[49]

Not as Tiresome

One Sunday evening, while Mrs. Taft was recovering from her stroke, she played bridge with Taft's brother Charles and some friends, while the President read Omar Khayyam. "I abstain," Taft remarked, "on account of public sentiment, but I don't think it is one whit as wicked as spending one's evening reading Omar." "Nor as tiresome," cried Mrs. Taft amid laughter.[50]

Approval

One day in August 1910, after Taft had approved some requests his private secretary, Charles D. Norton, and his military aide, Archie Butt, had brought him, Mrs. Taft sighed: "Will, you approve everything—everything Mr. Norton brings you, everything Captain Butt brings to you, everything everybody brings you." "Well, my dear," chuckled Taft, "if I approve everything, you disapprove everything, so we even up on the world at any rate." "It is no laughing matter," said Mrs. Taft. "I don't approve of letting people run your business for you." "I don't either, my dear," said Taft, "but if you will notice, I usually have my way in the long run." "No you don't," persisted Mrs. Taft. "You think you do, but you don't."[51]

CHAPTER 26

THE WILSON WIVES

Ellen Axson Wilson, 1860–1914
Edith Bolling Wilson, 1872–1961

Anatomy is destiny, said Freud, and Wilson seemed to agree. The sexes, he insisted, differed psychologically as well as physically: men were bold, courageous, and forthright, while women were modest, gentle, and sensitive. Wilson didn't regard women as intellectually inferior, but he did think their intelligence was largely "sympathetic" and that it served as a supplement and stimulus to men's more rigorous view of things. When he read Charles Egbert Craddock's *The Prophet of the Great Smoky Mountains* (1885), he was impressed by the book's powerful style, but on learning that Charles Egbert Craddock was the pen-name of Mary Noailles Murfree, he decided the book was of the "sympathetic sort" after all.[1] Wilson's beliefs were not the falsifiable kind. He came around to supporting woman suffrage in his later years, but continued to find the idea of an independent career woman distasteful.

Wilson thought of women as men's helpmates, and, except for the "manly" kind, who went off on their own, he adored them. He was married twice and had three daughters, and his relations with the women in his household were always thoughtful, generous, and affectionate. He also developed several close friendships with attractive women outside the family that meant much to him. In an age when gentlemen adjourned after dinner for cigars, brandy, and serious talk, leaving their wives, presumably, to idle chatter, Wilson preferred the company of "cultivated and conversable" women.[2] After an animated exchange with a lively lady he met at a reception one day, he told his wife Ellen: "I feel, after being with her, that I am stepping out of an *aurora borealis* into the common unprismatic light."[3] But Ellen, his first wife, was an *aurora borealis* for him, too; and so was his second wife Edith. Both Ellen and Edith were his close companions as well as sexual partners and both shared fully in his life and work. They saw eye to eye with him, too, on women's role as homemakers and helpmates.

Wilson's first wife, Ellen Axson, who was born in Savannah, Georgia, in 1860, was a Presbyterian preacher's kid, like Wilson himself, and, as he perceived from the outset, an unusual young woman: bright, attractive, talented, and energetic. When he saw her in church one Sunday morning in April 1883, while in Rome, Georgia, on business, he was almost immediately smitten. "What a bright, pretty face," he recalled thinking; "what splendid, mischievous laughing eyes! I'll lay a wager that this demure little lady has lots of life and fun in her!" He called on her father at once, interrupted a solemn conversation about church attendance to ask for an introduction, and not long after meeting her, commenced a "fast and furious courtship."[4] In September 1883 she agreed to an engagement; but the marriage was delayed while Wilson finished graduate work at Johns Hopkins and got a teaching position at Bryn Mawr, and Ellen spent a year studying painting at the Art Students League in New York City. Ellen had a feeling she might have become quite good if she had continued with her painting, but she gladly gave it up as a serious pursuit in order to marry the devoted young man from Atlanta whom she loved and thought destined for great things. "Can you keep a secret?" she said to her brother Stockton just after she got engaged. "He is the greatest man in the world and the best."[5] Until the marriage in Rome in September 1885, she exchanged fervent love letters with him almost daily.

For thirty years Ellen was the "polar centre" of Wilson's life.[6] She was a conscientious housekeeper; she handled the family finances, took cooking lessons to improve her skills, watched her husband's diet, cultivated the garden, and saved money by sewing clothes for the children and herself. She was an indispensable partner too; she proofread her husband's articles and books with him and, later, when he got into politics, went over speeches with him and even made suggestions for them. She was a stimulating teacher as well. Though Wilson had a doctorate from Johns Hopkins and she was only a graduate of the Female Seminary in Rome, Georgia, she was able to teach him a lot about fields in which he was woefully deficient when they first met: art, architecture, and literature. She helped him, too, as he struggled to master the German language, and she taught the children to read and provided them with religious instruction. Above all, she was Wilson's quiet counselor; she talked over problems with him, gave him advice, and was always ready to boost his spirits when he became depressed. Sometimes, when he was angry and made reckless statements, she would say, "Oh, Woodrow, you don't mean that, do you?" "Madam," he would then respond, with mock gravity, "I ventured to think I did until I was corrected."[7]

Wilson knew exactly how crucial Ellen was for his life. "My love for you," he once told her, "released my real personality, and I can never express it perfectly in either act or word away from you and your immediate inspiration. . . . Love unlocks everything within me that is a pleasure for me to use. I never used my mind, even, with satisfaction till I had you." He did, it is true, develop a deep affection for Mrs. Mary Allen Hulbert Peck, a bright and lively woman, separated from her husband, whom he met in 1907 and with whom he exchanged fervent letters for several years. But he seems to have kept the relationship within bounds, saw to it that Mrs. Peck became Ellen's friend, too, and eventually came to regret the temporary infatuation. "These letters," he later wrote of his exchanges with Mrs. Peck, "disclose a passage of folly and gross impertinence in my life. I am deeply ashamed and repentant. Neither in act nor even in thought was the purity or honor of the lady concerned touched or sullied, and my offense she has generously forgiven. Neither was my utter allegiance to my incomparable wife in anyway by the least jot abated. She, too, knew and understood and has forgiven, little as I deserved the generous indulgence."[8]

The Peck affair, Ellen Wilson once confessed to Dr. Cary Grayson, the family doctor, was the only occasion for unhappiness in all her years with Wilson. She knew he adored attractive and vivacious women, and she liked all of his favorites and shared his friendships with them. She shared his work with him, too, first in the groves of academe (Bryn Mawr, Wesleyan, and Princeton), and then, after he became Governor of New Jersey, in the wider world of politics. As Governor, Wilson was soon being talked of for the presidential race in 1912, and Ellen went into action on his behalf: scanned newspapers, entertained Democratic leaders for dinner, became his audience when he rehearsed speeches designed to make his views known nationally, and sent him letters full of encouragement and advice when he was on the road.

In March 1912, Wilson went to Atlanta for a conference and, after he left, William Jennings Bryan, still a power in the Democratic party, turned up in Princeton to give a speech. Ellen at once telegraphed her husband, urging him to return, and he succeeded in getting back just in time to hear the Great Commoner speak. Afterward, the Wilsons took Bryan out for dinner and he was "captivated" by Mrs. Wilson. Later, when a friend complimented her for "playing good politics" in arranging the Wilson-Bryan get-together, she seemed surprised. "My dear," she said calmly, "it was only good manners."[9] She helped out with Bryan again a few months later, when the Democratic Convention was meeting in Baltimore. As the delegates assembled, Bryan took on the party regulars in a fight for control of the Convention and sought the aid of Wilson and other presidential hopefuls. Wilson was inclined to straddle the issue, on the advice of his campaign advisers, but Ellen said firmly, "There must be no hedging," and he smiled and agreed, "You are right, of course." In the end he backed Bryan and the progressives; and, with Bryan's help, went on to win the nomination.[10] In November, he became the first Democrat since Grover Cleveland to win the national election. "My dear," Ellen said quietly when news of her husband's victory arrived election night, "I want to be the first to congratulate you."[11] Wilson once said that without his wife and his father he would never have made it to the White House.

Wilson's inaugural address on March 4, 1913, thrilled Ellen. As he was speaking, she suddenly got up and walked over to the rostrum to listen, even though he had run over the speech with her beforehand. The "big, garish 'White House,'" as she called it, did not intimidate her; she went on, as always, trying to make life as smooth and comfortable as she could for her husband.[12] "I must make believe very hard," she told a friend, "not that I am a different kind of woman,— in *some* respects—not *all,* thank Heavens."[13] Neither she nor her husband took any interest in Washington's little world of wealth and fashion, but as First Lady she presided over the dinners, teas, and receptions expected of her and did so charmingly.[14] She continued helping her husband; she gave advice on speeches, discussed issues (she read up on Mexico for him during a crisis in U.S.-Mexican relations in 1914), and attended conferences with him. But she also got involved in activities of her own: attended meetings of social workers, as a member of the Board of Associated Charities, toured government departments and had restrooms for women installed where they were lacking, and, though a segregationist like her husband, got interested in improving conditions among impoverished blacks in the nation's capital. Shortly after entering the White House, she took several Congressmen on a tour of the city's slum areas and persuaded them to sponsor a bill (called the "Alley Bill" or "Mrs. Wilson's Bill") providing for decent housing for the slum-dwellers. "I wonder," she once mused, "how anyone who reaches middle age can bear it if she cannot feel on looking back that, whatever mistakes she may have made, she has on the whole lived for others and not for herself."[15]

In March 1914, Mrs. Wilson had a bad fall in her bedroom. After several weeks she was able to get out of bed and even make a few public appearances, but she remained weak and listless after that and without appetite. In June, when Dr. Grayson put her to bed again, it was clear there was no hope; she was suffering from Bright's disease and tuberculosis of the kidneys. Wilson spent hours by her bedside, talking to her, trying to get her to eat, reading to her, or simply watching her while she slept, and praying for recovery. When one of his daughters asked him if the United States might become involved in the war that was developing in Europe, he put his hand over his eyes and murmured, "I can think of nothing—nothing, when my dear one is suffering."[16] Shortly before the end, on August 6, she whispered to Grayson, "Doctor, if I go away, promise me that you will take good care of Woodrow." Wilson was holding her hand when she stopped breathing. "Is it all over?" he asked; and when Grayson nodded, he cried, "Oh, my God, what am I going to do?"[17] For months afterward he was filled with guilt for having involved her in the hurlyburly of politics, and at times he even longed for death himself. But he buried himself in his work, poured out his heart in letters to close friends, and, at Dr. Grayson's suggestion, played golf, took automobile rides and trips on the presidential yacht, and gradually came out of his deep depression.

Seven months after Ellen's death, Wilson met Edith Bolling Galt, a handsome, well-to-do widow of forty-three, whose husband, a Washington jeweler, had died in 1908. The meeting was quite by accident. Mrs. Galt was a friend of Wilson's cousin, Helen Woodrow Bones, then helping out as White House hostess; and one afternoon in March 1915 the two came to the White House for tea. Just as they got out of the elevator on the second floor, they ran into the President, returning from the golf course with his friend Grayson. Edith was embarrassed; her boots were muddy from the walk she and Helen had just taken; but she observed that Wilson's golfing attire was "*not* smart," while she was at least wearing a splendid tailored outfit.[18] Wilson joined the tea party, was enchanted by his cousin Helen's friend, and before long was having her over to dinner at the White House. Mrs. Galt, he learned, had been born in Wytheville, Virginia, in 1872, the daughter of William Holcombe Bolling, a Circuit Court Judge, had helped run her husband's jewelry store for several years after his death, and, while she had only two years of formal schooling, had learned from her father to love the classics the way Wilson did.

The President began courting Edith right after their first meeting. He took her on walks and automobile rides and for trips on the presidential yacht. He read to her, too, the way he had to Ellen, and showered her with letters in which he made no secret of his rapidly developing affection for her. From almost the beginning, moreover, he discussed affairs of state with her, as he had with Ellen, and even had a private telephone line set up between the White House and her home on Twentieth Street, so they could be in constant touch. Early in May he announced his love and Edith was taken by surprise. "Oh, you can't love me," she cried, "for you don't really know me; and it is less than a year since your wife died." "Yes, I know you feel that," said Wilson, "but, little girl, in this place time is not measured by weeks, or months, or years, but by deep human experiences; and since her death I have lived a lifetime of loneliness and heartache."[19] Edith refused to commit herself at this time, and afterwards Helen Bones reproached her for breaking Wilson's heart. In September, however, to Helen's delight. Edith agreed to an engagement.

Not everyone in the White House circle shared Helen's—and the three Wilson daughters'—enthusiasm for the match. Some of Wilson's political advisers were afraid that remarrying so soon after his first wife's death might alienate the public and jeopardize his chances for re-election in 1916. In an effort to block or at least postpone the marriage, they decided to make use of the "Peck's Bad Boy" story that had surfaced during the 1912 campaign: that Wilson had had an affair with Mary Peck that lasted several years and that his wife had been shattered when she learned about it. There were rumors, Wilson's associates told him, that Mrs. Peck (who took her former name, Mary Allen Hulbert, when her divorce became final in 1912) was furious over Wilson's decision to marry Mrs. Galt rather than her, and that she was threatening to publicize the "love letters" Wilson had written her and was even thinking of selling some of the letters.

The story about Mrs. Peck came as a shock to Wilson. He simply couldn't believe that his old friend would engage in blackmail (and the blackmail story was, in fact, made out of whole cloth), but he was deeply disturbed by the thought that Edith might get involved in the kind of mudslinging that had disfigured the 1912 campaign. After he heard about the alleged threat, he sought out his fiancée at once and assured her that his relations with Mrs. Peck, with whom he had corresponded for years, had been entirely platonic, and that she had been Ellen's friend too, and the three of them had frequently socialized together. When he offered to release Edith from the engagement, however, she insisted on standing by him. "I am not afraid of any gossip or threat," she told him, "with your love as my shield. . . ."[20] Wilson was overjoyed and he hesitated no longer. Early in October he wrote out an announcement of their engagement to be issued to the press and the response to the news was mostly favorable. The wedding—a quiet service in Edith's Washington home—took place in December, and afterward Wilson and his bride boarded the train for a honeymoon in Hot Springs, Virginia. Edmund Starling, a Secret Service man who accompanied them, was amused to see the President dancing a little jig in his private car the next morning and singing, "Oh, You Beautiful Doll! You Great, Big Beautiful Doll!"[21]

As First Lady, Edith Wilson soon learned how to shake hands at receptions with a minimum of discomfort. "You put your middle finger down," Wilson instructed her, "and join the index and ring finger above it. In that way people can't get a grip. Your hand just slides through theirs."[22] Like Wilson's first wife, however, Edith Wilson did a lot more than serve as hostess for official functions. She supervised the household, handled her husband's finances, watched his diet, saw that he got proper rest and recreation, accompanied him when he campaigned in 1916, and, above all, provided him with a sympathetic audience whenever he wanted to talk about the problems facing him as President in a war-torn world. To bring her close to the center of things Wilson taught her the secret code he used for communicating with his representatives abroad; and soon she was decoding incoming messages for him and putting his outgoing messages into code.

The entry of the United States into World War I in April 1917 increased Edith Wilson's activities. She got out her sewing machine to make clothes for the soldiers; she also donned a Red Cross uniform and put in regular appearances at soldiers' canteens in Washington to hand out coffee and sandwiches. She became "The Shepherdess" too: arranged to put some sheep on the White House lawn to keep it close cropped and had the wool they produced sold at auction to benefit the Red Cross. Throughout the war she continued, as ever, to be the President's closest confidante. "From the first," she recalled, "he knew he could rely on my prudence, and what he said went no further."[23] At times she joined her husband and a group of his advisers known as "The Inquiry" to discuss war aims.

When the war ended in November 1918, Mrs. Wilson supported her husband's decision to attend the peace conference in Paris to ensure the achievement of his objectives; and Wilson insisted she accompany him. After their arrival in Europe in January 1919, the Wilsons made a triumphant tour of France, Italy, and Great Britain (where Edith refrained from curtseying when she met the Queen), and then Wilson went into a huddle with Allied leaders to fashion a treaty ending the war which would include his proudest creation: a League of Nations designed to uphold world peace. While Wilson labored at Versailles during the day, his wife visited soldiers' hospitals, rehabilitation centers, factories, and canteens; and at night she enjoyed hearing her husband talk about the day's proceedings and rake some of his fellow delegates over the coals. At the same time she worried about his health; the long and at times stormy hours he was putting in at Versailles steadily wore him down and at one point he came down with influenza. But his spirits soared when the treaty was finally completed, with his beloved League an essential part of it. "Well, little girl," he exclaimed, as their train left Paris in June 1919, "it is finished, and, as no one is satisfied, it makes me hope we have made a just peace; but it is all in the lap of the gods."[24]

The gods proved refractory. So fierce was the opposition to ratification of the Versailles Treaty in the U.S. Senate that Wilson decided to go on a tour of the country to promote popular support for the League of Nations. His wife strongly opposed the tour; her husband seemed so frail and worn after his labors in Paris that she wondered whether he could survive a grueling speech-making trip to the West Coast. But

Wilson was insistent. "I must go to the people," he told her. "Only they can prevail."[25] In the end she yielded; and when he set out from Washington by train in late August she went along to watch over him. In city after city she listened, at times spellbound and at times with tears in her eyes, to his eloquent plea for U.S. membership in the organization he had created for bringing peace to the world at long last, after centuries of strife and bloodshed. But she was increasingly alarmed by his rapidly deteriorating condition: the splitting headaches, the insomnia, the disgestive upsets, and the asthmatic attacks. "Let's stop," she cried at one point. "Let's go somewhere and rest. Only for a few days. . . ." "No!" said her husband stolidly. "I have caught the imagination of the people. They are eager to hear what the League stands for. I should fail in my duty if I disappoint them."[26] But his suffering became intolerable after he spoke in Pueblo, Colorado, late in September, and she called a halt to the tour and insisted he return at once to Washington. On October 2, three days after he was back in the White House, he suffered a massive stroke that left him paralyzed on one side.

With the President's collapse, Mrs. Wilson faced a dilemma no other First Lady had ever confronted: should she insist that her husband resign his office or should she find some way of helping him carry on with his responsibilities as President until he recovered from his illness? Mrs. Wilson sought the advice of her husband's doctors, and they all counseled against resignation. The President's mind was as good as ever, Dr. Francis Dercum, the neurologist, told her, and there was every reason to think he would make a good recovery so long as she shielded him against "every disturbing problem" while he was convalescing. "How can that be," she asked, "when everything that comes to an Executive is a problem? How can I protect him from problems when the country looks to the President as the leader?" "Madam, it is a grave situation," admitted Dercum, "but I think you can solve it." Then he outlined a plan of action. "Have everything come to you," he advised; "weigh the importance of each matter, and see if it is possible by consultations with the respective heads of the Departments to solve them without the guidance of your husband. In this way you can save him a great deal. But always keep in mind that every time you take him a new anxiety or problem to excite him, you are turning a knife in an open wound. His nerves are crying out for rest, and any excitement is torture to him." "Then," said Mrs. Wilson, "had he better not resign, let Mr. Marshall [Vice President Thomas Marshall] succeed to the Presidency and he himself get that complete rest that is so vital to his life?" "No," said Dercum, "not if you feel equal to what I have suggested. For Mr. Wilson to resign would have a bad effect on the country, and a serious effect on our patient. He has staked his life and made his promise to the world to do all in his power to get the Treaty ratified and make the League of Nations complete. If he resigns, the greatest incentive to recovery is gone; and as his mind is clear as crystal he can still do more with even a maimed body than any one else. He has the utmost confidence in you. Dr. Grayson tells me he has always discussed public affairs with you; so you will not come to them uninformed." Dercum convinced Mrs. Wilson; she resolved to follow his suggestions.[27]

Thus began what Mrs. Wilson called her "stewardship." Wilson, she said, "was first my beloved husband whose life I was trying to save," and "after that he was President of the United States."[28] To protect her husband she kept everyone, even the highest officials, out of his bedroom; to enable him to continue as President she became his intermediary with Senate leaders, diplomats, secretaries, advisers, and Cabinet heads. Every day she went carefully over the papers coming to his office for action, sifted out the most important ones, made summaries of them, and then took them to his bedside, and relayed his instructions afterwards to the appropriate officials. As Wilson got better, he was able to dictate notes to her to send to his associates; at times, too, he instructed her to get in touch with Senator Hitchcock and the other Senators leading the fight for ratification of the Versailles Treaty to convey his ideas on what strategy they should follow in the Upper Chamber. "I, myself," Mrs. Wilson wrote in her memoirs, "never made a single decision regarding the disposition of public affairs. The only decision that was mine was what was important and what was not, and the *very* important decision of when to present matters to my husband."[29]

Mrs. Wilson's determination to carry on with the Presidency during her husband's illness produced considerable criticism, particularly among the Republicans. Some people complained about the "Mrs. Wilson regency" and "Petticoat Government"; others sarcastically called her the "Presidentress" and said her husband was now "First Man."[30] But Mrs. Wilson had her defenders too. "All the more credit to her!" exclaimed Dolly Gann, sister of Kansas Republican Senator Charles Curtiss. "I am glad there was a woman in the White House who knew how to take the reins and use authority when it was passed upon her."[31] There was something ironic, however, about a Southern gentlewoman who vigorously opposed woman suffrage (even Wilson came reluctantly to support it) wielding power in the U.S. Government the way no other woman ever wielded it.

When it came to the major issue of the day—the bitter fight over ratification of the Versailles Treaty—Mrs. Wilson faithfully transmitted her husband's instructions to the Senators leading the fight for adoption. There were to be no concessions, she told them, to critics of the Treaty, like Henry Cabot Lodge of Massachusetts, who refused to accept America's membership in the League of Nations without a set of fairly stiff reservations. On November 19, 1919, however, when Senator Hitchcock told Mrs. Wilson that unless the administration accepted the Lodge reservations the Treaty would surely fail of ratification, she was deeply upset by the prospect of defeat, and decided, for the first time, to try to influence her husband's policy. "For my sake," she pleaded with him, "won't you accept these reservations and get this awful thing settled?" "Little girl," said Wilson gently, taking her hand, "don't you desert me; that I cannot stand." Then he added: "Better a thousand times to go down fighting than to dip your colours to dishonourable compromise."[32] Mrs. Wilson at once regretted that she had ever asked him to do something he regarded as "manifestly dishonourable" and resolved never to do it again. She went out to tell Hitchcock to stand firm against the Lodge reservations and, later, when the Treaty failed to pass the Senate, she sadly reported the news to her husband. Wilson said nothing for a moment or two and then told her: "All the more reason I must get well and try again to bring this country to a sense of its great opportunity and greater responsibility."[33] In March 1920, the Senate voted again on the Treaty—first with, and then without, reservations—and both times failed of passage. It was not until 1945, more than two decades after Wilson's death, that the United States finally made Wilson's great dream a reality, when it joined the United Nations, dedicated to checking aggression and keeping the nations of the world at peace. Mrs. Wilson lived to see the fruition of her husband's hopes and dreams.

Not long after the November vote on the Treaty, Republican Senator Albert B. Fall of New Mexico visited Wilson's bedside with Senator Hitchcock, ostensibly to discuss relations with Mexico. Wilson called the visitation a "Smelling Committee," for Senator Fall, a foe of the Treaty, was really there to see whether there was any substance to rumors that the President had lost his mind. "Well, Mr. President," said Fall piously after some desultory talk, "we have all been praying for you." "Which way, Senator?" chuckled Wilson, forcing a strained smile from the New Mexican. Then, observing that Mrs. Wilson was busily taking notes on the interview, Fall exclaimed: "You seem very much engaged, madam." "Yes," she said sharply, "I thought it wise to record this interview so there may be no misunderstanding or misstatements made."[34] Afterward, Senator Fall acknowledged to reporters that the President was mentally fit.

Wilson never became well again but he did make some improvement. He was eventually able to leave his bed and get around in a wheelchair and then, with help from others, walk a bit. In April 1920 he met with his Cabinet for the first time since his stroke, but Mrs. Wilson interrupted the meeting after a half-hour or so, when she thought it was taxing her husband's strength. As Wilson's health improved, he began toying with the idea of seeking the nomination in 1920 so he might continue his fight for the League; but Mrs. Wilson firmly vetoed the idea. She also killed another idea of his: to call on the Senators who had voted against the Treaty to resign office and seek re-election on the basis of their position on the League on Nations.

When the United States returned to "normalcy" under Warren G. Harding in 1921, the Wilsons retired to a house they had purchased on S Street and Wilson talked of practicing law again and of writing a book on government. His wife arranged a law partnership for him, but he finally decided he was not up to the task. As for the book, he got no further than the dedication: "To E.B.W., [who] has shown me the full meaning of life. Her heart is not only true but wise; her thoughts are not only free but touched with vision; she teaches and guides by being what she is; her unconscious interpretation of faith and duty makes all the way clear; her power to comprehend makes work and thought alike easier and more near to what it seeks—Woodrow Wilson."[35] In November 1923, he gave a brief radio talk, with the help of his wife, about the League of Nations, but he was clearly failing by then. He died in February 1924, at sixty-eight; and when his wife learned that the Senate had put Henry Cabot Lodge on a committee to attend the funeral services, she wrote to tell him bluntly he was *persona non grata,* and he agreed to stay away.

After Wilson's death, his wife went into seclusion for a year or so and then began appearing in public again. She traveled a great deal after that and made a special point of attending meetings of the League of Nations whenever she was in Europe. She also devoted much of her time to honoring the memory of her late husband; she helped Ray Stannard Baker on his multi-volumed biography of Wilson, became director of the Woodrow Wilson Foundation, founded to promote international understanding, helped make Wilson's birthplace in Staunton, Virginia, a national shrine, and served as a member of the Woodrow Wilson Centennial Committee in 1956. She was on hand, too, whenever a new road, bridge, or building was dedicated to her husband; and in 1939 she published her memoirs, hoping to correct the inaccuracies she detected in the books about Wilson written by his former associates.

Through the years Mrs. Wilson remained a loyal Democrat, though she had reservations at times about her party's policies. She attended the Democratic Convention in Houston, Texas, in 1928, and gave a speech there; and in 1932 she was in Chicago for the Convention that nominated Franklin D. Roosevelt, and even did some campaigning for him that year. When FDR launched the New Deal, however, she had misgivings about his welfare programs and doubted whether her husband would have approved of them. But she was a frequent visitor to the White House during the Roosevelt and Truman years; and she was a special guest at John F. Kennedy's inauguration in January 1961. Later that year she was anxious to attend the dedication of the Woodrow

Wilson Bridge over the Potomac on the 105th anniversary of her husband's birth, but she never made it. She died quietly on December 28, at eighty-nine, on Wilson's birthday.

In her later years Mrs. Wilson was frequently asked: "Really now, didn't you run the White House in 1919?" Her response was invariably one of astonishment. How could she ever have dreamed of taking over? Wasn't her husband the greatest man in the world?[36]

* * * *

Brown Dress

When Wilson was teaching at Princeton, his wife Ellen tried to economize to make ends meet and once innocently remarked that her clothes cost her only about forty dollars a year. "I should like to meet Mrs. Wilson," a woman headed for a faculty reception once told English professor Bliss Perry's wife. "Mrs. Wilson will probably be there," said Mrs. Perry, "and she will be wearing a brown dress." "How do you know she will wear a brown dress?" asked the woman. "Because," said Mrs. Perry, "her best dress *is* brown." There were snide remarks about Mrs. Wilson's wardrobe. "Mrs. Wilson," one mean-spirited faculty wife is supposed to have said, "every fall you look sweeter in that brown dress."[37]

Head

In the Wilson home at Princeton there were several statues on pedestals, including the Apollo Belvedere and the Winged Victory. When one of the Victory's wings broke, the man who came to mend it told Ellen Wilson, "Give me the head, ma'am; I can put that back on too, so you wouldn't hardly notice it." She was too polite to correct him. "I'm sorry," she said gently, "but I haven't the head. I wish I had."[38]

Periodic Comma-Cal Conversations

Ellen Wilson helped her husband read proof on his books and articles and sometimes, after spending all morning going over a manuscript, they would carry on a long conversation at lunch in proofreading style. "The soup comma my dear comma is delicious semi colon Maggie is an excellent cook period No wonder exclamation You taught her period," he would say and she would answer, "Thank you comma Woodrow period."[39]

Boasting

When Wilson was at Princeton, his children—Margaret, Jessie, and Eleanor—were enormously proud of him and liked to brag about him to their friends. One morning Ellen Wilson came into the nursery looking very serious. "Your father has been told that you are boasting about him," she said. "Is it true?" Eleanor denied it and Jessie thought a moment and then said, "I haven't, mother." Then Margaret burst out with, "Well, mother, I do—I tell them a few things." Mrs. Wilson then sat down, gathered the children around her and said, "I understand how you feel, but it does embarrass your father, so after this we four will have a beautiful time talking about him as much as we like when we are alone, but not to other people."[40]

Concentration

Ellen Wilson had formidable powers of concentration. The children loved to stand around while she was reading and watch their father try to attract her attention. It became a kind of game. First Wilson would make signs, then he would clear his throat, then speak softly to her, and finally start raising his voice. "Were you speaking to me?" Mrs. Wilson would finally say, looking up, as the children dissolved into laughter. "We thought we were," Wilson would say gently.[41]

Governor's Husband

One June day, when the Wilsons were vacationing on the New Jersey coast, a little boy came wandering into their cottage and Mrs. Wilson gave him some cake. When he got home the boy told his parents he had been to see the Governor and the Governor had given him some cake. But, he added, the Governor's husband wasn't at home.[42]

Fundamentally and Superficially

When he was Governor of New Jersey Wilson began to be talked of as a presidential possibility in 1912, but on speaking tours he repeatedly said he was not thinking of the Presidency. "Please, please don't say again that you are not thinking of the Presidency," Ellen Wilson wrote him worriedly. "All who know you well know that this is true fundamentally; but superficially, it can't be true and it gives cynics an opening they seize with glee."[43]

Don't Have Any

Shortly after Wilson received the nomination for President at the Democratic Convention in 1912, a woman reporter approached Ellen Wilson and asked why she never wore jewels. "Have you some sort of moral prejudice against jewelry, Mrs. Wilson?" she asked. Ellen, who had stinted on her attire so her husband could have the books he wanted and the children could study art and music, murmured something noncommittal, but the reporter was persistent. "But why, Mrs. Wilson?" she asked. "No," said Ellen Wilson finally, "I have no prejudice against [jewelry]; we just haven't any." The simplicity of her answer brought tears to the eyes of her daughter Eleanor.[44]

Put One Over on Teddy

Ellen Wilson was extremely pleased when her husband decided to break precedent and deliver his tariff speech to Congress in person shortly after he became President. She was even more pleased by the great applause his personal appearance before Congress elicited. "That's the sort of thing Roosevelt would have loved to do, if he had thought of it," she said delightedly afterward. "Yes," laughed Wilson, "I think I put one over on Teddy."[45]

Goner

When Pat McKenna, White House doorkeeper first saw Edith Galt with President Wilson, he cried: "She's a looker!" Added another White House aide: "He's a goner!"[46]

Typo

In 1915, when Wilson was courting Edith Bolling Galt, the *Washington Post* ran an article in which the following typographical error appeared: "The President spent much of the evening entering [entertaining] Mrs. Galt." The *Post* made every effort to recall the edition with the misprint, but several copies managed to hit the newstands. Wilson's enemies made use of the typo to back up their charge that he was a womanizer.[47]

The Bishop's Wife

Although Edith Galt's friends wanted her to have a White House wedding, she decided on a quiet wedding in her own little home with only close relatives present. And when she asked an Episcopal Bishop to perform the ceremony, she told him she couldn't invite his wife to be present since she was asking no one outside the immediate families, and the Bishop assured her that he and his wife understood the matter perfectly. Two days before the wedding, however, the Bishop arrived in Washington with his wife and told Mrs. Galt that they were sailing for England a few days after the wedding and that it would cause his wife "much chagrin to acknowledge to her titled friends there that she had not been asked to the marriage of the President where her husband had officiated."

The Bishop's change of mind deeply offended Mrs. Galt. She wrote him a frosty note explaining again why she couldn't include his wife in the wedding party and she concluded that "my only course is to excuse you from performing the ceremony." Wilson was amused when she read him her letter over the telephone. "Why not wait and think it over?" he suggested. "No," said Mrs. Galt firmly, "this letter goes to him right now. I will postpone our wedding rather than be bludgeoned into a thing of this kind." "Yes," Wilson finally agreed, "I was afraid of that. But, after all, the poor fellow has enough to stand with a wife like that." In the end, the rector of St. Margaret's Episcopal Church, where Mrs. Galt had a pew, and Wilson's pastor in the Central Presbyterian Church in Washington, shared the ceremony.[48]

Thank God for the Chance

"Let's lean on the rail instead of walking," Wilson told Edith one night in June 1915 when they were cruising down the Potomac on the *Mayflower,* "as I want to talk to you." Then he told her he had just received a letter from William Jennings Bryan saying he couldn't conscientiously continue as Secretary of State because he thought Wilson's policy was leading to war. Bryan was still popular among Democrats, especially among Democratic doves, and Wilson was reluctant to let him go, but when he asked his wife what she thought the effect of the resignation would be, she unhesitatingly replied: "Good; for I hope you can replace him with someone who is able and who would in himself command respect for the office both at home and abroad."

To clinch her point, Mrs. Wilson told her husband about a play she had seen the previous year. In it, a young woman eloped with a boorish young lover in an old-fashioned coach which broke down at the country estate of an attractive bachelor. The elopers had to beg hospitality for the night. The next morning, the host, already smitten by the young bride-to-be, surprised the lovers in the midst of a violent quarrel. The woman tried to end the spat with a kiss but her boy friend sullenly rejected it. "Kiss her, sir," cried the host impatiently; and then, in an aside to the audience, he added, "and thank God for the chance!" Mrs. Wilson's advice about the Bryan resignation, she said, was the same: "Take it, sir, and thank God for the chance." Wilson smiled and then went on to discuss a possible replacement for Bryan.[49]

Mrs. Peck's Letters

President Wilson's warm friendship with Mrs. Mary Hulbert (the former Mrs. Peck), shared by his first wife Ellen, eventually produced numerous stories, all of them false, about his efforts to suppress supposedly incriminating letters he had written her. One was that Louis Brandeis received his appointment to the U.S. Supreme Court as a reward for stealing or buying the letters and destroying them. Other stories named other associates of Wilson as involved in the enterprise. One day, Cleveland H. Dodge called on Colonel Edward M. House and said: "I've just learned that I paid Mrs. Peck one hundred and twenty-five thousand dollars for her letters." "When were you informed that?" asked House. "This morning," replied Dodge. "I beat you," laughed House. "I learned the day before yesterday that I paid her one hundred and fifty thousand dollars!"[50]

Pessimist

From the outset Mrs. Wilson was sure her husband was going to lose his bid for re-election in 1916. The masses were for him, she thought, but the Republicans had more money to put into the campaign than the Democrats did and the European war had created serious problems for his administration. A few nights before the election she lay awake thinking about these things, and when Wilson came into her bedroom to talk about a speech he had been working on she told him she had been making plans for their retirement after the election. "What a delightful pessimist you are!" cried Wilson. Then he told her: "One must never court defeat. If it comes, accept it like a soldier, but don't anticipate it, for that destroys your fighting spirit." On November 5, when the outcome of the election was uncertain, Wilson joined his wife in thinking he had lost. It wasn't until a few days later, when all the returns were in, that his victory was assured.[51]

Edith Wilson and the Colonel

In December 1917, to co-ordinate the operation of the railroads in wartime, Wilson prepared a special message to Congress asking for authority to place them under governmental control and sent a copy of his address to his adviser, Colonel E.M. House, for suggestions. When House arrived to talk things over with Wilson, the latter was in a conference, so Mrs. Wilson invited him to have tea with her while waiting for the President. As they sat down before the fire, House took out his copy of Wilson's proposal and argued forcefully against it. "Well," said Mrs. Wilson, when he finished, "we will have an interesting session after dinner, when you take this up with Woodrow." "Suppose you tell him about it first," said House, "and let him think it over." "Very well," agreed Mrs. Wilson, "I will." House, a White House guest at the time, went off to dress for dinner.

Shortly after House left, Wilson ended his conference and came in to see his wife. "Oh," he said, when he saw the tea-things, "you and House have been having a talk. What does he think of my message?" "He doesn't like it at all," Mrs. Wilson told him; she went on to summarize House's objections to the railroad plan. "Oh, I am sorry," said Wilson thoughtfully; "but I will be glad to have him go into his reasons; that is really where he will give me help." To Mrs. Wilson's astonishment, when House went into a huddle with Wilson after dinner, he said he was in complete agreement with the President's plan. "Why, Colonel," she exclaimed, "you couldn't! You said—" "Yes, I know I did," he interrupted, "but I have changed my mind." Mrs. Wilson was appalled that House failed to stick to his guns in her husband's presence; she decided that he was a "yes man" when he was with the President, but a critic behind his back. She never really trusted him after that.[52]

Free from Nerves

For the opening of the Third Liberty Loan President Wilson and his wife went to Baltimore. When they reached the beautifully decorated grandstand after a long drive through the city, Mayor Preston took Mrs. Wilson aside and began whispering to her. "Mrs. Wilson," he said, "I want to apologize to you for the absence of my wife. I have received anonymous letters saying the stand will be blown up today, so I would not let her come. Of course *I* had to." "Certainly, Mr. Mayor," said Mrs. Wilson, more amused than frightened, "in such circumstances she is most excusable." "I knew you would understand," said the Mayor, wiping his brow, "but I did not tell the President because I did not want to make him nervous, as we may be blown up at any moment." Mrs. Wilson was tickled by the Mayor's assumption that the President's wife was "free from nerves." But, with "no credit to myself," she recalled, "that happened to be the case."[53]

Looking the Part

After the United States entered World War I, Mrs. Wilson joined a Red Cross unit, donned a uniform that made her look a bit frumpish, and spent afternoons in a Washington canteen handing out candy and cigarettes to soldiers passing through town. One hot day in July, a soldier came up to her and said, "Am I right in thinking this is Mrs. Wilson?" "Yes, you are," said Mrs. Wilson, "and I am mighty glad to tell you how proud the President is of you boys." The soldier then brought some of his buddies over to meet the First Lady. But one of the doughboys, a tall, raw-boned fellow from the West, took a look at Mrs. Wilson, and then drawled: "Well, buddy, you can't string me. That's not the wife of the President of the United States." "Oh," interposed Mrs. Wilson, "you don't think I look the part?" "I certainly do not," he said frankly. "Well, I agree with you," laughed Mrs. Wilson; "but when you come back from France, if you will come to the White House I will do my best to 'look the part' and give you such a warm welcome it will convince you."[54]

Mrs. Wilson and the Ships

When Edward N. Hurley, chairman of the Shipping Board, asked Mrs. Wilson to rename the eighty-eight German ships in U.S. ports which the United States took over when it went to war, she thought it would be an easy task. And at first it was. She renamed the *Vaterland,* the largest ship, *Leviathan,* and the smallest, the *Minnow,* and she retained the names of ships the Germans had named after American Presidents, like *George Washington.* But the other ships posed a problem, and so did the new ships American factories were turning out by the hundreds. Every time she named a ship after an American city, river, lake, or mountain, and checked *Lloyd's Register,* she found the name had already been used. Finally she procured an Indian dictionary from the Library of Congress and began assigning Indian names to the ships. As a descendant of Pocahontas, she took special pleasure in picking euphonious Indian names for the vessels. But she worked hard. "Trivial as it may seem," she wrote, "as the War went on, and other duties increased, this work of naming the ships became a genuine burden. I would finish one list, take a deep breath, and send it in to Mr. Hurley and receive, with his note of thanks, a longer list of new and nameless ships." But Secretary of the Navy Josephus Daniels wrote to tell her that "when women decide to become cabinet officers the Secretary of Navy's portfolio should be assigned to you. Indeed, there seems to be a fitness in this for in the Navy we always call a ship 'she.' "[55]

Broke the Law

One day Secretary of the Interior Franklin K. Lane forwarded a letter he had just received to Mrs. Wilson. "Honorable Sir," said the writer, "I am a descendant of Pocahontas and the Powhatan Indians. President Wilson married Mrs. Galt whose name was Bolling. She is an Indian, a descendant of Pocahontas and Powhatan." Then the writer got to his point. "President Wilson," he said, "has broken the greatest law in the United States. He has served wine and liquor to Mrs. Wilson and she is an Indian, and the law says no one shall give or treat or bootleg or sell or blind tag liquor to an Indian et cetera—you can read it yourself. Now, Secretary of the Interior, I wish you to get out a warrant for President Wilson's arrest immediately according to the laws of the United States. I was given you as the proper one to push this proper prosecution at once." The letter ended by identifying the writer as "an educated Indian," whose cousin was a judge, and informing Lane he would "have proper attorneys represent my side at the hearing."[56]

Greatest Mistake

One evening late in October 1918, Mrs. Wilson went into her husband's office and found him pecking away at the typewriter. When she came in, he pulled a sheet of paper from the machine and handed it to her. "Tell me what you think of that," he said. It was an appeal for the election of a Democratic Congress the following month to show confidence in the President's leadership and make his work easier. Mrs. Wilson was upset by the message. Although the Republicans had been supporting the war effort enthusiastically, Wilson seemed to be implying that they couldn't be trusted. "Please don't release this, Woodrow," she begged him. "You know you don't feel this way, it doesn't represent your own views." "You're right, it doesn't," admitted Wilson. "But I promised [party leaders] that I would do it and I feel I must." "I beg you not to," cried Mrs. Wilson. "It is not like you and I feel it will have a very bad effect." "I promised," said Wilson stubbornly. Mrs. Wilson's instincts proved correct. Wilson's appeal for a Democratic Congress infuriated the Republicans and led many independent Republicans who might have voted Democratic to vote the straight Republican ticket. The result that the Republicans gained control of Congress in the 1918 elections and Republican leaders like Henry Cabot Lodge were able to sabotage Wilson's program for the post-war world. "That message," Mrs. Wilson said after the election, "was the greatest mistake Woodrow ever made."[57]

Can't Afford

In January 1919, when the Wilsons toured Europe, just before the opening of the Versailles Peace Conference, one of the ladies at a Buckingham Palace dinner asked Mrs. Wilson if she was a Quakeress. "No," Edith told her. "Oh," said the lady, "I thought you were because you wear no tiara." "I would wear one," said Mrs. Wilson amusedly, "but, you see, my husband can't afford to give me one."[58]

Mrs. Wilson and the Duchess

When the Wilsons were in Paris, a haughty French duchess, whose title was one of the oldest in the country, told the Serbian minister, M. Vesnitch, that she had no desire to meet them because they were only "ordinary Americans with no claim to aristocracy or title." "Oh, Duchess," cried Vesnitch, who knew Mrs. Wilson had Pocahontas in her family background, "there you are wrong, for Mrs. Wilson is directly descended from a princess, and of the only aristocracy in America; her grandmother seven times removed was the great princess Pocahontas." "Why, this is very important," cried the Duchess, "and I knew nothing of it at all. Of course I must go and call on Mrs. Wilson at once and give an entertainment in her honor."

When Vesnitch told Mrs. Wilson about the Duchess, she was filled with mirth and told Wilson she thought she would send the Duchess a picture postcard of Washington's Powhatan Hotel (where her mother was staying) and tell her "it's my ancestral palace, now used as a hotel for war workers." "Please wait," laughed Wilson, "until I finish this job over here before you play jokes on any French lady."

A few days later the American Ambassador's wife gave a reception for Mrs. Wilson at which the Duchess was present and when she introduced the Duchess to the President's wife, the latter curtsied, as if Mrs. Wilson was of royal descent. She then invited her and the President to dinner. When Mrs. Wilson said the President was too busy to accept dinner invitations and that she herself never dined out without him, the Duchess said she was "desolated." Then suddenly she smiled and cried: "Oh, but I learn that you visit hospitals." "Yes," said Mrs. Wilson, "I do." "*Voilà!*" she cried. "My house is a hospital, so you will come now?" "I will with great pleasure do that," said Mrs. Wilson.

When Mrs. Wilson arrived at the Duchess's chateau a few days later, she was greeted by the seven-year-old Duke in black velvet, holding pink apple blossoms, by a Cardinal in gorgeous red robes, and by the Duchess and the ladies on her staff wearing nurses' uniforms. As Mrs. Wilson stepped from the limousine, the little Duke knelt on one knee, kissed her hand, and presented the blossoms. Then the Cardinal raised his hand in silent benediction and the Duchess introduced Mrs. Wilson to her friends. And at the tea, to Mrs. Wilson's surprise, she and the Duchess were the only ones seated. Everyone else had to stand in the presence of royalty.[59]

Hiding Place

In February 1919, when it came time for Wilson to present the draft of the Covenant of the League of Nations to a plenary session of the Peace Conference, Mrs. Wilson was eager to hear him speak, for she knew the League was his greatest love. There was a problem; the meeting was for members only and the public was barred from the Foreign Office where the delegates were to assemble. But Mrs. Wilson asked Admiral Cary Grayson, the President's physician, whether it would be improper for her to ask Georges Clemenceau, the French Premier who was to preside over the meeting, if there was any way she could be admitted. Grayson, who wanted to be there too, encouraged her to approach "the Tiger," as he was called.

Before getting in touch with Clemenceau, however, Mrs. Wilson asked her husband if he had any objections. "In the circumstances," he told her, "it is hardly a request, it is more a command, for he could not very well refuse you." "That being the case," she said, "then I shall certainly make it." "Wilful woman," teased Wilson, "your sins be on your head if the Tiger shows his claws." "Oh, he can't, you know," she returned; "they are always done up in grey cotton [gloves]."

When Grayson broached the subject to Clemenceau, the Tiger's eyes twinkled. "It is impossible," he said. Then he went on: "Almost impossible. For so lovely a lady nothing is quite impossible." Then he suggested that Grayson and the President's wife hide in the little antechamber off the Room of the Clock, where the meeting was to be held, and keep the heavy curtains in front of it tightly closed. "For God's sake," he added, "don't let anyone see her, for if you do, the other wives will tear me apart."

On the day of the historic meeting, Mrs. Wilson and Dr. Grayson arrived early at the French Foreign Office, sneaked into the antechamber before any of the delegates appeared, sat down on the two chairs there, and pulled the crimson curtains shut. There they waited as the delegates filed in and Clemenceau called the meeting to order. When Wilson stood up to speak, after an hour or so of preliminary business, Mrs. Wilson forgot at once how hot and stifling it was in the little room. She got a look at him too; for since all the delegates were watching Wilson as he spoke, she and Grayson ventured to part the curtains a bit and peek out at him. "It was a great moment in history," Mrs. Wilson wrote later, "and as he stood there—slender, calm, and powerful in his argument—I seemed to see the people of all depressed countries—men, women and little children—crowding round and waiting upon his words. He rarely made long speeches, and this was no exception. I could see it had made a great impression, and I longed to go to him and tell him all I felt."

When the delegates had given Wilson a big ovation and adjourned the meeting, Mrs. Wilson and Dr. Grayson came out of their hiding place and went to join Wilson outside. "Are you so weary?" she asked when she saw him waiting for her in his car. "Yes," he answered. "I suppose I am, but how little one man means when such vital things are at stake."[60]

Fire and Smoke

In the fall of 1919, when Wilson was touring the country with his wife to win support for the League of Nations, he decided to look up his old friend Mary Hulbert in Los Angeles. Mrs. Hulbert had a doleful tale to tell Wilson and his wife. People had tried to buy the affectionate letters Wilson had written her, she said, and had put pressure on her to testify at impeachment proceedings against the President, and had even searched her rooms looking for material they could use against him during the 1916 campaign. Wilson was appalled that his old friend had been put through so much misery on his account. "God," he cried, "to think you should have suffered because of me." Mrs. Wilson decided to lighten the mood by a little teasing. "Where there's so much smoke," she smiled, "there must be some fire." "Then, perhaps," Mrs. Hulbert is said to have retorted, "you *were* [the German minister] von Bernstorff's mistress!"[61]

Sorry

During the celebration of the Woodrow Wilson Centennial in 1956, Mrs. Wilson went to Staunton, Virginia, Wilson's birthplace, to hear *New York Times* columnist Arthur Krock deliver a speech in Wilson's honor. Krock's address was mainly laudatory, but he did criticize Wilson for not having taken any prominent Republicans, like William Howard Taft and Elihu Root, with him on the peace delegation to Paris after World War I. When Krock ran into Mrs. Wilson at the luncheon afterward, he was afraid she might be unfriendly because of his critical remarks, but she seemed quite unruffled. A few days later, he met her at another Centennial meeting. "You must have been very gratified at Staunton on the occasion of the centennial of President Wilson's birth," he said. "It was a very great occasion." "Yes, it was," said Mrs. Wilson calmly. "I'm so sorry you could not have been with us."[62]

Common Thief

Mrs. Wilson was frequently a guest in the White House after FDR became President. At one dinner, FDR asked her: "How do you like these wine glasses, Mrs. Wilson? I got them off the *George Washington* for five cents apiece." Mrs. Wilson said she admired them and then told FDR that when the *George Washington* was being broken up at the end of World War I, Secretary of the Navy Josephus Daniels offered Wilson the beautiful desk he had used on the crossing. "I can't take it," Wilson said. "It belongs to the government." "No it doesn't," said Daniels. "It was a personal gift to you from Rodman Wanamaker." Wilson then said if that was the case he would like the desk very much. But a few days later Daniels called back to say the desk had disappeared. "We never found out what happened to it," Mrs. Wilson told FDR. Roosevelt (who had been Assistant Secretary of the Navy under Daniels) started laughing. "I can tell you," he said. "I have it. And what's more, you're not going to get it away from me." Mrs. Wilson then looked him straight in the eye and, to his amusement, exclaimed: "You're nothing but a common thief!"[63]

Memoirs

When Edith Wilson read Colonel House's books about the Wilson administration, she was so angry that she sat down and started writing her own memoirs. "But I never expected to publish it," she said. "I was writing the truth for my own satisfaction." But after her friend Bernard Baruch read the manuscript he said, "You must publish this book." "It's not good enough," she replied. "I don't know how to write." "It is good enough to have kept me up until one o'clock this morning," Baruch said. "Mark James has been helping me with a book. I'll send him to you." "I don't think he can help," said Mrs. Wilson. "I can't publish it." "I command you to," said Baruch. So she let Pulitzer Prize biographer Marquis James read the manuscript. "I can't change it," James said after looking it over. "It's yours and I would spoil it. Just leave out a few of the nasty remarks you made about Colonel House and some of the others." After making a few revisions, Mrs. Wilson went on to publish *My Memoir* in 1939 and was gratified by its warm reception.[64]

Nixon-Lodge Sticker

Mrs. Wilson was delighted when John F. Kennedy included a chapter on her husband in his book *Profiles in Courage,* and she gave him her warm support when he ran for President in 1960. One day, when she happened to notice that a man who called on her had a Nixon-Lodge sticker on his car, she told him emphatically: "Don't you ever park a car with that thing on it in *my* driveway again!"[65]

Didn't Dare

In October 1961, two months before her death, Mrs. Wilson was the star guest at a White House ceremony when JFK signed a law setting up a commission to plan a memorial to Woodrow Wilson. "You just sit right there," said Kennedy, handing her a pen, "and then we'll sign it." "I didn't dare ask you for it," said the still strong-minded former First Lady somewhat meekly, to general laughter.[66]

CHAPTER 27

FLORENCE KLING HARDING

1860–1924

Florence Kling Harding was a folksy First Lady. Just before the White House reception following her husband's inauguration in March 1921, she noticed the servants pulling down the window-shades to keep people from peeking in and cried: "Let 'em look if they want to! It's their White House."[1] As First Lady, she tried to mingle with "the people" as often as she could; sometimes she came downstairs when visitors were touring the White House, introduced herself, and started showing them the sights. After the wartime restrictions of the Wilson administration, the White House under the Hardings became refreshingly accessible to the public again.

Mrs. Harding's health was poor much of the time, but she stood in receiving lines for hours and soon developed "White House feet." At the first New Year's reception she shook hands for over two hours, and when a friend expressed pity, exclaimed, "Oh, now, my hand is good for two hours yet," went on to shake close to seven thousand hands that day, and suffered a swollen hand for days afterwards without complaint.[2] She was "simple, natural, and old-fashioned," a close friend observed, whenever she appeared at White House functions.[3] But she took special pleasure in visiting World War I veterans in the Walter Reed Army Hospital and taking them flowers from the White House gardens. "I was in a hospital for eight months with an open wound that had to be dressed twice a day," she once explained, "and I know what hospital life means to a patriot."[4] Her proudest entertainment: a big

Veterans Garden Party on the White House lawn. When the maid laid out a new outfit for the occasion, she cried, "I'll just wear this old hat instead of my new one, because the boys are accustomed to it, and, as soon as they see it, they know where I am."[5] The boys adored her.

Mrs. Harding didn't charm everyone. Mrs. Wilson, who met her just before leaving the White House, was shocked by her gaucheries. The White House's Elizabeth Jaffray (housekeeper whom Mrs. Harding quickly fired and then rehired) was also repelled at first by the plain lady from Ohio. Jaffray thought Mrs. Harding was a kind of "unpolished emerald" who "could never quite accustom herself to the greatness and importance of her position."[6] The Longworths—Nicholas, Ohio Congressman, and Alice Roosevelt, his wife—socialized with the Hardings but were unimpressed. Alice though Harding was pleasant, but "just a slob"; and she wasn't much taken with Mrs. Harding, either. "She was a nervous, rather excitable woman," observed T.R.'s daughter, "whose voice easily became a little high-pitched, strident. . . . She usually spoke of Mr. Harding as Warren Harding. It is impossible to convey her pronunciation of the letter *r* in print. Something like Wurr-r-ren Ha-rr-ding."[7]

If Harding was "Wurr-r-ren" to his wife, she was the "Duchess" to him. But he wasn't the only one to employ that rather austere appellation; all of his cronies did too. Sometimes, though, they also called her "ma" and "boss." Some people thought she really was the boss, but though she did a lot of nagging, she didn't always prevail. "I have only one real hobby—," she once acknowledged, "it's my husband."[8] Harding's career came first with her, and she was a big help to him in many ways. She gave shrewd advice on occasion, accompanied him on speechmaking tours, and faithfully protected his public image. Harding wanted to "chuck" all ceremony when he became President, but she saw to it that he did the right things when he moved into the Executive Mansion. She kept him from chewing tobacco in public (cigars were permissible), hied him off to church as often as she could, and tried, with less success, to prevent the assignations she suspected took place when she was not around. The Hardings, after all, represented small-town values to the great American Electorate and it was important to keep up appearances. As Harding announced in more than one speech, the heart of America lay in its farms, villages, and small towns. There dwelt "normalcy."

Marion, Ohio, was Florence Harding's hometown. Born there in 1860, she attended the local schools and then went off to the Cincinnati Conservatory of Music for a year. Florence was several cuts above Warren socially; her father, Amos Kling, banker and realtor, was the richest man in town, while Warren's father, an impecunious physician, was said to have a black ancestor or two in his background. When Harding first met his wife-to-be, she was a plain-looking, somewhat gawky, and headstrong young woman who was just as bent on having her way in the world as her father was. She was also "a mite wild."[9] To her strait-laced father's immense displeasure, she liked to go roller-skating with young men from the other side of the tracks; and he locked her out of the house more than once when she came home too late at night to suit him. In 1880, she eloped with Henry ("Pete") DeWolfe, son of the local coal-dealer, and her father was outraged; Pete was a likable but lazy lad, with a bent for the bottle, and the marriage quickly fell apart. When Florence returned to Marion, penniless, with a baby boy, after Pete deserted her, her father refused to help out. She ended by leaving the baby with her grandparents and becoming a piano teacher to support herself. She was a thirty-year-old grass widow when she started seeing Warren Harding, the twenty-five-year-old editor of the *Marion Star*.

The Hardings probably first met at the town's roller-skating rink. But according to a popular (though unsubstantiated) story, Florence was shopping with a girl friend when she saw Warren, standing on a street corner, for the first time. "Who is that handsome young fellow?" she is said to have asked. "Warren Harding of the *Star*." "Do you know him? I'd like to meet him." "He doesn't come to our house," said her friend doubtfully. "But maybe I could get word to him to be at the Baptist supper Wednesday."[10] Florence's father was furious when his daughter took up with the young journalist. One day, when he met Harding in the courthouse, he let loose a flood of profanity, called him a "nigger," and threatened to blow his head off. But Florence ignored her father, as she had when she took up with Pete, and continued seeing Warren. She also began overseeing him; if he sneaked out of town to see an old girl friend elsewhere, she was usually waiting for him at the station when he got back. Once he spied her from the train window and got off on the other side of the tracks, but she caught up with him. "You needn't try to run away, Warren Harding," she shouted. "I see your big feet."[11]

Florence wasn't the romantic type, but she was probably in love with the town's young Lothario. As for the easygoing Warren, he liked Flossie well enough, and, when it came to domineering women, he was used to that, for his mother had ruled the Harding roost when he was a boy. He was grateful, too, for admiration and affection from a woman whose social position, unlike his, was assured; and in the end he agreed to marriage. Florence's father boycotted the wedding (though her mother sneaked in to watch the ceremony), which took place on July 8, 1891, and it was seven years before he spoke to his daughter again, and fifteen years before he permitted himself to set foot in his son-in-law's home.

Soon after the marriage, Florence began helping out with the *Marion Star*, a weekly newspaper which Harding developed into a daily. "I went down there," she later recalled, "intending to help out for a few days and I stayed fourteen years."[12] She took over the circulation department, organized and trained (and sometimes spanked) the newsboys, and saw to it that the papers were regularly paid for as well as delivered. She handled advertising, too, while her husband concentrated on writing editorials and winning good will for the paper and himself. At noon each day, she hopped on her bicycle and headed home to prepare lunch for Warren. People working for the *Star* respected Mrs. Harding, but resented her penny-pinching. "She is nice to everybody," reported one observer, "knows how to run things, too. Runs her house; runs the paper if necessary; runs Warren; runs everything but the car, and she could run that if she wanted to."[13] She never ran Warren—or the *Marion Star*—as much as people thought she did, but it was not for want of trying. She made every effort, people noticed, to be "unostentatiously but unshrinkingly" at her husband's side at all times.[14] She once told a friend she had learned that she should "always have reserve food in the icebox and never let my husband travel without me."[15]

Harding did a lot of traveling. As public speaker and editorial writer, he had learned to platitudinize with energy and enthusiasm and he was in much demand as booster and glad-hander at community club luncheons and local Republican rallies. He also went on the Chautauqua lecture circuit for a time and thoroughly enjoyed spouting obviosities; he would have enjoyed himself even more had his wife not tagged (and nagged) along and wrangled with managers over travel expenses. In 1898 he entered politics with the warm approval of his wife, was elected to the Ohio Senate that year, and in Columbus met Harry M. Daugherty, political wheeler-dealer who thought Harding looked "like a President" and soon became his friend and manager.[16] With Daugherty's help, he was elected Lieutenant Governor in 1902; and on election night, there was a big victory banquet in his honor in the state capital. But when members of the *Marion Star* gathered to congratulate him on his return home, Mrs. Harding, knowing that they expected an end-of-the-year pay hike, announced coldly: "You men needn't look for a raise this time. That little show cost us thirteen hundred dollars."[17] But she was pleased nonetheless, and, like Daugherty, was convinced her husband had a great future in politics. In 1910, however, Harding ran for Governor of Ohio and lost, and decided to withdraw from politics for a while and go back to running the *Marion Star*.

In 1914, when Ohio's Senator T. E. Burton decided not to seek re-election, one of the *Star*'s staff members cornered Mrs. Harding and asked: "What's the matter with W.G. pointing for the Senate?" "Do you think he could get it?" asked Mrs. Harding excitedly. "Why not?" said the man. "The Republicans are getting together. He's got a lot of friends." "It costs such a lot of money to live in Washington," sighed Mrs. Harding. "Oh," said the man, "it isn't so expensive. And the paper isn't doing badly." "If he was a corporation lawyer and could pick up a lot of business on the side, I'd say yes," said Mrs. Harding thoughtfully. "But he couldn't do anything there. No, I don't know as we can afford it yet."[18] But when she learned that Daugherty was eager to have her husband run, she joined him in putting pressure on Harding to enter the contest. His victory that November, by a narrow margin, exhilarated her. She went to Washington with him in March 1915 with great expectations.

Mrs. Harding was proud of her position as wife of a U.S. Senator. She enjoyed mingling socially with Nicholas and Alice Longworth, fellow Ohioans, and through them, getting to know millionaire Edward B. McLean and his wife Evalyn Walsh, owner of the fabulous Hope diamond. She attended poker games with her husband at the Longworths' and the McLeans' and, since she was no gambler herself, acted as bartender. "Duchess," one of the players would say periodically during long evenings, "you are laying down on your job," and she would get up and mix whiskey and soda for the group.[19] When she became seriously ill with kidney trouble in 1918, Mrs. McLean sent her flowers, visited her regularly, and listened sympathetically to her troubles; and after she recovered, introduced her to the "smart set" in Washington. Sometimes, after hobnobbing with people in the world of fashion, Florence would ask Evalyn: "What did they say of me?"[20] She began keeping a little red book with the names of people in it who she felt had snubbed her. She was deeply hurt because the Longworths never deigned to set foot in her house, but her husband couldn't have cared less.

When Harry Daugherty began promoting Harding for the Republican presidential nomination in 1920, Mrs. Harding was at first cool to the idea. "I like being a Senator's wife," she announced.[21] But Daugherty soon won her over and she joined him in persuading her reluctant husband to enter the contest. "Harding was very fond of her and held the profoundest respect for her judgment. . . ," according to Daugherty. "If she backed our candidate, he would make the fight. If she opposed it, the issue would be doubtful."[22] But Harding did poorly in the quest for delegates in Ohio and Indiana and decided to pull out of the race. When he started to call his Ohio headquarters to announce his withdrawal, however, Mrs. Harding rushed over, snatched the phone out of his hand, and cried: "Wurr'n Harding, what are you doing? Give up? Not until the convention is over. Think of your friends in Ohio!"[23] Harding stayed in the race.

One afternoon in February 1920, Mrs. Harding visited Madam Marcia, popular Washington astrologer, with three other Senators' wives, and was so intrigued by the hocus-pocus she encountered that she returned alone, incognito, in May and asked the seeress to cast the horoscope for a man born on November 2, 1865 at 2:00. On subsequent visits, Madam Marcia told her the man in question was a great statesman and would become President of the United States some day, but would never live to complete his term. He was, moreover, "sympathetic, kindly, intuitive, free of promises and trustful of friends, enthusiastic, impulsive, perplexed over financial affairs." For good measure, she added what was common knowledge in Washington: "Many clandestine love affairs; inclined to recurrent moods of melancholia."[24]

Mrs. Harding was well aware of her husband's spells of moodiness. Precisely what she knew about his "clandestine love affairs," however, it is hard to say. By the time she consulted Madam Marcia, Harding had two affairs going on at the same time. The first was with Carrie, wife of his good friend James E. Phillips, a Marion businessman; he had fallen in love, apparently for the first time in his life, with Carrie in 1905 and was still deeply attached to her. The second affair was with Nan Britton, a young Marionite who had gotten a crush on him when she was in high school, taken the initiative, and succeeded in getting him to visit her regularly in New York City to engage in what she called "love's sweetest intimacy."[25] Until 1920, Mrs. Harding seems not to have known about Carrie; the Hardings and the Phillips saw a great deal of each other in Marion through the years and traveled abroad together a couple of times without either Florence or Jim realizing what was going on. But Harding saw Carrie secretly whenever he could, and when he couldn't he exchanged fervent love letters with her in which he was "Constant" and she was "Sis."[26] As for Nan, she had made her infatuation with Harding so obvious in Marion (she liked to write, "I love Warren Harding," on the school blackboard) that Mrs. Harding had regarded her with suspicion from the beginning.[27] But she seems not to have known about the baby girl Nan had in October 1919 and for whom Harding made generous provision.

Neither of Harding's dalliances came to light during the 1920 presidential campaign; nor does Harding himself seem to have feared that they would. At one point before Republican leaders chose him as their candidate ("the best of the second-raters") at the nominating convention in June, Colonel George Harvey took him aside and asked him bluntly whether there was anything in his background that might prove embarrassing in the campaign. Harding retired for about ten minutes to think things over. He surely must have thought about his extramarital activities; possibly he also thought of the old charge that the Hardings had Negro blood in them. But he had won three elections without having his personal life dragged in the mud and probably counted on the same thing in 1920. In any case he ended by assuring Harvey that there was nothing to worry about and his backers went into action to gather enough delegates to put him over the top at the Chicago Convention.

Mrs. Harding had mixed feelings about her husband's nomination. She knew that both her husband's and her own health were wretched;

he had heart trouble and she suffered from kidney ailments. She was also aware of the fact that being President had almost killed Woodrow Wilson, and wondered whether her husband (who had gone to the Battle Creek sanitarium several times to recover from nervous exhaustion) could take the pressure of the highest office in the land. Possibly, too, the Washington's soothsayer's prediction about a truncated term of office bothered her, for she was a superstitious woman. "I am content to bask in my husband's limelight," she told a reporter when she first arrived in Chicago, "but I cannot see why anyone should want to be President in the next four years." Then she suddenly added: "I can see but one word written over the head of my husband if he is elected and that word is Tragedy."[28] But when her husband received the nomination on the tenth ballot she was in transports; and she even did something she usually avoided as she grew older: posed for the photographers.

When the convention adjourned, Mrs. Harding threw herself with enthusiasm into her husband's front-porch campaign: greeted delegations to Marion jovially, spoke to women's groups, helped Harding convince voters that they were "just folks like themselves," and went with him on a short speaking tour later in the campaign.[29] But Harding's managers took no chances; they paid the Phillipses off (for Jim had found out about Carrie) with $20,000 plus a monthly stipend and got them out of the country for the duration. And when a pamphlet appeared just before election day announcing that one of Harding's ancestors was "a West Indian Negro of French stock," they quickly suppressed it with the help of the Wilson administration. Harding wanted to deny the pamphlet's allegations publicly, but Mrs. Harding persuaded him to keep quiet.[30] On election day, her husband rolled up huge popular and electoral majorities and went on to become one of the nation's most popular Presidents.

The White House didn't change the way the Hardings lived appreciably, though both came to find it oppressive. Harding went on, as before, uttering the kind of glitzy generalities that Americans adore (even intellectuals, if they like the utterer). He also continued the poker games; his cronies now assembled on the second floor of the Executive Mansion and the Duchess still served the drinks. (There was no liquor of course at public functions on the first floor, for Prohibition was now the law of the land.) Harding continued seeing Nan, too, after he became President; he sneaked her in and out of the White House with the help of friendly Secret Service men and enjoyed blissful moments with her in the coat closet adjoining his main office. One afternoon, Florence almost caught him at it, but, forewarned, he managed to get Nan out of the building just in time.

Mrs. Harding may have been brighter—certainly she was shrewder—than her husband, but she was far less generous. She insisted on cutting a passage from one of his speeches that seemed friendly to the

League of Nations, for she hated Woodrow Wilson's creation.[31] She also opposed giving a pardon to socialist leader Eugene Debs, who had been jailed for making anti-war speeches during World War I. But Harding pardoned him anyway. Since he was a fellow Midwesterner, Harding invited him to the White House for a chat after he got out of prison and found he liked him very much.

Some of the people Harding liked ended by betraying him. In 1922 and 1923, evidence of graft and bribery among friends he had appointed to high office began coming to light and soon threatened to destroy his Presidency. Harding was plunged into despair; he found it almost impossible to sleep at night. His wife became plagued by anxiety; she even had a Secret Service agent stationed outside her bedroom to protect her. The Surgeon-General, Charles Sawyer, an old Marion friend, finally recommended a change of scene and the Hardings decided on a trip to Alaska and the West Coast that summer. On June 20, 1923, they boarded a special train, with Dr. Sawyer and a naval doctor to take care of Harding and with a trained nurse for Mrs. Harding. Harding insisted on making speeches en route and he was warmly received whenever he appeared in public. But his spirits remained low and he was increasingly listless. At one point he told his associates: "In this job I am not worried about my enemies. It is my friends that are keeping me awake nights." A few days later he asked reporters what a President should do whose friends had betrayed him.[32] Mrs. Harding was feeling too low herself to make the effort to bolster his spirits.

In Seattle, Harding became seriously ill and was rushed to San Francisco and put to bed, with local doctors as well as his own physicians in attendance. Dr. Sawyer had announced that the President was suffering from ptomaine poisoning, but his diagnosis was wrong; all four doctors agreed that he had really suffered a heart attack. The end came quickly and quietly on August 3. Mrs. Harding was reading him an article in the *Saturday Evening Post* filled with kind remarks about his policies and he was enormously pleased. "That's good," he told her at one point. "Go on, read some more."[33] Moments later he was dead of an apoplectic stroke. Mrs. Harding was frantic at first and then steeled herself. "I will not break down," she kept telling herself and soon calmed down.[34] The night before the funeral she went to the East Room of the White House where her husband lay in state and sat quietly there for a long time. "No one can hurt you now, Warren," she finally said, and then went over and picked a little bouquet from the flowers sent for the funeral and placed it on his coffin.[35]

After the funeral, Mrs. Harding stayed in Washington for a few months going through her husband's papers and weeding out letters she thought might embarrass him. Then she returned to Marion, went into seclusion except for daily trips to her husband's grave, and placed herself under Dr. Sawyer's care. Sawyer's sudden death while she was with him came as a terrible shock; and soon after, on November 21, 1924, she died herself of chronic nephritis and myocarditis and was buried beside her husband in his memorial tomb in Marion.

Mrs. Harding was mercifully spared two books: Nan Britton's *The President's Daughter* (1927), a tell-all exposé, filled with details about her involvement with the President; and Gaston B. Means's *The Strange Death of President Harding* (1930), a sensational potboiler by a convicted perjurer filled with lies, including the suggestion that Mrs. Harding had poisoned her husband. Mrs. Harding was undoubtedly "a strange and rather difficult woman," as her housekeeper Elizabeth Jaffray discovered, but she was deeply devoted to her husband though she knew he often strayed from the fold.[36] The night before his funeral she sent for two newspapermen whom he had praised shortly before his death and told them to go in and look at the President's body. "I know what some of his critics were saying," she said quietly. "They charged that he was weak. I want you to look at those firm lips of his and see. I look at them and know they show he was not weak. I know that he had strength and courage."[37]

* * * *

America

Before the King and Queen of Belgium came to Washington, many Washington society women took lessons in the gentle art of curtsying to prepare themselves for a reception at the White House but not Mrs. Harding, then a Senator's wife. When she was introduced to Queen Elizabeth, she held her hand out and looked her straight in the eye. "She was in our country," Mrs. Harding explained afterward. "If I were being received by royalty in its own country, I should try to conform to the customs of the country. But this was in America."[38]

Watch Me

Just before Mrs. Harding became First Lady, she is said to have asked a Senator: "Who was the most successful Lady of the land?" "Oh, I don't know," said the Senator. "Mrs. Cleveland or Dolley Madison, I suppose." Mrs. Harding then tapped her shoulder. "Watch me!" she told him.[39]

Interview

At the Republican Convention in 1920, Harding's campaign manager Harry Daugherty was pleased by the way Mrs. Harding handled herself when approached by reporters. "You know my husband ought to win this fight," she told one reporter before the balloting began. "He has a winning way about him that has always disarmed enmity. He can differ sharply with a man—but always without offending him— ." "I see you like him," laughed the reporter. "Oh, I know you'll think I'm boasting," she said, "but I have only one fad. The only fad I've had for the last twenty-six years. And that is my husband. It's old-fashioned, I know. But that is the way I feel about it." "They say Hiram Johnson [California Senator] will bolt the Convention if he is not nominated—," said the reporter. "Do you think so?" "Of course not," smiled Mrs. Harding. "I know Senator Johnson well. He's a fine fellow, though he does make a great noise at times. His bark is much worse than his bite. He's just trying to scare the delegates into voting for him." "Think you'd like the White House?" asked the reporter. "I would not," said Mrs. Harding, frowning. "We've a lovely home in Washington and many warm friends. Being a Senator's wife suits me. It's pleasanter, quieter, its problems never heartbreaking." Daugherty thought Mrs. Harding's interview was nigh-perfect and was convinced it helped her husband obtain the nomination.[40]

Excitement

As the balloting proceeded at the Republican Convention in 1920, with Harding in the lead, Daugherty began to worry about Mrs. Harding, who was following proceedings from the gallery. He knew she had a weak heart; he also knew she had opposed her husband's quest for the nomination. What if the excitement proved too much for her? "If she should suddenly die in the wild scene that would follow his nomination," Daugherty began thinking, "he would feel that he had killed her." As the delegates prepared for the tenth ballot (the one that put Harding over the top), Daugherty hurried up the stairs and into Mrs. Harding's box.

When Daugherty reached Mrs. Harding, he found she had taken her hat off, was humped forward in her chair, and in her right hand was gripping two enormous hat pins in vogue at the time. "It's terrible, isn't it?" said Mrs. Harding when she saw him. "What?" he murmured. "All this wild excitement," she said. "This yelling and bawling and cat-calling. I can't follow it." "I didn't think you would," Daugherty told her, "but something's going to happen down there in a few minutes that may shock you if you don't look out—" "What do you mean?" she cried. "I've come up here to ask you to keep cool," said Daugherty. "That's a good joke, I must say," she snorted. "It's a hundred and ten in this place and you advise me to keep cool!" "You know what I mean," said Daugherty. "Something's going to happen on the next ballot and I want you to be prepared for it—" "What?" she wanted to know. "We have the votes," Daugherty told her, leaning closer. "Your husband will be nominated on the next ballot." At these words, Mrs. Harding gave a sudden start and fairly leaped from her chair. As she did so, she drove both hat pins deep into Daugherty's side. He sprang back, felt the blood running down his leg and into his shoe, but tried hard to ignore it. "Remember now," he went on weakly, for Mrs. Harding seemed unaware of what had happened, "no excitement. Keep your head and laugh at the antics below—" "I will," she promised. Daugherty felt his way down to the floor of the Convention, the blood swishing in his shoe, and caught hold of a chair to steady himself. And as the hall exploded in cheers with Harding's victory, he tottered back to his hotel, fearing the worst, and looked down fearfully at his shoe. To his relief, it was perspiration, not blood, trickling down there. Mrs. Harding's hat pins hadn't hurt him a bit; he had just become overexcited.[41]

Frightful Picture

After Harding's nomination for President, his managers began preparing a biographical pamphlet and sought a good picture of his wife to go in it. But Mrs. Harding proved stubborn; she hated having her picture taken and flatly refused to pose for the photographer. When Daugherty finally went in desperation to see her, she told him the only picture she had was an old one he couldn't possibly use. Daugherty insisted on seeing it and she laughingly brought out a silly-looking old picture, taken when she was a young woman, perched on a bicycle. Daugherty quickly grabbed the picture, put it in his pocket, and cried: "Well, if you will not have a picture taken, I will use this one." "You wouldn't use that thing, would you?" she exclaimed in disbelief. "I will," said Daugherty. "Unless you give me a better one. We cannot finish the booklet without a picture of you. It's ready to go to the press, and I'll take this picture home and use it unless you have another one made immediately." Mrs. Harding tried to laugh it off as joke, but when she saw that Daugherty was deadly serious, she caved in and agreed to pose for a new picture. And during the campaign, she was a good sport about it. "All right. Come on boys," she said when photographers surrounded her. "I always take a frightful picture and I hate this. But I know you've got to do it."[42]

Not a Bit of It

For some reason Mrs. Harding took a dislike to Vice President Coolidge and his wife. When the widow of a Missouri Senator offered her house and grounds for an official residence for the Vice President and a bill was introduced into Congress authorizing acceptance of the gift and making an appropriation for the upkeep, Mrs. Harding snapped: "Not a bit of it, Not a bit of it. I am going to have that bill defeated.

Do you think I am going to have those Coolidges living in a house like that? An hotel apartment is plenty good enough for them." The bill was defeated.[43]

Instructions

A few days before the Hardings left for California in June 1923, Mrs. Harding took Secret Serviceman Edmund W. Starling aside to discuss plans for the trip. "I want you to promise me something," she said. "Wherever we are to stop, I want the doctors, General Sawyer and Captain Boone, as close to the President's room as possible. If they can be put in the adjoining suite, I would appreciate it. At any rate, I want to be informed of their room number in each place. I am also taking a trained nurse with us." Starling knew that Mrs. Harding was just recovering from a serious illness so he nodded obligingly. But she looked sharply at him and exclaimed: "You understand?" "Perfectly," he said. "Are you *sure* you understand?" she said doubtfully. Then she added: "It is not for myself that I want this done, but for Warren." "I understand," he said in some surprise. "I will follow your instructions."[44]

CHAPTER 28

GRACE COOLIDGE

1879–1957

The first time Grace Goodhue saw Calvin Coolidge, he was in his underwear. This was on a summer day in Northampton, Massachusetts, in 1904. She was watering flowers at the time, happened to glance at the house next door, and saw the thirty-two-year-old Calvin standing by the window in his Union suit, with a brown derby perched on his head, busily shaving. She burst into laughter at the sight; and when he heard her laugh and looked out of the window, she quickly turned her head away in embarrassment. But he soon contrived an introduction through a mutual friend and began an assiduous courtship. He wore the hat while washing and shaving, he explained when they met, to keep an unruly forelock out of his face. She accepted the explanation; she also agreed to picnics, rides, walks, and talks with him. He soon proposed.

Grace and Cal were an unlikely pair; her friends found it all quite unbelievable. She was warm, friendly, outgoing, gregarious, and cheerful; he was quiet, austere, deliberate, uncommunicative, and sometimes glum. But they both came from Vermont, had college degrees, and shared the granite-like values—strive and succeed, sink or swim, work and win, pluck not luck—of their region. She had grown up in Burlington, where her father was a steamboat inspector on Lake Champlain, and had studied at the University of Vermont before commencing work in Northampton as a teacher in the Clarke Institute for the Deaf. He had been born and raised in Plymouth, Vermont, where his father was a storekeeper, had attended Amherst College, and then settled in Northampton to begin law work. The two Vermonters quickly discovered they agreed on the two-spheres theory of sexual relations: man's world was that of the provider while woman's was that of the home-maker.

For all his diffidence, young Cal tried hard to come out of his shell while courting the twenty-six-year-old Grace. She was a good skater and though Cal had never been on the ice, he dutifully acquired a pair of skates and joined her and her friends at a skating party on the Connecticut River. But after stumbling around for a while, he suddenly gave up, took off the skates, slung them over his shoulder, and said he "guessed we'd better go home," and he never skated again.[1] He was a poor dancer, too, and had no taste for either the theater or music, both of which Grace loved. But she enjoyed the hikes and chats, and as Cal put it years later: "From our being together we seemed naturally to come to care for each other."[2] When friends teased Grace about her beau's taciturnity, she laughed and said she was used to silence at the Clarke Institute. "Having taught the deaf to hear," quipped one of her friends, "she may now inspire the dumb to speak."[3]

Cal's folks liked Grace. "That's a likely-looking gal, Calvin," his grandmother told him. "Why don't you marry her?" "Mebbe I will, Grandma," said Coolidge.[4] One day he showed up in Burlington when Grace was away and called on her father. "Up here on some law business, Mr. Coolidge?" said Captain Goodhue uncertainly. "No," said Coolidge, "Up here to ask your permission to marry Grace." "Does she know it?" asked Goodhue in surprise. "No," said Coolidge, "but she soon will."[5]

When Coolidge proposed, he wasted no words. "I am going to be married to you," he informed Grace one day; and she promptly accepted.[6] Her mother didn't like Cal and tried to get her daughter to wait a year to think things over, but Cal was persistent and Grace stuck by him. They were married in the Goodhue parlor by a Congregational minister on October 4, 1906, and then entrained for a two-week honeymoon in Montreal. But after a week there, Cal announced: "Grace, gotta be going back home." When she asked why, he said: "Running for school committee, gotta go back and make a speech."[7] So they returned at once and after occupying temporary quarters for a while, eventually settled in one half of a two-family house on Massasoit Street that was to be their home until Coolidge became President. Coolidge lost the contest for school committeeman, but won all subsequent elections: as Mayor of Northampton, member of the Massachusetts legislature, Lieutenant-Governor and then Governor of Massachusetts, and finally as Vice President and then President of the United States. His "hobby," he once told an interviewer, was "holding office."[8]

Grace Coolidge quickly learned that politics was not her sphere. One evening, early in the marriage, she heard that her husband was going to speak at a church gathering and got ready to go with him. When she appeared at the door with her hat and coat on, he asked: "Where are you going?" "I just thought I'd go out to hear you talk," she answered. "Better not," he advised.[9] So she stayed home that night and it was years before she heard him speak in public. But she seems not to have minded being excluded from his work. She threw herself cheerfully into her tasks as wife and mother; with the help of a maid, she cooked the meals, washed the clothes, kept the house neat and clean, knitted and darned stockings for the boys, Calvin, Jr., and John, read to them, and even played baseball with them, for her husband was usually too busy for games. One afternoon, Coolidge came home with a well-worn old laundry bag and tossed it on the table. When she opened it, she found it contained fifty-two pairs of men's socks full of holes. She went uncomplainingly to work at once mending them, but later asked Coolidge whether he had married just to get his socks darned. "No," he said solemnly, "but I find it mighty handy."[10]

Coolidge did, of course, play his part at home. He stoked the furnace, supervised the household budget, and on occasion acted as babysitter so his wife could visit friends in the evening. But one night when he was at home and one of the boys fell out of bed, he called his wife at once on the party line. "Grace?" he cried. "What is it, Calvin?" she asked. "Hop home," he directed, and she returned at once.[11] In Northampton, people said, "Grace Coolidge always made the best of things."[12]

Coolidge thought his wife was a creditable homemaker, but he fussed about her cooking for a time because he was a finicky eater and had digestive troubles. She soon learned to serve the kind of food he liked, but was no great shakes at first with baking, as she freely admitted. To tease her, Coolidge sometimes dropped her biscuits on the floor and stamped his foot as they landed to emphasize the weight. He kidded her about her pie crust too. The first time she made apple pie the crust was so tough that neither of them could get the pie down at dinner. But when two friends dropped by for a visit that evening, Coolidge insisted they have some dessert and out of politeness, they managed to get through the pie. When they finished it, he twinkled, "Don't you think the Road Commissioner would be willing to pay my wife something for her recipe for pie crust?"[13]

Coolidge joshed his wife about her gullibility as well as about her baking. Once a salesman talked her into buying a family medical book they could hardly afford and she had such misgivings right after making the purchase that she put the book in an inconspicuous place on the parlor table without saying anything about it. A few days later, however, when she started leafing through it, she discovered her husband had written on the flyleaf: "This work suggests no cure for a sucker."[14] But Mrs. Coolidge did some teasing of her own. She liked to imitate her husband's nasal twang—he really did pronounce the word, cow, in four syllables—and teased him about his terse way of putting things. But she knew just when she could get away with mimicking him, and he was always amused by her performances. The first time she heard him speak in public, though, she was so tickled by his rustic twang that she hid behind a pillar to conceal her mirth.

Economy was Coolidge's watchword. He used a party line for telephoning, acquired utensils at low prices when the Norwood Hotel went out of business, denied himself a car, and did little traveling or entertaining. But for all his penuriousness, he had one extravagance: he wanted his wife to wear pretty clothes and never complained about the cost. He liked to go window-shopping with her, too, and she liked having him along. He helped her pick out dresses, wraps, hats, and shoes, and she thought he had pretty good taste. She was impressed, too, with his patience. "Much more than I have," she admitted. "Often I would be willing to quit—willing to take anything just for the sake of quitting; but Mr. Coolidge never gives up until he finds just what I want."[15] Once she saw him take one of her hats out of the closet, measure the crown with a string, and slip the string into his vest pocket. A few weeks later, when they left for a summer visit in Vermont, he presented her with an elegant new hat and she was touched by his thoughtfulness. But the hat was scarcely suitable for summer in the country, and though she loyally wore it whenever she went out, she realized, to her chagrin, that "the neighbors thought I was 'high-hatting' them."[16]

As Coolidge moved up the political ladder, his salaries increased, but he continued to be frugal. While serving in the Massachusetts legislature and even after he became Lieutenant-Governor of the state, he simply rented a room in the Adams House in Boston to stay in during the week and returned to Northampton by train (milk-train, not express) every weekend to be with his family. After his election as Governor in 1918, however, his friend Frank W. Stearns, wealthy Boston Department store-owner, urged him to rent a house, move his family to Boston, and have his wife preside there in a style befitting the Governor's wife. But Coolidge continued as before; he let his family stay in Northampton and merely rented an additional room in the Adams House for his wife to stay in when she visited him. "Mr. Coolidge may be Governor of Massachusetts," Mrs. Coolidge explained, "but I shall be first of all mother of my boys."[17] Boston's elite—the Henry Cabot Lodges, for instance—scorned such pennypinching, but Mrs. Coolidge loyally defended her husband. "Although my husband has moved up, it makes no difference in our mode of living," she declared. "Why should it? We are happy, well, content. We keep our bills paid and live like everyone else."[18] She did attend her husband's installation as Governor, however, enjoyed leading the grand march at the inaugural ball, and impressed people with her poise and graciousness. But she was upset when some friends she had invited to the inaugural ceremony were denied admission because of a foul-up in arrangements. "I guess you'll have to grin and bear it," one of the inauguration officials told her. "Evidently I have to bear it—," she retorted, "but I won't grin!"[19]

As Governor's wife, Mrs. Coolidge charmed everyone when she appeared in public and was soon creating a great deal of good will for her husband. Frank Stearns was ecstatic. "One of [Coolidge's] greatest assets is Mrs. Coolidge," he told a friend. "She will make friends wherever she goes, and she will not meddle with his conduct of the office."[20] Coolidge was fully aware of the way his wife was winning friends for him and he was proud of her appearance too. "I wish you were here to see the dresses my wife has," he wrote his stepmother early in 1920. "Folks who see them know why I cannot pay very high rent."[21]

Coolidge continued as Governor to exclude his wife from his political world and never dreamed of discussing his work with her. As a result there is no way of knowing what she thought when he received nationwide attention for his peremptory statement during the Boston police strike in 1919: "There is no right to strike against the public safety by anybody, anywhere, anytime." But she followed his rise to national prominence with interest and was excited when he began to be mentioned for the Republican nomination for President in 1920. When the Republicans met in Chicago in June to pick their candidate, she was with her husband in the Adams House as he received telegrams and telephone calls keeping him posted on developments. The announcement of Harding's nomination on the tenth ballot produced momentary disappointment; Coolidge was chagrined that he hadn't done better himself in the balloting, and he got up and went out for a long walk when he heard the news. Shortly after he rejoined his wife, the phone rang and he picked it up, listened a moment, and then announced: "Nominated for Vice President." "You don't mean it?" cried his wife, thinking he was joking. "Indeed I do," he assured her. "You are not going to take it, are you?" she exclaimed. "I suppose I shall have to," he said glumly.[22] The victory of the Harding-Coolidge ticket in November left her with mixed feelings. "More hotel life, I suppose," she sighed.[23] But the New Willard Hotel in Washington proved pleasanter than Boston's Adams House.

Coolidge's elevation to the Vice Presidency in 1921 revolutionized his—and his wife's—style of living. Not only did Coolidge himself acquire high status as a substitute for the President on state occasions; his wife, too, became an important person in Washington. Mrs. Coolidge was now Second Lady of the Land; for the first time in her life she was able to share in her husband's work, appearing beside him at dinners, dedications, and ceremonial demonstrations, and taking precedence wherever she went. The Coolidges soon found themselves dining out four or five nights every week, attending luncheons, teas, benefits, balls, and receptions, and associating with politicians, diplomats, and celebrities from all over the world. To Mrs. Coolidge's surprise, her husband seemed to enjoy the busy social life, at least at first, and she facetiously told Frank Stearns that he was "becoming quite a 'social butterfly.' "[24] Coolidge almost never refused a dinner invitation; as he once explained: "Got to eat somewhere."[25] But he was strict about hours; 10:00 p.m. was his bedtime, and when it came time to leave, he always got up and announced, "Grace, we're leaving," even if she was enjoying herself. Once she stayed late at a tea party and he telephoned her. "Grace, I've come home," he declared. "You come home, too." The deadline for teas was 6:00 p.m.[26]

But for all his socializing in Washington, Coolidge remained as close-mouthed as ever. At dinners he was so reserved that people often forgot he was there until he rose to leave; one hostess said that every time he opened his mouth a moth flew out. Once, at an afternoon tea, a society woman rushed up to Mrs. Coolidge and exclaimed: "I'm so delighted. I am going to have the honor of sitting beside your husband at dinner tomorrow evening . . . ." "I'm sorry for you," laughed Mrs. Coolidge. "You'll have to do all the talking yourself."[27] She was amused by the stories about "Silent Cal" that started circulating in Washington at this time and even passed a few of them along herself. Her favorite was the one about the woman who turned to the Vice President at a dinner party and said, "Oh, Mr. Coolidge, you are so silent. But you must talk to me. I made a bet today that I could get more than two words out you." "You lose," said Coolidge. According to Mrs. Coolidge, the exchange actually took place.[28]

When the Coolidges first arrived in Washington, the Cliff Dwellers there tended to look down on them at first as rustic boors. "How do you suppose they will adapt themselves to using an automobile?" sneered one socialite.[29] But Mrs. Coolidge quickly blossomed in the new environment; at her first reception, observers noted that she "kept her wits at the end of her tongue."[30] She was, to be sure, socially inexperienced when she became Second Lady, but she had years of experience in keeping conversations going at home, and she was soon shining at the dinners and receptions she attended with her husband. It was her charm, wit, gaiety, and gift for repartee, everyone agreed, that made her husband's public appearances a success. "She gave everyone a sense of ease and enjoyment because she was so richly endowed with *joie de vivre* herself," observed Mrs. Henry W. Keyes. "I doubt if any Vice-Presidential hostess has ever wrung so much pleasure out of Washington or given so much in return. Everybody liked her, and because she went everywhere and did everything she became a familiar as well as a popular figure. She is the one women in official life of whom I have never heard a single disparaging remark in the course of nearly twenty years."[31]

Even the abrasive Alice Roosevelt Longworth praised her; she liked the way Mrs. Coolidge was "amused by all the official functions and attentions, yet was always absolutely natural and unimpressed by it all."[32] When Mrs. Coolidge returned to Northampton for a short visit in May 1921, there was a local headline: "Mrs. Coolidge, Unchanged, Finds It Good To Be Home." Immensely pleased, Coolidge wrote his father: "Grace is home as you may know from the paper. She is wonderfully popular here. I don't know what I would do without her."[33]

When Coolidge suddenly became President upon Harding's death in 1923, he thought he could "swing it," but his wife had misgivings about the role she was expected to play as the President's wife. "We all know Calvin will make it," she told Mrs. Dwight Morrow, who had sent her words of encouragement. "I have been somewhat doubtful of my own ability, but if you say I can come through I know I can."[34] She did more than come through; she came through marvelously. *"She is very nice,"* cried Chief Justice William Howard Taft shortly after she became First Lady.[35] "She's chuck plumb full of magnetism," chortled Will Rogers.[36] Her "genuine interest" in people, noted one White House guest, and "her *kindness* . . . cast a sort of glow around her wherever she went."[37] The White House staff liked her as much as the public did; housekeeper Elizabeth Jaffray found her "warm, friendly and talkative, a champion smiler," and Chief Usher "Ike" Hoover reported that members of the President's household said she was "ninety percent of the Administration."[38] When a foreign diplomat was asked whether he understood what she had said to him at a reception, he exclaimed: "That is not necessary. To look at her is gladness enough."[39] Before long people were comparing her to Dolley Madison; Will Rogers called her "Public Female Favorite No. 1."[40]

Mrs. Coolidge's schedule was always jampacked; she received callers by appointment, rode in parades, planted trees, laid cornerstones, did Red Cross work, supported fund-raising for the deaf, stood with her husband in receiving lines, and attended countless luncheons and dinners. Partly because the Federal Government, under a new law, began paying expenses for the President's official entertaining, the Coolidges had more house guests during their six years in the White House than any previous administration. Some of their guests: the Prince of Wales, the Queen of Rumania, Will Rogers, Charles A. Lindbergh, the Rockefellers, and movie stars Douglas Fairbanks and Mary Pickford. Mrs. Coolidge, as always, carried off each occasion nicely, filling in for her husband's long silences and putting guests at ease by her warmth and good humor. She was troubled, however, by the way her husband scheduled appearances for her without saying anything about them until the last minute. "Calvin, look at me," she said one morning at breakfast as her husband was submerged in his newspaper. "I find myself facing every day a large number of engagements about which I know nothing, and I wish you'd have your Secret Service prepare for me each day a list of the engagements for the coming week, so that I can follow it." "Grace," he said coming out from behind his newspaper, "we don't give out that information promiscuously."[41] After that, she adopted a policy of having coat, hat, and gloves handy at all times for last-minute calls on her time. One day, when "Ike" Hoover asked her whether she was going with the President on a certain mission, she told him: "I do not know; but I am ready."[42]

Diligent though she was about her White House responsibilities, Mrs. Coolidge continued to remain severely aloof from politics. She never discussed public issues with her husband, made no attempt to influence him on policies and appointments, gave no speeches, refused all interviews, and never talked politics with guests. "If I had manifested any particular interest in a political matter," she once reflected, "I feel sure I should have been properly put in my place."[43] Curiously, though, Coolidge intruded a great deal on what was regarded as his wife's special sphere of influence: the household. He insisted the housekeeper send menus each morning to him as well as to his wife, carefully went over White House accounts in an effort to save money, prowled around the kitchen, looked in on his wife's mail, and went over guest lists she drew up to cross off the names of people he didn't like. He also continued to lay down the law when it came to her personal life; he forbade dancing in public, driving a car, taking a plane flight, having her hair bobbed, or wearing culottes for hiking in the country. Shortly after she became First Lady, Mrs. Coolidge visited the White House stables, decided to take up horseback riding, acquired a riding outfit, and went out to Fort Meyer for her first lesson. But when reporters got wind of it and newspapers came out with headlines the next morning ("Mrs. Coolidge Takes Up Riding"), Coolidge told her at breakfast: "I think you will find that you will get along at this job fully as well if you do not try anything new."[44]

At times, Mrs. Coolidge chafed under the regimen. "Being wife to a government worker," she once mused, "is very confining."[45] She found it difficult to submerge her own identity in the persona of First Lady. "This was I, and yet not I—," she recalled years later, "this was the wife of the President of the United States and she took precedence over me; my personal likes and dislikes must be subordinated to the consideration of those things which were required of her."[46] When her husband became cranky or sullen or unreasonable or gloomy, she usually passed it off with a smile or laugh or a quip or a tease. But sometimes she became seriously upset, and took a long walk or plunged into vigorous knitting to keep herself in control. "Many a time, when I have needed to hold myself firmly," she confessed, "I have taken up my needle. It might be a sewing needle, knitting needles, or a crochet hook—whatever its form or purpose, it often proved to be as the needle of the compass, keeping me to the course."[47]

But Mrs. Coolidge never remained angry for long. She knew that her husband adored her and that in his own quiet way he tried to make his love clear to her just about every day. "I have scant patience with the man of whom his wife says, 'He never gave me a cross word in his life,'" she once confessed. "It seems to me he must be a feckless creature. If a man amounts to much in this world, he must encounter many and varied annoyances whose number mounts as his effectiveness increases. Inevitably comes a point beyond which human endurance breaks down, and an explosion is bound to follow."[48] Mrs. Coolidge's was essentially a healthy-minded temperament that refused to dwell on the darkness; there is no reason to doubt that she was in good spirits most of the time. She sang and whistled when she was at work, and won the nickname "Sunshine" from the White House staff. (Coolidge was "Smiley," but the sobriquet was sardonic.) "There is a song from one of our not new musical comedies about girls," Mrs. Coolidge once wrote, "which says something about 'the short, the fat, the lean, the tall; I don't give a rap, I love them all.' This is the way I feel about people, and I have been fortunate in being placed where I had an opportunity to gratify my taste by meeting great numbers of them."[49]

Coolidge's—and his wife's—enormous popularity produced a triumphant re-election as President in 1924 and considerable speculation in political circles about his intentions in 1928. But for a long time mum was the word with the taciturn Vermonter. When Will Rogers visited the White House in May 1927, he said to Mrs. Coolidge in a loud stage whisper at one point, "I wish you would tell me if the President is going to run again," and she told him, "You find out if you can, and let me know."[50] Then, suddenly, in the summer of 1927, her husband issued a terse statement, "I do not choose to run for President in 1928," from his summer place in the Black Hills of South Dakota that took everyone unawares. "That was quite a surprise the President gave us this morning," Kansas Senator Arthur Capper remarked to Mrs. Coolidge at lunch that day. When she asked what he meant and he told her, she exclaimed: "Isn't that just like the man? He never gave me the slightest intimation of his intention. I had no idea!"[51] Later, though, she confessed: "I am rather proud of the fact that after nearly a quarter of a century of marriage, my husband feels free to make his decisions without consulting me or giving me advance information concerning them."[52] But Mrs. Coolidge wasn't the only one left out; her husband didn't even give his close friend Frank Stearns any inkling of what he planned to do. "He and I," said Mrs. Coolidge of Stearns, "were in the same class as recipients of Presidential confidence. Many a time we have put our heads together and figured out that two and two made four only to learn later that we had been adding the wrong numbers."[53]

Mrs. Coolidge had no regrets about leaving the White House. She was never bowled over by the glamor of being First Lady and she was looking forward to being a private person again. She was worried, moreover, by her husband's health and thought he was ready for a quiet life. "Please treat me rough when I get home," she wrote a Northampton friend, "and kick me about a bit so I'll realize I'm human."[54] Asked about her husband's plans for retirement, she admitted, "I haven't any idea what he has in mind." But she added: "I always did like the unexpected and am waiting with intense interest the next jump."[55]

But there were no more "jumps" for the Coolidges when they returned to Northampton after attending Herbert Hoover's inauguration in 1929. Coolidge took up his law practice again, did articles for various magazines on governmental problems, published his autobiography, and wrote a daily syndicated newspaper column for a time, while his wife resumed her housework, continued her Red Cross activities, and served as chairman of the board of trustees for the Clarke School for the Deaf. She wrote poetry, too, and one of her poems, dedicated to the memory of Calvin, Jr., who had died of blood poisoning at sixteen during the White House years, appeared in *Good Housekeeping* in 1929. Some years before, when Boston University awarded her a honorary LL.D., she said it should have been a D.D.: "Doctor of Domesticity."[56] In 1931, an advertising agency offered Coolidge $75,000 for the use of her name, but he turned the offer down flatly: "She wouldn't like it."[57] Late in 1932, he wrote her his last love letter while in New York on business: "I have thought of you all the time since I left home."[58] In January 1933, he died of coronary thrombosis at sixty-one.

"I am just a lost soul," Mrs. Coolidge confided to a friend after her husband's death. "Nobody is going to be believe how I miss being told what to do. My father always told me what I had to do. Then Calvin told me what I had to do."[59] But her life as a widow changed little and rumors of a remarriage were unfounded. She helped raise an endowment for the Clarke School, traveled to Europe, kept in touch with son John and his family, and followed the World Series—she was known as No. 1 fan of the Boston Red Sox—with excitement. When she was asked about political views upon her return to Europe in August 1936, she declared: "Well, we all are interested in politics—we should be. But I am not actively interested."[60] She did join the interventionist organization, the Fight for Freedom Committee, however, just before Pearl Harbor, and after the United States entered World War II, sponsored war-bond sales and campaigned for the salvage of fats, metals, and waste paper. After the war, she broke her rule about interviews on one occasion and talked briefly to a reporter about herself. "I was born with peace of mind," she declared. "It is a matter of inheritance, training, and experience."[61] When her health began failing in the early 1950s and she had to stop going to Boston for her beloved baseball games, she was delighted when the President of the American League sent her a bouquet of flowers to console her for having to miss the World Series.

On July 9, 1956, Mrs. Coolidge died of heart failure at seventy-eight and was buried beside her husband in Plymouth. He had already paid tribute to her in his *Autobiography* (1929): "For almost a quarter of century she has borne with my infirmities and I have rejoiced in her graces."[62]

✲ ✲ ✲ ✲

Missing

Grace Goodhue soon learned that economy was one of Coolidge's watchwords. Early in her acquaintance with him, he expressed the desire to go on a picnic which she and some of her friends at the Clarke School were planning. After consulting her friends, she told him he could go along if he provided the lunch. He readily agreed, and brought chicken sandwiches, strawberry shortcake, and a dozen macaroons for the outing. At the picnic, the little group quickly devoured the sandwiches and shortcake and then moved in on the macaroons. After the meal, Coolidge began picking up the dishes and getting the leftovers together. Then he began counting the macaroons, and, to the amazement of the picnickers, insisted that each person tell him how many macaroons he or she had eaten. Half a macaroon, he explained, was missing. And "for it," Mrs. Coolidge remembered, "no account has ever been made."[63]

Human

In the summer of 1905, when Coolidge was courting Grace Goodhue, he went to Burlington for a visit and she invited him to accompany her on a visit to a friend of hers who lived a few miles out of town. Coolidge reluctantly agreed to go; he wasn't very sociable. But he slicked himself up for the occasion; he wore a dark-blue serge suit, new and perfectly tailored, a black derby hat, and patent-leather oxford shoes tied with the widest and shiniest pair of silk shoelaces Grace had ever seen. Just before they got into the carriage, he put a whisk broom in the back. Arriving at the friend's place, he assisted Grace from the carriage to the doorstep and then drove the horse into the back and hitched him to a large ring in the corner of the barn. Then he took the whisk broom from the back of the buggy, brushed the dust and horsehairs from his clothing, and rejoined Grace. Every move, his wife recalled, was so deliberate that to her he was like "a small boy performing some hated task because his mother has asked him to do it as a favor to her."

Once inside the house and introduced to the friend, Coolidge sat down on the very edge of a large sofa in the parlor and lapsed into silence. The friend was normally a talkative person, but the conversation on this occasion was sporadic and halting, and the man on the sofa remained severely aloof. Suddenly he got up, smiled, and announced: "We'll be going now." Then he went out to get the horse. "My land, Grace," cried the friend, "I'd be afraid of him!" As they drove home, young Grace chided her fiancé. "Now why did you act like that?" she wanted to know. "She thinks that you are a perfect stick and said she'd be afraid of you." Declared Coolidge imperturbably: "She'll find I'm human."[64]

Education

Sometimes Mrs. Coolidge got the impression her husband didn't think much of her education. Shortly after the wedding, he asked her one evening out of the blue when Martin Luther was born. When she said she had no idea whatsoever, he cried: "Didn't they teach you anything where you went to school?"[65]

Governor's Chair

In 1904, Northampton celebrated its 250th anniversary, and one evening was devoted to a reception for the Governor and his Council, given by the Daughters of the American Revolution. Grace Goodhue accompanied Coolidge to the City Hall, where the reception was held, and after strolling around a bit, they sat down in two comfortable vacant chairs. They hadn't been there long before a charming hostess approached them, said the chairs were reserved for Governor and his wife, and persuaded them to leave. Fourteen years later, when Coolidge was elected Governor of Massachusetts, he turned to his wife and cried: "Well, Grace, even the Daughters of the American Revolution can't put us out of the Governor's chair now."[66]

Piano

One June, when Coolidge was Mayor of Northampton, his wife took a group of Northampton high school girls on a sight-seeing tour of Washington that included the White House. When the group entered the East Room, Mrs. Coolidge was so taken by the gorgeous gilded piano the Roosevelts had installed there that she stepped over to touch it and was roughly pushed away by the guide. "Some day," she said afterward, "I'll come back here, open that piano and play it, and he won't order me out." Years later, when she was First Lady, she recalled the episode and got in the habit of giving the piano a little kick whenever she needed to let off steam.[67]

Junior

The first Coolidge boy, John, was born on September 7, 1906, and the second on April 13, 1908. For the second boy, the Coolidges hadn't agreed on a name, and, since it was Easter-time, Mrs. Coolidge temporarily called him "Bunny." But she kept urging her husband to decide on a name and he kept putting it off. Finally she told him the time had come to pick a name. "Well, my dear, I agree with you," he said, "but I thought that before we called him Calvin, I'd see if he knows anything." So Calvin, Jr., it was.[68]

Musical Speech

When Coolidge was Lieutenant-Governor of Massachusetts, he had occasion to give a talk at a dinner in Boston celebrating the 100th anniversary of the Chickering Piano Company. Mrs. Coolidge attended the dinner, for she loved music, and was amazed to hear her quite unmusical husband talk knowledgeably about the great composers and their masterworks. After the dinner, when she laughingly asked him where he had acquired all the musical information, he looked embarrassed and quickly changed the subject. Mrs. Coolidge decided the Chickering speech was the only ghostwritten speech her husband ever delivered. Years later, she observed, he omitted it from a collection of speeches he approved for publication.[69]

Vice President

One day, when Mrs. Coolidge was to join her husband for a conference in Washington, she got into such a last-minute rush that she arrived at Continental Hall without her ticket. When she announced she was Mrs. Coolidge, the doorman looked blank. "What's your husband's first name?" he asked. "Calvin," she said. "What's his business?" asked the doorman doubtfully. "He's Vice President." "Vice President of what?" he wanted to know. But he finally let her in.[70]

Like Abigail

Mrs. Coolidge knew that supercilious Bostonians like the Lodges were disdainful of her husband's parsimony and that they had low expectations for her as First Lady. "I don't know what they expect me to do," she told a friend soon after entering the White House. "Hang my wash in the East Room, like Abigail Adams?"[71]

Coolidge Whistle

Mrs. Coolidge could whistle very well, but her husband couldn't. When he wanted the dogs to come, he blew a little whistle he had acquired and, according to "Ike" Hoover, blew it "like a locomotive." Once when he had mislaid the whistle and was trying to whistle for the dogs himself, his wife cried: "What's the matter, poppa, don't your teeth fit tonight?"[72]

Speech

When she was in the White House, Mrs. Coolidge tried hard to avoid publicity. "It has been my unbroken policy not to see newspaper writers or give interviews to anyone," she told a friend. "At the word interview spoken or written, my ears go up and my chin out." But once, when she was giving a luncheon for some newswomen under the giant magnolia tree Andrew Jackson had planted in memory of his wife Rachel, someone asked for a speech. Without a protest, she simply raised her hands and went on to give a five-minute talk in the sign language she had learned at the Clarke School. Her guests were delighted.[73]

Handkerchief

At one of Mrs. Coolidge's tea parties in the Red Room, one of the Cabinet wives had a slight cold and periodically took a handkerchief out of her purse to use. But while she was telling a story, she absent-mindedly put one of the tea napkins into her bag. "You may as well bring it right out into the open—," Mrs. Coolidge teased her. "I saw you putting it in your bag!" The woman shamefacedly held up the White House napkin. But, to Mrs. Coolidge's chagrin, it was full of holes, "to cry shame on my vigilance as a housekeeper."[74]

Exhausted

One evening, when the Coolidges and some friends were taking a cruise on the *Mayflower,* Mrs. Dwight D. Morrow and Mrs. Frank B. Kellogg sat next to the President at dinner when he was in one of his laconic moods and they simply couldn't get him to say anything. The next morning when Mrs. Coolidge joined him in the dining saloon for breakfast he queried: "And where are my fair ladies?" "Exhausted by your conversation of last evening!" she told him.[75]

Contrast

When Howard Chandler Christy arrived at the White House to do a portrait of Mrs. Coolidge with Rob Roy, her white collie, beside her, he asked to wear a red gown for color contrast. But her husband objected; he wanted her to wear a white brocaded satin gown which he liked very much. "If it's contrast you want," he told her, "why not wear white and paint the dog red?"[76]

Open Door

In July 1924, five years after the death of Calvin Jr., Mrs. Coolidge wrote a poem, "The Open Door," eventually published in *Good Housekeeping*, which revealed her deep religious bent.

You, my son
Have shown me God.
Your kiss upon my cheek
Has made me feel the gentle touch
Of Him who leads us on.
The memory of your smile, when young,
Reveals His face,
As mellowing years come on apace.
And when you went before,
You left the gates of Heaven ajar,
That I might glimpse,
Approaching from afar,
The glories of His grace.
Hold, son, my hand,
Guide me along the path,
That, coming,
I may stumble not,
Nor roam,
Nor fail to show the way
Which leads us home.[77]

Mrs. Coolidge's Hike

In June 1927, Coolidge established his summer White House in the Black Hills of South Dakota, with his office in a schoolhouse in Rapid City and his residence in a game lodge eight or ten miles farther up in the hills. Shortly after settling in there, Mrs. Coolidge set out for a morning's hike with James Haley, young secret serviceman, as guide. Taking a downhill trail, she walked with a quick, long stride, and swiftly put distance behind her. But, as was her habit, she also stopped frequently to pick wild flowers. At one o'clock, her punctual husband arrived home for lunch and found she had not returned. When he learned she had left shortly after nine that morning, he began pacing the porch nervously, and as time passed became increasingly upset. At 2:15, just as he was about to organize a searching party, Mrs. Coolidge, quite fatigued from her long uphill return, appeared and cried gaily: "Hello, papa! Sorry to keep you waiting!" But Coolidge was not in any cheerful mood, and he took her at once into the house. No one knows what he said to her, but there was much speculation. The *Boston Herald* had a headline: "WIFE'S LONG HIKE VEXES COOLIDGE; PRESIDENT PACES PORCH AS FIRST LADY HITS 15-MILE TRAIL." So did the *Boston Post:* "FIRST LADY ALMOST LOST; PRESIDENT WORRIED, ON POINT OF FORMING SEARCH PARTY JUST AS MRS. COOLIDGE RETURNS. The *Boston Globe* had this to say: "WIFE'S DELAY TAXES COOLIDGE'S PATIENCE. SHE GOES OFF ON LONG HIKE AND LUNCHEON GETS COLD. PRESIDENT SITS ON PORCH AN HOUR WAITING FOR HER TO EXPLAIN." A few days later the *Globe* had another headline: "HALEY OUT AS MRS. COOLIDGE'S ESCORT."[78]

President's Widow

In 1936, Mrs. Coolidge took a trip to Europe with her friend Mrs. Florence B. Adams and was unrecognized until she reached Switzerland. Mrs. Adams had written ahead for reservations in a hotel in the little Swiss town they were visiting and when she and Mrs. Coolidge arrived, the hotel people, knowing the widow of a former American President was going to stay here, greeted them with great deference. "Mrs. Adams," said one of the clerks, as Mrs. Coolidge went off to look at their rooms, "would you sign for yourself and Mrs. Lincoln?"[79]

Romance

The Coolidges lived happily together for twenty-eight years, but when Mrs. Coolidge was asked, toward the end of her life, how she had come to marry her husband, she said: "Well, I thought I would get him to enjoy life and have fun but he was not very easy to instruct in that way." Once a reporter asked her about their early "romance," and she looked at him, smiled, and said facetiously: "Have you ever met my husband?"[80]

CHAPTER 29

LOU HENRY HOOVER

1874–1944

Lou Henry Hoover once told an interviewer that she majored in geology at college, but had "majored in Herbert Hoover" ever since.[1] Majoring in Hoover meant, from the outset, placing her own considerable organizing talents at his service whenever he needed them. In China during the Boxer Rebellion, she mobilized the women in Tientsin's foreign settlement for defense while he worked with the men; in London, at the outbreak of World War I, she joined him in caring for American tourists stranded in Europe, and she did her part for Belgian Relief as well; and in Washington, she worked closely with him when he became Food Administrator for the Wilson administration and then Secretary of Commerce for Harding and Coolidge. Her "controlling impulse," noted one observer when she was a Cabinet wife, "is to help and second her husband in every small or big thing in life."[2] "Mrs. Hoover," declared a *New York Evening Post* writer shortly after Hoover's election as President in 1928, "has been a feminine counterpart of her husband, his mental, physical, and spiritual equal through their lives."[3]

The collaboration began early. The Hoovers met as undergraduates at Stanford University in 1894, when Lou, the daughter of a Monterey banker, was a freshman, and Herbert, an orphan, was a senior, and quickly discovered they had a lot in common. They were both born in Iowa in 1874, grew up with a fondness for hiking and camping in the great outdoors, came to California about the same time, and picked geology as their major field of interest at Stanford. They first met, appropriately enough, in Professor John Caspar Branner's geology laboratory one afternoon, when Branner was showing Lou some rock specimens, with his assistant, the shy, boyish-looking Herb, standing nearby. "Hoover brought them in from the field," Branner told her. "They've been called carboniferous but I'll eat my hat if they aren't precarboniferous." Then he turned to his prize student and asked: "Isn't that your opinion, Hoover?" Hoover blushed and remained silent; but a little later, when Branner left the room, he got to talking with Lou, and the two were soon friends.[4] Lou's sorority sisters teased her when she started seeing Hoover regularly, for he was a Quaker and opposed to fraternities as undemocratic. But she didn't mind making friends with a "barbarian," as non-Greeks were called, and she was soon joining him on field trips and going to campus parties with him, too, though he wasn't a particularly good dancer. On one outing, when he started to help her over a fence, she astonished him by leaping over it herself. She was a bit of a tomboy and better at sports—baseball, basketball, skating, archery—than Hoover was. He preferred fishing, and later, medicine ball. Hoover went riding with her but lacked her love of horses. She was an excellent horsewoman.

In May 1895, Hoover graduated from Stanford with a degree in geology and mining engineering and went off to work in various mines in the Southwest before securing a position with a London firm to develop a gold mine in Australia. Before leaving the country in October 1897, he and Lou, now a junior, became informally engaged; and after he left, she told her sorority sisters she planned to get married when her fiancé had a job that would keep him in one place long enough to make a home. But Hoover never did settle down for long. He did so well at developing efficient methods for mining gold in Australia that late in 1898, he was offered the position of chief engineer for the Chinese Engineering and Mining Company, with headquarters in Peking. Overjoyed by the boost in his fortunes, Hoover sent Lou a long cablegram from Australia proposing marriage as soon as he got home and a honeymoon en route to China. She at once cabled her acceptance; she had graduated from college by this time and was teaching in public school in Monterey; and she was filled with excitement by the thought of adventures abroad. She started reading up on China.

Lou, originally an Episcopalian, wanted a Quaker wedding, for she had adopted Hoover's faith. But there was no Friends Meeting in Monterey; there was no Protestant minister in town either and no time to locate one. In the end, a Catholic priest whom Lou had befriended agreed to perform the ceremony with the permission of his bishop. The wedding took place on February 10, 1899, shortly after Hoover returned from abroad, and the following morning the Hoovers took the train to San Francisco and then boarded a ship for China, with a pile of books about the Celestial Empire to study on the passage to China.

The Hoovers reached Tientsin in March 1899, rented a house in the foreign compound there, and began planning expeditions into the countryside in a search for coal and other minerals. Mrs. Hoover worked closely with her husband: mapped the geology of the regions he planned to visit, translated material about Chinese mining available in European sources, catalogued the mining laws of the world for him to give Chinese officials, and helped write her husband's reports. "We are not working just for Mr. Hoover," one of the young engineers on his staff observed, "but for Mrs. Hoover too."[5] She went on expeditions with her husband, whenever she could, by canal boat, pack mule, and on horseback, and prepared for the trips by reading up on the history and mythology of the places they were to visit. When expeditions were too difficult for her to accompany, she remained in Tientsin studying Chinese with a native teacher and in time became fairly proficient at the language. Hoover himself mastered only about a hundred words, but ever afterward, he recalled, his wife "kept that hundred words in use between us by speaking Chinese to me on *sotto voce* occasions."[6] In her spare time, Mrs. Hoover also began collecting ancient Chinese porcelain.

The Hoovers blossomed in China. Hoover succeeded in locating valuable coal deposits in northwestern China and some gold mines, too, and impressed all his associates with his energy, skill, diligence, and resourcefulness. Despite their Quaker faith, the Hoovers enjoyed the high status they held in both Chinese and foreign circles, and became accustomed to entertaining American and foreign diplomats as well as young engineers on Hoover's staff in their Tientsin residence. From the outset they employed what Hoover called "the necessary multitude of Chinese servants," including houseboy, maid, cook, gardener, and rickshaw boys to deliver messages for them and to haul them around town.[7] The Hoovers were bothered by the "squeeze," that is, the way servants (and Chinese officials too) padded expense accounts and threw in extra charges for non-existent services. That the "squeeze" was a way of compensating for dirt-cheap wages—the Hoovers paid $108 a month for ten servants—seems never to have occurred to them. But they found it pleasant having flocks of underlings to command and were amused by their supposed comicalities. Sometimes, when Mrs. Hoover told the cook simply to make omelets for dinner, he had a way of crying, "You catch 'em," and then serving the usual four- or five-course dinners anyway.[8]

In June 1900, the Boxer Rebellion—an uprising of fervent Chinese patriots against foreigners in China—shattered the Hoovers' routine. In Tientsin, as well as in Peking, the Boxers began attacking and killing both foreigners and Chinese Christians and it was necessary for the foreign communities in those cities to organize for defense until Western troops arrived to put down the revolt. During the four-week siege of Tientsin, Hoover helped erect barricades around the foreign compound and organized all the able-bodied men into a protective force to man the barricades. Mrs. Hoover went to work too: helped convert the foreign colony's club into a hospital, took her turn nursing the wounded, rationed food and planned meals, and served tea every afternoon to those on sentry duty. She made her daily rounds by bicycle, keeping as close to the walls of the buildings as she could to escape bullets; but once a bullet punctured one of the tires of her bicycle and brought her to the ground. Like her husband, though, she remained calm and efficient throughout the crisis and even seemed to enjoy the excitement. One afternoon, when she sat home playing solitaire to rest up from her work at the hospital, a shell suddenly burst nearby and she ran to the back door to see what had happened and found a big hole in the backyard. A little later a second shell hit the road in front of the house and she went out to see what damage it had done there. Then came a third shell; this one burst through one of the windows of the house and demolished a post by the staircase. Several reporters covering the siege rushed into the living room to see if she was all right and found her back at the card table. "I don't seem to be winning this game," she remarked coolly as they came in, "but that was the third shell and therefore the last one for the present anyway. Their pattern is three in a row." Then she got up and suggested: "Let's go and have tea."[9]

With the arrival of Western troops and the lifting of the siege in August, Hoover instructed his wife to leave at once for Japan while he wound up his work in China. "You will go out," he told her. "I will not," she said. "You will!" he cried. "I will not go until you go!" she persisted. "Hum-m-m," he grumbled, gave up, and started to leave. "All right, Bertie," she called after him with a smile. "All right, Lou," he returned.[10] There were "two chins in the Hoover family," Hoover's engineers liked to say.[11] A few days after the crisis was over, Mrs. Hoover was amused to read an obituary about her in a Peking newspaper. "There were three columns of it, too!" she told friends in the United States. "I was never so proud in my life."[12]

Mrs. Hoover was even prouder when her husband was offered a junior partnership in a British mining firm with interests all over the world and took her to London to live in November 1901. To oversee the company's far-flung mining operations, Hoover traveled far and wide—to China, Burma, India, Egypt, Australia, Western Europe—and took his wife with him wherever he went, on passenger lines, tramp steamers, tugboats, trains, cars, buggies, horses, and camels. When the babies arrived—Herbert, Jr., in 1903, and Allan Henry in 1907—he took them along, too, in baskets, even though they were only a few weeks old on their first trips. In 1908, Hoover struck out on his own; he set himself up as an independent consulting mining engineer, with headquarters in New York and San Francisco as well as London, and was soon swamped with assignments. The journeys abroad—to supervise the finances and operations of some twenty mining companies engaging his services—continued; and the Hoovers soon amassed a great fortune. In their elegant mansion in London, Mrs. Hoover got in the habit of entertaining her husband's associates on a grand scale; she also extended her hospitality to British and American officials, foreign dignitaries, and celebrities like Joseph Conrad and H. G. Wells as well. For the fun of it, the Hoovers spent their spare time translating a famous Latin treatise on mining and metallurgy, Agricola's *De Re Metallica* (1556), with Mrs. Hoover handling the Latin and her husband tracking down the scientific references. The translation was published in 1912 in a limited, deluxe edition and met a warm reception among classical scholars as well as professional scientists. Mrs. Hoover also helped her husband out with his textbook, *The Principles of Mining* (1909), which grew out of a series of lectures he gave at Stanford and Columbia. She was particularly insistent that he include a chapter on character-building for young engineers and the importance of professional ethics.

In the fall of 1914, the Hoovers booked passage to the United States on the *Lusitania,* but when they called a few days later to confirm their reservation, the steamship clerk exclaimed: "Don't you know there is a war on?"[13] The outbreak of World War I transformed the Hoovers' life; it took them out of the world of business and placed them squarely in the public arena. Hoover himself was soon devoting all his energies to taking care of the thousands of American tourists stranded in Europe with the coming of war, and Mrs. Hoover toiled diligently at his side. While Hoover dug up money and credit for the tourists and booked passage for their return to the United States as soon as he could, Mrs. Hoover handled the women and children; she saw to their food and lodgings, and, while they remained in London, arranged sight-seeing tours for them in the city and trips to "cathedral towns" and to Shakespeare and *Lorna Doone* country. One elderly woman for whom they succeeded in finding space on a ship bound for America demanded a written guarantee that it wouldn't be sunk by submarines, so they obligingly gave her one.

Hoover was soon involved in Belgian Relief and Mrs. Hoover threw herself energetically into that work too. Not only did she help out at the Belgian Relief headquarters in London and visit Belgium with her husband to ascertain its needs; she also returned to the United States to make speeches in San Francisco and Los Angeles to raise money and supplies for the cause. She succeeded, too, in persuading the Rockefeller Foundation to provide free shipping for American food and clothing destined for Belgium. "She is the most capable woman alive," exclaimed a Los Angeles schoolteacher after hearing one of Mrs. Hoover's Belgian Relief speeches.[14] But after the war, when Belgium's King Leopold presented her with the Cross Chevalier, Order of Leopold, she exclaimed: "What is there to say? I have done nothing extraordinary, nor anything more than a woman should do for the man she loves. I have been deeply interested in Mr. Hoover's work and have tried to be of whatever assistance I could. My chief hobbies are my husband and our sons."[15]

When Mrs. Hoover returned to London after several weeks of speech-making on behalf of Belgian Relief in the United States, she plunged into work for the British as well. She helped organize the American Women's Hospital at Paignton, to care for wounded soldiers, and became active in the Women's Committee for Economic Relief, which, among other things, established knitting and sewing factories to employ British women thrown out of work by the war to make clothes for British servicemen. "Mrs. Hoover has a real Yankee temperament and personality and will stand no nonsense . . . ," chortled the *Washington Post* in a series of articles about her work appearing in February 1915. "When she takes a thing in hand it is as good as done."[16] But as in China during the Boxer Rebellion, Mrs. Hoover was thrilled by the excitement of life in wartime London. One night, after the Germans started dropping bombs on London from Zeppelins, a bomb landed with a great explosion near the Hoover house and Mrs. Hoover and her husband rushed up to the boys' room only to find the beds empty. After a frantic search, they finally located the lads on the roof of the house watching with fascination the searchlights scanning the sky and the dogfights in the air. The Hoovers joined the boys and they all watched a Zeppelin being brought down in flames north of London.

When the United States entered World War I in April 1917, President Wilson picked Herbert Hoover as the logical man to head the Food Administration, set up to conserve food for the war effort; and in May, the Hoovers transferred their war activities from London to Washington. As Food Administrator, Hoover devised various strategies for encouraging the voluntary conservation of food and fuel at home so the United States could provide American and Allied troops with what was needed on the Western front. Mrs. Hoover pitched in to help with her usual energy. She welcomed members of her husband's staff for mealtime conferences at her home and sat in on the discussions; she also organized the women who came to Washington to work for her husband into the Food Administration Girls' Club and saw to it that members secured a dormitory to live in and one of the best cafeterias in town in which to eat. In speeches and interviews, moreover, she helped her husband enlist housewives as well as restaurants and food suppliers in the great conservation effort. Before long, the word, Hooverize, came into circulation; it meant to save on food by cutting down on the use of wheat, meat, and sugar and carefully cleaning one's plate at mealtimes. In her own home, Mrs. Hoover "hooverized" with enthusiasm; she stopped serving beef and pork, substituted corn and rye for wheat, and used honey and molasses instead of sugar. She also made up wheatless and meatless recipes for American housewives to try out. And when she began her long association with the Girl Scouts in 1918, she placed special emphasis on the cultivation of war gardens by the youthful members of the young organization.

When the war ended in November 1918, Hoover made plans to resume his engineering business but was soon called back to public service again, to administer relief and economic reconstruction for the war-devastated nations of Europe. To Mrs. Hoover's disappointment, there were fewer opportunities for her to get involved in her husband's work as "Food Regulator of the World" after the war; and she remained in the United States most of the time while her husband was at work in Europe. She did, though, help arrange special banquets—with plain food, tin utensils, and an empty high chair for the Invisible Guest, a hungry child—to raise money for European relief. At a fund-raising dinner in New York City in December 1920, she succeeded in persuading a thousand guests to pay $1,000 a plate for the cause. For the rest, she concentrated on drawing up plans for a hilltop residence for the family on the Stanford campus, which she designed as a kind of "Hopi house" with flat roofs and wide terraces.

In March 1921, when Hoover became Secretary of Commerce in the Harding Cabinet, Mrs. Hoover cheerfully abandoned work on her dream house in Palo Alto and went to work developing an old colonial house in Washington into a splendid residence where she could extend hospitality to her husband's associates and to friends they had made all over the world who visited the nation's capital. The Hoovers had done a lot of entertaining ever since their Tientsin days, but in Washington they fairly outdid themselves. There were guests for breakfast as well as for lunch and dinner at the Hoover place on S Street. For Hoover, meals were in part occasions for getting to know his colleagues better and even transacting business; for Mrs. Hoover, they were part of her tireless effort to make her extremely reserved husband more of a social creature. She never succeeded in socializing Hoover; but for all his reticence, he seemed to like being surrounded by people at mealtimes. An apostle of efficiency, he was proud, too, of the adroit way his wife met her responsibilities as a Cabinet wife. One evening, two Senators and their wives arrived at the Hoover mansion while the Hoovers were at the dinner table with guests. "You know," Hoover whispered to his wife, "I invited them this morning and forgot to tell you." Mrs. Hoover laughed, sent the original guests into the library, and told her husband, "You meet the Senators and keep them busy for a while." Then she had the cook raid the ice-box and a little later presided calmly over an entirely new dinner for the enlarged group.[17]

For all the entertaining, the Hoovers, as Quakers, refused to take society in Washington seriously; they rebelled, too, at some of the rules governing social intercourse there. The wives of high Washington officials were expected to call in person on each other during the social season or leave their cards; but Mrs. Hoover refused to spend four or five afternoons a week in what she regarded as wasteful activity and soon persuaded the other Cabinet wives to abandon the old custom. Instead of circulating in prestigious social circles, moreover, she spent her energies furthering the work of the new women's organizations appearing on the landscape: the Girl Scouts, the League of Women Voters, the Campfire Girls, the General Federation of Women's Clubs. "I believe that even after marriage it is possible for a woman to have a career," she announced in a speech in St. Louis in 1926 at a Girl Scout conference. When asked what she thought of the woman who spent all her time taking care of her home, she exclaimed: "I think she is lazy. The modern home is so small there is little work to do. The baby? It isn't a baby for long. There is no reason why a girl should get rusty in her profession during the five or six years she is caring for a small child."[18] People tend to universalize their own individual experiences when they generalize about things, and Hoover was doubtless thinking about her own life when she spoke to the Girl Scout leaders in St. Louis.

In speeches to women's groups, Mrs. Hoover urged involvement in politics as well as in careers of their own along with homemaking. "Women should get into politics," she told a meeting of Republican women in Philadelphia in May 1923. "They should take a more active part in civic affairs, give up some of their time devoted to pleasure for their duty as citizens. Whether we are wanted in politics or not, we are here to stay and the only force that can put us out is that which gave us the vote. The vote itself is not a perfect utility. It is perfected in the way it is used."[19] When Hoover ran for President in 1928, however, she was reluctant at first to accompany him on campaign trips. But Hoover was insistent. "I need you there," he told her. "And who else should receive all the bouquets?" In the end she went along with him and even made brief speeches at times from the rear platform of his campaign train. After being presented with some flowers by a group of well-wishers in Palo Alto, she announced gaily: "I enjoy campaigning because my husband makes the speeches and I receive the roses!"[20] After Hoover's victory in November, she accompanied him on a goodwill tour of South America during which her facility in Spanish made friends for them wherever they went.

With her husband's inauguration as President in March 1929, Mrs. Hoover took her superabundance of energy, enterprise, and efficiency to the White House as the nation's foremost chatelaine. She filled the building with art objects from all over the world, transformed two rooms into historic places filled with furniture from the Monroe and Lincoln periods, and planned Camp Raridan, a little village of log cabins in the

Virginia hills, as a summer retreat for the President and his associates, which, her son Allan remarked, called for "roughing it in perfect comfort."[21] She continued, nay, stepped up, the pace of entertaining she had taken up as a Cabinet wife. There were guests for all the meals; and there were teas, too, sometimes two of them the same afternoon. Asked to characterize the Hoover administration, Ava Long, the new White House housekeeper, once exclaimed: "Company, company, company!"[22] At the end of their first three years in the White House, the Hoovers had dined alone only three times, each time on the occasion of their wedding anniversary.

The entertainments were lavish. Mrs. Hoover insisted on "the best of everything" regardless of cost, and her husband cheerfully picked up the check for any excess of the amount authorized by Congress for official entertaining.[23] "They set the best table that was ever set in the White House," according to Chief Usher "Ike" Hoover.[24] But the White House staff found that the proclivity of the Hoovers to add last-minute guests—sometimes scores of them—to their luncheon and dinner lists kept them in a state of continual crisis. At one big party, five hundred people turned up instead of the two hundred planned for; and the White House servants frantically cleaned out all the grocery stores in the neighborhood (and also unpacked picnic boxes filled with goodies for a trip to Virginia) in order to get enough food together for the crowd. One morning, at 11:00 a.m., housekeeper Ava Long went shopping downtown, with the understanding that there were to be only four extra people for lunch; when she got back at noon, Mrs. Hoover notified her there would be forty people for lunch at one. Mrs. Long at once had an emergency conference with Katherine, the cook, and in record time the latter assembled all the leftovers in the ice-box and managed to produce enough croquettes for all the guests by one o'clock. Afterward, a foreign dignitary asked for the recipe and Ava Long picked the name, "White House Surprise Supreme" for the concoction. "Only with the help of a perfectly trained and good-natured staff," she wrote after leaving the White House, "was it possible to make the domestic wheels of the White House move smoothly and silently."[25]

The Hoovers wanted the White House wheels to move as unobtrusively as possible. People on their staff, in fact, felt at times that the President and his wife would even have preferred invisibility. Hoover kept such a distance from the servants—he hardly ever looked at them and almost never spoke to them—that they referred to him among themselves as "His Majesty."[26] Mrs. Hoover was perforce less remote, but in her own way more imperious. She developed signals as a substitute for words to keep the wheels moving; putting her hand to her hair meant the servants were to announce dinner; touching her glasses indicated they were to clear the table. She also insisted that the butlers stand in perfect silence at meals and refrain from showing any interest in the table conversation. "After each service," reported Alonzo Fields, the head butler, "I took my position and stood at attention. To smile at some joking remark, overheard by chance, would mean the last time I would go into the dining room. I approached the table to change services in a dignified, alive manner, but never rushing or hurried. There could be no scraping of plates. The tinkle of silver on china would invite an immediate report to the head man."[27]

Hoover's shyness with people not in his own class probably explains his inability to communicate with the White House help; and his wife's passion for efficiency, which she shared with her husband, surely had something to do with the severe regime she imposed on all the servants. The Hoovers accepted the Quaker doctrine of spiritual equality all right; and were devoted to American democratic ideals too. But they had absorbed more of the imperial outlook of the European upper classes from their experiences in China and Britain than they realized. Mrs. Hoover was, to be sure, concerned with the welfare of the thirty-two servants she commanded in the White House; she gave them presents, looked after their health, and even paid the college bills of one of the maids. But *noblesse oblige* didn't quite make up for *hauteur*. "I found it much easier to admire the Hoovers than to like them," confessed Ava Long. "Finer people never lived. But the President and Mrs. Hoover rarely broke through the barrier between those who serve and those who are served."[28]

The Great Crash of 1929 and the Great Depression that followed killed the Hoover Presidency. Mrs. Hoover felt keenly the mounting criticism of her husband as the economy went into a tailspin; and, as always, supported his efforts to meet the newest crisis with heartfelt fervor. She began making her entertainments simpler; she wore cotton dresses in public to encourage the domestic textile industry; and in speeches to women's groups she echoed her husband's views: that the Depression originated in European, not American, conditions after World War I, and that the proper response was for people to tighten their belts, continue working hard, and carry on with "activities that are essential" until the economy righted itself.[29] She also urged American women to increase their efforts to help the needy. "My plea is that our most important duty is to find when, how, and where people need help," she announced in a radio broadcast from the White House in November 1931. "The winter is upon us. We cannot be warm, in the house or out, we cannot sit down to a table sufficiently supplied with food, if we do not know that it is possible for every child, woman, and man in the United States also to be sufficiently warmed and fed."[30] She made generous donations herself to charitable organizations.

In 1932, Mrs. Hoover accompanied her husband again on a tour of the country in his bid for re-election and once more made little speeches for him from the rear platform of his train. She was devastated by

Franklin D. Roosevelt's victory in November and, like her husband, was convinced that the Democrats had grievously misled the American voters in what she thought was an unfair campaign against the President. Just before FDR's inauguration, however, she invited Eleanor Roosevelt to the White House and graciously showed her around the place. But when Mrs. Roosevelt asked to see the kitchens, she stopped, drew herself up, and exclaimed: "I'm sorry, but the housekeeper will have to show you the kitchens. *I* never go into the kitchens."[31] On her last day in the White House, she stopped by the elevator for a moment to take leave of her maid. "Maggie," she said, tears in her eyes, "my husband will live to do great things for his country."[32]

As an ex-President, Hoover missed being away from the center of things, but Mrs. Hoover took to retirement in her beloved house on the Stanford campus with pleasure. She went on hikes and bike rides, had reunions with her sons and grandchildren, served a term as president of the Girl Scouts, continued speaking on occasion to women's groups, helped out with Community Chest drives, and, above all, enjoyed entertaining the Stanford faculty and students at her home. "It's a beautiful evening," she would telephone the director of a girls' dormitory at the college. "Will you bring some of the girls up to enjoy the view? I have 65 coffee cups."[33]

With the outbreak of World War II, Hoover became involved again in European relief activities and made the Waldorf-Astoria Hotel in New York the headquarters for his work and a second home for him and his wife. As during World War I, Mrs. Hoover assisted her husband in his efforts to help European refugees, and after the United States entered the war she went to work for the Red Cross. On January 7, 1944, she attended a concert in New York, returned to the hotel to rest up a bit before joining her husband for dinner, collapsed suddenly, and died a few minutes later of a heart attack. Hoover lived on until 1964. In his memoirs, which appeared in the 1950s, he says little about her, but that was probably the way she would have wanted it. "One of Mrs. Hoover's chief characteristics," declared the *Memphis Scimitar* right after her death, "was her ability to be of great aid to her husband, yet remain completely in the background."[34]

* * * *

Not Snooty

One day Stanford's geology professor, John Branner, told Hoover that his wife wanted him and his friend Bob McDonnell to come to dinner Saturday night. "She's asked some girls," he said, handing them some slips of paper, "and these are the ones she would like you to bring." As soon as he left the room, Hoover turned to McDonnell and asked: "Who'd you draw?" "Grace Diggles," said McDonnell. "Who are you taking?" "Lou Henry," replied Hoover with disappointment, for he had been a bit intimidated by her at their first meeting. "I didn't go much for her. What do you say we trade?" "O.K. by me," said McDonnell, "one's as good as the other." When they told Branner about the trade, he wasn't sure his wife would approve, for she planned her dinners carefully, but suggested all four come to the house at the same time.

When the foursome arrived at the Branner house Saturday night, in the midst of a heavy rainstorm, Lou Henry handed her raincoat and umbrella to McDonnell, while Hoover took care of Grace Diggles's things. Mrs. Branner was surprised by the switch in partners but said nothing when she greeted them at the door. When it came time to go into the dining room, however, she went back to her original plans. "Mr. McDonnell," she said sweetly, "will you please take Miss Diggles in to supper." So Hoover ended up at Lou Henry's side after all. During the evening, McDonnell glanced over from time to time expecting to see a look of misery on his friend's face. Instead, he saw Hoover totally absorbed in his companion. Lou Henry had succeeded in breaking through Hoover's shy diffidence almost at once. She shared her enthusiasm for the Branners with him, got him to talking about geology with her, and impressed him with her zeal for learning. He was pleased, too, to discover that they both came from Iowa and had moved west about the same time. And he was delighted to learn that she loved the outdoor life as much as he did. "Say," he told McDonnell at the end of the evening, "Lou Henry's all right. She's not a bit snooty like I thought she was."[35]

Easy

During the siege of the American embassies in Peking at the time of the Boxer Rebellion, Mrs. Hoover and the other women served meals behind a barricade of sugar barrels and rice bags. "Cooking was easy," she said later, "and so was marketing. All we had to do when we were hungry was to tap the barricades."[36]

Reunion

When Mrs. Hoover was in Tientsin with her husband, she kept a cow in order to supply her baby with milk. One day, unfortunately, the cow disappeared, and its little calf was even more disconsolate than Mrs. Hoover was. One of the Chinese lads working for her, however, had a suggestion for locating the lost animal. Why not take the "cow's pup," as he called the calf, and walk him through the settlement? Then, when mother and child smelled each other, they would start mooing and there would be a family reunion. Mrs. Hoover liked the idea; so one quiet night, she and some of her friends started on their search with the "cow's pup" in tow. Since they suspected German soldiers, they headed first for the stockade where the German cavalry was encamped. To their delight, as soon as they reached the stockade the "cow's pup" began making noises and there was an appropriate response from inside the stockade. Mrs. Hoover then approached the German sentry, explained the problem in her best German, and requested the return of her cow. For a minute or two, the sentry didn't seem to understand her request. Then it dawned on him; she wanted to reunite the calf outside with the cow inside. No sooner said than done. He grabbed the calf by the halter, pulled him inside the stockade, closed the door firmly behind him, and shouted: *"Danke schön!"* Mrs. Hoover loved telling the story.[37]

Intermission

One night, when the Hoovers were living in London, Mrs. Hoover entertained some American friends at the opera. It was Wagner's four-part Ring Cycle and that meant a long session, commencing at four in the afternoon and, after an hour out for dinner, continuing on far into the night. Mrs. Hoover met her friends for the first part of the program, and for the intermission saw them to dinner at the Savoy. Then she disappeared and did not return until the rise of the curtain after dinner. Her guests were curious; where did their hostess go between curtains? Later they learned; she had gone home to hear her children say their prayers.[38]

Beautiful House

In April 1928, after Hoover had become a leading contender for the Republican nomination for President, a reporter visited Mrs. Hoover at her Washington home, spoke of the beauty of the place, and then, hoping to sound her out on politics, said that the next time he saw her, he might be in a still more beautiful house. He was thinking of the White House, of course, but Mrs. Hoover carefully avoided going out on a limb. "Yes," she laughed, "you *must* come to see us in California!"[39]

Tea Party

In June 1930, Mrs. Hoover entertained the wives of Congressmen at tea and included Mrs. Oscar DePriest, wife of a black Congressman from Chicago, on the guest list. Although she arranged the tea on two different days and included Mrs. DePriest with a group of women who had no prejudices, Southern segregationists accused her of "defiling" the White House. "Mrs. Herbert Hoover," stormed a newspaper in Mobile, Alabama, "offered to the South and to the nation an arrogant insult yesterday when she entertained a Negro woman at a White House tea. She has harmed Mr. Hoover to a serious extent. Social admixture of the Negro and the white is sought by neither race. The Negro is entitled to a social life, but that the two races should intermingle at afternoon teas or other functions is inadmissable." Similar editorials appeared in newspapers in Houston, Austin, Memphis, and Jackson, Mississippi.

But Mrs. Hoover had her defenders too. The *Washington Post, New York Times, Chicago Tribune,* and *Boston Evening Transcript* all praised her. Even a newspaper in Bristol, Virginia, had kind words for the First Lady. "The President of the United States is President of all the people, white, black, red, or yellow," wrote the Bristol editor. "The First Lady entertained a Negro at the White House as a courtesy from one branch of government to another. Mrs. Hoover is internationally minded. Politically she put into practice the brotherhood of man and religiously the fatherhood of God, even if the individual is an image carved in ebony." The President backed her up too. The following week he invited a black professor from Tuskegee College for lunch at the White House.[40]

CHAPTER 30

ELEANOR ROOSEVELT

1884–1962

A few weeks after Franklin D. Roosevelt's victory over Herbert Hoover in the 1932 election, Mrs. Hoover offered to show the President-elect's wife around the Executive Mansion. The two agreed on a morning for the visit and then Mrs. Hoover asked where she should send the White House limousine and military aide to pick up Mrs. Roosevelt. To her surprise, the latter refused both the car and the aide; she planned, she said, to walk over from the Mayflower Hotel after breakfast. The State Department's chief of protocol, FDR's cousin, was horrified when he heard about it. "But Eleanor darling," he cried, "you can't do that! People will recognize you! You'll be mobbed!"[1] His protests were in vain; when the time came, Eleanor Roosevelt walked quietly over to the White House with a friend and after touring the place returned to her hotel on foot too. The White House staff—and Mrs. Hoover—didn't know quite what to make of it.

Eleanor Roosevelt, like her husband, was a great precedent-breaker. Right after becoming First Lady, she insisted on running the White House elevator herself instead of having the Chief Usher operate it for her. When the servants began rearranging the furniture for her soon after she moved in, she pitched in and started helping out to speed things up. At an early reception, she greeted the guests at the door instead of waiting until they were all inside and then descending the staircase in grand style to meet them. More important: shortly after her husband's inauguration, she began holding press conferences of her own for the women reporters. Newsmen scoffed at the idea at first, but Mrs. Roosevelt proved to be such good copy that they were soon envying their colleagues on the distaff side. "Washington had never seen the like—," exclaimed Bess Furman in a story about Mrs. Roosevelt for the Associated Press, "a social transformation had taken place with the New Deal."[2] There was criticism, of course, as well as praise for Mrs. Roosevelt's departures from the customary ways of First Ladies; but it became quickly clear to everyone that the new President's wife was in her own way going to be just as activistic as the new President himself.

People who came to know Mrs. Roosevelt when she was First Lady found it hard to believe that she had once been painfully shy, insecure, and even inarticulate, and that as a young woman she had been so apolitical that she hadn't even favored votes for women. The Eleanor Roosevelt story is one of extraordinary growth in knowledge and understanding, and it contains a great deal of pain and suffering. Mrs. Roosevelt, like her husband, was born to privilege; but in neither case did privilege prove to be a prison, for both of them succeeded in transcending the parochialism of the upper-class circles in which they were reared and in developing broad interests and sympathies as adults. In Eleanor Roosevelt's case a restless intellectual curiosity and a desire to be of some use in the world played a major role in liberating her; but her own sorrows also made her extremely sensitive to the griefs that all people everywhere experience.

Eleanor Roosevelt's parents were what the present age calls "beautiful people": her mother, Anna Hall, an attractive, even glamorous figure in New York society, and her father, Elliott Roosevelt, younger brother of Theodore Roosevelt, a handsome and dashing man-about-town. But Eleanor, who was born in New York in October 1884, did not inherit their good looks. Though by no means the ugly duckling she thought she was, she was rather plain-looking as a child and couldn't help sensing her mother's disappointment that she wasn't prettier. "She is such a funny child," her mother told people, "so old-fashioned that we always call her 'Granny.' "[3] Her mother "tried very hard," Mrs. Roosevelt recalled, "to bring me up well so my manners would in some way compensate for my looks, but her efforts only made me more keenly conscious of my shortcoming."[4]

With her father, things were different; he called Eleanor his "Little Nell," showered her with affection, and became the center of her childhood world. Unfortunately, he was away a great deal, either on alcoholic binges or seeking cures for his alcoholism in sanitaria in this country and abroad. Eleanor always came alive when he turned up at home and treasured the affectionate letters sent her between visits; but in his absence, she was shy, withdrawn, full of fears, and starved for affection and appreciation. When she was only eight, her mother died of diphtheria, and two years later she lost her father too. In 1892, she and her two brothers went to live with their grandmother, Mrs. Valentine Hall, a rich New York widow who had once been a society belle like Eleanor's mother. Unfortunately, Grandmother Hall had had such trouble bringing up her own children that she decided to exercise strict discipline with the three grandchildren. Her principle, Mrs. Roosevelt remembered, was that "no" was easier to say than "yes"; and the governesses she hired to take care of Eleanor were equally demanding. But Eleanor did learn to ride, studied French, did a lot of reading and writing on her own, and became filled with a desire to excel in her studies in order to win the approval of her elders. Her "real education," however, did not commence, she said, until 1899, when her grandmother sent her to England to study when she was fifteen.[5]

In England, Eleanor Roosevelt was enrolled in a finishing school called Allenswood, located in a suburb of London, run by a remarkable Frenchwoman, Marie Souvestre, who asked for creative thinking as well as disciplined behavior from her students. Mlle. Souvestre was something new for Eleanor: a political liberal as well as a religious freethinker. At Allenswood, Eleanor soon found herself in a far wider world than the one she had known at home and she responded to it with eagerness and excitement. Mlle. Souvestre quickly recognized Eleanor's talents and singled her out for special attention. Not only did she encourage Eleanor's zeal for learning; she also taught her how to dress attractively and behave with poise in company. On holidays, she took her to the Continent with her as traveling companion and even made her handle the travel arrangements on occasion. Under Mlle. Souvestre's tutelage the once rigid and inhibited Eleanor threw off her fears, developed self-confidence, established warm relations with her classmates, and emerged two years later a thoughtful and gentle young woman with a great deal of quiet charm.

Eleanor returned to New York when she was eighteen for her "coming out." She was at the horse show in Madison Square Garden in November 1902, which opened the social season, and attended the big Assembly Ball at the Waldorf-Astoria in December with the other debutantes; she also appeared at countless luncheons, teas, dinners, late suppers, and dancing parties during the winter season. Mlle. Souvestre had warned her against taking any of this too seriously, but she needn't have worried, for Eleanor didn't really enjoy it very much. She considered herself a poor dancer; she reproached herself, too, for not being a society belle the way her mother and grandmother had been. "She wasn't a belle by any means," one of her escorts recalled years later. "She was too tall for most of the young men. But," he added, "she was an interesting talker. And she was always gracious and pleasant."[6]

One young man who found the nineteen-year-old Eleanor very much to his liking was the twenty-one-year-old Franklin Roosevelt, a distant cousin, who came down from Harvard to take in the parties. She had met him before, when they were children, but it had never occurred to her that they would ever become seriously involved as young adults. Soon, however, he was squiring her around town regularly; and one weekend in November 1903, when they went up to Massachusetts to visit Eleanor's brother Hall, a student at Groton, he proposed to her. With her help, he told her, he "would amount to something some day." "Why me?" she is reported to have exclaimed. "I am plain, I have little to bring you."[7] But he was insistent; and after searching her own feelings, she decided she loved him and agreed to an engagement. When Franklin told his mother, a widow who had long centered her life on her only son, she was stunned, and quickly arranged a trip to the Caribbean for him, hoping he would get over his infatuation. But "Franklin's feelings did not change," Eleanor noted with satisfaction, and in the fall of 1904, their engagement was made official.[8] By this time, Franklin had graduated from Harvard and was studying law at Columbia University, while Eleanor was teaching classes for slum youngsters at a settlement house in New York under the auspices of the Junior League. The wedding took place on March 17, 1905, with Eleanor's uncle, President Theodore Roosevelt, giving away the bride and stealing the show while he was at it.

Eleanor Roosevelt's marriage ended her tentative ventures into self-realization. Sara Roosevelt, her domineering mother-in-law, simply took over after the honeymoon; she secured a place in New York near her own for the young couple to live in, purchased furniture for it, and hired maids and servants for them. It was "Yes, Mama," and "No, Mama," for Eleanor, with cooks to handle the meals and nurses to care for the babies, when they came, and with Eleanor increasingly a supernumerary, she felt, in her own home. She even gave up her settlement work; serving on charitable boards was enough *noblesse oblige* for the conventional young society matron she was becoming. At times, though, she was overwhelmed by a feeling of uselessness; and one occasion, to FDR's astonishment, she broke down into tears as she sat before her dressing table. "I was beginning to be an entirely dependent person . . . ," she reflected later, "I was not developing any individual taste or initiative. I was simply absorbing the personalities of those about me and letting their tastes and interests dominate me."[9]

In 1910, when FDR decided to run for the New York State Senate, Eleanor was mildly interested and even went to hear one of his campaign speeches. She was shocked, however, when he came out for woman suffrage, for she "took it for granted," she recalled, "that men were superior creatures, and . . . knew more about politics than women. . . ."[10] But since FDR was a suffragist, she, too, as a dutiful wife, became one; and when her husband, to everyone's surprise, won his election as a Democrat in the overwhelmingly Republican Dutchess County, she went to Albany to establish a home there for him during sessions of the state legislature. In Albany she was free of her mother-in-law for the first time in her marriage and found it exhilarating to manage the household (and the children) herself. "I had to stand on my own feet now," she wrote later, "and I think I knew that it was good for me. I wanted to be independent. I was beginning to realize that something within me craved to be an individual."[11] In London, some years earlier, an English lady had asked her to explain the difference between the state and federal governments in the United States and she had felt abjectly ignorant. In Albany, she began attending meetings of the legislature and finding American government and politics stimulating. She also began meeting new kinds of people: professional politicians and newspapermen who were far more interesting than the society people she knew in New York. In June 1912, FDR took her to the Democratic Convention in Baltimore that nominated Woodrow Wilson for President and she found the proceedings highly amusing. This was, she thought, another "step forward."[12]

In 1913, FDR became Assistant Secretary of the Navy in the Wilson administration and his wife continued her political education in Washington. She made the social calls expected of government wives, entertained senior naval officers, inspected navy yards with her husband, and joined him in socializing with a group of lively Washington officials who liked to talk politics and eat the scrambled eggs (her specialty) served at Sunday evening gatherings. When the United States entered World War I, she took up Red Cross work, knitted clothing for the servicemen, and spent many hours at the Railroad Canteen serving snacks to the soldiers passing through town. At the same time, she spent more time with the children (Anna, James, Elliott, Franklin, Jr., and John), trying, not always successfully, to avoid the strict discipline she had known as a child without succumbing to the easy indulgence her mother-in-law encouraged with them.

During the war, Mrs. Roosevelt made valiant efforts to conserve food but things didn't turn out exactly as she had planned. In July 1917 the Food Administration picked the Roosevelt household as "a model for other large households," and the *New York Times* sent a newswoman around to interview Mrs. Roosevelt about her food-saving methods. "Mrs. Roosevelt," the reporter wrote afterward, "does the shopping, the cooks see that there is no food wasted, the laundress is sparing in her use of soap, each servant has a watchful eye for evidence of shortcomings on the part of others; and all are encouraged to make helpful suggestions in the use of 'left overs.'" The *Times* writer ended with a quotation from Mrs. Roosevelt: "Making ten servants help me do my saving has not only been possible but also highly profitable." The *Times* story, not surprisingly, produced a great deal of mirth in Washington and elsewhere; and FDR gleefully joined in the teasing. "All I can say is that your latest newspaper campaign is a corker," he told his wife, "and I am proud to be the husband of the Originator, Discoverer and Inventor of the Household Economy for Millionaires! Please have a photo taken showing the family, the ten cooperating servants, the scraps saved from the table and the hand book. I will have it published in the Sunday *Times*. . . ." "I do think it was horrid of that woman to use my name in that way," moaned Mrs. Roosevelt, "and I feel dreadfully about it because so much is not true and yet some of it I did say. I will never be caught again that's sure and I'd like to crawl away for shame."[13]

The following year Mrs. Roosevelt experienced a far greater humiliation than the *Times* story about her food-conservation efforts. In September 1918, when FDR returned from a trip to Europe with pneumonia, she started handling his mail for him while he was ill; and one day, to her horror, she came across some letters revealing that for the past year or so he had been deeply involved with Lucy Page Mercer, the attractive young Maryland woman who had been serving as her own social secretary since 1914. The discovery almost wrecked the Roosevelt marriage. "The bottom dropped out of my own particular world," she told a close friend years later, "and I faced myself, my surroundings, my world, honestly for the first time. I really grew up that year."[14] Mrs. Roosevelt offered FDR his freedom, but when he thought over the consequences of a divorce—the effect on the children, on Lucy Mercer, a devout Catholic, and on his own political future—he decided not to accept it. Instead there was a reconciliation; Mrs. Roosevelt forgave her husband and he in turn agreed to break off relations with Lucy and never see her again. But Mrs. Roosevelt was never quite the same after that. Though she continued to be devoted to her husband (and he to her), she seems to have decided, once for all, that she would no longer be the "patient Griselda" she thought that she had been ever since her wedding day.[15] She would distance herself from her mother-in-law, for one thing; and, for another, though she would continue helping her husband in his work in any way she could, she would also strike out on her own and follow her own bent, too, as much as she could.

Mrs. Roosevelt found several outlets for her energies when the war ended in 1918. Her concern for the unfortunate, in abeyance since her youthful settlement-work days, led her to visit Washington's Naval Hospital regularly, with gifts and words of cheer for the patients there, and to undertake a campaign to persuade government officials to improve conditions among the shell-shocked veterans housed in St. Elizabeth's Hospital. She began, too, at long last, to take an interest in feminism; in October 1919, she attended a meeting in Washington of the International Congress for Women Workers and began learning something about working conditions among women in industry. At the same time, she started studying politics, with the help of her husband and his friend Louis Howe, an Albany newspaperman who had become FDR's secretary in 1913 and was convinced FDR had a great political future. When FDR received the Democratic nomination for Vice President in 1920, she went on campaign trips with him, publicly announced her support of the League of Nations, one of the issues of the campaign, and, at Howe's request, began giving advice on her husband's campaign speeches. The defeat of the Democrats that year took FDR back to business and law work in New York City, but his wife continued her education in politics. She took classes in typing and shorthand, joined the League of Women Voters, and took over responsibility for preparing monthly reports on Congressional legislation for the League. FDR enjoyed giving her advice on parliamentary tactics. "Eleanor," he told her just before a League meeting she hoped to influence, "you be there early and sit up front. Just as soon as the report is read, you get up and move that it be tabled. That motion is not debatable." She did as he instructed. "It worked," she exalted afterward. "You should have seen their jaws drop!"[16]

In August 1921, when the Roosevelts were staying at their summer place on Campobello Island, New Brunswick, FDR, who was now thirty-nine, was suddenly struck down by poliomyelitis. It was a severe attack; and despite intensive treatment, he was never able to walk again without leg braces and canes. His mother wanted him to abandon public life and retire to Hyde Park and the life of a country gentleman, but his wife, backed by Louis Howe, strongly opposed her. Dr. George Draper, the orthopedic specialist brought in on the case, also urged that FDR live as active a life as he could; otherwise, he warned, he would sink into hopeless invalidism. FDR himself was eager to remain active and in the end, his mother had to yield. By the spring of 1922, FDR was appearing in his office in New York every day and getting around on a wheelchair. "His illness," Mrs. Roosevelt reflected years later, "finally made me stand on my own two feet in regard to my husband's life, my own life, and my children's too."[17] She invited Howe to stay at the Roosevelt home in order to help her keep FDR interested in public affairs.

Howe soon put Mrs. Roosevelt to work. Under his prodding, she joined the Women's Division of the Democratic State Committee of New York and took charge of fund-raising. She also learned to drive and began taking people to the polls on election day. She got in the habit, too, of bringing people she met to the house to meet FDR—state officials, party workers, social workers—in order to keep him abreast of things. With Howe's encouragement, she even ventured into speech-making. She was so nervous at first that she had a way of giggling in the middle of speeches, but Howe soon broke her of this habit. "Have something you want to say," he instructed her, "say it and sit down."[18] Gradually she acquired self-confidence in public and was soon in great demand as a speaker. In an article entitled, "Why I Am a Democrat," she spoke for her husband as well as herself when she wrote: "On the whole, the Democratic Party seems to have been more concerned with the welfare and interests of the people at large and less with the growth of big business interests."[19] But in the midst of all her activities, she reassured her husband: "I'm only being *active* till you can be again. . . ."[20] This wasn't quite the case. She was finding politics as exhilarating as her husband did; she was also building a life of her own.

In 1924, the "Franklin-Eleanor team," as it came to be called, turned up at the Democratic Convention in New York that summer. Mrs. Roosevelt presented several women's planks to the Resolutions Committee, and FDR, appearing on crutches, nominated New York Governor Al Smith for President in a stirring speech that everyone agreed was the highlight of the convention. In 1928, FDR nominated Smith again at the Democratic Convention in Houston; and after the delegates had picked the "happy warrior," as FDR called him, as their candidate, Smith persuaded FDR to run for Governor of New York. In 1928, as in 1924, Mrs. Roosevelt took an active part in the campaign; she worked at party headquarters in New York and also made campaign speeches for Smith. Smith lost heavily to Hoover in November, but FDR won handily in New York. Shortly after the election, a newswoman called on the Governor-elect's wife and was shown into the library. Moments later, Mrs. Roosevelt rushed into the room with her arms full of books and pamphlets, tossed them on the sofa, and sat down to begin the interview. She had just returned from teaching at Todhunter, a private school for girls, she explained. The reporter also learned that Mrs. Roosevelt was running a furniture factory at Val-Kill, near Hyde Park, editing the *Women's Democratic News,* traveling to Albany periodically to push for progressive state social legislation, and serving on the boards of several civic and charitable organizations. "Putting it mildly," she wrote afterward, "if Mrs. Roosevelt continues with all her present activities after she moves to Albany, she will be the busiest woman in official life today."[21]

As Governor's wife, Mrs. Roosevelt was busier than ever. She continued teaching, writing, editing, and running the Val-Kill enterprise; but she also began visiting state institutions—prisons, insane asylums, hospitals—with her husband. It was difficult for FDR to walk, so he got in the habit of driving around the grounds with the administrators of these institutions and sending his wife inside to see how they were actually being run. At first, her reports, she freely admitted, were unsatisfactory. She would tell FDR what was on the menu for the day and he would ask: "Did you look to see whether the inmates actually were getting that food?" She soon learned to look in the cooking pots on the stoves to see whether the contents corresponded to the menus, to observe whether the beds were too close together, indicating overcrowding, and to probe into the attitudes of the patients toward their supervisors.[22] In an interview in 1930 on a "Wife's Job," she told a reporter that being a man's partner was even more important than being a mother and a housekeeper. "Today," she declared, "we understand that everything else depends upon the success of the wife and husband in their personal partnership relation."[23] The Roosevelts had long since ceased to share the same bedroom; but the close partnership they developed after FDR became a cripple was to last until his death.

In 1930, FDR easily won his bid for a second term as Governor and at once became a leading contender for the presidential nomination in 1932. His wife wasn't particularly anxious to have him run for President, but she was willing to help him if that was what he wanted to do. "If polio did not kill him," she told a reporter who wondered about his health, "the Presidency won't."[24] She didn't always agree with her husband on the issues. An ardent prohibitionist, she split with him over the 18th (Prohibition) Amendment which he wanted repealed. She was also disappointed when he publicly repudiated the League of Nations in order to win the support of the Hearst newspapers for his candidacy. But she shared his major objective—to make life better for the average man, woman, and child—and was heartened by the vigorous measures he took in New York to cope with the crisis posed by the Great Depression. During the 1932 campaign, she helped writers who were preparing FDR biographies, worked diligently in the Women's Division of the Democratic party, supplied her husband with material for speeches, and at times joined him on the campaign train. And on election night, she even sipped a little champagne to celebrate his victory over Hoover. But she looked forward to the White House with trepidation; she was afraid it would imprison her in an endless round of ceremonial activities.

Mrs. Roosevelt needn't have worried. She soon found that the opportunities to help her husband and to serve the public were far greater in Washington than in Albany and that she was going to be busier than ever in the White House. And the public soon learned, even before FDR's inauguration, that the new First Lady was going to be quite unlike her predecessors in the White House. The *Nashville Tennessean* took a look at her life as Governor's wife in Albany and then predicted that she would make even more of a stir in the nation's capital than Theodore Roosevelt's lively daughter Alice. "It begins to look," crowed the editor, "as if Anna Eleanor Roosevelt is going to make Alice Roosevelt Longworth look like Alice-Sit-by-the-Fire."[25] Mrs. Roosevelt did of course do the traditional things; she presided graciously over all the teas, luncheons, receptions, and State dinners expected of her, and even added special parties for women's groups to her crowded schedule. But, unlike America's other First Ladies, she also became heavily involved in her husband's policies and programs; and she managed to carry on as an independent writer and lecturer at the same time.

Mrs. Roosevelt didn't waste a minute after the inaugural ceremony. Soon she was inspecting Washington's slums, reporting her findings to the press, and urging Congressional action to clean them up. And when the bonus army veterans, who had been forcibly ejected from Washington by the Hoover administration, returned to beg Congress to vote funds for their World War I pensions, she went out, at FDR's suggestion, with friendly greetings and plenty of coffee and joined them all in singing "There's a Long, Long Trail." ("Hoover sent the army," remarked one veteran; "Roosevelt sent his wife."[26]) Meanwhile, she followed the framing of all the major New Deal programs—the National Recovery Act (NRA), Agricultural Adjustment Act (AAA), Civil Works Administration (CWA), Work Projects Administration (WPA)—with keen and critical interest and made useful suggestions for all of them. She insisted on consumer representation and equal pay for women in the NRA codes; she persuaded AAA officials to use surplus farm goods to feed the hungry instead of destroying them in the hope of producing higher farm prices with scarcity; she had frequent conferences with CWA administrator Harry Hopkins on work projects for the unemployed, especially for women and young people; and later on, when WPA replaced CWA, she succeeded in having projects for unemployed professionals—writers, painters, musicians, actors—included in the program.

Mrs. Roosevelt was conscientious about her mail too. To her astonishment, she received over 300,000 letters in 1933, some of them filled with venom, but others containing words of praise and many of them pleas for help. Mrs. Roosevelt dispensed with the old form letters; she insisted on taking all but the anonymous hate letters seriously. She had her staff sort and classify the letters and handle them accordingly: forward appeals for jobs to the appropriate departments and find some way of responding to appeals for money (including donations of her own) from people in desperate need. She answered the most informative letters personally and passed some of them on to her husband. But she was puzzled by some of the communications. One woman wrote to say she wanted to adopt a baby and asked Mrs. Roosevelt to find one for her; then she wrote to say that when Mrs. Roosevelt got her the baby, she would need a cow; then she pointed out that if she had both the baby and the cow, she would need an icebox for the milk. Another woman wrote to tell about her ailments—fallen arches, hemorrhoids, "female troubles," hernia, poor liver—but added: "Thank goodness, it doesn't all hurt at once."[27]

But Mrs. Roosevelt spent only part of her time conferring with New Deal officials and answering letters in her White House office. She also became the White House's first great peregrinator. "My missus goes where she wants to," FDR once chortled, "talks to everybody, and does she learn something!"[28] To the delight of the struggling young airline industry, she frequently traveled by air, sometimes on junkets of her own, but often on quests for information her husband needed. During her first year as First Lady, she traveled 38,000 miles by plane, train, bus, and automobile; in her second year, it was 42,000 miles. Once, in her early White House days, she hopped into her little roadster with her friend Lorena Hickok and drove to Quebec for a three-week vacation; when she got back, FDR began quizzing her closely about life in Maine—how the fishermen there lived, what the farms were like, how the houses were built, what kinds of schools and churches she saw in the villages—and she quickly decided that "this was the only way I could help him" outside of supervising the White House staff.[29] She soon got in the habit of observing conditions carefully wherever she went—whether on vacations, on lecture tours, or even on visits to members of the family—and then answering with some precision her husband's probing questions when she got back to Washington. As the President's "eyes and ears," she inspected coal mines, sharecropper camps, mental hospitals, Puerto Rican slums, and of course numerous New Deal projects, frequently on direct assignment from her husband. One day she went off to look over a Baltimore prison without telling FDR beforehand; and when he called for her a little later and her secretary said, "She is in prison, Mr. President," he exploded in laughter. "I'm not surprised," he cried, "but what for?"[30]

From the beginning, Mrs. Roosevelt had her favorite projects. One of them was Arthurdale, a subsistence homestead community she helped organize and finance after visiting the coal mining town of Scott's Run, West Virginia, and being horrified by the terrible living conditions she encountered among the unemployed miners there. Established with government assistance in 1933, Arthurdale became an attractive community of fifty families supporting itself by farming, with a school, town hall, post office, and shop for local crafts, which Mrs. Roosevelt enjoyed visiting. The community didn't prosper the way she had hoped it would, but she was convinced that "the human values were most rewarding."[31] Another project she conceived and helped develop was the National Youth Administration (NYA), a government agency established in 1935, which provided employment for young people out of work and gave high school and college students part-time jobs so they could continue with their education. NYA was Mrs. Roosevelt's "pet Government agency," *Life* reported in 1940; she "functions as a kind of inspector-general of NYA projects."[32]

When NYA was in its formative stages, Mrs. Roosevelt saw to it that Mary McLeod Bethune, the black educator, was placed on the advisory board to ensure the participation of young blacks in NYA programs. She had once been indifferent to the plight of blacks as America's second-class citizens; but by the time she entered the White House, she had developed keen sympathies for them and, as First Lady, she became their special champion (along with Secretary of the Interior Harold L. Ickes) in the Roosevelt administration. She soon befriended Walter White, leader of the National Association for the Advancement of Colored People (NAACP), as well as Mary Bethune, and had frequent discussions with both of them about the fate of blacks under the New Deal. She also entertained blacks at the White House, visited black schools and communities, and tried to persuade her husband to support an anti-lynching bill and the abolition of poll taxes (which served to disfranchise black voters). For political reasons, FDR refrained from publicly endorsing these measures; but he never objected to his wife's activities on behalf of blacks, even though some of his own closest associates were critical.

In 1939, Mrs. Roosevelt decided to attend the organizational meeting of the Southern Conference on Human Welfare, attended by black as well as white delegates, in Birmingham, Alabama. When she arrived at the auditorium and found it segregated by law into black and white sections, she made a quiet, but stunning, gesture of protest: she slowly edged her chair into the center aisle separating the races as the meeting proceeded. A few weeks later, when the Daughters of the American Revolution refused to allow the distinguished black singer Marian Anderson to perform in Constitution Hall, the only auditorium in Washington large enough to accommodate the crowds expected to attend the concert, Mrs. Roosevelt decided, after giving the matter careful thought, to resign from the organization. She then helped Secretary Ickes arrange an outdoor concert for Marian Anderson on the steps of the Lincoln Memorial. Afterward Mary Bethune said she "came away almost walking in the air."[33] Others were not so happy. Some people (not all of them Southerners) were outraged by the First Lady's efforts to help the blacks and began circulating vicious racial jokes about her. They also charged that she was encouraging the formation of "Eleanor Clubs" among blacks to work for the replacement of black women in the kitchen by whites. American blacks, not surprisingly, came to adore Mrs. Roosevelt; and black voters, traditionally Republican, began shifting in large numbers to the Democratic party in part because of her work.

While answering her mail, taking an active interest in Federal and state projects, and seeing to it that New Deal officials did not overlook the nation's least privileged citizens, Mrs. Roosevelt also carried on a busy writing schedule. She had begun writing for magazines in the 1920s, took pride in her career as a journalist, and was reluctant to abandon it after becoming First Lady. But here, as elsewhere, she soon became busier than ever. In 1934 she began writing a monthly column, "Mrs. Roosevelt's Page," for the *Woman's Home Companion* on a variety of subjects, ranging from gardens and recreation to education, maternal mortality, and old age. And in 1942, she launched another monthly column, "If You Ask Me," in the *Ladies' Home Journal* (which she moved to *McCall's* in 1949), in which she gave disarmingly simple, commonsensical, and judicious answers to questions about religion and sex as well as about some of the famous people she had met.

Mrs. Roosevelt was a newspaper columnist too. She did a daily column called "My Day," beginning in 1935, which dealt with her activities as First Lady and was carried in scores of newspapers around the country. Once FDR took a crack at newspaper columnists at a press conference and a reporter reminded him of his wife. "She is in an entirely different category," said Roosevelt; "she simply writes a daily diary."[34] In time, however, "My Day" became more than a chatty little report on shopping trips, family reunions, and Christmases at the White House; in it, Mrs. Roosevelt began discussing New Deal projects that interested her, taking positions on social issues, and commenting on major world events. Sometimes FDR himself used the column as a trial balloon; and politicians got in the habit of reading it carefully for clues to the President's intentions.

In the mid-thirties, Mrs. Roosevelt turned autobiographical; at the suggestion of friends, she took time out to write her memoirs. When she completed the manuscript, FDR went over it carefully and made editorial suggestions, and after revisions Harper published it in 1937 with the title *This Is My Story.* The book was a critical as well as a popular success and quickly became a best-seller. Even Alice Longworth, who enjoyed disparaging her cousin as a soft-headed do-gooder, was impressed by it. "Have you read it?" she wrote a friend. "Did you realize Eleanor could *write* like that? It's perfect; it's marvelous; she can *write* . . . all at the highest pitch."[35] By the time *This Is My Story* appeared, the First Lady had a radio program of her own and was on the lecture circuit, too, discussing such topics as youth problems, progressive education, the relation of the individual to the community, and life in the White House. She gave all of the money she earned from her writing and lecturing to charitable organizations.

One morning FDR phoned his wife. "What's this I hear?" he cried. "You didn't go to bed at all last night?" It was true, she confessed; she had been working on her mail the night before and suddenly noticed it was getting light, and decided it wasn't worth the trouble going to bed at all.[36] FDR was amused—and impressed—by his wife's indefatigability, and so were thousands of Americans. But there were critics too. Some of them thought she should spend more time in the White House. "Instead of tearing around the country," one irate citizen wrote her, "I think you should stay at home and personally see that the White House is clean. I soiled my white gloves yesterday morning on the stair-railing. It is disgraceful."[37] Others thought she should spend more time with her family. But Mrs. Roosevelt, in fact, always rushed off posthaste to be with her children in emergencies: when Franklin, Jr., had an auto accident in Virginia; when James had a serious operation in Minneapolis; when Anna had her first baby in Seattle; when Elliott was contemplating divorce in California; and when John's wife was about to have her first baby in Massachusetts. FDR, at any rate, was proud of his wife. He liked to quote her: "My missus says . . . , "My missus told me . . ." He kept her portrait on his office door, moreover, and once exclaimed to Frances Perkins: "That's just the way Eleanor looks, you know—lovely hair, pretty eyes. And she always looks magnificent in evening clothes, don't she?"[39] But some people thought she had too much influence on the President for the country's good.

Mrs. Roosevelt always minimized her influence. "I never urged on him a specific course of action, no matter how strongly I felt," she insisted, "because I realized he knew of factors in the picture as a whole of which I might be ignorant." Still, she made many suggestions to him during the New Deal years. She brought people to him—sharecroppers, garment-workers, young radicals—she thought he ought to meet; passed on ideas and proposals to him which people had broached to her; left letters, articles, and clippings by his bedside for him to read before retiring at night; and engaged in long and frequently lively discussions with him about politics, Congressional legislation, the state of the Union, and foreign affairs. Sometimes FDR tried out ideas on her which he was still mulling over in his mind and played the devil's advocate to rouse her interest. One evening he took a position on an issue with which he knew she disagreed and argued so forcefully for it that she became vehement as she marshalled all the arguments she could muster against it. The next day, to her surprise, he coolly presented the precise position she had taken the day before in a conversation at tea with the American Ambassador to Great Britain. "Without giving me a glance or the satisfaction of batting an eyelash," she wrote later, "he calmly stated as his own the policies and beliefs he had argued against the night before! To this day I have no idea whether he had simply used me as a sounding board, as he so often did, with the idea of getting the reaction of the person on the outside, or whether my arguments had been needed to fortify his decision and to clarify his own mind."[40]

Sometimes, in fun, FDR exaggerated his wife's influence. "Never get into an argument with the Missus, you can't win," he once told Lorena Hickok. "You think you have her pinned down here . . . but she bobs up right away over there somewhere! No use—you can't win!"[41] FDR did of course frequently win; his wife's opinion was only one among many opinions he took into consideration before making final decisions on vital issues. But he always listened to her with interest and respect. Assistant Secretary of Agriculture Rexford Tugwell was impressed by the interchange. "No one," he said, "who ever saw Eleanor Roosevelt sit down facing her husband, and holding his eyes firmly, say to him,

'Franklin, I think you should . . . ,' or, 'Franklin, surely you will not . . . ,' will ever forget the experience."[42]

For all of her absorption in the work of the New Deal, Mrs. Roosevelt was eager to leave the White House at the end of her husband's second term and she thought he felt the same way. But the outbreak of war in Europe in 1939 inevitably produced pressure on him to run for a third term, and though she would have preferred retirement, she loyally supported his decision to run again in 1940. She also helped him out at the Democratic Convention that year. When the delegates assembled in Chicago in July to renominate the President, many of them were angry and resentful; they felt that FDR was taking them too much for granted and they balked at rubber-stamping his choice of Secretary of Agriculture Henry A. Wallace (whom they regarded as an impractical ultra-New Dealer) as his running mate. "This convention is bleeding to death," Harold Ickes telegraphed Roosevelt, who was following the proceedings in Hyde Park. "Your reputation and prestige may bleed to death."[43] FDR refused to go to Chicago himself to pacify the delegates, for he was hoping for a "draft," but he finally suggested that his wife make an appearance there. "You know Eleanor always makes people feel right," he said. "She has a fine way with her."[44] In the end, his wife saved the day. Though lukewarm about Wallace herself, she flew out to Chicago (and piloted the plane to New York for a while), spoke simply but movingly to the delegates about the need for harmony in a time of crisis, and succeeded in mollifying them. After her talk, the convention gave Wallace the vice-presidential nomination, as Roosevelt wanted. A little later, he phoned his wife to tell her she had done "a very good job," and newspapers reported: "MRS ROOSEVELT STILLS THE TUMULT OF 50,000."[45]

During World War II, Mrs. Roosevelt accepted her first official government position, that of deputy-director of the Office of Civilian Defense (OCD). For her, "defense" involved much more than fire engines, hospitals, bomb shelters, and warning systems; it involved morale, and morale, she insisted, depended on the extent to which a community provided its citizens with decent housing, nursery schools, homes for the elderly, and recreational facilities. "Real defense," she announced, "is making, day by day, a way of life which we would gladly do all we can to preserve . . . ."[46] The New Dealish bent she gave OCD produced so much outrage among conservatives in Congress and in the press that she finally resigned her position. But she continued to press for her humanitarian ideals elsewhere. She tried to persuade the State Department to let more refugees into the country, pressed hard for fair treatment of the blacks in the armed forces and in war industries, and visited Japanese-American relocation camps in Arizona to show her sympathy for American citizens of Japanese descent who had been hastily (and unjustly) evacuated from the West Coast after the attack on Pearl Harbor. She could hardly conceal her disappointment when her husband announced that "Dr. Win-the-War" had replaced "Dr. New Deal" after Pearl Harbor, for she was anxious to preserve and even expand New Deal social programs during and after the war.

World War II by no means curtailed Mrs. Roosevelt's travels; she continued to be a perpetual peregrinator. In 1942, at FDR's suggestion, she flew to Great Britain (with the code name, Rover, impishly assigned to her, she thought, by her husband) to visit American fighting men there; in 1943 she flew to the South Pacific on a similar mission; and in 1944 she toured American bases in the Caribbean and Central America. When she was in the South Pacific she was eager to visit Guadalcanal, an island American forces had wrested from the Japanese after one of the bloodiest battles of the war. Admiral William F. Halsey was not about to let her go there; he knew the island was still being bombed daily by the Japanese and he also knew that FDR didn't really want her to place herself in danger there. But she looked so crestfallen when he refused her request that he relented and promised to postpone his final decision until she had toured some of the other islands in the area and visited Australia and New Zealand. He followed her activities as she made the rounds in her Red Cross uniform, with something approaching stupefaction. "I marvelled at her hardihood, both physical and mental," he later reported; "she walked for miles, and she saw patients who were grievously and gruesomely wounded. But I marveled most at their expressions as she leaned over them. It was a sight I will never forget." He finally took her to Guadalcanal and never regretted doing so. "She alone," he said after the trip, "had accomplished more good than any other person, or any group of civilians, who had passed through my area."[47] Mrs. Roosevelt chatted pleasantly with the G.I.s wherever she went, signed thousands of autographs, took messages home to their families for them, and also entered into a correspondence with some of them after returning to the United States.

One trip Mrs. Roosevelt didn't get to make was to Yalta. When FDR was making plans for his conference there with Churchill and Stalin in February 1945, she wanted to go along, partly because she was eager to see the Crimea and partly because she was concerned about his health. But FDR told her there were to be no wives there; instead, he would take daughter Anna along as confidential secretary, since Churchill was bringing his daughter with him. While FDR was in the Crimea, Mrs. Roosevelt kept him informed about political developments in Washington and he sent her brief but affectionate letters from Yalta. Upon his return to Washington, she told him she was disappointed that Estonia, Latvia, and Lithuania, once independent nations, but absorbed by Russia in 1939, had not been given their freedom, but FDR asked her: "How many people in the United States do you think would be willing to go to war to free Estonia, Latvia, and Lithuania?"[48] She finally decided that he had done the best he could at Yalta, with the Red Army already occupying Eastern Europe and the United States pressing Russia to enter the war against Japan after the defeat of Germany.

On April 12, 1945, a few weeks after reporting to Congress on the Yalta Conference, FDR suddenly died of a massive cerebral hemorrhage while resting up in Warm Springs, Georgia, from his exertions. Mrs. Roosevelt was attending a charitable benefit in Washington at the time, and when she was called back to the White House and told the news, she exclaimed: "I am more sorry for the people of this country and of the world than I am for ourselves."[49] When Vice President Harry Truman, hastily summoned from the Capitol, arrived, she put her arm on his shoulder and said gently, "Harry, the President is dead." Truman was speechless for a moment and then said, "Is there anything I can do for you?" "Is there anything *we* can do for *you?*" she returned. "For you are the one in trouble now."[50] To her four sons, in uniform in various parts of the world, she sent a cable: "DARLINGS: FATHER SLEPT AWAY THIS AFTERNOON. HE DID HIS JOB TO THE END AS HE WOULD WANT YOU TO DO."[51]

When Mrs. Roosevelt flew down to Warm Springs to arrange for the return of her husband's body to Washington, she was in for another shock. She learned that Lucy Mercer Rutherfurd, now a middle-aged but still attractive widow, had been with FDR when he died and that he had been seeing her from time to time after he became President. She also found out to her anger that her daughter Anna had occasionally invited Mrs. Rutherfurd to the White House when she was out of town. Mrs. Roosevelt had it out with Anna, was cool to her for a few days, and then seemed to have put the matter out of her mind. In her memoir of the White House years, *This I Remember* (1949), she says nothing about it, of course, but there are passages in the last part of the book which undoubtedly reflect the deep hurt she still felt about her husband's infidelity. She had an "almost impersonal feeling" about FDR's death, she recalled, and in trying to understand why this was so, she says that it may have grown out of the fact that "much further back I had had to face certain difficulties until I decided to accept the fact that a man must be what he is" and that "life must be lived as it is. . . ." She then reflects: "All human beings have failings, all human beings have needs and temptations and stresses. Men and women who live together through long years get to know one another's failings; but they also come to know what is worthy of respect and admiration in those they live with and in themselves." And of her partnership with FDR in the White House, she had this to say: "When I went to Washington I felt sure that I would be able to use opportunities which came to me to help Franklin gain the objectives he cared about—but the work would be his work and the pattern his pattern. He might have been happier with a wife who was completely uncritical. That I was never able to be, and he had to find it in other people. Nevertheless, I think I sometimes acted as a spur, even though the spurring was not always wanted or welcome. I was one of those who served his purposes."[52] Despite the bleak conclusion, the ambivalences are all there: human beings have their weaknesses, but she isn't quite reconciled; she had great respect and admiration for her husband, but perhaps it wasn't entirely to his credit that he seemed to prefer adulation to disagreement; she was happy to be his partner in the White House, but the New Deal was essentially his enterprise, not hers, and she was merely his instrument. After all the years of teaching, writing, lecturing, advising on government policies, and working for a kinder social order, she seems to be saying, she hasn't really achieved genuine self-fulfillment.

Mrs. Roosevelt was to mellow in the years following her husband's death. The time came indeed when she could say that he had been just as progressive in his social vision as she had been. But she would have been shocked by the speculations two of her sons engaged in after her death about her relations with their father. In a book appearing in 1974, Elliott Roosevelt depicted her as so sexually inhibited that FDR was driven, almost in spite of himself, to have an affair with Lucy Mercer, and later on, with Marguerite ("Missy") LeHand, his pretty young White House secretary.[53] But James Roosevelt begged to differ with his brother Elliott. In a book he published in 1976, he agreed that his mother had been repressed (he quotes her as telling Anna, "Sex is an ordeal to be borne"), but he doubted very strongly that his father was able to have sexual relations after his bout with polio. FDR's romance with Missy LeHand, if that's what it was, and his relations with Lucy Rutherfurd while he was President, were, he insisted, perforce "affairs of the spirit," not of the flesh. On one point, though, James agreed with his brother: like Woodrow Wilson, FDR enjoyed the company of attractive and vivacious young women and was able to forget the cares of office in their company.[54] Judging from the testimony of the Roosevelt brothers, it wasn't so much that Mrs. Roosevelt disagreed with her husband on certain issues or that she was too critical; the problem was that she was always too earnest. She lacked the light touch; she simply couldn't be frivolous. And that was hard on FDR, who needed to kid around from time to time. Mrs. Roosevelt admitted as much in her later years.

"The story is over," Mrs. Roosevelt told reporters when she left the White House.[55] She was wrong; after a few weeks of stock-taking she was back at work again: writing, lecturing, teaching, and working hard for causes dear to her heart: civil liberties, racial equality, national health insurance, the extension of social security, government-subsidized housing for the poor, and federal aid to education. She was able to speak more frankly now, with the restraints of the White House gone, and she was also free to participate more actively in politics. She served on the executive board of the NAACP, helped found the Americans for Democratic Action (ADA), a liberal social-action group, and actively supported Adlai Stevenson's bid for the Presidency in 1952 and 1956. Though she always spoke gently and without dogmatism, she was as controversial as ever. To racists, she was a "nigger lover"; to reactionaries, she was a "Communist." When she came out against Federal aid to parochial schools because she thought it violated the church-state separation clauses of the Constitution, Francis Cardinal Spellman exploded in wrath, called her anti-Catholic, and accused her of "discrimination unworthy of an American mother." Defending her position in "My Day," she pointed out: "The final judgment, my dear Cardinal Spellman, of the worthiness of all human beings is in the hands of God."[56] Of Westbrook Pegler, whose attacks on her as "La Boca Grande" became increasingly pathological with the passage of time, she said mildly: "The poor man needs to be able to say disagreeable things. What else can he write about now?"[57]

Mrs. Roosevelt's first love in her later years was the new United Nations organization, to which President Truman appointed her a delegate in December 1945. As a member of the UN's Human Rights Commission she played a major role in drafting a Universal Declaration of Human Rights and getting it adopted by the UN General Assembly. And at sessions of both the Human Rights Commission and the General Assembly, she impressed delegates from all over the world, including the Russians, with her mastery of debating techniques as well as of the substantive issues. Arthur Vandenberg, former Republican Senator who served on the American delegation with her, had feared she would be soft-headed and naive, but her down-to-earth realism quickly impressed him. "How did you get along with Eleanor?" a reporter asked him after they had worked together a while. "I've said a lot of mean things about Mrs. Roosevelt," responded Vandenberg, "but I want to tell you that I take them all back. She's a grand person and a great American citizen."[58] Even the Russians, who frequently clashed with her at the UN, came to respect her. Her formula for dealing with them was simple and straightforward: "Have convictions; be friendly; stick to your beliefs as they stick to theirs; work as hard as they do."[59] In 1957 she had a lively exchange with Soviet Premier Nikita Khrushchev at his Yalta villa about the relative merits of capitalism and communism, and when she was about to leave, he asked whether he could tell the press that they had had a friendly conversation. "You can say," she told him, "that we had a friendly conversation but that we differ." Khrushchev grinned and then exclaimed: "Now! At least we didn't shoot at each other!"[60]

In 1953, when Dwight D. Eisenhower became president, Mrs. Roosevelt resigned her UN post and spent the next few years working for the American Association for the United Nations, giving speeches around the country to win support for the UN, and circling the globe in an effort to learn as much as she could about all the other nations, old and new, on the planet. When John F. Kennedy became President in 1961, he made her a UN delegate again; and her appearance on the floor of the General Assembly for the first time since 1953 produced a tremendous ovation. Kennedy also appointed her to the advisory board of the new Peace Corps (so reminiscent of her beloved NYA) and asked her to chair the President's Commission on the Status of Women. And when the "Bay of Pigs" invasion to overthrow Fidel Castro's Communist regime ended in failure, she served on the "Tractors for Freedom" committee which exchanged American tractors for the soldiers captured in Cuba.

Mrs. Roosevelt was now ailing; a rare form of blood disease was slowly wasting her away. She tried hard to keep up with her crowded schedule—writing, lecturing, committee work—but wasn't always able to do all the things she loved doing. "I can't work," she announced sadly one day when Elenore Denniston arrived to help out with the new book she was writing. "I don't understand it." Then she apologized: "And you have come so far."[61] In the fall of 1962, she had to go to the hospital for tests, but chafed under the constrictions there and insisted on going home to die. "I'll tell you about it just once," she said to friends; "then we won't mention it again. But one thing I learned. I'm not afraid to die."[62] Years before, when Edward R. Murrow asked whether she believed in a future life, she had told him: "I believe that all you go through here must have some value, therefore there must be some *reason.* And there must be some 'going on.' How exactly that happens I've never been able to decide. There is a future—that I'm sure of. But how, that I don't know. And I came to feel that it didn't really matter very much because whatever the future held, you'd have to face it when you came to it, just as whatever life holds you have to face it in exactly the same way. And the important thing was that you never let down doing the best you were able to do. . . ."[63]

Mrs. Roosevelt always tried to do her best, though she never really thought, deep down, that it was good enough. At times, in her sorrow, she sneaked off to Rock Creek Park in Washington to contemplate the statue of Grief which Henry Adams had erected in honor of his wife after she committed suicide in a fit of depression. But she always snapped out of it and went back to work for a better world. "The future is literally in our hands to mold as we like," she declared in her last book, *Tomorrow Is Now,* published a few months after her death at seventy-eight on November 7, 1962. "But we cannot wait until tomorrow. Tomorrow is now."[64]

* * * *

Watchful Eye

In March 1915, when the Roosevelts went to San Francisco to attend the opening of the Panama Pacific Exposition, they shared a hotel suite with their friends, Assistant Secretary of State William Phillips and his wife. One morning when the four were having breakfast together, Mrs. Roosevelt asked her husband whether he had received a letter from a certain person. "Yes," said FDR as he drank his coffee. "Have you answered it, Dear?" she asked. "Yes," he said. "Don't you think you should answer it now?" she persisted. "Yes," said FDR. And he did answer it right after breakfast. "I gathered," said Phillips afterward, "that the letter might never have received a reply without the watchful eye of his wife."[65]

Hideously Uncomfortable

One evening in the winter of 1920, the Roosevelts went to a party at the Chevy Chase Club in Washington and FDR enjoyed it so much he decided to stay on when his wife left. When Mrs. Roosevelt, bored by the party, got home, she found she had forgotten her latchkey. Instead of seeking help, she simply settled down on the doormat and leaned against the wall, feeling sorry for herself and angry with her husband. Hours later, Roosevelt finally showed up, flushed with wine, women, and song, and his wife rose "like a wraith," according to Alice Longworth, to confront him. "But darling, what's happened?" cried FDR. "What are you doing here?" "Oh, I forgot my key," she told him. "But couldn't you have gone to the Adolph Millers? You could have spent the night there, or you could have gone to Mitchell Palmer's house. . . ." "Oh, no," said Mrs. Roosevelt, "I've always been told never to bother people if you can possibly avoid it." "You must have been hideously uncomfortable," said FDR. "Well," sighed Mrs. Roosevelt, "it wasn't *very* uncomfortable." Years later, looking back on the incident, Mrs. Roosevelt regretted the way she had made FDR "feel guilty by the mere fact of having waited" in the vestibule that night.[66]

Nothing Else To Do

In the early 1920s, Mrs. Roosevelt started getting interested in politics, and before long she was traveling around New York state, organizing women's groups and campaigning for equal opportunity for women in the Democratic party. Once when she and another volunteer worker called on a county chairman in the southern part of the state, the latter's wife told them he was not at home. "All right," said Mrs. Roosevelt, who knew the chairman opposed women in politics and was probably hiding out from her, "we will just sit here on the steps until he comes." They sat down, and after an hour or so of waiting, the wife reappeared and told them that she had no idea when her husband would return. "It doesn't matter," said Mrs. Roosevelt with a smile. "We have nothing else to do. We will wait." After another hour or so, the chairman capitulated. He came out to the porch, looking somewhat embarrassed, and talked to Mrs. Roosevelt and her coworker.[67]

Identity

On the night of Roosevelt's election as President in November 1932, Mrs. Roosevelt was found weeping in a room at the end of a corridor in the hotel where FDR and his associates had been getting election results. Asked what was wrong, she sighed: "Now I'll have no identity."[68]

Mrs. Roosevelt and the New York Police

After the attempt on FDR's life in Miami in February 1933, Roosevelt wanted to ask the Secret Service to assign a man to protect his wife. "Don't you dare do such a thing," Mrs. Roosevelt said firmly. "If any Secret Service man shows up . . . and starts following me around, I'll send him right straight back where he came from." But Colonel Edmund Starling, head of the presidential detail, brought the matter up time and again with Roosevelt and Louis Howe. He was particularly worried about the way Mrs. Roosevelt drove herself around unescorted.

Local police found Mrs. Roosevelt just as refractory as the Secret Service did. In March 1933, when she went up to New York to visit the headquarters of the Women's Trade Union League, she found four policemen in front of the building. "What are you all doing here?" she asked. "We're here to guard Mrs. Roosevelt," said one of the patrolmen. "I don't want to be guarded," she cried; "please go away." "We can't do that," said the patrolman, somewhat embarrassed; "the captain placed us here." Mrs. Roosevelt then went in the building, called Louis Howe, who got in touch with Police Headquarters, and a few minutes later the police captain appeared. "Please take them all away," Mrs. Roosevelt directed, after thanking him for his concern. "No one's going to hurt me." "I hope not," said the captain doubtfully, as a crowd gathered. "Make your mind easy on that subject," Mrs. Roosevelt assured him, and he finally withdrew the detail. "Americans are wonderful," said Mrs. Roosevelt afterward. "I simply can't imagine being afraid of going among them as I have always done, as I always shall."[69]

Barn Door

Since Mrs. Roosevelt refused to have a bodyguard, the Secret Service decided to provide her with a gun to protect herself and talked her into taking lessons in how to handle it. After considerable resistance, she finally agreed to go to the FBI firing range for target practice. After she had been there a few times, however, FBI Chief J. Edgar Hoover became alarmed. "Mr. President," he told FDR, "if there is one person in the U.S. who should not carry a gun, it's your wife. She cannot hit a barn door." Later on, Earl Miller, a Roosevelt bodyguard with whom she formed a close friendship, taught her how to shoot the gun, and she began carrying it, unloaded, in the glove compartment of her car. Mrs. Roosevelt enjoyed Miller's company so much that there were rumors of an affair. But as Miller once put it: "You don't sleep with someone you call Mrs. Roosevelt."[70]

Solved Problem of Living

In an interview of April 1933, Cissy Patterson, publisher of the *Washington Herald,* asked Mrs. Roosevelt how she was able to move through "these cram-crowded days of hers with a sure, serene, and blithe spirit" and without the "sick vanity" and "wounded ego" that afflicted most people. "You are never angry, for instance?" she remarked. "Oh, no," agreed Mrs. Roosevelt, "I really don't get angry . . . You see I try to understand people." "But when you were young," said Mrs. Patterson, "were you free like this? So free—free of yourself?" "No," said Mrs. Roosevelt. "When I was young I was very self-conscious." Then how had she achieved such self-control? "Little by little," Mrs. Roosevelt explained. "As life developed, I faced each problem as it came along. As my activities and work broadened and reached out, I never tried to shirk. I tried never to evade an issue. When I found I had something to do—I just did it. Really I don't know—" When Mrs. Patterson suggested the First Lady was "a complete extrovert," the latter just glanced up over her knitting needles, and said nothing. Mrs. Patterson finally gave up trying to get Mrs. Roosevelt probe her inmost psyche. She concluded the new First Lady had "solved the problem of living better than any woman I have ever known."[71]

Meals

When women reporters asked Mrs. Roosevelt how it happened that the opinions she expressed in her daily column, "My Day," coincided so frequently with the President's views, she told them: "You don't just sit at meals and look at each other!"[72]

Iodine

Not long after Mrs. Roosevelt settled in the White House, she engaged Sheila Hibben, an authority on cooking, to spend a few days teaching the White House cooks various special recipes. One day Mrs. Hibben arrived in Mrs. Roosevelt's room a bit early, and as she picked up a book to look at while waiting for the First Lady, Mrs. Roosevelt's German shepherd, Major, bit her in the ankle. As she let out a yell, Mamie, the black maid rushed in, followed a moment later by Mrs. Roosevelt. "What's the matter?" cried Mrs. Roosevelt. "Major's done bit Mrs. Hibben!" cried Mamie. "Mamie," said Mrs. Roosevelt, equal to any emergency, "that settles it. From now on, we will have iodine kept in this room."[73]

Niece

In the summer of 1933, Mrs. Roosevelt took a trip to Quebec with AP reporter Lorena Hickok (with whom she seems for several years to have had an infatuation), and while there stopped to look at a little church by the water and got to talking with the village priest. "Are you any relation to Theodore Roosevelt?" he asked, when she mentioned her name. "I was a great admirer of his." "Yes," smiled Mrs. Roosevelt, "I am his niece." She was delighted to find that "no recent history seemed to have penetrated to this part of the world."[74]

Poor Man

One day, as she rode through Tarrytown, New York, Mrs. Roosevelt's motorcycle escort was thrown off his vehicle and slightly injured. With great concern, Mrs. Roosevelt accompanied him to the hospital and stayed there with him until she was certain that he was all right. A few weeks later, she was driving through the city again and spotted the man with his head still in bandages. When he recognized her, he got off his motorcycle and started toward her, but she jumped out of the car and started running toward him. "I will not have that poor man running to me," she exclaimed.[75]

Guess Who's Coming to Dinner

Whenever FDR refused to make appointments to see people his wife thought he should see, she invited them to dinner at the White House. One evening she invited National Youth Administration head Aubrey Williams to dinner so he could talk with the President. FDR liked Williams, but was surprised when he appeared at the dinner table. During the dinner, Mrs. Roosevelt mentioned a young Chinese student she had just met and urged FDR to meet him so he could quiz him about China. When FDR said he was too busy, Mrs. Roosevelt insisted he take time out to talk with the lad. "Send your Chinaman to the State Department," suggested FDR. "Franklin," protested Mrs. Roosevelt, "you know perfectly well that the State Department is on the other side politically so far as China is concerned and it will be useless in the matter." But the President was stubborn; he said he just wasn't going to waste his time talking to her Chinese student. "Well," said Mrs. Roosevelt finally, "then I will have him to dinner as my guest." At this point, Williams exploded in laughter. He ruefully confessed that he was there that evening for the same reason.[76]

Talking to a Man

"I was talking to a man the other day," Mrs. Roosevelt began a speech to a women's group, "and he said. . . ." When she made her point, one women, impressed with what she had said, wanted to know: "Who was that man you talked to?" "Franklin," Mrs. Roosevelt told her.[77]

Applause

Mrs. Roosevelt became adept at answering critical questions after her lectures. In Akron, Ohio, a hostile member of the audience asked her: "Do you think your husband's illness has affected your husband's mentality?" "I am glad that question was asked," said Mrs. Roosevelt calmly. "The answer is Yes. Anyone who has gone through great suffering is bound to have a greater sympathy and understanding of the problems of mankind." The audience applauded her loudly.[78]

Snowdrift

Mrs. Roosevelt went to FDR's office one winter day when he was surrounded by newsmen to say goodbye before leaving to attend Cornell Week. Roosevelt looked out at the falling snow and told her to telephone him if she got caught in a snowdrift. "All right," she said, as she left, "I will telephone you from a snowdrift." "And she would too!" FDR told the newsmen gleefully.[79]

Can't do a Thing

Once Mrs. Roosevelt asked her husband whether her advocacy of the anti-lynching bill introduced into Congress might hurt his efforts to get Southern votes for his rearmament program. "You go right ahead and stand for whatever you feel is right," he assured her. When she hesitated, he added: "Well, I have to stand on my own legs. Besides, I can always say I can't do a thing with you."[80]

Getting Things Ready

"Eleanor," Louis Howe once said to Mrs. Roosevelt, "if you want to be President in 1940, tell me now so I can start getting things ready."[81]

Losing Heads

When Mrs. Roosevelt was in London during World War II, she had dinner with Prime Minister Winston Churchill and his wife and soon got into an argument with Churchill about Spain. When she said she thought the United States and Britain should have done something to help the Loyalists, who were anti-fascist, during the Spanish Civil War, Churchill said he had been for the Franco government until Germany and Italy intervened on the side of the Franco fascists. But he also told Mrs. Roosevelt that the Spanish Loyalists had come under control of the Stalinists and if they had won, people like Mrs. Roosevelt and himself would have been the first to lose their heads. When Mrs. Roosevelt said that losing heads was unimportant when it came to big issues, Churchill exclaimed: "I don't want you to lose your head and neither do I want to lose mine." At this point, Mrs. Churchill leaned across the table and said, "I think perhaps Mrs. Roosevelt is right." "I have held certain beliefs for sixty years," said Churchill, quite annoyed, "and I am not going to change now." Mrs. Churchill got up at once and the dinner—and the argument—was over.[82]

Gentle and Sweet

When China's Madame Chiang Kai-shek visited Washington in February 1943, Mrs. Roosevelt, who met her first, found her sweet, gentle, delicate, and full of charm, and in talking about her, made FDR quite eager to meet China's First Lady. Madame Chiang spent several days at the White House and, to Mrs. Roosevelt's chagrin, turned out to be tough and demanding. One evening FDR got to talking about the trouble he was having with labor leader John L. Lewis and suddenly turned to Madame Chiang and asked: "What would you do in China with a labor leader like John Lewis?" Without a moment's hesitation, Madame Chiang lifted her beautiful little hand and slid it quietly across her throat. "Well," said FDR to his wife later, "how about your gentle and sweet character?"[83]

Change

After FDR's inauguration for a fourth term in January 1945, his physical energy began declining and he became increasingly less inclined to engage in the vigorous give-and-take with his wife that he had once welcomed. One night he and his old friend Harry Hooker got to talking about compulsory military service for young men after the war and it was clear they both strongly favored it. But Mrs. Roosevelt found the idea of peacetime conscription distasteful and began arguing heatedly against it. Suddenly she realized that her husband was becoming extremely upset and she dropped the subject as soon as she could. When Hooker took her to task afterward for upsetting her husband, she resolved never to do it again. FDR, she reluctantly admitted, "was no longer the calm and imperturbable person who, in the past, had always goaded me to vehement arguments when questions of policy came up. It was just another indication of the change which we were all so unwilling to acknowledge."[84]

Banging the Gavel

During the meetings of the United Nations' Human Rights Commission, which Mrs. Roosevelt chaired after World War II, the Russians did a lot of filibustering. One Russian delegate named Pavlov was particularly good at it. "The words rolled out of his black beard like a river," Mrs. Roosevelt recalled, "and stopping him was difficult, indeed." Finally she developed a method for blocking the flow. She watched his face closely as he jabbered on, and then, when he paused to take a breath, she banged her gavel loudly and reprimanded him for wasting the Commission's time.[85]

Wide Awake

When Mrs. Roosevelt was a member of the U.S. delegation to the United Nations, she enjoyed entertaining UN members at her apartment in New York and at Hyde Park too. Once a young State Department man was roasting hot dogs on the terrace fireplace at Hyde Park and conversing with a man from New Zealand. In a low voice, the New Zealand delegate asked whether he was right in thinking that sometimes during long speeches Mrs. Roosevelt went to sleep. "She does indeed," laughed the American, "and so do I. About three o'clock in the afternoon it happens. An alternate is supposed to wake me, and then I wake up Mrs. Roosevelt by handing her a note about something. Not long ago I was jerked out of a nice nap by news that the television machine was about to focus on our delegation. I hastily sent the fearful message to Mrs. Roosevelt and she scribbled back, 'I know it and am wide awake.' "[86]

No Picket Line

When Mrs. Roosevelt was scheduled to speak at a university in the Southwest, a student who was a member of the ultra-rightist John Birch Society approached the chancellor and announced: "We intend to picket Mrs. Roosevelt's lecture tonight." "Dear me," sighed the chancellor, "Do you know her subject?" "No," said the young Birchite. "But she is a controversial figure and a university is no place for controversial ideas." "She is speaking on how best to fight Communism," the chancellor told him. "Well?" said the student puzzledly. "Well, you see," the chancellor went on, "with your ideas of guilt by association, if you picket her for speaking against Communism, you will have aligned yourself, by your own logic, on the side of Communism." That night there was no picket line.[87]

Military

Once Mrs. Roosevelt entertained some Russians in New York and on July 4th, took them out to see the parade. "Military?" they asked, when a group of people went by in uniform, for they were used to seeing great displays of military strength in Red Square on national holidays in their own country. "Boy scouts," Mrs. Roosevelt explained. Another group marched by in uniform. "Military?" they asked again. "The volunteer fire department," said Mrs. Roosevelt. At length, a car went by, containing four middle-aged war veterans, stuffed into uniforms they had long grown too stout to wear in comfort. "Military," announced Mrs. Roosevelt triumphantly.[88]

CHAPTER 31

BESS W. TRUMAN

1885–1982

Bess W. Truman, complained reporters, was "a hard person to write about" when she became First Lady.[1] "After nine months in the White House," observed *Newsweek* in January 1946, "Mrs. Harry S. Truman was so little known, even in Washington, that she did her Christmas shopping alone and unnoticed in capital department stores."[2] But that is exactly what she wanted; she appeared to have a profound passion for anonymity.

Mrs. Truman shrank from publicity. While her husband strode prominently on the public stage as U.S. Senator, Vice President, and then President of the United States, she remained resolutely in the background. Right after becoming First Lady in April 1945, she promptly discarded Eleanor Roosevelt's activist policy; she refused to hold press conferences, refrained from expressing her opinions in public, and gave only succinct answers to questions submitted in writing by women reporters anxious to find out what she was like. Her favorite words, it seemed, were "no comment."[3] Once, when her secretary told her a reporter wanted to know what she was going to wear to tea that day, she exclaimed, "Tell her it is none of her damn business."[4] Her policy with reporters was plain and simple: "Just keep on smiling and tell 'them' nothing."[5]

One day, shortly after Harry Truman became President, he came upon his wife burning some papers in the fireplace and asked what she was doing. "I'm burning your letters to me," she announced. "Bess," cried Truman, "you oughtn't to do that." "Why not?" said Mrs. Truman. "I've read them several times." "But think of history," wailed Truman. "I have," said Mrs. Truman firmly and went on with her work. But she didn't get to destroy all the letters.[6] And in the hundreds that survive there is plenty of plain talking, common sense, good humor, and deep affection.

Elizabeth Virginia ("Bess") Wallace was Harry's one and only. He liked to tell people it was love at first sight when he caught a glimpse of her one morning in Sunday School at the Presbyterian church in Independence, Missouri. The year was 1890; he was only six and she was five. She had golden curls, he recalled, and beautiful blue eyes; and he was sure she was the prettiest girl in the world. "I was too backward to look at her very much," he said later. "And I didn't speak to her for five years."[7] He finally struck up an acquaintance with her in fifth grade, and after that they were friends until they graduated from high school together in 1901. It was a great day for Harry whenever he got to carry her books home from school for her; it was even better when they met to study Latin together with one of his cousins as tutor. But they were only friends and not even close friends at that. Bess was a tomboy; she rode, skated, climbed trees, played baseball and tennis, and could whistle through her teeth as well as any boy in town. But Harry had poor eyes and wore eye-glasses ("Four Eyes") and preferred reading books and playing the piano to engaging in sports. Once he improvised on the piano for Bess; and when he started rippling arpeggios across the keyboard one of Bess's girl friends recognized what he was doing right away. "It's Bessie's ice cream whistle!" she exclaimed.[8]

Bess's mother Madge Wallace didn't take much to young Harry. The Trumans, after all, struggled to make a living—Harry had to work in Clinton's Drug Store as a boy—while the Wallaces stood high in Independence's pecking order, chiefly because Madge's father, George Porterfield Gates, was a wealthy flour miller and lived in one of the finest houses in town. But the "queenly" Madge didn't give the Trumans much thought; they moved to Kansas City shortly after Harry graduated from high school and it never occurred to her, or to Bess, for that matter, that they would ever see him again.[9] The Trumans came on hard times—Harry's father lost his life savings speculating on grain futures—and Harry gave up the idea of college, took jobs with the Santa Fe Railroad and a Kansas City bank, and then ended up working on his grandfather's farm in Grandview. But the Wallaces weren't doing so well either. Bess's father worked as deputy county recorder (issuing marriage licenses) for a time and then as a customs officer in Kansas City, but never earned what he needed to support his family the way he wished. His debts mounted and he took to drink; and one night in 1903, after an importing business he launched went broke, he got in the bathtub with a revolver and killed himself. Mary Paxton, next-door neighbor, went over to console the eighteen-year-old Bess and found her walking around in the backyard with her face set and her hands clenched. She joined her and they walked wordlessly together until dawn.

Bess never spoke of her feelings that night. For the rest of her life, in fact, she avoided even mentioning her father's name to anyone. Years later, when her daughter Margaret learned about the suicide by accident and went in shock to her father about it, the latter exclaimed: "Don't you *ever* mention that to your mother!"[10] Bess Truman's deep-seated reticence and her dislike of publicity, Margaret Truman came to believe, grew in part out of fear that her father's name might be dragged in the mud by her husband's political enemies. It was Margaret Truman's belief, too, that the suicide shaped Bess's attitude toward marriage. "She saw," according to Margaret, "that her mother's way of loving her father, the passive, tender but more or less mindless love of the genteel lady, was a mistake. It failed to share the bruises, the fears, the defeats a man experienced in his world. It left him exposed to spiritual loneliness. If she ever found a man she could trust . . . Bess Wallace vowed she would share his whole life, no matter how much pain it cost her. She rejected absolutely and totally the idea of a woman's sphere and a man's sphere."[11]

In 1904, Bess went with her mother and three younger brothers to live with her grandparents in their fine mansion on North Delaware Street. The following year Grandfather Gates sent her to Bartow, a finishing school in Kansas City, where she made excellent grades, won a shot-put contest, and became a star on the basketball team. She had plenty of suitors, for she was something of a belle, and did a lot of socializing, but she was in no hurry to get married. She seemed happy playing bridge, being active in the Needlework Guild, having tennis games, going to parties with her friends, riding the handsome black horse her grandfather gave her, and driving his Studebaker (the first in Independence) around town. She also became in effect head of the Wallace family; her mother depended heavily on her for companionship and her brothers needed looking after too. Then, one day in the summer of 1910, Harry Truman, now a dirt farmer, suddenly stepped into her life again. He was in town to visit his aunt and some cousins; and when his aunt said she needed to return a cake plate belonging to Mrs. Wallace, Harry seized the plate "with something approaching the speed of light" and took off.[12] Two hours later he returned with a big smile on his face. "Well, I saw her," he announced triumphantly. He added that he was going to see her again.[13] Thus began his long courtship of his boyhood friend.

Courting Bess wasn't easy. Truman worked long hours on his grandparents's farm in Grandview; and, when he could get away, trips to Independence by train and streetcar took over an hour each way. But he went as often as he could, showered Bess with letters, and took satisfaction in knowing that Bess enjoyed his company, though her mother looked down her nose on him as a dirt farmer. Harry himself sometimes felt the social distance keenly; after losing money in some get-rich-quick investments, he apologized to Bess for not doing better in the world and confessed that the Trumans simply lacked the Midas touch. "I have been so afraid you were not even going to let me be your good friend," he told her. "To be even in that class is something."[14] He had enough self-confidence, however, to propose in June 1911, in a lengthy letter that was by turns serious and jocular. "If the drought searing the countryside went on much longer," he started off by saying, "water and potatoes will be as much of a luxury as pineapples and diamonds." Then he segued into his proposal: "Speaking of diamonds, would you wear a solitaire on your left hand should I get it?" He went on to regret his clumsiness; if he were "an Italian or a poet," he said, he would use all the "luscious language of two continents" to convince her of his devotion, but he was just "a kind of good-for-nothing American farmer." But he added: "I've always had a sneakin' notion that some day maybe I'd amount to something."[15] Bess turned him down, as he had more than half expected, but he was grateful that she didn't make fun of his proposal. "You see I have never had any desire to say such things to anyone else," he told her. "All my girl friends think I am a cheerful idiot and a confirmed old bachelor. They really don't know the reason nor ever will."[16]

Gradually Harry made progress. Bess accepted his picture; she extended a standing invitation to visit her on Sundays; she let him call her "Bess"; she visited him in Grandview a couple of times; she went on long walks and on fishing trips with him and enjoyed teasing with him; she gave him her picture; and in November 1913 she even said that if she ever got married, he would be the man. But even after they considered themselves unofficially engaged, Truman continued to be self-deprecatory. He could hardly believe it, he wrote her, that "the best girl in all the universe" cared for "an ordinary gink" like him. "How does it feel," he asked, "being engaged to a clodhopper who has ambitions to be Governor of Montana and Chief Executive of the U.S.?"[17]

Bess's mother remained cool. But after Truman got enough money together to buy a second-hand Stafford (an open touring car) in which to squire Bess around, she seemed to thaw out a bit. And then, finally, in the summer of 1917, seven long years after Truman had first proposed, Bess asked her mother to announce their engagement. But at this point Harry demurred. The United States had just entered World War I; and Harry, a member of the Missouri National Guard, was commissioned a first lieutenant in the U.S. Army, and convinced Bess it would be better to wait until the war was over before getting married. In April 1918, after several months organizing and training an artillery regiment (and writing Bess at least once every day), Truman went overseas, was in combat in France, and returned to the United States in April 1919, with the rank of major. In June, a few weeks after his discharge, he and Bess were married in an Episcopalian service in Independence, honeymooned in Chicago and Port Huron, Michigan, and then took up residence in the Gates mansion where Bess had lived since 1904.

In November 1919 Truman teamed up with one of his army buddies to open a haberdashery in Kansas City. Bess helped out; she planned the advertising, took inventory, and went over the books at night. When the business failed in the post-war depression, she supported her husband's decision to pay off the debts instead of declaring bankruptcy, even though it meant cutting down on expenses for years afterward. She also supported his decision to go into politics with the help of Tom Pendergast, Democratic boss in Kansas City whose nephew Truman knew in the army; but she looked askance at first at some of the rough-and-ready types Truman began bringing home when he ran for county judge in 1922. One day, a flamboyant politician with a bulbous red nose picked up little Margaret (born in 1924) to give her a kiss and the latter reached out and pinched his nose. "Served him right," sniffed Mrs. Truman as the man yelled in pain and hastily put the baby down.[18] She always felt "poky" when Truman was out of town on business, but kept her eye on his office when he was away, sent him newspaper clippings, and exchanged daily letters with him.

In 1934 Pendergast asked Truman to run for the U.S. Senate, and, as Truman told his wife afterward, "I said yes."[19] He knew she had doubts about his moving up the political ladder, with all the public exposure it involved, but he also knew he could count on her loyal support. During the campaign she appeared on the platform with him at rallies but refrained from making any speeches, even brief ones, herself. "A woman's place in public," she told Ethel Noland, "is to sit beside her husband, be silent, and be sure her hat in on straight."[20] She had mixed feelings about her husband's victory that fall. "Of course I'm thrilled to be going to Washington," she told a reporter after the election. "But I have spent all my life here on Delaware Street and it will be a change."[21]

Mrs. Truman found she liked Washington better than she had expected. She found it easy to take care of the little furnished apartments Truman rented during sessions of Congress, and did a lot of sightseeing at first with her mother (who was always part of the Truman household) and Margaret. She didn't like Washington's cleaning establishments, however, and since postal rates were low those days, sent the family laundry to a place in Kansas City to be handled. She went to work in her husband's office soon after arriving in Washington: went over his mail, signed letters, and, later on, did research and gave advice on speeches. In 1941, after Truman began his second term as Senator, he put "the Boss," as he called his wife, on the payroll at $4,500 a year; and when some Republicans tried to make something of it, he told reporters, "She earns every cent I can pay her. She helps me with my personal mail, helps me with my speeches and with my committee work. I don't know where I could get a more efficient or willing worker."[22] People never doubted Truman's basic honesty and integrity, even though his patron, Tom Pendergast, was jailed in 1939 for bribery and income-tax evasion.

With the coming of World War II, Truman proposed, helped organize, and headed a Senate committee investigating war contracts and did such an efficient job of saving the taxpayers' money that he soon won national attention and began to be mentioned for the vice-presidential nomination in 1944. Truman rejected the idea out of hand, "I talked it over with Bess," he told friends who wanted him to try for the nomination, "and we've decided against it. I've got a daughter and the limelight is no place for children."[23] When his friend Tom Evans argued with him about it, he finally told him he didn't want "to drag a lot of skeletons out of the closet." "Well, now wait a minute . . . ," said Evans, surprised. "I didn't know you had skeletons. What are they?" "The worst thing," said Truman, "is that I've had the Boss on the payroll in my Senate office and I'm not going to have her name dragged over the front pages of the papers and over the radio." "Well, Lord," cried Evans, "that isn't anything too great." He went on to tell Truman he knew of other Senators and Congressmen who had their wives and relatives on the payroll and assured him there was nothing wrong about it so long as they all put in an honest day's work, as he knew Bess was doing.[24] But Truman continued resistant; and it wasn't until President Roosevelt put direct pressure on him at the time of the Democratic Convention that he finally gave way. Truman's concern, his daughter later speculated, was not so much about having his wife on the payroll as it was about the possibility that the Republicans might drag his family, including Bess's father, in the mud during the campaign.

At the Democratic Convention in July 1944, the delegates picked Truman as FDR's running mate on the second ballot, and the reluctant candidate then took his disappointed wife through the noisy crowd up to the platform afterward to acknowledge the cheers. As he stood there, grinning and gesticulating, Mrs. Truman turned discouragedly toward him and sighed: "Are we going to have to go through this for the rest of our lives?"[25] To reporters crowding around her when the convention adjourned she confessed she had been opposed to the nomination but was now "almost reconciled."[26] Back in Kansas City a few days later to do some shopping, she got on the elevator in a department store and at once attracted the attention she so much disliked. "This is Mrs. Truman," shouted one woman. Another woman stared at her as if she were a "department store dummy," and then sneered, "Why, she's wearing seersucker." Bess pretended not to notice either of them, but later she exclaimed: "I wonder if they thought a vice-presidential candidate's wife should be dressed in royal purple!"[27] Incidents like this, though, were the least of her husband's worries. He continued to fear that his enemies would bring his wife and "Mother Wallace" into the campaign and he was anxious to guard their privacy. "This is going to be a tough, dirty campaign," he told Margaret, "and you've got to help your dad protect your good mama."[28] Fortunately, his fears proved groundless. Mrs. Truman never became an issue in the campaign, even though she appeared on the platform with her husband when he was on the campaign trail in the fall. On election night she went to bed before the returns were in and seemed more resigned than jubilant when she learned that the Roosevelt-Truman ticket had triumphed at the polls.

Life changed considerably for the Trumans after the inauguration in January 1945. They were now expected to make the social rounds in Washington and attend countless luncheons, dinners, teas, receptions, and cocktail parties, as guests of honor and as surrogates for President and Mrs. Roosevelt. Sometimes they went to three cocktail parties and a big dinner party all in one day. Neither Truman nor his wife enjoyed gadabouting in this fashion; they were still essentially plain-spoken and unassuming Missourians who found the pomposities and pretentiosities of Washington's social set essentially ridiculous. But to Truman's chagrin he looked ridiculous himself on one occasion. One evening he attended a party at the National Press Club and agreed to play the piano for the reporters. As he sat down at the Steinway grand and began to play, suddenly the sultry young Hollywood actress Lauren Bacall went up, got onto the top of the piano, and went into a slinky pose; and though Truman turned his head away at once, photographers snapped a picture and it appeared in newspapers and magazines all over the country. When asked afterward what his wife thought about it, Truman said ruefully, "She said she thought it was time for me to quit playing the piano."[29]

FDR's sudden death in April 1945 seems to have taken both the Trumans by surprise, despite rumors of poor health (vigorously denied by White House spokesmen) ever since 1944. Soon after Truman was summoned to the White House and told about FDR's passing, he telephoned his apartment and Margaret answered. "Let me speak to your mother," he said quietly. "Are you coming home to dinner?" his daughter asked. "I'm going out." "Let me speak to your mother," repeated Truman. "I only asked you a civil question," said Margaret, a bit hurt at his peremptory tone. "Margaret," cried Truman impatiently, "will you let me speak to your mother!" Extremely upset at this point, Margaret handed the phone to her mother; and moments later Mrs. Truman was in tears as she heard the news.[30] A little later she stood next to her husband when he took the oath of office as President and looked like "a woman in pain."[31]

Shortly after Mrs. Truman became First Lady, she acceded to a request by women reporters to hold press conferences the way Eleanor Roosevelt had been doing since 1933. Then she canceled the conference the last minute and decided to go her own way in the White House. When the reporters protested, she told them: "I am not the one who is elected. I have nothing to say to the public."[32] She did agree to entertain reporters at tea, and even attend their luncheons, but she insisted everything was off the record. She also agreed to answer questions submitted in writing; but her answers were invariably succinct and singularly uninformative. "Friends of Mrs. Truman insist that the country will love her when it comes to know her," declared *Newsweek* in January 1946. "But it seems very doubtful that Bess Truman, intent on being herself and upon seeing that her family has some privacy even in the White House, will ever give the country that chance."[33]

Despite her insistence on privacy, Mrs. Truman did the entertaining required of her (mercifully reduced, to be sure, by the mourning period for FDR and food shortages during and after the war) and even critics praised her dignity and affability. She had "a strong tennis arm," she bragged, for the handshaking; she had a good memory, too, for names and faces, and pleased guests by knowing who they were.[34] At one dinner party, she sent her husband a note that the quiet woman next to him whom he was ignoring was Dr. Lise Meitner, the distinguished atomic scientist. On another occasion, when a fiercely anti-Truman politician approached in the receiving line and Truman dared his wife to trip him up, Mrs. Truman whispered back: "Shush. Remember, he is a guest in our house."[35] She was cordial to Washington's socially pretentious, too, even though she shared her husband's disdain for them. "She made sure that the snooty women were well treated," said Truman. "That's something I wouldn't do. . . . I couldn't talk to them."[36]

While she was in the White House, Mrs. Truman tried hard to keep things in the family the way they had been she became the First Lady. She handled White House bookkeeping and oversaw expenditures; supervised the daily menus; looked for cobwebs and dust in the rooms; continued to address her own Christmas cards; got on the phone for long chats with friends in Independence; played ping-pong with Margaret in the basement; drove her own car when shopping; listened avidly to baseball games over the radio; entertained the Independence Tuesday Bridge Club at the White House; chatted with her husband evenings and listened to him play the piano. But like Truman she regarded the Executive Mansion as the "Great White Jail."[37] She did enjoy being waited on by butlers and servants (and, like Truman, treated them with kindness and consideration), but not much else. "We are not any of us happy to be where we are," she wrote a friend, "but there's nothing to be done about it except to do our best. . . ."[38] In 1946, someone asked her: "If it had been left to your own free choice, would you have gone into the White House in the first place?" Her answer: "Most definitely would *not* have."[39] She did admit there were "enjoyable spots" to being First Lady, but added that "they are in the minority." After she had been First Lady for three years and was asked how she liked it in the White House, she murmured, Oh, so-so."[40]

For a time, Mrs. Truman, who had long enjoyed being her husband's partner, felt as though she were being left out of things. Just before her husband went to Potsdam for a Big Three conference in July 1945, he telephoned her in Independence, where she was spending a few days, and found her extremely upset by the way he had been ignoring her ever since becoming President. "I am blue as indigo about going," he wrote her afterwards. "You didn't seem at all happy when we talked. I'm sorry if I've done something to make you unhappy. All I've ever tried to do is make you pleased with me and the world. I'm very much afraid I've failed miserably. But there is not much I can do now to remedy the situation." But he did try to remedy it. When he got to Potsdam he wrote and telephoned her every day and kept her informed of the work he was doing, just as he had when he was Jackson County judge and Senator from Missouri.[41]

Once back in Washington after Potsdam, however, Truman plunged into his work again and was soon inadvertently neglecting his wife. He was too busy, it seemed, to talk over problems with her, as in the old days, or discuss his decisions, major and minor, with her and listen to her suggestions. She had once been his informal, but indispensable, partner, whose advice he eagerly sought; she was now, she increasingly felt, a mere outsider. Sometimes, she thought, her husband seemed hardly aware of her existence. In December 1945 her resentment came to a head. When Truman flew out to Independence on Christmas Day to join his family for Christmas there, he found his wife in a stormy mood. "So you've finally arrived," she exclaimed when he turned up. "I guess you couldn't think of any more reasons to stay away. As far as I'm concerned, you might as well have stayed in Washington." There was a big fight, and when Truman returned to Washington on December 27, he was still so angry that he wrote his wife a blistering letter and mailed it by special delivery to Independence that night. But the following morning, he had grave misgivings, got on the phone, and called Margaret. "I want you to do something very important for me," he told her. "Go over to the post office and ask to see Edgar Hinde [the postmaster]. Tell him to give you a special delivery letter that I mailed to your mother, yesterday. It's a very angry letter and I've decided I don't want her to see it. Burn it."[42]

Margaret did as her father requested; and he wrote again and toned down his feelings. He was still upset and couldn't quite bring himself to apologize for slighting her during his first few months as President, but he did try to get right with her again. He began with a bit of self-pity. "Well," he wrote, "I'm here in the White House, the great white sepulchre of ambitions and reputations. I feel like a last year's bird nest which is on its second year." He went on to express disappointment at their reunion on Christmas Day. "You can never appreciate," he said, "what it means to come home as I did the other evening after doing at least one hundred things I didn't want to do and have the only person in the world whose approval and good opinion I value look at me like I'm something the cat dragged in. . . . I wonder why we are made so that what we really think and feel we cover up." He then reminded her of the sudden responsibilities thrust upon him as President and regretted his inability to take his new position in stride. "This head of mine," he sighed, "should have been bigger and better proportioned. There ought to have been more brain and a larger bump of ego or something to give me an idea that there can be a No. 1 man in the world. I didn't want to be. But, in spite of opinions to the contrary, *Life* and *Time* say I am." But if the Presidency was that important, he concluded, then "you, Margie and everyone else who may have any influence on my actions must give me help and assistance, because no one ever needed help and assistance as I do now. If I can get the use of the best brains in the country and a little bit of help from those I have on a pedestal at home, the job will be done."[43]

Truman's letter seems to have restored harmony; he and his wife seem to have been back in harness soon after the Christmas crisis of 1945. Truman began briefing her on the international situation, went over programs like the Marshall Plan with her, and discussed appointments, speeches, and Congressional legislation with her. In his later years he said she had been his "chief adviser" while he was President and "a full partner in all my transactions—politically and otherwise."[44] He also said she had been in on all his major decisions as President. "I discussed all of them with her," he told Marianne Means. "Why not?"[45] Her advice, he said, was indispensable. "Her judgment was always good," he recalled. "She never made a suggestion that wasn't for the welfare and benefit of the country and what I was trying to do. She looks at things objectively, and I can't always. . . ."[46]

Truman may have exaggerated his wife's influence on his work as President, but there is no doubt that she was again in on things after the first few hard months of neglect. "Bess never hesitated to try to influence Harry Truman's decisions," according to Margaret Truman. "But she never attempted to control him. . . ."[47] She evaluated people for him, made suggestions for speeches, and expressed her opinions frankly about his policies. But her influence seems to have been mainly broad and indirect rather than focused on specific issues. She listened sympathetically to him as he talked about policies, plied him with questions that forced him to think things through and, above all, permitted him to let off steam in private when he was upset, but strongly urged restraint in public. "She never nagged him," averred Matt Connelly, Truman's appointments secretary. "Once he made a decision, whether or not she agreed with it, she accepted it."[48] But she was never reconciled to his outbursts of profanity in public. Whenever he called someone an S.O.B. or told him to go to hell, he knew he faced a severe dressing down at the hands of his wife. The White House servants got used to hearing her tell him, "You didn't have to say that!"[49]

Mrs. Truman didn't want her husband to run for another term in 1948. Like many other people, including their best friends, she doubted that he could beat Republican candidate Thomas E. Dewey that year, and she longed for retirement in Independence in any case. Once Truman threw his hat into the ring, however, she cheerfully agreed to help out and hoped for the best. During the campaign she avoided speech-making, as always, but joined her husband, with Margaret, for the barn-storming tour of the country he undertook in the fall. Truman delivered several major addresses en route; but he was at his best when making short, salty, off-the-cuff remarks from the rear platform of his train as it stopped in town after town on its way across the country. The crowds gathering to hear him at each stop enjoyed the barbs he directed at Dewey and the Republicans; but they also liked the way he brought his wife and daughter into the picture. "Howja like to meet my family?" he would cry after making a few remarks. Then, as the crowd roared its assent, he introduced his wife as "the boss" and Margaret as "the boss's boss." Neither Bess nor Margaret enjoyed the ritual (and he abandoned it after Bess finally threatened to leave the train if he did it once more), but the crowds loved it.[50]

During the campaign Mrs. Truman joined the small group of advisers which met every night to plan the next day's strategy. She also went over her husband's speeches with him, saw that he ate properly and got enough rest, and supplied ailing reporters with aspirin and sewed buttons on coats for people on Truman's staff. Throughout, she was deeply moved by her husband's serene confidence in victory, although all the experts forecast a Dewey landslide, but she continued doubtful herself. "Does he really believe that he'll be elected?" she asked Clark Clifford in disbelief a few weeks before the election.[51] But she shared his joy when he confounded the experts by winning in November; and on election night, after recalling that Republican Congresswoman Clare Boothe Luce had once called her an "ersatz First Lady," she said triumphantly: "I wonder if Clare Boothe Luce will think I'm real, now!"[52] A few days later, when White House usher J. B. West came into her office to offer congratulations, she held up a copy of *Time,* with Dewey's picture on the cover, and exclaimed: "Well, it looks like you're going to have to put up with us for another four years!"[53]

With the outbreak of the Korean War in 1950 and the rise of McCarthyism, Truman's second term turned out to be even stormier than the first; and when there was talk of his running again in 1952, Mrs. Truman put her foot down. She simply couldn't survive another four years in the White House, she told her husband, and she doubted strongly that he could either. Truman agreed. On March 29, 1952, he announced his intention of retiring at the end of his second term; and afterward his wife, said a friend, "looked the way you do when you draw four aces."[54] But when Adlai E. Stevenson proved reluctant at first to be the Democratic candidate that year, there was more talk about a Truman renomination. "Matt," said Truman one day to his appointments secretary, "do you think the old man will have to run again?" Connelly pointed to Mrs. Truman's picture behind the presidential desk and asked, "Would you do that to her?" Truman was silent for a moment and then said, "You know, if anything happened to her what would happen to me?" "All right," said Connelly, "I think you've thought about it."[55] In the end Stevenson accepted his party's nomination and lost to Dwight D. Eisenhower in November.

Mrs. Truman could hardly wait to get out of the White House and back to her home in Independence. To her surprise, the streets of her beloved hometown were jampacked with people waiting to greet her and her husband when they arrived there by train. "If this is what you get for all those years of hard work," she told Truman, "I guess it was worth it."[56] But leaving the public stage was even more worthwhile for her. She settled happily down to quiet domesticity, and, except for helping her husband on his memoirs, put politics behind her and concentrated on cooking meals, taking care of the house, keeping up with Margaret (who was married in 1957 and eventually had four sons), following baseball on radio and TV, and reading detective stories. When Merle Miller traveled to Independence to interview Truman for a television documentary, Mrs. Truman balked at being included. "I have no desire to have my voice recorded for posterity," she told him.[57]

As former president, Truman continued to make news. He did some lecturing as well as writing; he also delighted reporters (and the public) by his peppery comments on public issues while taking his morning walks. But Mrs. Truman did her old duty: tried to keep her impulsive husband in check. When the Democratic Convention met in 1956 to nominate Stevenson for a second time, Truman turned up in Chicago with a big button proclaiming his support for New York Governor Averell Harriman; and he persisted in his anti-Stevenson drive long after it was clear he had lost. Deeply disturbed by his behavior at the convention, Mrs. Truman sought out their old Kansas City friend Tom Evans. "Tom," she cried, "can't you do something to stop Harry? He's making a fool of himself." It was the only time Evans ever saw her in tears.[58]

When 1960 rolled around, Mrs. Truman persuaded her husband to stay away from the Democratic Convention in Los Angeles that summer. But after John F. Kennedy received the nomination, Truman volunteered to make some speeches for him and went on the road again. In San Antonio that fall he announced that anyone who voted for Richard M. Nixon ought to go to hell. After the speech, Richard Donahue found the former President pacing the floor of his hotel room. "The Madam," he explained, had just called to bawl him out. "If you can't talk politer than that," she told him, "you come right home."[59] It was his last campaign; his health began failing in the late 1960s.

When Truman died in 1972 at eighty-nine, his wife put a letter he had written to celebrate their thirty-eighth anniversary on the desk in their upstairs bedroom, and spent a little time every morning rereading it before going down to breakfast. It was a simple letter—not much more than a listing, year by year, of the major events in their life together from 1920 to 1958—and it was signed, "Your no account partner who loves you more than ever!"[60] The passing of her husband didn't change Bess Truman's life appreciably. She continued to read a lot, watched sports on television, kept up with Margaret and her grandchildren, and guarded her privacy. When she turned ninety, one of Margaret's friends came by with a tape recorder to quiz her about an ancient Independence scandal and Mrs. Truman told her: "You might as well put that away. I will not say one word to you or anyone else on that subject."[61] She did, however, appear in public in 1976 when President Gerald Ford and his wife Betty came to Independence for the dedication of a statue of her husband. But arthritis and failing eyesight kept her increasingly confined after that; and when a man who had for years been shooting pigeons off the roof of the house called to say he was getting too old to do it any more, she sniffed: "Too old! You don't know the half of it!"[62] Told that the Gallup Poll had listed her among the top twenty women most admired in the United States, she said frankly, "I don't know why."[63]

Mrs. Truman outlived all the other First Ladies. Edith Wilson lived to eighty-nine; and Mary Scott Harrison, Benjamin Harrison's second wife, lived to ninety. Bess W. Truman died in October 1982 at the age of ninety-seven. She probably would have liked the *New York Times* headline: "BESS TRUMAN IS DEAD AT 97: WAS PRESIDENT's 'FULL PARTNER.' "[64]

* * * *

Never Guests

When Truman was nominated for Vice-President in 1944, Republican Congresswoman Clare Boothe Luce reported that his wife was on the Senate payroll, and though Truman explained all the work she was doing in his office, Luce persisted in calling her "Payroll Bess." Truman never forgave her. After he became President, *Time-Life-Fortune* owner Henry R. Luce asked him why he never invited his wife to the White House. "I have been in politics for thirty-five years," said Truman, "and everything that could be said about a human being has been said about me. But my wife has never been in politics, and she has always conducted herself in a circumspect manner. No one has a right to make derogatory remarks about Mrs. T. Your wife has said many unkind and untrue things about her. And as long as I am in residence here, she'll never be a guest in the White House. . . ."

Another *persona non grata* was New York Congressman Adam Clayton Powell, Jr. In the fall of 1945, when Mrs. Truman took tea with the Daughters of the American Revolution after the organization had barred his wife, pianist Hazel Scott, from performing in Constitution Hall, Powell began calling her "the Last Lady of the Land." From then on Truman kept Powell off the White House guest lists, even for the annual Congressional receptions. The Trumans supported civil rights for blacks; but in this instance they failed to realize the importance of symbolic acts the way Eleanor Roosevelt did in 1939. "Truman," said his friend Harry Vaughan, "never forgave anyone who cast the slightest slur on his women folk." He "apparently felt the insult to his wife," concluded J. B. West, "was worse than the insult to Powell's wife and to American Negroes."[65]

Complains to Husband

During Harry Truman's first week in the White House, Eddie McKim, the chief White House secretary, found some stenographers working on the thousands of letters that Eleanor Roosevelt had received after her husband's death. "So this is 'My Day,' " he cried. "Mrs. Roosevelt is no longer riding the gravy train. Stop it!" And he fired the women and ordered the work stopped. When Mrs. Truman heard about it, she was furious and complained to her husband. The next morning the two women resumed their work on Mrs. Roosevelt's mail and McKim was transferred to another position.[66]

Messy

Soon after the Trumans moved into the White House, FDR's grandson, young Johnny Boettinger, arrived to look for some things he had left there when visiting his grandparents. Truman insisted on helping him locate his stuff and the two began making the rounds of the White House closets. When the boy parted the dresses hanging in one of Mrs. Truman's closets and looked on the floor in the back, he cried: "Her closets are as messy as Grandma's!" Truman laughed delightedly, told his wife about it later, and liked to tease her about it whenever she implored him to be neater.[67]

Sit Down

When Mrs. Truman visited her bridge club in Independence for the first time after becoming First Lady, all her friends stood up when she entered the room. "Now stop it, stop it this instant," she cried. "Sit down, every darn one of you!"[68]

Teasing Truman

On November 27, 1945, Margaret Truman attended the opening of the Metropolitan Opera in New York with her mother, and when the two entered the box just before the curtain rose, the orchestra burst into the "Star-Spangled Banner" and people in the parquet turned around to level their opera glasses at the First Lady and her daughter. Margaret felt elated by all the attention and held her head high. Suddenly she caught her mother's warning eye. "She gave me a sardonic glance," Margaret recalled, "and I realized that I was preening, and would probably never hear the last of it from the teasing Truman."[69]

Manners

When the newspapers made a big thing out of Margaret Truman's dancing several times at El Morocco with a gentleman famous for his marriages and love affairs, Mrs. Truman called her on the telephone, made small talk for a few minutes, and then sprang her question: "How did you happen to be dancing with So-and-So the other evening?" "He asked me," said Margaret. "I couldn't be rude." "I should think once would be enough to take care of manners," said Mrs. Truman. "He was such a good dancer," said Margaret. "Do you have to do it again?" her mother wanted to know. The man had, in fact, been calling Margaret ever since the night in the New York nightclub, but her mother's remonstrance strengthened her determination never to see him again.[70]

Pay For It

Early in 1947, Margaret Truman made her debut as a singer in a radio broadcast from New York and to her delight received a check for $1,500 for her performance. When she got back to Washington, she decided to celebrate by buying the best mink scarf she could find. The saleslady gave her a big smile when she told her casually to charge the bill to Mrs. Harry S. Truman. "Oh no you don't" interposed Mrs. Truman. "What?" cried Margaret in surprise. "You bought it, you pay for it," exclaimed her mother. "You're making your own money now."[71]

Swing That Bottle

When Mrs. Truman was called on to christen an Army C-54 hospital plane, someone forgot to score the ceremonial champagne bottle so it would shatter easily, and when she swung the bottle against the plane it didn't break but put a dent in the aluminum fusilage. Mrs. Truman tried again, and then again and again, but all she succeeded in doing was make the dent larger. Finally one of the mechanics rushed over, grabbed the bottle, held it against the plane, and smashed it with a wrench. The champagne spurted out and drenched the First Lady. "Fine thing for a shot-put champion," teased Margaret afterward. Truman said she should have pushed back her hat, spat on her hands, and let fly a mighty swing the way she did when playing sandlot baseball as a kid. Retorted Mrs. Truman: "I'm sorry I didn't swing that bottle at you!"[72]

Battle Royal

At dinner one night the Trumans had watermelon for dessert. Suddenly Truman flipped a watermelon seed off his thumb over to his wife and she quickly shot one back. Daughter Margaret then gleefully joined in the Three Stooges routine and soon a watermelon-seed battle royal was raging. When the butler came to remove the plates he hastily retreated from the shower of seeds and waited outside, nearly bent double with laughter, until the battle ended.[73]

Guns Smoking

Mrs. Truman got fun out of pretending to be grouchy first thing in the morning. When she joined her family for breakfast at eight she growled playfully at her husband or twitted them about their behavior the night before. The White House staff enjoyed the mock-serious routine. "Is she wearing two guns this morning?" Chief Usher J. B. Fields would ask the butler who served breakfast. "Just one today," he might answer; or, if the First Lady was doing particularly well, he would say: "Both guns smoking!"[74]

Insult

When the Russian diplomat Nicolai V. Novikov learned that for a White House dinner he was to be seated near representatives of two Baltic countries Russia had taken over, he pretended to be ill and turned down the invitation at the last minute. Truman was enraged; he considered it an insult to his wife. The next day he summoned Undersecretary of State Dean Acheson to his office and told him he wanted

Novikov declared *persona non grata* and thrown out of the country. "Why?" cried Acheson in astonishment. "He insulted Mrs. Truman by turning down that invitation at the last second," stormed the President. "I'm not going to let anyone in the world do that." Truman's secretary, Matt Connelly, quickly got Mrs. Truman on the line. "Let me talk to him," she said.

Mrs. Truman told her husband to calm down and suggested discussing the matter with Acheson. "I'm talking with him now," said Truman and handed the phone to Acheson. "His critics will have a field day," Mrs. Truman told the Undersecretary. "We've already given them too much ammunition." "What do you—er—suggest?" asked Acheson. "Tell him you can't do anything for twenty-four hours, something like that," she said. "By that time he'll be ready to laugh about it." At this point Acheson decided to convey Mrs. Truman's message to her husband in words of his own. "Above himself—yes," he murmured, glancing at the President. "Too big for his britches—I agree with you. Delusions of grandeur." Truman grabbed the phone. "All right, all right," he cried. "When you gang up on me I know I'm licked. Let's forget all about it."

When Truman put the receiver down, he reached for a photograph on his desk which his wife had given him when he left for France during World War I. "I guess you think I'm an old fool," he said, "and probably I am. But look on the back." And Acheson read the inscription written by Mrs. Truman years before: "Dear Harry, May this photograph bring you safely home again from France—Bess."[75]

Gray

One day President Truman drove past a billboard in Washington advertising the Broadway musical, *Gentlemen Prefer Blondes* and remarked to his companion, columnist Leonard Lyons: "*Real* gentlemen prefer gray."[76]

Friendly Rector

When columnist Drew Pearson began suggesting people he thought Truman should fire, the latter was roused to wrath. At a dinner for the Reserve Officers Association he departed from the text of his speech to exclaim, "Everybody is telling me who I should have on my staff and in my Cabinet. No s.o.b. is going to dictate to me who I'm going to have!" When he returned to the White House his wife was waiting up to give him a severe lecture on dignity in the Presidency. A few days later Bill Hassett, his correspondence secretary, came hurrying into his office with the news that the rector of a large Washington church had just told the press that under similar provocation he might have said just what the President had said. "I just wish," cried Truman, "that rector would go talk to my wife!"[77]

No More Press Conferences

When Arkansas Senator J. William Fulbright became critical of the administration's policies, Truman lost his temper and called Fulbright an "overeducated s.o.b." He also told reporters that Fulbright's report on the Reconstruction Finance Corporation was "asinine." Afterwards he confessed to his staff that "the madam thinks I shouldn't have any more press conferences."[78]

Stick of Butter

When the Trumans entered the White House, the imperious housekeeper Evelyn Nesbitt (who had forced poor FDR to eat sweetbreads day after day) was still ruling the roost. One day Mrs. Truman needed some butter to take to a potluck luncheon with her bridge club, and stopped by the White House kitchen to pick it up. "Oh, no!" exclaimed Mrs. Nesbitt. "We can't let any of our butter go out of the House. We've used up almost all of this month's ration stamps already." But Mrs. Truman was not to be put off. She summoned Howell G. Crim, J. B. West's predecessor as Chief Usher, and told him: "Our housekeeper tells me I can't take a stick of butter from the kitchen!" "Why, of course you can," he sputtered. "She is entirely out of order!" "Then I think it's time to find a new housekeeper," said Mrs. Truman. A little later Mrs. Nesbitt found herself on the way out.[79]

Seven Years

In March 1952, Mrs. Truman took the women's press corps on a tour of the White House, which had just undergone extensive repairs. It was an election year and one of the reporters asked if she would like to spend four more years in the White House. "That is a question to which you are not going to get a yes or no out of me," she said frankly. "But could you stand it if you had to?" persisted the reporter. "Well," laughed Mrs. Truman, "I stood it for seven years." To her immense relief, however, her husband decided not to run again.[80]

Air Conditioning

When Truman was in the hospital, the superintendent wanted to install air conditioning in his room. But the former President declined; no one else in the hospital had air conditioning and he didn't want to receive special privileges. Several days later, however, an old friend reprimanded him. "It may be all right for you to do without air conditioning," he said, "but what about Mrs. Truman sitting here day after day in this insufferable heat? Why are you being so stubborn, and why don't you let the hospital put air conditioning in?" Within an hour an air conditioning unit was put in the room.[81]

Sounds Like It

On a trip to Europe, when the Trumans were in Italy, they visited Salerno, where Truman had fought as an army officer during World War I. The next day a newspaper quoted him as having complained about the "squirrel-headed generals" who directed the battle there. Truman insisted he had been misquoted, but his wife exclaimed: "Well, it certainly *sounds* like something you'd say!"[82]

Chrome

After leaving the White House, the Trumans acquired two cars. Since Mrs. Truman wouldn't let her husband drive her new Chrysler, Truman bought himself a Dodge coupe. He hadn't had it long, however, before he scraped all the chrome off one side when he hit a very narrow gate in the back of the house in Independence. His wife was triumphant; it proved she was right in keeping him away from her car. But about two weeks later she called his office sounding crestfallen. "Anything the matter?" he asked. "Well," she said sheepishly, "I hate to tell you this, but the truth is I was coming through the gate this morning and scraped the chromium off the side of my car."[83]

Not the Same

One morning, when Merle Miller was interviewing Truman for a television program, the former President put on some records of Glenn Gould and Arthur Rubinstein playing Beethoven's *Appasionata* Sonata. "I can play that," he finally announced, turned off the record-player, went over to the piano, sat down, and played a few halting bars. "Now, Harry," interposed Mrs. Truman, who was standing near the piano, "we all know it's not the same." He gave her a big smile.[84]

Library Card

After Truman's death, his wife did a lot of reading, but she always insisted on getting in line at the library and checking out her books like everyone else. One day she discovered she did not have her card. The librarian said it didn't matter. "Yes, it does," said Mrs. Truman. "I'm no different from anyone else. If I don't have a card, I can't take out these books." The librarian finally persuaded her to let him check them out on his card.[85]

CHAPTER 32

MAMIE DOUD EISENHOWER

1896–1979

One day, less than a month after her wedding, Mamie Eisenhower watched in astonishment as Dwight Eisenhower, then a second lieutenant, strode into their quarters at Fort Sam Houston around lunchtime, went over to the piano, and began assembling the gear he had stowed behind it when they first moved in. "You're not going to leave me this soon after our wedding day, are you?" she wailed as he went about his task. "Mame," said Ike, going over to put his arm around her, "there's one thing you must understand. My country comes first and always will. You come second."[1]

Mamie never forgot Ike's words. "It was quite a shocker for a nineteen-year-old bride," she said years later.[2] But there were other shockers. Being an Army wife, she soon learned, not only meant periodic separations from her husband; it also called for continual moving from post to post and trying to put a home together in one new setting after another. "I've lived in everything from shacks with cracks to palaces," Mrs. Eisenhower once declared; in one year she recalled, she and her husband lived at seven different Army posts.[3] Worse than the searing separations and the periodic peregrinations, however, was the nagging fear that beset her when she knew Ike was in harm's way in some distant place. She was never entirely reconciled. "It used to anger me," she said toward the end of her life, "when people would say, 'You're an Army wife, you must be used to Ike being away.' I never got used to his being gone. He was my husband. He was my whole life."[4]

Mamie knew next to nothing about Army life when she met and married Dwight D. Eisenhower, a 26-year-old junior officer fresh out of West Point, in 1916. She knew even less about the kind of austerity that was to face Army officers like Eisenhower during the inglorious 1920s when the United States was lazily at peace with the rest of the world. Mamie wasn't exactly a spoiled brat, but she was surely reared in comfort. Her father, John Doud, did so well with his meat-packing firm in Boone, Iowa, where she was born in 1896, that he was able to retire at thirty-six, move to Denver, and spend summers in the mile-high city and take his family (four daughters in all) to San Antonio every year for the cold months.

As a young woman Mamie was accustomed to maids, servants, and cooks, as well as elegant automobiles, pleasant vacation trips, a nice weekly allowance, and even an electric coupe ("Creepy") to ride around town in if she so desired.[5] She went to dancing school, of course, and after finishing grade school, attended Miss Wolcott's, a "very katish" finishing school for the daughters of Denver's best families.[6] Mamie was "handable," her mother said, and therefore nice to have around, both at home and in school; and though she learned nothing about cooking (why should she?), she did learn from her parents how to keep accounts and run a household.[7] Pert, chirrupy, and gregarious as a young woman, she went to countless movies, dances, and picnics, and entertained her friends at parties by running off popular tunes (she played by ear) on the piano for them to sing by. She had plenty of boyfriends and went on dates several times a week, but remained engagingly uninvolved. When the Douds left for San Antonio in the fall of 1915, one of her Denver beaux said plaintively, "Suppose you fall in love with some guy in San Antone?" "I won't do that," she assured him. "There's nobody in Texas that I specially like."[8] Shortly afterward her path crossed Dwight D. Eisenhower's.

Mamie first met Ike in October 1915. She was visiting Fort Sam Houston, just outside of San Antonio, one Sunday afternoon, and sitting in front of the Officers' Club with Lulu Harris, wife of Major Hunter Harris, and several other women, when Eisenhower, Officer of the Day, came out of his Bachelor Officers' Quarters, across the street, to start inspecting the guard posts. "Ike," cried Mrs. Harris, "won't you come over here? I have some people I'd like you to meet." "Sorry," said Eisenhower, "I'm on guard and have to start an inspection trip." "Humph!" said Mrs. Harris in an aside to Mamie. "The woman-hater of the post!" To Ike she exclaimed: "We didn't ask you to come over to *stay*. Just come over here to meet these friends of mine." At this, Ike walked across the street to say a polite hello and was at once glad he did. Mamie "attracted my eye instantly," he recalled; she was "a vivacious and attractive girl, smaller than average, saucy in the look about her face and in her whole attitude." Mamie's first reaction on seeing Ike was:

"He's a bruiser." As he approached she thought: "He's about the handsomest male I have ever seen."[9] After the introductions, Ike asked her to make the rounds with him and she consented, even though she didn't especially enjoy walking, and they took to each other at once. The following day, when she got home from a fishing trip, the maid told her a "Mr. I-something" had been calling all afternoon.[10] When the phone rang again, she answered it; it was Ike asking "Miss Doud" to go dancing that evening. She had a date; he suggested the next night and she had a date then too. In the end Ike settled for an evening four weeks away. Before ringing off, though, Mamie remarked: "I'm usually home about five. You might call some afternoon." Ike responded: "I'll be there tomorrow."[11]

Ike was at the Douds' the following day; and most days after that one too. If Mamie was out, he stayed anyway, chatted with her parents, and got along with them swimmingly. When he proposed to Mamie on St. Valentine's Day in 1916, she accepted at once and he gave her a replica of his West Point ring. Her parents heartily approved the match, though her father warned her about the hardships she faced as an Army wife and was pleased when she assured him she could meet them bravely. There was some talk of postponing the wedding until November, when Mamie turned twenty, but with the United States headed for war with Germany and the likelihood that Ike might be sent abroad if war came, they decided not to wait. The wedding took place at the Doud mansion in Denver on July 1, 1916 (the day Ike became a first lieutenant), with a Presbyterian minister officiating. "For heaven's sake, Ike," cried Mamie's father just before the ceremony, "take it easy and sit down!" "I can't," said Ike, who was garbed in a starched, snow-white dress uniform. "Why not?" asked Doud. "Spoil the crease in my pants," said Ike.[12] After the wedding, there was a two-day honeymoon in Eldorado Springs, in the mountains west of Denver, and then Ike and Mamie took the train to Abilene, Kansas, to visit Ike's family.

In Abilene, Ike and Mamie had their first quarrel. One afternoon Ike went out to play poker with some of his old buddies there; and when suppertime rolled around he was still at the game. Against the advice of Ike's mother, Mamie decided to call him up. When she reached him, he told her he was sorry to be so late, but that he never left a poker game when he was behind. "You don't understand, honey," he cried, when she insisted he return at once. "I can't quit now. I'm a loser." "Come now," ordered Mamie, "or don't bother to come at all," and she slammed down the receiver. "That will fetch him," she told Ike's mother. "Don't get too upset if it doesn't," murmured the latter. Ike didn't appear until 2:00 a.m. "It took a little while longer than I thought," he announced, pleased by his winnings, "but. . . ." Mamie exploded at this point; and the two went upstairs, closed the bedroom door, and quarreled until dawn.[13]

Ike and Mamie soon made up; and in the years that followed they went on to build a good marriage for themselves. But there were more "spats," as Mamie called them, not all of them minor ones, in a marriage that lasted over half a century. "There were a lot of times when Ike broke my heart," Mamie once told Julie Eisenhower, wife of grandson David. "I wouldn't have stood it for a minute if I didn't respect him. It was the kind of thing where I respected him so much, I didn't want to do anything to disappoint him."[14] Ike respected Mamie, too, of course, and was rarely disappointed in her. His letters to her through the years when he was away were always affectionate; and to celebrate their fortieth wedding anniversary he had the words, "I love you better today than the day I met you," engraved on a ruby heart as a surprise gift for her.[15] Mamie felt the same way about Ike.

But if Mamie adored Ike, she was never unassertive. Ike learned soon after the wedding that she was, in her own way, as strong-willed a person as he was. But he also gradually came to realize that she was about all he could have hoped for in an Army wife and probably more. She made his life the center of the marriage; gave him crucial support whenever he needed it; never complained about being excluded from his professional work and never interfered with it; and learned to perform with élan the social skills expected of the wife of a junior officer who was eager to impress his superiors. Mamie's finishing school had prepared her to put her husband first; her mother's tireless solicitude for her father also shaped Mamie's behavior as a wife. But Mamie found it easy to focus her life on Eisenhower's because she admired as well as loved him, and shared his ambitions for an outstanding military career. "I knew from the day I married Ike that he would be a great soldier," she told a reporter after Ike became famous. "He was always dedicated, serious and purposeful about his job. Nothing came before his duty. I was forced to match his spirit of personal sacrifice as best I could. Being his wife meant I must leave him free from personal worries to conduct his career as he saw fit."[16]

One way Mamie eased Ike's burdens was to take charge of the family finances. She was good at this; she did all the shopping, for clothing as well as food, and managed nicely (if a bit penuriously) on the small salary Ike earned as a junior officer in peacetime. When it came to cooking, she was less sure of herself. "I was a cooking-school dropout," she joked, after taking cooking at the "Y" as a young bride and learning little more than how to make mayonnaise.[17] (She already knew how to make fudge.) Mamie exaggerated her culinary deficiencies, for in time she became competent enough, but she was willing to accord primacy to Ike ("the world's best cook"), for it was a hobby with him and he was proud of his vegetable soup and steaks. "Ike cooks anything better than anybody," she told people, "that's why I hate to work hard over a meal."[18]

If Ike helped with the meals, Mamie instructed him in the social graces. She had an aristocratic flavor about her; she "could have married *anybody,*" wailed a Denver friend when she chose the homespun Ike rather than a wealthy socialite as her mate.[19] "She takes full credit," averred her son John, "for smoothing the edges off the rough-and-ready Kansan and for teaching him some of the polish that later put him in good stead." She became an accomplished hostess, moreover, and charmed Ike's fellow officers, and their wives, too, by her entertainments. At Fort Sam and other posts Ike and Mamie gave such good parties—buffet suppers, card games, songfests ("Now, Ike, no bellowing")—that their place was called "Club Eisenhower."[20] The ease with which Mamie moved in Army circles won a great deal of good will for Ike among the top brass as well as among junior officers.

But Mamie was not faultless; she committed her share of gaffes in the early days. In March 1918, when Ike was appointed to command the Tank Corps at Camp Colt, near Gettysburg, Pennsylvania, and went on ahead to report for duty, Mamie soon got busy packing the furniture they had required in San Antonio. But when one of her friends saw what she was doing, she expressed surprise. "You'll always be on the move," she said. "Don't clutter up your life with a lot of *things.* Travel light." She suggested selling the furniture to the new arrivals at Fort Sam and then buying new things in Gettysburg. Mamie took her advice, and to Ike's utter astonishment, disposed of everything for ninety dollars. "How could I have been so gullible?" lamented Mamie in retrospect. "I know I was young, but not *that* young; sure I could have applied simple arithmetic and figured out that $90 could never replace another apartment. What I exchanged for nine $10 bills cost originally over $900. It's a wonder Ike didn't wring my neck." Ike was in fact more upset by the way she disposed of his two civilian suits. He had ordered them custom-made right after graduating from West Point, to wear when off-duty, and he was proud of them. But Mamie regarded them as "horrid eyesores" and sold them for $10. Ike blew up when he heard what she had done, but soon cooled off.[21]

Ike was more amused than angry by the boner his wife pulled when they were stationed at Camp Meade, south of Baltimore, a couple of years later. One day, shortly after they had moved into their barracks apartment there, Newton D. Baker, the Secretary of War, arrived at Camp Meade for an inspection, and happened to stop by the Eisenhower apartment when Ike was on duty in the field. Mamie invited him in and chatted pleasantly with him for a minute or two, and then Baker asked: "What does your husband do best, Mrs. Eisenhower?" "Oh," she said brightly, "he plays an awfully good game of poker." Afterward, Ike groaned when he heard about the exchange. "What possessed you to say such a thing to the Secretary, Mamie?" he sighed. "I took it for granted he knew you were a good soldier," said Mamie.[22]

Eisenhower was a good soldier all right, but his career seemed to be foundering. He was disappointed that he didn't get overseas during World War I, and discouraged by his unchallenging assignments and slow advance in rank after the war. His consolations were Mamie, who always cheered him up, and their little boy, Icky (christened Doud Dwight Eisenhower), on whom he showered affection. The companionship of his wife and child more than made up for the frustrations of his military life. It helped, too, when Mamie began going out of her way to cultivate some of his own interests. At Camp Colt, she knew he enjoyed visiting the nearby Gettysburg battlefield and started going along with him. She hadn't the slightest interest in Civil War battles when she began taking the tours, but under Ike's lively tutelage she soon became fascinated by the subject.

In January 1921 life turned suddenly bleak for the Eisenhowers. Just before Christmas their beloved Icky came down with scarlet fever (contracted, apparently, from the maid, a Gettysburg girl who had just recovered from the illness) and after a few days in the post hospital died, when he was only a little over three years old. Icky's death devastated Mamie and almost produced a nervous breakdown in Ike. For months afterwards they tormented themselves with self-reproach for not having taken better care of the boy; and at times they even seemed silently to blame each other for what had happened to him. "Neither ever forgot," observed a Gettysburg friend, "not even a half century later."[23] It was a long time before either of them could talk about Icky to their friends.

The loss of Icky hurt the Eisenhower marriage. By the time the Eisenhowers arrived in Panama, a year later, for Ike's new assignment with General Fox Conner in the Canal Zone, their relations had become distant and strained. Ike buried himself in his work; and Mamie, distraught by the heat, the uncomfortable quarters, and the isolation of Camp Gaillard, spent long hours pouring out her troubles to General Conner's wife Virginia. Finally Mrs. Conner told Mamie bluntly that she would lose her husband if she didn't start using her feminine charms on him. "You mean I should *vamp* him?" cried Mamie. "That's *just* what I mean," said Mrs. Conner firmly. "Vamp him!"[24]

Mamie didn't exactly copy Theda Bara, the famous Hollywood Vamp of the 1920s, but she did begin dressing more attractively; she also had her hair redone and added the bangs across the forehead that were such a big hit with Ike (and, later, the American public). Most important, she started spending more time with Ike and less with Mrs. Conner, and gradually succeeded in restoring the bonds that had formerly held the two so closely together. "Mamie did win back her husband," said Mrs. Conner triumphantly years later, "and probably saved the marriage."[25] The birth of another boy, John Sheldon Doud, in 1922, was also a big help. "John did much to fill the gap that we felt so poignantly and so deeply every day of our lives since the death of our first son," Eisenhower wrote years later. "While his arrival did not, of course, eliminate the grief that we still felt—then and now—he was precious in his own right and he did much to take our minds off the tragedy."[26] Mamie couldn't help being overprotective of her new son; and she apologized for her "smother love" when he reached maturity.[27] But Ike, for his part, reined in his feelings; he could never bring himself to be as playful and openly affectionate with Johnny as he had been with Icky. As an adult John Eisenhower remembered his father as an austere, though devoted, figure.

In 1924 the Eisenhowers left Panama, to Mamie's relief, and Ike proceeded to carry out a series of assignments, including the General Staff School in Fort Leavenworth in Kansas, and the Army War College in Washington, which won him increasingly favorable attention from the Army's highest-ranking officers. But when he agreed to go to Manila in 1935 as assistant to General Douglas MacArthur, military adviser to the Philippines Government, Mamie was stunned. "I built myself up for a letdown," she said in later years. "I wasn't counting on finding anything delightful or delicious in such a hot place. I guess it was because the bad dream of three years in Panama was my only comparison."[28] She insisted on remaining in Washington a year to see Johnny through eighth grade and then resignedly joined her husband in Manila in 1936. When Ike took his hat off to greet her at the pier she was in for a shock: he was completely bald. As she gasped, Ike quickly explained that he cropped his hair to keep cool. "Crop—what?" she exclaimed in disbelief.[29]

Manila turned out better than Mamie expected. The heat bothered her at first; and so did a stomach disorder that kept her in bed much of the time. But a gall-bladder operation in 1938 put her on her feet again; and the elegant, air-conditioned suite in the Manila Hotel to which Ike was eventually assigned made life in the tropics bearable, even pleasant, for her. Before long she had established a new "Club Eisenhower" in Manila and was hosting a series of teas, dinners, and mah-jongg and bridge parties that delighted her Filipino as well as American guests. When President Manuel Quezon awarded Eisenhower the Distinguished Service Cross of the Philippines in the fall of 1939, he handed Mamie the medal after his speech eulogizing Ike. "You pin it on," he told her, "for you helped him to earn it."[30]

In June 1941, Eisenhower became Chief of Staff of the Third Army in San Antonio; and in July he and Mamie arrived on their old stamping grounds at Fort Sam Houston on the day of their 25th wedding anniversary. Mamie liked being there again; her husband was now a colonel, with better quarters and more perks than the ones they had known in the old days. But like most Americans she was keenly aware of the deterioration of relations with Germany and Japan and uneasily shared her husband's belief that the United States would sooner or later enter the Second World War. On December 7, 1941, she was sitting by the radio, knitting and listening to a football game, when a voice suddenly broke in: "Pearl Harbor—Japanese planes attacking!" She hastily turned up the volume, heard the announcer cry, "Battleships in flames!", and then rushed into the bedroom where Ike was taking a nap. As she entered, the bedside phone rang and Ike reached groggily for it. "The Japs have hit us," he told Mamie when he put down the receiver. "That's it—that's war."[31] A few days later General George Marshall, Army Chief of Staff, summoned him to Washington.

World War II brought Eisenhower fame and glory. For Mamie, the war years—"my three years without Eisenhower"—meant loneliness and misery.[32] With Ike abroad from March 1942 until June 1945, she was now an "army widow" with a vengeance. She was proud of his eminence—first, as Commanding General of U.S. Forces in Europe and then as Supreme Commander of Allied Expeditionary Forces landing in Normandy—but feared for his safety. Her health, still precarious from her years in the tropics, took a tumble after Ike left for London, and at first she spent much of her time bedridden in her little apartment in the Wardman Park Hotel in Washington. At length she roused herself from her lethargy. "I finally told myself I couldn't carry on that way," she said later. "I came to feel that God was not going to let anything happen to Ike until he had done what he was intended to."[33] So although the colds, headaches, insomnia, stomach trouble, and dizzy spells (produced by an inner-ear difficulty) continued, she plunged into Red Cross work, went to parties where she thought she could create goodwill for Ike, served as hostess for the Stage Door Canteen, and conscientiously answered all the letters she received from the mothers and wives of servicemen in Ike's command.

But Ike was never long out of Mamie's thoughts. "I'm worried sick about him," she told a reporter.[34] She clipped newspaper stories and magazine articles about him to paste in a big scrapbook, feasted her eyes on her seven photographs of him in the bedroom, showered him with gifts for Christmas, birthdays, and wedding anniversaries, and, above all, wrote him cheery, chatty letters in which she tried to conceal her anxieties. To her delight, Eisenhower suddenly showed up in Washington in January 1944 on a short leave. There was a reunion in the Wardman Park, a trip to West Point to see John, a senior in the U.S. Military Academy, and then a vacation in a cottage in White Sulphur Springs, West Virginia. Just before Ike left, Mamie told him, "Don't come back again till it's over, Ike. I can't stand losing you again."[35] Later she wrote: "I said goodbye to him and thought my heart would break."[36]

Mamie's unhappiness during the war was compounded by the rumors drifting back to Washington about her husband's love affair with

Kay Summersby, an attractive young Englishwoman who became one of his jeep drivers shortly after he arrived in London in 1942. Mamie couldn't bring herself to believe the stories, but they were persistent, and she finally mentioned them in letters to her husband. Ike vehemently denied the allegations; he repeatedly assured her that he had "never *been in love with anyone but you,*" and urged her to stop worrying. "You must realize," he wrote her in February 1943, "that in such a confused life as we lead here all sorts of stories, gossip, lies etc can get started without the slightest foundation in fact. I don't even let my people tell me what they are—my poor brain is sufficiently burdened with things that are true. So I want you to know that you can smile at anything—I'm trying to do my duty, every day, and my only hope is that this war will be over quickly so I can go over with you, minute by minute, everything that's happened to either of us since I last [saw] you."[37] Mamie tried to forget the Summersby tale—"Of course, I don't believe it," she told a friend, "I know Ike"—but it bothered her nonetheless.[38] Ordinarily a light drinker she began overindulging on occasion, like many other army wives, when she was at parties. But when someone in her family warned her about what was happening, she quickly snapped out of it and went back to her usual one-drink custom. But a new story began making the rounds: that Mamie was an alcoholic. The fact that she stumbled at times because of inner-ear difficulties seemed to lend credence to the charge of alcoholism.

The rumors about Mamie's alcoholism were clearly false. What about the story that Ike had a love affair with one of his jeep drivers during World War II? For years the Summersby story (which pursued Ike into the White House) rested on mere hearsay. Then, a few years after Ike died in 1969, it seemed to receive confirmation from some remarks made by Harry S. Truman (who had taken a strong dislike to Ike in the heat of the 1952 campaign) in an interview with writer Merle Miller in 1961 and 1962. In *Plain Speaking: An Oral Biography of Harry S. Truman* (1973), appearing a year after Truman's death, Miller reported that Truman had told him that in June 1945 Eisenhower took steps to divorce Mamie and marry Kay, but that Army Chief of Staff George C. Marshall wrote him a blistering letter, threatening to "bust him out of the army" if he persisted, and Ike at once abandoned his plans and eventually broke off with Kay.[39]

The Truman statements created a sensation. Even Kay Summersby, living in New York and dying of cancer, seemed surprised. She had published her memoirs, *Ike Was My Boss,* in 1948 and in it mentioned neither marriage proposal nor romance. But the Truman revelations inspired her to engage a ghostwriter to update her memories; and the ghostwriter, borrowing liberally from *Ike Was My Boss,* produced a book entitled *Past Forgetting: My Love Affair with Dwight D. Eisenhower* (1976), which recalled assignations with Ike that had apparently been past remembering in the 1948 book. In the new book, Kay acknowledged that she hadn't known of Ike's desire to marry her until she read Miller's book; she also revealed that on the three occasions when Ike tried to make love to her he couldn't pull it off. ("Oh Kay," he was quoted as sighing, "I'm sorry. I'm not going to be any good for you.")[40] Kay's book, which appeared after her own death in 1974, inspired the American Broadcasting System to produce a six-hour mini-series in May 1979, centering on Ike's love for Kay; and a paperback novel, *Ike,* based on the TV show, appeared as well. By this time, however, Ike's son John had obtained permission from his mother to publish a collection of Ike's wartime *Letters to Mamie* (1978) that showed beyond a shadow of doubt that Ike had never wavered in his deep affection for his wife at any time during World War II.[41]

Ike's wartime letters to Mamie probably tell all; he liked Kay but he loved Mamie. "She is a very popular person in the whole headquarters and everyone is trying to be kind," he wrote Mamie, right after Kay's fiancé, Captain Richard Arnold, was killed by a land mine in North Africa in 1943. "But I suspect she cannot long continue to drive—she is too sunk!"[42] He tried to help her cope with her tragic loss by adding secretarial work to her duties as chauffeur. But she was only one of his drivers when he was in Europe. Sergeant Leonard D. Dry was his chief chauffeur and two other members of the British Women's Auxiliary Corps also drove for him. All four drivers agreed, when interviewed after the publication of *Past Forgetting,* that if there was a romance between Ike and Kay it was one of the best kept secrets of the war. Ike's naval aide, Harry Butcher, who was with Eisenhower much of the time during the war, also strongly doubted the Summersby story. "If a romance had been brewing between Ike and Kay," he declared, "I could hardly have missed it."[43] And according to Ike's orderly, Michael J. McKeogh, the episodes in Kay's book about Ike's fruitless efforts to make love were without foundation in fact. "I woke the General in the morning," he told an interviewer, "and I never went to bed until my boss was in bed at night. There was never anyone in the General's bed at night. There was never anyone in the General's bed but the General. Never, at any time all during the war, wherever we were quartered."[44] The Truman story must surely be put down to the faulty memory of an aging former President who was still rankling over Ike's charges in the 1952 campaign that the Truman administration was riddled with corruption.

When World War II ended, the world was Ike's oyster. Not only was he America's most popular war hero; he was also held in high esteem throughout Europe. "Imagine how I felt," Mamie told an interviewer years later, "hearing presidents, prime ministers, kings, and generals praising Ike to the whole wide world. I was so happy I could hardly breathe."[45] But she couldn't help teasing her husband about his fame.

After he received a big ovation at a victory celebration in New York at the end of the war, she went up to him and said mischievously: "May I touch you?"[46] But Ike kept his head. When he received an honorary degree at Columbia University in 1947, Mamie rose with the rest of the audience as he walked down the aisle, but he stepped out of the line, took her hand, and said earnestly: "Don't you ever stand up for me, Mamie."[47] The following year he became president of Columbia University and in 1951, Supreme Commander of NATO forces in Europe. But his entry into politics in 1952 was probably inevitable.

Mamie wanted Ike to stay out of politics; she was afraid it would kill the privacy she cherished. But as her husband, pressed hard by Republican leaders, was trying to decide whether to go for the nomination in 1952, she carefully refrained from giving any advice. "Honey," she told him, "it's your decision. My job will be the same as always—to take care of you and our home."[48] After he entered the presidential lists, however, she soon found herself heavily involved, too, and to her surprise, discovered there was something "exhilarating" about politics.[49] "If you are going to help Ike," a Republican National Committeewoman told her just before the Republican Convention in Chicago in June, "you must get into the picture." "I will try," promised Mamie, "but you know me."[50] She did try; and she succeeded beyond her own expectations. She mingled amiably with the delegates in Chicago, remembered names and faces, and charmed everyone with her friendliness and good humor. "I don't understand politics," she said repeatedly. But a reporter who saw some solemn delegates go into a room to meet her and emerge with big smiles on their faces, told her: "You're the best politician of the lot!"[51]

After Eisenhower received the Republican nomination, he launched a campaign tour—eight major speeches and seventy-seven whistlestop talks—that featured his wife the way Harry Truman had featured Bess in 1948. Borrowing from Truman, Ike gave short speeches from the rear platform of his campaign train as it moved around the country and always finished his remarks with the announcement: "And now I want you to meet my Mamie." Then, as the people gathered around the train cheered lustily, his wife (who, unlike Bess Truman, enjoyed the ritual) smiled and waved and then leaned over to shake hands and sign autographs.[52] One day, just at dawn, the Eisenhower train arrived at a little station in North Carolina and the Eisenhowers found two or three hundred people already gathered to greet them. "Come on, Honey," cried Ike gaily, still in his dressing gown. "Let's say hello to them!" "Like this?" said Mamie doubtfully; she had just gotten out of bed. "Sure!" said Ike. "You're pretty as a picture." So Mamie went to the car door, wearing a robe and with her hair in curlers, and stood with her husband as he talked to the crowd. "*She's* going to get him elected," a woman in the crowd exclaimed. Three press association photographers were covering the tour, but only one of them was up early enough to record the event. "The other two boys are in bad trouble," Jim Hagerty, one of Ike's campaign advisers, remarked at breakfast a little later. "They're apt to get fired." "We can fix that," said Mamie at once. "I'll get my hair in curlers again and we'll restage it." The pictures of the Eisenhowers in bathrobes, which appeared in newspapers everywhere, were "dynamite," according to Hagerty. "Mamie is a property!" exulted one leading Republican. "Mamie must be worth at least 50 electoral votes," wrote *New York Times* columnist James Reston.[53]

Mamie helped with speeches as well as platform appearances. She listened as Ike read drafts of all his major speeches aloud to her as they traveled around the country and offered suggestions. Sometimes, though, she was a little too frank for him. On one occasion he started reading a speech and hadn't gone very far before she interrupted. "Ike," she said, "you can't say that. It's not in character." Eisenhower looked up, a bit irritated, and then continued reading. Again she stopped him. "Ike," she repeated, "that simply isn't *you!*"[54] After she made a few more interruptions, Ike suddenly lost his temper, threw the manuscript down, and stalked out of the room. But he ended by discarding the speech. (After little "spats" like this Ike gave Mamie formal, not folksy, introductions to whistlestop crowds: "And now I want you to meet Mrs. Eisenhower.")[55] Usually, though, Mamie liked his speeches; and whenever she expressed approval he got a big grin on his face. "Don't change a word, Ike," she told him once when a speech of his had moved her to tears.[56]

During the 1952 campaign Mamie was conscientious about answering the letters she received from Ike suporters, but there were so many she had to have the help of two secretaries. "I get up with letters," she once sighed. "I go to bed with letters. I guess I'm what you would call a well-lettered woman."[57] She was doing it all, though, for Ike, and that made it easy. "After thirty-six years of marriage," observed *Newsweek,* "her face still lights up like a bobby-soxer's when she talks about Ike."[58] Told that a carefully conducted poll had picked her husband as the greatest American, she smiled and exclaimed: "They didn't have to go to all that trouble. I could have told them."[59] She was moved to tears at times by the enthusiasm Ike roused during the campaign. "It's inspiring," she said, "the way people look at him with such affection, interest, and hope."[60] When a reporter asked whether she would have let Ike run for President if she had known how deeply involved she would become in the campaign, she exclaimed: "Why this comes naturally. I've been training for it for 36 years. When you're in the Army, you get used to chasing after your husband."[61] She told another reporter: "The bigger the crowds, the more people I met, the happier I became. Seeing thousands and thousands of people adoring Ike, believing in his leadership, kept me cloud-high all the time."[62] She wept on election night when Ike went over the top.

At his inauguration on January 20, 1953, Eisenhower went out of his way to give Mamie special attention. For one thing, he walked over to her after taking his oath of office, clasped her shoulders, and kissed her on the cheek. For another, he insisted that she, rather than Vice President Richard Nixon, ride back to the White House with him after the inaugural ceremony to view the inaugural parade. Some observers expected Mrs. Eisenhower to be innovative, too, and they looked forward to an activistic First Lady somewhat in the manner of Eleanor Roosevelt. "There were indications . . . ," noted *Newsweek,* "that the First Lady did not intend to remain in the quiet obscurity which had marked Mrs. Bess Truman's tenure. As the President's wife, she hoped to move beyond Washington's social life and to be a figure in her own right."[63]

Mamie disappointed Washingtonians. She and her husband turned out to be just as big homebodies as the Trumans were. One Washington hostess, asked if she had entertained the Eisenhowers in her home, admitted, "No, not yet." Then she added: "But nobody else has either."[64] Ike and Mamie preferred spending their free evenings staying home reading, playing cards, and watching movies and television to making the social rounds in Washington. "For years my evenings have been somebody else's," Ike told a friend. "That's the way it was, and I never resented it. But at last I've got a job where I can stay home nights, and, by golly, I'm going to stay home."[65] As for Mamie: "I never intended to be anything but Ike's wife." She liked to say she had "only one career, and his name is Ike."[66]

Mamie stayed in bed most mornings, but not because she was indolent. After breakfast she transformed the big Eisenhower bed into a kind of command post and held earnest conferences with the Chief Usher, the housekeeper, and the secretaries to plan the day's schedule and go over the mail. She was efficient and thrifty as manager of the household; she hunted for bargains in the morning papers and sent the servants scurrying around town to purchase food and supplies at good prices. She planned menus, seating arrangements (with her and Ike seated next to each other instead of on opposite sides of the table), and floral decorations for luncheons and state dinners; and she bristled if anyone intruded into her domain. One day, as she was going over the menu for a luncheon with Chief Usher J. B. West, she exclaimed: "What's this? I didn't approve this menu!" "The President did, two or three days ago," West explained. "I run everything in my house," frowned Mamie. "In the future all menus are to be approved by *me* and not by anybody else.!"[67]

J. B. West found Mamie "imperious" at times. Though usually breezy and informal in dealing with the White House staff, she also revealed "a spine of steel," according to West, forged by years of military discipline. "As wife of a career army officer," observed West, "she understood the hierarchy of a large establishment, the division of responsibilities, and how to direct a staff. She knew exactly what she wanted, every moment, and exactly how it should be done. And she could give orders, staccato crisp, and detailed, and final, as if it were she who had been a five-star general. She established her White House command immediately."[68] She ordered the servants to use the service, not the family elevator; she prohibited them from walking through the White House proper to get from one wing of the building to the other; and she insisted on correct nomenclature. J. B. West was Mister West, not West; and his subordinates were to be addressed by their first names. "Mrs. Eisenhower knows every single thing that goes on in this house," said one of the butlers, "who's here, what they're doing, and why."[69]

Yet if Mamie was strict and demanding as mistress of the White House she was also generous and thoughtful. She lavished praise on the staff whenever White House entertainments went off well. She also took a personal interest in her employees and spent much time talking to them about their families, homes, and health, and always sent flowers whenever any of them (or members of their families) became ill. She had a "birthday calendar," too, and saw to it they all had presents and a cake, baked in the White House kitchen, to celebrate their birthdays. "She wants to be everybody's godmother," said one White House staffer.[70] Christmas was special too. For her first Christmas in the White House she spent days assembling and wrapping gifts for everyone on the staff. "Well, I've finally done it," she said triumphantly when all the presents were placed under the big tree in the West Hall. "It's been my desire all my life, to be able to give a Christmas gift to *everyone* who works for me!"[71] In July 1955 she and Ike threw a big party for the entire White House staff at the farm they had purchased and remodeled in Gettysburg, Pennsylvania. It was their first permanent home and they used it for weekends and holidays during the White House years.

Mamie rarely discussed issues or politics with her husband. "Ike took care of the office," she once said. "I ran the house."[72] Sometimes he talked over appointments with her, for he considered her a shrewd judge of people. On rare occasions he even asked her opinions about policies. "Let me try this out on Mamie," he once said in the midst of a White House conference on the cost of living, "she's a pretty darn good judge of things."[73] Mostly, though, her work was running the White House, and she delighted her husband by her skill. "I personally think," he wrote years late, "that Mamie's biggest contribution was to make the White House livable, comfortable, and meaningful for the people who came in. She was always helpful and ready to do anything. She exuded hospitality. She saw that as one of her functions and performed it, no matter how tired she was. In the White House, you need intelligence and charm—to make others glad to be around you. She had that ability."[74]

For a time it looked as though Mamie would be First Lady for only four years. In September 1955, Ike suffered a coronary thrombosis while vacationing in Colorado, and she was convinced he should reject demands of Republican leaders that he run for a second term in 1956. He made a good recovery, however, and his doctors decided he would do better keeping busy than going into retirement, and she finally came around. "She felt, and said," Ike wrote in his memoirs, "that it would be best for me to do exactly whatever seemed to engage my deepest interest. She thought idleness would be fatal for one of my temperament; and consequently, she agreed that I should listen to all my most trusted advisers and then make own decision. She said she was ready to accept and support me in that decision no matter what its nature."[75] Ike presented her with a "military medal" (a jeweled medallion) for having taken such good care of him when he was in the hospital; and in February 1956 he told her: "I've made up my mind. I'm going to run."[76] She interposed no objections at this point. "I just can't believe," she told friends, "that Ike's work is finished;" he "still had a job to do."[77] He won re-election easily in 1956, but a year later suffered a mild stroke. Then she had misgivings. "I'm not so sure," she said worriedly, "we're ever going to be able to live in Gettysburg."[78] But Ike recovered quickly and remained in good health and spirits until the end of his second term in 1961.

The Gettysburg farm house in which the Eisenhowers lived in retirement was Mamie's "dream house." She had supervised its construction and loved the quiet life they lived there in the first house of their own. Life was simple and easy in Gettysburg. Ike hunted, fished, painted, oversaw the dairy farm, and worked on his memoris; Mamie ran the household, as always, and was as economical as ever: saved discount coupons, shopped for specials in the local supermarkets, and insisted on taking her place in the line at the counter. She also watched TV with Ike in the evening and, above all, took good care of him; she watched his diet, insisted on proper rest and exercise, and saw that he dressed properly when going outdoors. "Whenever Ike went away," she said, "the house sagged. When he came home, the house was alive again."[79] In July 1966 they celebrated their Golden Wedding Anniversary, and for the occasion Mamie tried to look her best for her "boy friend," as she called Ike. "For any marriage to be successful," she told a reporter, "you must work at it. Young women today want to prove something, but all they have to prove is that they can be a good wife, housekeeper, and mother. There should be only one head of the family—the man."[80] But Mamie was never as submissive as she sounded in this pronouncement, nor did Ike ever want her to be.

In April 1968, Ike suffered another heart attack and was taken to the Walter Reed Hospital in Washington, and spent his last months there with Mamie at his side. "We've had a wonderful life together," he told her a few days before he died. Just before the end he told son John: "Be good to Mamie."[81] After his death at seventy-eight in March 1969, Mamie returned to Gettysburg to live, but the house now "sagged" badly. "When Ike died," she told Julie Eisenhower, grandson David's wife, "the light went out of my life."[82]

Eisenhower dominated Mamie's thoughts until her own death at eighty-two in September 1979. "I miss him every day," she told friends;[83] and she told her son, "I feel like I'm fighting a one-person battle to keep his name alive."[84] Her rare public appearances in her last years centered on her husband. She visited his grave in Abilene, Kansas, every year to celebrate his birthday; she went to West Point for the dedication of a student center named Eisenhower Hall; she was on hand in Norfolk for the dedication of the super-carrier *Dwight D. Eisenhower;* she attended graduation exercises regularly at Eisenhower College, a liberal-arts institution founded in 1965 in Seneca Falls, New York, and helped raise money for the institution too. She also enjoyed going over to the campus of Gettysburg College, not far from her farm, to see the statue of Eisenhower there. "I always speak to him when I pass it," she said. "During the winter I don't like to see the snow covering his head."[85] There were buildings named after her too. At Eisenhower College the main academic building was called the Mamie Doud Eisenhower Hall. And there was a Mamie Eisenhower Pavilion at the Eisenhower Medical Center in Palm Springs, California, and she attended the dedication.

The student upheavals of the late 1960s and early 1970s bewildered Mamie; nor could she understand the rapidly developing feminist movement. When asked in 1973, "Do you think women need 'liberating'?" she replied, "No—but that is my own opinion."[86] The abbreviation, Ms., puzzled her. "I never knew," she told an interviewer, "what a woman would want to be liberated from."[87] She took the two-spheres view of the sexes for granted. "When Ike came home, he came home," she once said. "He left all his work at the office. I never went to his headquarters when he was an army officer. I only went to his White House office four times—and I was invited each time." She once summed it all up: "I was Ike's wife, John's mother, the children's grandmother. That was all I ever wanted to be. My husband was the star in the heavens. It wasn't that I didn't have my own ideas, but in my own era, the man was the head of the household . . . Ike and I never did the same things . . . I knitted and he painted. There was never any competition between us."[88] She once impishly told an interviewer that the secret of her long and happy marriage to Ike was that they had "absolutely nothing in common!" She explained: ". . . even though Ike and myself didn't do the same things, we always enjoyed each other's company."[89]

To the end Mamie scoffed at the Kay Summersby story. She even calmly watched the ABC series about her on television in March 1979, a few months before her death, and found it amusing. "Mickey," she told Michael McKeough, Ike's old orderly, who lived near her with his wife, "it doesn't bother me one way or the other."[90] When one of her friends began deploring the show, she exclaimed: "Now don't you bother your pretty little head about it for one minute. We both know it just isn't true."[91]

Mamie never doubted Ike. Nor did she doubt herself. Discussing Lord Mountbatten with a friend one day, she said firmly: "If I had once thought that there was an iota of truth in the Kay Summersby affair, I would have gone after Mountbatten. And believe you me, my friend, I could have gotten him!"[92]

* * * *

Less Spectacularly

One evening young Mamie Doud went dancing with her Uncle Joel at a night spot in Denver. First they did waltzes and fox trots, but when the band began jazzing it up they got energetically into the spirit of things. Suddenly one of the uniformed attendants interrupted them. "The management," he told them, "requests that you dance a little less spectacularly."[93]

Protecting Mamie

Soon after Ike married Mamie he was made Provost Marshal of his post and had considerable trouble at first keeping the untrained soldiers there in tow. Since Mamie was one of the few women in camp and had to be alone a lot he gave her a .45 pistol for protection and showed her how to use it. A few weeks later he decided to try her out. "Mamie," he said, "let's see you get your pistol out—as if there were somebody trying to break in through the front door." Mamie went hunting for the pistol which she had hidden behind the piano, inside a bedding roll, under a stack of other possessions. She had hidden it so well and it took her so long to locate it that Ike concluded "she couldn't have gotten it out in a week, much less in a hurry." He decided to concentrate on making the camp safer.[94]

Non-Stop

During the great mobilization of 1917, Ike's regiment moved to Leon Springs, twenty miles out of San Antonio, leaving Mamie separated from her husband without any mode of transportation between the two places. But one Sunday, Mamie resolved to drive to Ike's camp. Though she had never driven a car before, she decided to use the little jalopy Ike had left behind even though it wasn't in the best of shape. She knew how to start the car and how to put it into forward gear and that was all. After phoning Ike in advance she began her twenty-mile trip early in the morning when she would have the whole road to herself. En-route she stopped for nothing—she didn't know how—not even a rail-road track crossing. Ike waited nervously for her at the entrance to the camp, and when he saw her coming down the hill he breathed a sigh of relief. As she approached, he heard her crying, "Ike! Get on, get on quickly—I don't know how to stop this thing!" As he ran toward the car, she yelled: "Jump on!" He leaped onto the running board, opened the door, hopped in, and took over. He spent the rest of the day giving her driving lessons.[95]

Mac Returns

When General Douglas MacArthur returned to Manila after a short absence, his staff planned a big welcome-home dinner for him. The Eisenhowers were in their quarters dressing for the occasion when Mrs. Eisenhower felt suddenly dizzy. "I feel queer, Ike," she said. "What is it?" "It's not you," Ike told her, "it's an earthquake. By golly, it's a hum-dinger!" When the shock, which sent furniture tumbling about the room, ended, Eisenhower, who had his arm around his wife, announced: "It's all over. We'd better finish dressing." Smiling weakly, his wife ex-claimed: "Does this always happen when Mac comes back?"[96]

Bats

When Eisenhower took up duty in Camp Gaillard in the Panama Canal Zone in January 1922, his wife developed an almost instant dislike for the place. The house in which they were quartered was old and flimsy, damp and mildewed from frequent thundershowers, and crawling with ants, mosquitoes, cockroaches, and other vermin. Worst of all were the bats. The second night there, Mrs. Eisenhower heard a swish overhead, put on the light, and covered her head with a sheet. "Ike!" she screamed. "There's a bat in the room. Kill it!" "I can't do that, honey," said Ike. "It's against the law." He went on to explain that, years before, the French had brought bats to Panama to kill the mosquitoes and that it was against the law to harm them. Mamie was not convinced. "Law or no law," she cried, "kill that bat!" So Ike did just that. He grabbed his dress sword, chased the intruder around the room and finally yelled: "Got him!" Ike's frantic efforts turned Mamie's fright into mirth. "Ike was so grim and earnest," she recalled, "I didn't dare laugh. His long leaps and fancy sword-play on the furniture was a riot. I kept thinking of Douglas Fairbanks, the great leaper and jumper and swordsman of the movies. Only Fairbanks didn't fight bats."[97]

Announcement

One evening in November 1942, Mrs. Eisenhower joined the Milton Eisenhowers for talk and games at their place in Falls Church, Virginia, but couldn't keep her mind on mah-jongg because the radio was on full blast. "Please turn that darn thing off," she begged, but to her surprise, her brother-in-law insisted on keeping it going. Then a solemn voice interrupted the music. "We interrupt this program for an important announcement." Came the news: "American and British troops under the command of Lieutenant General D. Eisenhower are landing at several points on the coast of North Africa."[98]

Tea

When Eisenhower was Chief of Staff after World War II his wife had occasion to entertain former British Ambassador Lord Halifax and his wife at tea. But just as she was beginning to pour the tea, Eisenhower said, "Would anybody rather have a drink?" "Well . . . ." said Halifax, hesitantly. "No you don't!" cried Mrs. Eisenhower. "I went to all sorts of trouble to get the sort of tea Ike told me he used to have with you in England. Now you take your tea, and after that you can have a drink."[99]

Invitation

Just before Inauguration Day, Mrs. Eisenhower received a beautiful embossed invitation to the inaugural ball. "What should we do about this?" she asked Ike. "Turn it down," he said with a straight face. "Tell them we've got another engagement."[100]

Archenemies

On Inauguration Day, Mamie stood for five hours in the White House stand watching people march by and finally sought respite. She found a chair in the rear, sat down, slipped off her shoes, reached down, and began rubbing her arches. When news photographers spotted her and leaped forward to get some pictures she said she didn't mind. Every woman in the country, she said, would sympathize with her, though she added, with a wink at her mother standing nearby, that perhaps she shouldn't be caught giving aid and comfort to her archenemies.[101]

Which General?

At a dinner in Washington to honor General George C. Marshall, former Ambassador Joseph C. Grew, who presided over the festivities, misspoke himself. "General Marshall," he announced, "wants nothing more than to retire to his Leesburg, Virginia, home with Mrs. Eisenhower." As the banqueteers roared with laughter, Grew hastily cried: "My apologies to the general!" But as the room became quiet, Mrs. Eisenhower piped up: "Which general?"[102]

Mrs. Ike

"I've got *my* man right here," Mamie told White House Chief Usher J. B. West, "where I want him!" She loved the White House and was proud of being First Lady. After her first reception she asked for a platform "so everyone can see me." So West arranged for carpenters to build a little platform which elevated her about a foot and enabled her to look her visitors in the eye as she shook hands. But when she used the platform with people during her second reception, she found it didn't work. "We almost lost me," she told West the next day. "They nearly jerked me off the platform." So the next time she stood on the landing of the grand staircase overlooking the Green Room and waved to people as they passed by. A few days later she tried another stance: on the bottom step facing the lobby, with four leading Republican women on the second step right behind her. But this didn't satisfy her either and she told West she wanted the women to move back a couple of steps. "They're so close," she said, "nobody knows who is *me!*"

Mrs. Ike, as the President called her, enjoyed being the center of attention outside as well as inside the White House. She liked to ride in the back seat of the big White House limousine and wave to people on the street when they recognized her. In parades, when she rode in a car behind the President's car, somebody usually yelled, "Where's Mamie?" to the President in the car ahead, and she gleefully rolled down the window, poked her head out, and shouted, "Here I am, here I am!" For State dinners she disapproved of the U-shaped banquet arrangements whereby Presidents sat across the table from their wives. "I don't think the First Lady should have her back to so many of the guests," she said, and arranged to sit next to her husband in the throne-like mahogany chairs at the head of the E-shaped banquet table.[103]

Tardy

Unlike his wife, Ike was punctual. One day Mrs. Eisenhower arrived later than usual for dinner. "Do you realize," cried Ike, "that you have kept the President of the United States waiting?" "Why no," said Mrs. Eisenhower demurely. "I've been busy making myself pretty for my husband."[104]

Outside Kiss

Once, when Jimmy Carter was President, he visited Mrs. Eisenhower in Gettysburg and, to her surprise, kissed her on leaving. "My Lord," she told her doctor, "I didn't know what to do. I hadn't been kissed by a man outside of the family since Ike died."[105]

CHAPTER 33

JACQUELINE BOUVIER KENNEDY

1929–

Soon after she became First Lady in 1961, Jacqueline Kennedy became a celebrity. Her youthful charm, good looks, aristocratic poise, whispery voice, and stunningly attired figure quickly captured the hearts of millions of Americans—and foreigners too—and made her every word and deed the object of intense scrutiny, analysis, discussion, and commentary. French fashion director Madame Ginette Spanier announced that the new White House chatelaine possessed "star quality and chic," and American fashion designers heartily agreed.[1] Anthropologist Margaret Mead declared that Mrs. Kennedy had "a special kind of presence—a combination of qualities that Americans have long admired in young stage and screen stars but have seldom hoped to find in the wives of famous men."[2] Other admirers racked their brains for appropriate appellations: superstar, long-stemmed American Beauty, Her Elegance, First Lady of Glamour, Queen of America, Porcelain Princess, America's Newest Star. By 1962 Mrs. Kennedy was being called the "First Lady of the Western World" and was inspiring torrents of adulatory articles in TV and movie magazines as well as in the more respectable organs of opinion. And long after her husband's tragic death in 1963 she continued, despite her efforts to avoid reporters, to make headlines.

Mrs. Kennedy never sought fame. She cherished her privacy and shied away from the publicity that went with being a politician's wife. "It's frightening," she once exclaimed, "to lose your anonymity."[3] A couple of weeks after John F. Kennedy's election as President a woman approached her and said, "You're Mrs. Kennedy, aren't you? I recognize you from your pictures." Sighed the President-elect's wife, "I know. That's my problem now."[4] At JFK's inauguration in January 1961 she told friends resignedly: "I feel as though I have just become a piece of public property."[5]

As First Lady, Mrs. Kennedy's graceful behavior in public quickly turned her into a major asset to the President. She was delighted to find herself helping her husband in this way but insisted on preserving an area of privacy in the White House for her family and on keeping the children, Caroline and John, Jr., out of the public eye as much as possible. Her own favorite First Lady was the uncommunicative Bess Truman. "Mrs. Truman was always just Mrs. Truman," she said admiringly. "Her central responsibility was to be Mrs. Truman."[6] Jackie (as she was popularly called) came to enjoy being First Lady of the Land, but she resolved to go on being Mrs. Kennedy too.

Being Mrs. Kennedy meant holding onto some of the independence, even perverseness, that she had developed as a child. Born to wealth and privilege in New York in 1929, she learned to move with ease and assurance amid the conventionalities of her class without surrendering her individuality. Her parents were what celebrity-mongers would one day call "beautiful people." Her father, New York stockbroker John Vernou ("Jack") Bouvier III, one of "the 400," was "drippingly handsome," and her mother, Janet Lee, was also an attractive and stylish presence in high society.[7] The Bouviers accustomed Jackie to elegant living on the family estate at East Hampton, Long Island, in the summer, and in their Park Avenue apartment in Manhattan during the winter, and saw to it that she had a comfortable childhood surrounded by maids, butlers, nurses, and chauffeurs. "Jackie," her father once told her, "you never have to worry about keeping up with the Joneses, because we are the Joneses. Everybody has to keep up with us."[8]

But Jonesville was no paradise. Jackie's parents quarreled incessantly over money and over her father's infidelities, and in 1936, to the eight-year-old girl's distress, they separated, and she found herself with divided loyalties. Jackie adored her father; he was full of fun, generous with money, and deeply devoted to her, and she couldn't help enjoying his company far more than she did that of her rather sedate and thrifty mother. In time, though, as her parents vied for her affection, she learned to play them against each other when she wanted money for clothes, trips, and pets. And later on, when her mother divorced Bouvier and married Hugh D. Auchincloss, wealthy Washington lawyer and stockbroker, in 1942, she started playing her father and "Uncle Hugh," as she called her stepfather, against each other and ended by enjoying the best of both worlds: vacations with her father and her beloved horses at East Hampton, Long Island, winters at Merrywood, the Auchincloss country estate in Virginia, and summers at Hammersmith Farm, another Auchincloss estate just outside Newport, Rhode Island.

Jackie's girl friends thought she was different: independent-minded, at times withdrawn, a bit rebellious. "Sometimes," they complained, "you have the feeling she isn't really there." They thought her willful at times, too, and called her "Jacqueline Borgia."[9] At New York's fashionable Chapin School, which she attended for a time, her mischievousness won her the reputation of being "the very worst girl in the school" and forced the head-mistress, Ethel Stringfellow, to call her on the carpet just about every day. At length Stringfellow won her over by talking to her about horses (which Jackie at times seemed to prefer to humans) and convincing her that even thoroughbreds have to be schooled before they can perform. "I mightn't have kept Jacqueline," she said later, "except that she had the most inquiring mind we'd had in the school in thirty-five years."[10] Jackie's mother had mixed feelings about her daughter. "I like to use the word 'original' in describing Jacqueline," she once said. "She was brilliant, with strong feelings about things, gifted artistically, and always good in her studies."[11] As a child, though, Jackie was unusual chiefly for her exceptional skill as an equestrienne.

There was much of course that was unexceptional about young Jackie Bouvier. Like the other girls in her set she took dancing and ballet lessons as a child, went to dog and horse shows, attended finishing schools that taught rich girls how to behave with decency and decorum, and made her debut as a teen-ager. When she "came out" socially in Newport at age seventeen, society columnist Igor Cassini ("Cholly Knickerbocker") singled her out for special attention in the *New York Journal-American*. "Queen Deb of the Year is Jacqueline Bouvier," he told his readers, "a regal brunette who has classic features and the daintiness of Dresden porcelain. She has poise, is soft-spoken and intelligent, everything the leading debutante should be."[12] But Jackie was more than a run-of-the-mill young debutante. She read widely in British and Continental literature, wrote stories and poems for her family, friends, and school chums, with drawings to go with them, and did creditable classwork in every school she attended: at Chapin, Miss Porter's in Farmington, Connecticut, and, later on, Vassar College, where she became interested in art history. At the Sorbonne, where she spent her junior year in college, she developed fluency in French and a lasting affection for French art and literature.

In 1951, while completing her college education at George Washington University in Washington, Jackie decided to enter the *Prix de Paris* contest sponsored by *Vogue*. The rules were stringent: contestants had to submit a personal profile, the layout for an entire issue of the magazine, four technical articles on high fashion, and an essay of five hundred words on "People I Wish I Had Known." There were over a thousand entrants in the contest but Jackie won first prize. The judges liked everything she submitted but were particularly struck by the people she picked to write about: Russian ballet impresario Serge Diaghileff, French poet Charles Baudelaire, and British writer Oscar Wilde. They were even more intrigued by the reasons she gave for her choices, "Baudelaire and Wilde were both rich men's sons," she wrote, "who lived like dandies, ran through what they had, and died in extreme poverty. Both were poets and idealists who could paint sinfulness with honesty and still believe in something higher. . . . Diaghileff possessed what is rarer than artistic genius in any one field, the sensitivity to take the best of each man and incorporate it into a masterpiece all the more precious because it lives only in the mind of those who have seen it and disintegrates as soon as he is gone."[13]

Jackie's prize consisted of a year's position at *Vogue,* six months in the New York office and six months in the Paris office, with the possibility of permanent employment if she showed real promise on the job. After thinking it over, however, Jackie decided to turn it down. Her mother thought she had been away from home long enough, for one thing, and, for another, Jackie herself seems to have had doubts about making good. "I guess I was scared to go," she said years later. "I felt then," she added, "that if I went back I'd live there forever, because I loved Paris so much. . . ."[14] But she got a trip to Europe anyway in the summer of 1951, and afterwards she and her sister Lee, who had gone with her, put out an amusing little book, *One Special Summer,* about their peregrinations on the Continent. In one place in the book the two girls revealed their "Dreams of Glory." Lee, wearing a long gown and a supercilious expression, pictured herself as "Carolina, Dugesa de Bronville," and Jackie, with regal costume and crown, appeared as "Jacqueline, Fille Naturelle de Charlemagne."[15] At twenty-one, apparently, Jackie thought it more fun to think of herself as the daughter of an emperor than as a successful editor at *Vogue.*

But the *Vogue* contest drew Jackie to journalism. Soon after returning from Europe she decided to look for a job on a newspaper, and *New York Times* columnist Arthur Krock, a friend of the family, offered to help. Late in 1951 he telephoned Frank Waldrop, editor of the *Washington Times-Herald,* on Jackie's behalf. "Are you still hiring little girls?" he asked. "Yes," said Waldrop. "Well, I have a wonder for you," said Krock. "She's round-eyed, clever, and wants to go into journalism. Will you see her?" Waldrop said he would and a few days later set up an interview. "Do you want to go into journalism," he asked Jackie, "or do you want to hang around here until you get married?" "No, sir!" cried Jackie, "I want to make a career!" "Well," said Waldrop, "if you're serious, I'll be serious, if not, you can have a job dipping things." "No, sir!" repeated Jackie, "I'm serious." Waldrop then told her to come back after the holidays and added: "Don't you come back to me in six months and say you're engaged!" "No, SIR!" said Jackie fervently.[16]

Early in 1952 Jackie went to work. She became the *Washington Times-Herald*'s "Inquiring Camera Girl" and, to prepare her daily column, began traveling around town to interview people, take their pictures, and write up the answers to questions she asked about matters of current interest. Waldrop was impressed by her conscientiousness. "She was a business-like little girl," he said admiringly, "nice, quiet, concentrated, obviously very, very earnest in wanting to be a professional. She was self-sufficient, good at listening, and she handled her job efficiently."[17] One of her first interviews was with Vice President Richard M. Nixon's wife Pat. And one of her best, later on, was with Nixon himself and with the junior Senator from Massachusetts, John F. Kennedy. Nixon gave a solemn answer to her question about the page boys in the U.S. Senate, but JFK said laughingly it might be a good idea if the Senators and the pages changed places now and then.

By the time of the interview Jackie knew JFK fairly well. A year or so earlier Washington newsman Charles Bartlett had brought her and Jack, then a Congressman, together for a dinner party at his house, hoping to play matchmaker, and he was delighted when they took to each other at once. But after a few dates the association languished; Jackie even became engaged for a time to a young New York stockbroker, John G. W. Husted (to the dismay of her mother, who didn't think he was rich enough). It wasn't until 1952, when JFK launched his bid for the U.S. Senate, that he and Jackie began taking each other seriously. Even then, Kennedy's Senate campaign kept him in Massachusetts much of the time and he saw Jackie only on his periodic returns to Washington. "It was a very spasmodic courtship," Jackie recalled. "He'd call me from some oyster bar up there with a great clinking of coins to ask me out the following Wednesday in Washington." Once she showed friends a postcard he had sent from Bermuda saying, "Wish you were here. Cheers, Jack," and commented ruefully, "And that was my entire courtship correspondence with Jack!"[18]

With his election to the U.S. Senate in November 1952, however, Jack began pressing his suit more vigorously. There were movies together, bridge games, dinners with friends, and occasional lunches together at the Capitol. At thirty-six, JFK was ready to get married; he wanted to move ahead in politics and knew that the American people liked their public officials to have wives and children. But Jackie held herself back at first, occasionally stood him up, and sometimes left town without telling him beforehand; a woman, her father had told her repeatedly, should always play hard to get and let the man take the initiative. In time, however, she fell deeply in love with the handsome and wealthy Senator from Massachusetts, and to a friend finally confessed: "What I want more than anything else in the world is to be married to him."[19] Jack, for his part, was intrigued by Jackie; she was bright, witty, attractive, and companionable and her reticences piqued him, for he was used to easy conquests. She had "class," too, unlike the starlets and models he knew; her Bouvier and Auchincloss connections meant that marriage would raise his own social status a notch or two. He was never starry-eyed about her—"I'm not the heavy lover type," he once confessed—but he liked and respected her and probably developed more affection for her than for any other woman in his life.[20] Jackie was aware of JFK's reputation as a womanizer, but refused to let it bother her. Her father, after all, was a rake, and so was JFK's father. Weren't all men like that? "I don't think there are any men who are faithful to their wives," she once said grimly. "Men are such a combination of good and evil."[21]

In May 1953, Jackie surprised Jack by taking off for London to cover the coronation of Queen Elizabeth II for the *Washington Times-Herald.* Jack read her witty reports and lively sketches for the newspaper with delight, but was somewhat dismayed by the letters she sent him about the parties she was attending and the young men she was meeting in London. "ARTICLES EXCELLENT," he cabled her, "BUT YOU ARE MISSED."[22] When she finally returned to Boston, loaded with books she had purchased for him, he was eagerly awaiting her in the airport. Twenty-four hours later they were engaged. "Aunt Madie," Jackie telephoned one of her relatives right afterward, "I just want you to know that I'm engaged to Jack Kennedy." When her aunt expressed pleasure at the news, Jackie went on: "But you can't tell anyone for a while, for it wouldn't be fair to the *Saturday Evening Post.*" "What," cried her aunt, "has the *Saturday Evening Post* to do with it?" "The *Post* is coming out tomorrow, with an article on Jack," Jackie explained. "And the title is on the cover. It's 'Jack Kennedy—the Senate's Gay Young Bachelor.' "[23] Shortly after the engagement Jack told his fiancée that when he met her at the Bartletts back in 1951 he had decided that when he was ready to get married she was the one. "How *big* of you!" she exclaimed.[24]

The engagement was announced late in June 1953, just before Jackie's twenty-fourth birthday, and the wedding took place in Newport on September 12. The Kennedy-Bouvier wedding was a big social event, attended by Kennedy politicos as well as Bouvier and Auchincloss socialites. There were ten bridesmaids, fourteen ushers, six hundred guests in St. Mary's Church, where Archbishop Richard Cushing officiated, and 1,700 guests at Hammersmith Farm for the reception afterward. When crowds of people pressed around the newly-weds as they emerged from the church after the ceremony, Jackie was upset and a bit frightened, but JFK, a seasoned politician, took it in his stride. Jackie's father had planned to give the bride away, but when he got slightly tipsy in his hotel room before the ceremony, Jackie's mother barred him from the church, and Hugh Auchincloss took his place. Later, when Jackie learned what had happened, she wrote her father a long and affectionate letter that moved him to tears. "Only a rare and noble spirit could have written that," he told friends.[25]

After honeymooning in Acapulco, the Kennedys settled down in Washington's fashionable Georgetown, and, to please JFK, Jackie took a course in American history at the Georgetown School of Foreign Service. But she didn't like it much; "American history," she decided, "is for men."[26] So was politics. A Republican (like the Bouviers and Auchinclosses) before she met JFK, she had never bothered to vote and remained practically apolitical for some time even after switching her allegiance to her husband's party. She once complained to a friend about "all those boring politicians" she had to meet through her husband and said she much preferred the company of artists, writers, and photographers.[27] "She breathes all the political gases that flow around us," JFK once remarked, "but she never seems to inhale them."[28] Still, she strove to be a good Senate wife. She translated French articles and books about Indochina for her husband, accompanied him on speaking tours, and arranged to send hot lunches to his Senate office to keep him from skipping meals. Above all, she got him to take more care with his grooming, educated him in good food, broadened his appreciation of the arts, modified his abrupt ways with guests who bored him, and sought to provide him with a place of comfort and relaxation at home away from his work. "I'm an old-fashioned wife," she told an interviewer, "and I'll do anything my husband asks me to do."[29]

But there were clashes. JFK was shocked by his wife's extravagances, and Jackie was bothered more than she thought she would be by his flirtations. Sometimes she sulked; and JFK, who found the tension intolerable, did everything he could to get her in a good mood again. He once compared himself to a straight line and Jackie to a wavy line intersecting it because of her ups and downs. But for all the differences in taste and temperament the two had much in common. "They were so much alike," their friend Lem Billings insisted. "Even the names—Jack and Jackie: two halves of a single whole. They were both *actors* and I think they appreciated each other's performances. It was unbelievable to watch them work a party. Jackie would be sitting with some old guy who'd almost nodded off and suddenly ask a question so filled with implied indiscretion that this old guy's eyes would almost pop out of his head. And for the remainder of the conversation he'd practically be married to her in intimacy. Jack was exactly the same way. Both of them had the ability to make you feel that there was no place on earth you'd rather be than sitting there in intimate conversation with them."[30]

In October 1954, JFK entered the Hospital for Special Surgery in New York for the first of two spinal operations and at one point was close to death. While he was in the hospital and during his long convalescence afterward, Jackie helped brighten his life. She read to him, played checkers and Monopoly with him, brought him amusing presents, and persuaded him to take up painting as a diversion. At one point she had actress Grace Kelly don a nurse's uniform, walk into his hospital room, and announce: "I'm the new night nurse." ("I must be losing it," sighed Kelly when JFK turned out to be too sick to recognize her.[31]) When JFK began work on his Pulitzer Prize-winning book *Profiles in Courage* (1956) during his confinement, Jackie pitched in to help: did research for him, helped organize material, and took down his notes. "This book," he announced in the preface, "would not have been possible without the encouragement, assistance, and criticism offered from the very beginning by my wife, Jacqueline, whose help during all the days of my convalescence I cannot ever adequately acknowledge."[32] JFK was back at his Senate desk in May 1955, but there was more sadness for the Kennedys that year: in her first pregnancy Jackie had a miscarriage.

In August 1956, Jackie, pregnant again, went to Chicago with her husband for the Democratic National Convention, but spent most of her time in the hotel room and missed his unsuccessful bid for the vice-presidential nomination. When the convention ended, JFK flew to the Riviera for a Mediterranean cruise and Jackie joined her mother at Hammersmith Farm to await the arrival of the baby. On August 23, she was rushed to the hospital, had an emergency Caesarean performed, and gave birth to a stillborn baby one month prematurely. When JFK heard the news he was reluctant at first to cut short his fun-making, but when a friend told him, "You better haul your ass back to your wife if you ever want to run for President," he caught the next plane home.[33] Rumors of an estrangement began circulating in Washington about this time; there was even a story, for which there is no evidence, that JFK's father paid Jackie a million dollars to stay with his son. The marriage was undoubtedly experiencing stresses and strains, but there is no reason to believe that either JFK or his wife was seriously contemplating separation. For their next wedding anniversary Jackie drew some amusing sketches entitled "How the Kennedys Spoil Wedding Anniversaries," reflecting on the troubles they had seen. One sketch showed JFK in a hospital bed with Jackie beside him; another showed Jackie in bed, with JFK at her side.[34] But children finally arrived—Caroline in November 1957 and John, Jr., in November 1960—and the marriage seems to have taken a turn for the better after that. Both the Kennedys loved children and became devoted parents. But JFK also realized children were a political asset and saw to it that the christening of their first-born received a great deal of publicity.

In January 1960 Kennedy announced his candidacy for the Democratic nomination for President and soon after entered his name in the primaries being held that year. Jackie offered to help but he was dubious at first about her political value. He thought she was too aristocratic for the average voter; she had too much status, he said, and not enough quo. "The American people just aren't ready for someone like you," he told her. "I guess we'll just have to run you through subliminally in one of those quick flash TV spots so no one will notice."[35] She was reduced to tears by his remark; but she was determined to be a good campaign wife. When a reporter asked her opinions about marriage, she gave an impeccable reply. "You do what your husband wants you to do," she declared. "My life revolves around my husband. His life is my life. It is up to me to make his home a haven, a refuge, to arrange it so that he can see as much of me and his children as possible—but never let the arrangements ruffle him, never let him see that it is work. . . . I want to take such good care of my husband that, whatever he is doing, he can do better because he has me. . . . His work is so important. And so exciting."[36]

Before long, though, Jackie was surprising JFK—and herself too—by the way she took to campaigning. Her first public appearances at her husband's side were enormously successful; she was, after all a political rarity: a presidential candidate's wife who was still young and beautiful and at the same time witty, charming, and intelligent. When she accompanied JFK on his rounds he sometimes told an aide: "As usual, Jackie's drawing more people than we are."[37] But she was soon doing more than simply appearing in public with her husband. She began writing a weekly column, "Campaign Wife," about her activities for circulation among JFK workers; she also gave press conferences, held fund-raising teas, invited leading Democratic women into her house for discussions, sponsored a telephone campaign soliciting the advice of women on the conduct of the campaign, and organized listening parties for the TV debates with Republican candidate Richard M. Nixon. Sometimes she even ventured to give a few speeches on her own when JFK had to be back in Washington. In one town she went into a supermarket, picked up the store's microphone, and announced: "Just keep on with your shopping while I tell you about my husband, John F. Kennedy." Then she talked briefly about his service in the U.S. Navy during World War II and the work he had done in Congress since 1946 and concluded: "He cares deeply about the welfare of this country. Please vote for him."[38]

In Wisconsin, Jackie talked to dairy farmers, and in West Virginia visited miners in their shacks and chatted with their wives. On appropriate occasions, too, she made little speeches in French, Spanish, and Italian, to the delight of foreign-born citizens, the irritation of Republicans, and the amusement of her husband. ("I assure you," he told one crowd, "that my wife can also speak English."[39]) "People say I don't know anything about politics," she told reporters at one point. "And, in a way, it's true. But you learn an enormous amount just being around politicians."[40] She seemed at long last to be inhaling the political gases around her. "It's the most exciting life you can imagine," she exclaimed, "always involved with the news of the moment, meeting and working with people who are enormously alive, and every day you are caught up with something you really care about."[41] Long before the campaign ended JFK had revised his opinions about his wife. "When we first married," he reflected, "my wife didn't think her role in my career would be particularly important. I was already in the Senate and she felt she could make only a limited contribution. Now, quite obviously, I'm in a very intense struggle—the outcome uncertain—and she plays a considerable role in it. What she does, or does not do, really affects that struggle. Since I'm completely committed, and since she is committed to me, that commits her."[42]

Jackie was pregnant again when the campaign got rolling and this, too, became a political asset. Crowds loved it when JFK referred to his wife's condition in speeches and predicted the baby would be a boy because that was what Jackie had told him. When someone remarked that it would be just like Jackie to have the baby on the eve of the election, to sway the voters, she exclaimed mock-seriously, "Oh, I hope not. I'd have to get up the next day to go and vote." Asked what she would do if her baby was born on Inauguration Day, she asked innocently, "When's Inauguration Day?" and was surprised by the laughter."[43] The baby—John, Jr.—didn't come until three weeks after JFK's victory that November, and the President-elect was on his way to Florida when he heard that Jackie had gone into labor. "I'm never there when she needs me," he exclaimed and at once headed back to Washington to be at her side.[44]

Before entering the White House, Jackie took a conventional view of her future role as First Lady. "I'd be a wife and mother first," she told reporters, "then first Lady." As always, she gave her husband precedence in the scheme of things. "I would do whatever he wanted . . . ," she declared. "I certainly would not express any views that were not my husband's. I get all my views from him—not because I won't make up my mind on my own, but because he would not be where he is unless he were one of the most able men in his party, so I think he's right."[45] But Kennedy wanted her to define her role as First Lady herself. "A man marries a woman, not a First Lady," he said. "If he becomes President, she must fit her own personality into her own concept of a First Lady's role. People do best what comes naturally."[46]

Jackie did not seem to take naturally to her new role. At any rate she did a lot of complaining about it at first. She compared herself to a moth hanging on a windowpane; she also said she felt as though she were in a fishbowl with all the fish outside looking in. "The one thing I do not want to be called is 'First Lady,' " she told Chief Usher J. B. West at the outset. "It sounds like a saddle horse. Would you notify the telephone operators and everyone else that I'm to be known simply as Mrs. Kennedy and not as First Lady?"[47] Before long, however, she found herself enjoying her new position. Though she never reconciled herself to the invasions of privacy that go with celebrity, she did discover that being the President's wife gave her opportunities to make use of her special skills and interests in ways that aided her husband as well as satisfied her own desire for self-expression.

After a few White House receptions and dinners Mrs. Kennedy became Washington's most popular hostess. From the beginning she discarded the stiff formality of traditional State dinners and turned them into informal but elegant affairs at which there was good talk and fun as well as good food and wine, and plenty of culture too. The guests might be artists, writers, and musicians as well as politicians and businessmen, and after dinner Isaac Stern or Pablo Casals or a Shakespearean troupe might perform for the guests. "Mrs. Kennedy," wrote a British journalist admiringly, "has substituted gaiety, informality and culture for the traditional stuffed shirt."[48] After one evening in the White House the *Boston Globe*'s Betty Beale announced: "That White House dinner should end for the time being any further question as to who is now the No. 1 hostess in Washington. Presidents' wives have always held this title in name, but Jacqueline Kennedy now holds it in fact, and there are no two ways about it."[49] Jacqueline Kennedy became No. 1 fashion leader, too, as the youthful "Jackie look" swept the nation, though she denied wanting to influence women's fashions. "I have no desire to," she insisted, a bit disingenuously. "That's about at the bottom of things I attach importance to."[50]

Mrs. Kennedy attached more importance, at least for a time, to redoing her new home. When she toured the White House a few weeks before her husband's inauguration, she was appalled by its drabness. "It looks like it's been furnished by discount stores," she sighed. "It looks like a house where nothing has ever taken place. There is no trace of the past."[51] Soon after becoming First Lady she launched an ambitious plan to make the Executive Mansion a "national historic object," which by its furniture, paintings, sculpture, and other furnishings would reveal something about the history of the Presidency from the earliest years. After persuading Congress to designate the White House as a national museum, she enlisted the aid of museum directors, historians, and art experts, formed committees to supervise the project, and began soliciting gifts of furniture (and funds for acquiring furniture) with historical as well as artistic value for the White House. She also rummaged through White House storerooms and was delighted by some of her findings. "When the First Lady gets on the trail of something," said one of her advisers, "she never gives up. If you have a bed that used to be in the White House, she'll have you sleeping on the floor before you know what happened."[52]

Mrs. Kennedy's objective in all this was "restoration," not "redecoration"; she wanted to restore to the White House's rooms the furnishings of earlier times—James Monroe's, Abraham Lincoln's—in order to remind Americans of their nation's past. When her work was finished, she prepared a *Historic Guide to the White House,* the first one ever published, which became immediately popular with tourists who began flocking to the White House in greater numbers than ever before to see what she had done. In February 1962, she also appeared on television, touring the White House, explaining the restoration, room by room, and answering questions posed by interviewer Charles Collingwood about White House history. Writer Norman Mailer found something meretricious in her performance ("One did not feel she particularly loved the past of America."), but most viewers—over 48,000,000—approved.[53] "Unlike past renovations of White House interiors," said *Time,* "Mrs. Kennedy's is likely to be lasting."[54] The President was impressed by the executive ability his wife had shown in the restoration project.

JFK was even more impressed by the development of his wife into what he called his "number one ambassador of good-will."[55] In Canada, where the two of them spent three days early in 1961, Jackie made an excellent impression whenever she appeared in public. "Her charm, vivacity, and grace of mind," declared one Canadian leader, "have captured our hearts."[56] In France, where the Kennedy's went in May for conferences with President Charles deGaulle, she was even more of a success. The French adored her; they liked her Bouvier French background, facility in their language, splendid attire, and youthful beauty, and gathered in huge crowds to greet her—*Vive Jackie, Vive Jackie, Vive Jackie!*—wherever she and her husband went. "It quickly became evident," wrote one reporter, "that the radiant First Lady was the Kennedy who really mattered."[57] At a luncheon for press correspondents the last day in Paris, JFK told the guests: "I do not think it altogether inappropriate to introduce myself. I am the man who accompanied Jacqueline Kennedy to France."[58] In Vienna, Jackie charmed Soviet Premier Nikita Khrushchev (who was having stormy sessions with JFK), and in London, where she and her husband dined with Queen Elizabeth, she was also a big hit. "Jacqueline Kennedy," declared the *Evening Standard,* "has given the American people from this day on one thing they had always lacked—majesty."[59] And so it went; everywhere she traveled, she took the natives by storm. In India (where she was called "Queen of America") and Pakistan, which she visited with her sister Lee in the spring of 1962, she created such warm feelings for her country that after her return to Washington the U.S. Information Agency produced a color film, "Jacqueline Kennedy's Asian Tour," for which she did the narration, to distribute around the world.[60]

JFK couldn't help being proud of his still essentially apolitical wife. "Kay," he told Kay Halle, a White House aide, "you have no idea what a help Jackie is to me, and what she has meant to me."[61] But for all her success as First Lady Jackie continued to insist that her primary interest was her family. "Jack's always proud of me when I do something like this," she said after the Asian trip, "but I can't stand being in front. I know it sounds trite, but what I really want is to be behind him and to be a good wife and mother."[62] During the Cuban missile crisis with Russia in the fall of 1962, when JFK asked if she wanted to move closer to the special underground shelter for the First Family, she said that if war came she would go to his office and share the consequences. "Jack appreciated her," their friend "Chuck" Spalding insisted, when doubts were raised about their compatibility. "He really brightened when she appeared. You could see it in his eyes; he'd follow her around the room watching to see what she'd do next. Jackie *interested* him, which was not true of many women."[63]

But there were persistent tensions. Jackie's spending habits continued to bother JFK. In 1961 her personal expenditures for clothes, paintings, expensive foods, and antiques amounted to over $105,000; and in 1962, when she was supposed to be economizing, they exceeded $121,000. JFK repeatedly remonstrated with his wife but it cannot be said he made much progress in the "Battle of the Budget." He was even more upset by her periodic retreats into seclusion and her insistence on exercising what she called "the PBO" (polite brush-off) on inappropriate occasions. "I can't stand those silly women," she said when she refused to attend a Congressional Wives' Prayer Breakfast.[64] JFK was furious when she boycotted the Distinguished Ladies' Reception held in her honor. "Jesus Christ," he stormed, "you can't do this!" But he ended by going himself in her place.[65] Jackie always pleaded family when she retreated into privacy. "If I were to add political duties," she explained, "I would have practically no time with the children."[66] She tried hard to give her children as normal a life as possible in the White House fishbowl. She read to them, taught them French, reserved playtime for them, and organized a nursery school so they could make friends their own age. "If you bungle raising your children," she once said, "I don't think whatever else you do well matters very much."[67]

More serious was the tension over JFK's continued philandering. Though he had vowed "to keep the White House white" before becoming President, he was soon straying far and wide from the marital path after his inauguration.[68] He had frequent meetings with Judith Campbell (mistress of Chicago Mafia capo Sam Giancana), a young beauty whom he had met through singer Frank Sinatra during the 1960 campaign; and also saw a lot of Mary Meyer, *Washington Post* editor Ben Bradlee's married sister-in-law. He enjoyed, too, the company of a couple of blondes from the White House secretarial pool whom he called "Fiddle and Faddle" (Jackie called them "the White House dogs"). There were rumors, moreover, about assignations with some of Hollywood's choicest sex queens—Angie Dickinson, Jayne Mansfield, Kim Novak, Marilyn Monroe—to say nothing of countless starlets with whom Sinatra put him in touch. Jackie tried to ignore her husband's behavior, but there is little doubt that she was deeply hurt by it all. Once she handed him a pair of panties she found in her pillow slip and cried: "Here, would you find who these belong to? They're not my size."[69] Perhaps to show her resentment, she began spending a great deal of time away from the White House: three or four days a week with her children in the horse and hunt country at Glen Ora, Virginia, frequent trips to New York City, Palm Beach, and Hyannis Port, and vacations with her sister Lee in Italy and Greece. Comedian Jimmy Durante's famous end-of-program valedictory, "And goodnight, Mrs. Calabash, wherever you are," was parodied in Washington: "And goodnight, Mrs. Kennedy, wherever you are."[70]

In August 1963, Jackie gave birth to a baby boy, Patrick Bouvier Kennedy, but he died of an infection after thirty-nine hours, and the loss seems to have brought her and JKF together again. "Jack and Jackie were very close after Patrick's death," according to Bill Walton. "She hung onto him and he held her in his arms—something nobody ever saw at any other time because they were a very private people."[71] A couple of weeks later JFK began making plans for a trip to Texas to do some politicking in preparation for the 1964 presidential campaign and asked his wife: "Maybe now you'll come with us to Texas next month?" "Sure I will, Jack," she said promptly.[72] "You're going to be very proud of me," she told Letitia Baldridge, her former social secretary. "I'm going to start campaigning."[73] In St. Louis a newspaper vendor looked at a headline about her forthcoming trip to Texas with the President and broke out into a big smile. "Now everything's going to be great," he told Charles Bartlett, the Tennessee editor who had brought the Kennedys together back in 1952. "This is a girl with a lot of savvy."[74]

In Texas, as elsewhere, Jackie attracted admiring attention wherever she and the President went. When they arrived at the Rice Hotel in Houston on November 21, the manager looked past JFK toward Jackie and cried, "Good evening, Mrs. President!"[75] Later on, when JFK asked his aide David Powers how the crowd greeting him compared with the one on his last visit to Houston, the latter told him: "Well, Mr. President, about as many turned out to see you as the last time. But there seem to be a hundred thousand more shouting, 'Jackie! Jackie!' "[76] "You see, Jackie," Kennedy said, "you really are important to me." With a big smile Jackie exclaimed: "I'm looking forward to campaigning with you in 1964."[77] When she learned that a group of Latin-American citizens was waiting downstairs in the hotel she asked her husband, "Would it be helpful if I went down and talked to them?" "It would indeed," JFK assured her. Her little speech in Spanish produced a big ovation and afterwards she cried: "Oh, it's wonderful, wonderful! I love it!"[78] When the presidential party left Houston for Forth Worth, Kennedy told Powers: "I'm glad Jackie is pleased with the trip. That's a definite plus, isn't it?"[79]

Fort Worth was just as friendly as Houston and once more Jackie was the center of things. When JFK came out of his hotel to address a crowd, people started clamoring, "Where's Jackie? Where's Jackie?" "Mrs. Kennedy is organizing herself," JFK explained. "It takes a little longer, but, of course, she looks better than us when she does it."[80] At a Chamber of Commerce breakfast he told the people gathered to honor him, "Two years ago I introduced myself in Paris by saying that I was the man who had accompanied Mrs. Kennedy to Paris. I am getting somewhat the same sensation as I travel around Texas."[81] Afterwards Jackie told him: "I'll go anywhere with you this year." "How about California in the next two weeks?" "I'll be there," she said. Beaming, JFK turned to Powers—"Did you hear *that?*"[82]

When JFK heard that the weather was going to be good in Dallas on November 22, he ordered the bubble top off his car. "I want all these Texas broads to see what a beautiful girl Jackie is," he told his aides.[83] Just before heading for Dallas, though, he showed his wife the full-page, black-bordered advertisement appearing in the *Dallas News* that morning accusing him of treason. "We're heading into nut country today," he told her. Then he added thoughtfully: "But, Jackie, if somebody wants to shoot me from a window with a rifle, nobody can stop it, so why worry about it?"[84] When the presidential limousine was ready to roll into Dallas, Powers gave Jackie instructions. "Be sure to look to your left, away from the President." he said. "Wave to the people on your side. If you both wave at the same voter, it's a waste."[85]

In Dallas, as in other Texas cities, the crowds gathered to greet the Kennedys were enthusiastic. "You sure can't say Dallas doesn't love you, Mr. President," exclaimed Texas Governor John Connally's wife. "No, you can't" grinned JFK.[86] Seconds later he was shot in the head by Lee Harvey Oswald firing from the sixth-floor window of the Texas Textbook Depository in Dealey Plaza. As Jackie saw part of her husband's head explode, she recoiled, scrambled onto the trunk of the car, clasped hands with the Secret Service man climbing onto the car from the street, and returned to her seat as the car started racing to the hospital. "My God, what are they doing?" she screamed. "My God, they've killed Jack, they've killed my husband. . . . Jack, Jack!"[87] At the hospital afterwards and on the plane where she sat by her husband's casket on the flight back to Washington she refused to change her blood-stained dress, even for the swearing-in of Vice President Lyndon B. Johnson as President. "I want them to see what they have done to Jack," she told LBJ.[88]

Back in Washington Mrs. Kennedy supervised arrangements for her husband's funeral on November 25 with what her mother called "a truly remarkable sense of the fitness of things." She chose St. Matthew's Cathedral for the funeral service and Arlington (rather than Boston) for the burial place, and afterwards received dignitaries from all over the world at the White House with a quiet charm and poise that made a tremendous impression on people, the world over, who watched the proceedings on television. "Have you ever seen anything like your cousin?" exclaimed France's Jean Monnet when he came across John Davis at the White House reception. "She was fantastic, wasn't she?" nodded Davis. "Extraordinary," cried Monnet. "Such strength I have rarely seen in all my years in public life. She was an example to the entire world." After the funeral Mrs. Kennedy appeared on television to thank the hundreds of thousands of people who had sent their condolences.[89] Before leaving the White House she placed an inscription on the mantel in the Lincoln bedroom she and JFK had occupied. "In this room lived John Fitzgerald Kennedy with his wife Jacqueline during the two years, ten months,and two days he was President of the United States."[90]

Out of the White House Mrs. Kennedy's popularity continued undiminished. She was, according to a Gallup Poll, the most admired woman in the world. So many tourists came by her Georgetown home hoping to catch a glimpse of her that in the fall of 1964 she moved to New York City, hoping to gain more privacy. A year after her husband's assassination a State Department official declared that she "could become, if she wanted to, the most powerful woman in the world."[91] Some Democrats wanted her to seek the vice-presidential nomination in 1964; others suggested she seek a Senate seat, accept an ambassadorial post, write a newspaper column, or start appearing regularly on television. But the former First Lady rejected all of these suggestions. She concentrated instead on building up the Kennedy legend.

In an article for *Life* not long after the assassination, Theodore White revealed that Mrs. Kennedy had come to think of the Kennedy years in the White House in terms of the mythical kingdom of Camelot. When JFK was a boy, she told White, he dreamed of performing valiant deeds like one of King Arthur's knights, and as President his favorite music was that from the famous Alan Jay Lerner-Frederick Loewe Broadway musical entitled *Camelot* (1960). "Don't let it be forgot," she told White, quoting lines from one of the *Camelot* songs, "that once there was a spot, for one brief shining moment, that was known as Camelot."[92] To canonize JFK the former First Lady persuaded President Johnson to change the name of Cape Canaveral to Cape Kennedy (in 1973 it was changed back), arranged a special tribute to JFK for delegates to the 1964 Democratic Convention, and worked hard planning the John F. Kennedy Memorial Library in Boston (opened in 1979). In a memorial issue on the first anniversary of JFK's death *Look* quoted her as saying, "Now I think that I should have guessed it could not last. I should have known that it was asking too much to dream that I might have grown old with him and see our children grow up together. So now he is a legend when he would have preferred to be a man. . . ."[93]

In 1965, Mrs. Kennedy approved the request of William Manchester to do a book on the assassination (*The Death of the President,* 1967) and consented to interviews. But in 1966, when his publisher sold serialization rights to *Look* she accused him of commercializing the assassination; she also objected to certain passages in the manuscript—JFK traipsing around in shorts before going to bed, Jackie peering into the mirror to see if she had wrinkles—and went to court seeking an injunction forbidding publication. In the end Manchester made some minor changes in the manuscript and agreed to turn an even larger share of the book's earnings over to the Kennedy Library than already stipulated, and the suit was dropped. "I think it is so beautiful what Mr. Manchester did . . . ," Jackie told the *New York Times.* "All the pain of the book and now this noble gesture of such generosity. . . ."[94]

Jackie herself was less generous. Though she received a Presidential widow's pension, a government allotment of $50,000 a year for office expenses, and an annual income of $200,000 a year from a trust fund set up by JFK, she continued as of old to spend lavishly on herself and felt obliged to become thoughtfully thrifty when it came to paying her employees. She resisted giving her secretary Mary Gallagher a salary raise ($12,000 out of her $50,000 government allowance), even though the latter neglected her family to serve her. She also balked at paying her maid, Providencia Paredes, for working overtime, though "Provo," as she was called, toiled far into the night for her. To save money, too, she suddenly dismissed Evelyn Lincoln, JFK's secretary, who had been spending long hours on the Kennedy correspondence, and when the latter complained bitterly, she said plaintively: "Oh, Mrs. Lincoln, all this shouldn't be so hard for you, because you still have your husband. What do I have now? Just the Library."[95]

In 1968, when her brother-in-law Robert F. Kennedy, with whom she had become very close after JFK's assassination, launched his campaign for the Democratic presidential nomination that year, she offered to help out. She was shattered when he was felled by an assassin in June after winning the California primary, and even contemplated leaving the country. "I hate this country," she told a friend after RFK's funeral. "I despise America and I don't want my children to live here anymore. If they're killing Kennedys, my children are number one targets. . . . I want to get out of this country."[96] The following summer she shocked her admirers around the world when she announced her engagement to Aristotle Socrates Onassis, wealthy Greek shipowner in his sixties, whose reputation as a businessman was none too good and whose rough and ready ways seemed so alien to those of the former First Lady. "Jackie," a friend told her, "you're going to fall off your pedestal." "That's better than freezing there," Mrs. Kennedy replied. "Tucky," she told her friend Nancy Tuckerman, "you don't know how lonely I've been."[97]

The marriage—a Greek Orthodox ceremony—took place in October 1969 on Onassis's private island of Skorpios, and the news that a prenuptial contract had provided her with $5,000,000 added to the general dismay over what she had done. "Jackie—," cried a London newspaper, "How Could You?" "America Has Lost a Saint," observed a West German newspaper. "All the World Is Indignant." France's *Soir* found the marriage "sad and shameful." Said a Kansas City housewife: "I feel almost as bad as when Jack was assassinated."[98] "We all thought she would marry someone like André Malraux," a Polish writer told Jackie's first cousin John H. Davis, "and here she goes off with a gangster."[99] But Davis did not share the surprise of Jackie's fans around the world over her choice of husbands. She had always liked older men, he pointed out, and she had learned from her parents to marry rich. The Onassis marriage, moreover, gave her the security and companionship she desperately needed after the second Kennedy assassination. "I think she married Onassis because he represented an anchor of security and strength," said former Kennedy official Roswell Gilpatric. "He could take her out of herself, take her away to other places, other life-styles."[100]

For a time the marriage was successful. The Onassises got along well and enjoyed each other's company and took pride in each other. Ari was proud of being married to a celebrity (he got into the *Social Register* after the wedding) and Jackie enjoyed being married to one of the richest men in the world. Then the relationship began going downhill. Onassis, like JFK before him, was appalled by his wife's profligacy and there were quarrels over money. The loss of his son in a plane accident in 1973 also hurt the marriage, for he began regretting he hadn't married a younger woman who could provide him with another son. His health was declining, too, and he decided to revise his will to reduce the amount Jackie would receive upon his death. Toward the end of his life, he was even contemplating divorce. In January 1974, he became seriously ill and entered the American Hospital in Paris; two months later he was dead. After the funeral Jackie issued a graceful statement about her second husband. "Aristotle Onassis rescued me at a moment when my life was engulfed with shadows," she said. "He meant a lot to me. He brought me into a world where one could find both happiness and love. We lived through many beautiful experiences together which cannot be forgotten, and for which I will be eternally grateful."[101]

In 1975 Mrs. Kennedy settled in New York again and started working for the Viking Press as a consulting editor. In 1977, however, when Viking made plans to publish a novel by a British writer which she thought exploited JFK's assassination, she handed in her resignation. In the spring of 1978 she joined the editorial staff at Doubleday and the writers with whom she worked found her shrewd, perceptive, conscientious, and imaginative. "She doesn't want to be known as the wife of the President of the United States or as the wife of one of the richest men in the worked," one of her friends observed. "She wants to be known as *herself*. Of course she's ambitious, and wouldn't settle for being anything second-rate. She wants to be important—and in her own right."[102] Asked to name her greatest achievement, Mrs. Kennedy once said softly: "I think it is that after going through a rather difficult time, I consider myself comparatively sane."[103]

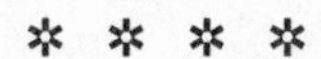

Lost

One day, shortly after the four-year-old Jackie went with her nurse to play in Central Park, the phone rang at home and when Mrs. Bouvier answered it, she found it was the police. "We have a little girl here," they told her. "We can't understand her name, but she knows her telephone number. Could she be yours?" When Mrs. Bouvier arrived at the police station she found Jackie chatting with the lieutenant. "Hello, mummy," cried the little girl when she saw her mother. The lieutenant explained what had happened. He found Jackie walking alone in the park, and when she saw him she stepped up to him and announced: "My nurse is lost."[104]

Chekhov

"Mummy," said the six-year-old Jackie one day. "I liked the story of the lady and the dog." She was reading Chekhov's short stories. "Did you understand all the words?" asked her mother, somewhat surprised. "Yes," said Jackie, "except what's a midwife?" Her mother tried to explain and then asked: "Didn't you mind all those long names?" "No," said Jackie, "why should I?"[105]

Says a Lot of Things

When Jackie was attending the Chapin School, she got into mischief a lot and was frequently sent to headmistress Stringfellow's office for disciplining. "What happens," her mother once asked her, "when you're sent to Miss Stringfellow?" "Well," said Jackie, "I go to her office, and Miss Stringfellow says, 'Jacqueline, sit down. I've heard bad reports about you.' I sit down. Then Miss Stringfellow says a lot of things—but I don't listen."[106]

Christmas Gift

When Christmas rolled around, Jackie asked the other photographers at the *Washington Times-Herald* what they would like as a gift and they all said they hoped it would be in a bottle. A few days later she presented them with a bottle, beautifully wrapped in a lovely box, with all their names listed on a card attached to it. Eagerly they unwrapped her gift. It turned out to be a quart of milk.[107]

Penny

"A penny for your thoughts," JFK once said to Jackie at dinner. "If I told them to you," said Jackie, "they wouldn't be mine, would they, Jack?"[108]

Waiter

Shortly after their marriage the Kennedys took a vacation in the south of France and had dinner on Aristotle Onassis's yacht, where Winston Churchill, one of JFK's heroes, was guest of honor. Wearing his best formal attire, Senator Kennedy tried hard to make a good impression, but Churchill practically ignored him. As he left the yacht disappointedly, Jackie looked at his dinner jacket and said: "Maybe he thought you were a waiter, Jack."[109]

Politics

Jackie tried hard but for some years simply couldn't get interested in politics. "The more I hear Jack talk about such intricate and vast problems," she said, "the more I feel like a complete moron." Once she sought Bobby Kennedy's advice and at dinner a little later asked her husband whether it was true "that confirming Albert Beeson to membership on the National Labor Relations Board has the effect of wrecking the Bonwit Teller clause of the National Labor Relations Act." JFK almost collapsed with laughter. "Jackie is superb in her personal life," he once told friends, "but do you think she'll ever amount to anything in her political life?" "Jack is superb in his political life," Jackie retorted, "but do you think he'll ever amount to anything in his personal life?" On another occasion he remonstrated with her for being so distant to some politically important people. "The trouble with you, Jackie," he said, "is that you don't care enough about what people think of you." "The trouble with you, Jack" returned Jackie, 'is that you care too much what people think of you."[110]

Campaigner

When JFK ran for the Senate in 1968, Jackie spoke briefly at an Israel bond rally in Boston and made a speech in French to Cercle Français in Worcester. But when she accompanied two of her husband's campaign workers to a rally in the parish hall of a French Catholic Church in Chicopee, no one showed up and JFK's advisers were in despair. But Jackie saved the night. She walked over to the rectory, introduced herself in French to the pastor and sat with him a half-hour or so drinking tea and talking to him in French. The following Sunday, according to JFK's associates, at every mass the priest had good things to say about JFK.[111]

Ruination

Once, when Jackie wore an especially stylish hat during the 1958 Senate race, JFK exclaimed: "What are you trying to do, ruin my political career? Take it off!"[112]

Southie

During the Wisconsin primary in 1960, Jackie appeared at a rally one night in Kenosha and tried to warm the crowd up until her husband appeared. To get things going she suggested singing a song. Her choice: "Southie Is My Home Town." The Kenoshans looked bewildered; they had never heard of the Boston Irish number. When JFK arrived he winced when he heard about his wife's gaffe.[113]

Acapulco

Before Los Angeles was picked as the site for the Democratic Convention in 1960, a leading Democrat turned to Jackie at a dinner party and asked if she had some thoughts about a suitable city. With a straight face she said: "Acapulco."[114]

Hairdo

During the 1960 campaign, Jackie received many letters saying she should change her hair style to one more fitting for the wife of a presidential candidate. Some people even sent her haircombs. "Oh, Mary," JFK once said to his wife's secretary, Mary Gallagher, one day, "how is the mail running on Jackie's hair?" "Heavy," Gallagher told him. Jackie finally issued a statement. "All the talk over what I wear and how I fix my hair has me amused," she announced, "but it also puzzles me. What does my hairdo have to do with my husband's ability to be President?"[115]

Snappy

What some people called Jackie's "devil-may-care chic" came in for criticism during the 1960 campaign. "She looks too damn snappy," complained one Manhattan woman. "I just don't like women who look snappy." "Jackie," said a close friend, "you are too much of an individualist. If you get in the White House, you'll have to make some concessions." "Oh, I will," said Jackie. "I'll wear hats!"[116]

Resting Up

Three hours before the inaugural parade ended, Jackie Kennedy left the reviewing stand and went to the White House to rest up. She had a "Reception for Members of the President and Mrs. Kennedy's Families" to attend, following the parade, and five inaugural balls ahead of her that night. But she never got to the Kennedy-Lee-Bouvier-Auchincloss reception.

Mrs. Kennedy's relatives and in-laws gathered at the White House late that afternoon and waited. After a while they became restless and her mother, Janet Auchincloss, told them her daughter was "up in the Queen's bedroom trying to relax." When JFK arrived around seven he cried: "Where's Jackie?" "Oh, she's upstairs resting," Mrs. Auchincloss told him. "Well," said JFK, "I guess I better go on up there too," and disappeared. At length Michel Bouvier, one of Jackie's cousins, went up to see Jackie and returned to say: "She's just resting. . . . You know she's got to go to five goddamn balls this evening, don't you?" "But," said someone plaintively, "can't she come down even for a *minute?*" "Not even for a *second,*" said Bouvier. The party finally broke up without Jackie's making an appearance.

Years later Michel Bouvier revealed that when he went up to see Jackie, she was sitting in bed looking frightened. "Oh, Michel," she wailed, "I don't know what to do? Is everybody there?" "Just about everybody," he replied. "Oh, Michel, really, I can't go down, I can't go down, I really can't. I've got all those balls to go to and I've got to look good for them." One relative thought an attack of shyness explained her unwillingness to go down.

The new First Lady did attend the inaugural balls with her husband that night. But in the midst of the ball at the Armory, she "just crumpled," and returned to the White House, leaving JFK to attend the last two himself. JFK didn't get back to the White House until dawn, and Jackie spent her first night there alone.[117]

Thoughtfulness

JFK had a bad back and used a special rocking chair to ease the pain. When the Kennedy's moved into the White House, Mrs. Kennedy had the rocker upholstered and the cane back padded with foam rubber to increase its effectiveness. Headlines: "JFK IS ROCKED BY WIFE'S THOUGHTFULNESS."[118]

Glassware

After campaigning with JFK in West Virginia in 1960, Jackie, shocked by the poverty she encountered in the coal-mining areas, was anxious to do something for the state. To help out she began buying White House crystal from the Morgantown Glassware Guild, Inc., an important West Virginia industry. "Mrs. Kennedy . . . ," according to a White House release for May 20, 1961, "ordered the glassware out of gratitude and affection for West Virginia. . . ." The following year a well-known manufacturer of costly glassware offered to donate a complete set of expensive crystal to the White House, but Mrs. Kennedy demurred. She wanted to continue purchasing glassware in West Virginia "until they aren't poor any more" there. "The poverty hit me more" in West Virginia "than it did in India . . . ," she told a friend, "maybe because I just didn't realize that it existed in the U.S." She said she would "practically break all the glasses & order new ones each week," if "it's the only way I have to help them— ."[119]

Important Things

When the Kennedys were in France in May 1961, French couturiers invited Jackie to attend a private showing of their collections, but she declined because she thought American women would resent it if the First Lady spent an afternoon admiring gowns by Dior, Givenchy, and Chanel. Asked at a news conference why she had turned down the invitation to the fashion show, she said, "I have more important things to do." But this remark nettled French couturiers, so she hastily changed her plans for the dinner in honor of the Kennedys in the Palace of Versailles their second night in Paris. She had planned to wear another Cassini gown to the dinner, but she now decided to include a Givenchy bodice. French couturiers were pleased.[120]

In About Ten Years

"She is unique for the wife of an American president," André Malraux told French President Charles deGaulle after the Kennedys had visited France. "Yes, she's unique," agreed deGaulle. "I can see her in about ten years on the yacht of a Greek petrol millionaire."[121]

Ought To Be a Law

Mrs. Kennedy tried to shield little Caroline from the public, but she was good copy and her doings got into the newspapers anyway. Grumbled a Republican National Committeewoman: "There ought to be a law against lady politicians who are three years old."[122]

Decision-Maker

Whenever Mrs. Kennedy asked her husband whether Caroline should appear at some reception or whether she herself should wear a short or long dress, JFK would say: "That's your province." Yes," she would say, "but you're the great decision maker. Why should everyone but me get the benefit of your decisions?"[123]

Just as Bad

When she was in the White House Jackie spent as much time as she could with her two children. "I always imagined I'd raise my children completely on my own," she once told an interviewer. "But once you have them, you find you need help. So I do need Dr. Spock a lot and I find it such a relief to know that other people's children are as bad as yours at the same age."[124]

Charity

When Jackie learned that her husband donated his salary as President to charity, she cried: "Hey, how about donating some of it to me?"[125]

Introduction

Once, when Jackie was taking a French photographer on a tour of the White House, she opened the door to show him one of the offices where a secretary was busily at work and exclaimed: "And this is a young lady who is supposed to be sleeping with my husband!"[126]

Marvelous Gifts

One day Jackie's secretary Mary Gallagher found the First Lady busily perusing a picture catalogue for trading stamps. "Oh, Mary," cried Jackie, "do you know what I've just learned from Anne Lincoln? You know, all the food we buy here at the White House? Well, she told me that with the stamps the stores give us, we can trade them in for these marvelous gifts!"[127]

Other Things

In the spring of 1963, Harris Wofford, JFK's special assistant on civil rights, took civil-rights leader Martin Luther King, Jr., to the White House for a meeting with the President. Just as the two entered the little elevator near the White House's main entrance, Jackie Kennedy got on, wearing blue jeans and covered with dust. "Oh, Dr. King," she cried, after introductions, "you would be so thrilled if you could have been with me in the basement this morning. I found a chair right out of the Andrew Jackson period—a beautiful chair." "Yes—yes—is that so?" murmured King. "I've just got to tell Jack about that chair . . . ," the First Lady went on as the elevator reached the family quarters. "But," she added, "you have other things to talk to him about, don't you?" After she left King said to Wofford: "Well, well—wasn't *that* something?"[128]

Ninety-nine Things

"People have told me ninety-nine things that I had to do as First Lady," Mrs. Kennedy once remarked to Nancy Tuckerman, but added triumphantly: "and I haven't done one of them!"[129]

Wait in Line

When someone told JFK that Cuba's Ernesto ("Che") Guevara said that the one American he wanted to meet was Jackie Kennedy, he exclaimed: "He'll have to wait in line."[130]

Caviar

"She is not as American as apple pie," a friend once said of the First Lady. "She is as American as caviar."[131]

Movie Star

"He was our President," said White House staffer Letitia Baldridge after JFK's death, "but she was our movie star."[132]

Back in White House

In the spring of 1968, when JFK's brother Robert entered the presidential race, no one was more excited than Jackie Kennedy. "Won't it be wonderful," she cried, "when we get back in the White House again?" "What do you mean, *we?*" said Bobby's wife Ethel coldly.[133]

CHAPTER 34

LADY BIRD JOHNSON

1912–

One day Lady Bird Johnson located her "dream house" in Washington, rushed off to tell her husband about it, and found him talking politics with his administrative assistant John Connally. When she entered the room, LBJ looked up, listened for a moment, and then turned back to Connally. Suddenly his wife broke in. "I want that house!" she exclaimed. "Every woman wants a home of her own," she went on, "and all I have to look forward to is the next election." After she stormed out of the room, Connally advised Johnson to buy the house and LBJ did just that.[1] It was one of two times, he said later, that he saw his wife lose her temper. The other time was when they were out riding on the LBJ ranch and he impatiently gave her horse a sharp prod leading the startled animal to leap so suddenly that Lady Bird nearly lost her balance. "I'll manage my own horse," she said with an angry look that LBJ never forgot.[2]

Ordinarily Lady Bird was the soul of serenity. Nothing untoward—LBJ tantrums, family crises, mean attacks by LBJ's enemies—seemed to faze her. "You can't make her mad," her elder daughter Lynda Bird once declared. "She is a lot like Melanie in *Gone with the Wind,* except with more drive. I have never seen her lose her temper."[3] Younger daughter Lucy Baines agreed. "Mother is the calmest, most even-tempered person I knew," she said admiringly. "When Daddy, Lynda and I get excited or upset over something, she pulls us all together. She's really the knot of the family. She can live through any difficult, trying time and never get mad, or lose her temper. She has complete control over her emotions. The rest of us have tempers, but Mother is a calm soul who smooths us down, and makes all of us feel closer together."[4] The tempestuous LBJ was just as impressed by Lady Bird's equanimity as his daughters were. "As a companion she has no equal," he told an interviewer. "She's soft and kind and understanding. She's always willing to meet you more than halfway. If you want to ride and she wants to walk, she'll ride. If she wants a fire in the living room and you don't, she says sweetly, 'We'll have it some other time.' "[5]

Like her family, Lady Bird's friends—she had no enemies—could never say enough in praise of her. "That's the greatest woman I have ever known," said Speaker of the House Sam Rayburn, who knew her well. "She's good and she's kind and she doesn't have a mean thought."[6] Republican Congressman Charles Halleck's wife Blanche said she never met anyone who didn't like Mrs. Johnson. Her press secretary Elizabeth Carpenter said she was as "gentle and serene as the cypress trees" in East Texas, "quiet, deep-rooted, protected."[7] Democratic Congressman Hale Boggs's wife Lindy once paid high tribute to her and then confessed: "I make her sound like a combination of Elsie Dinsmore and the Little Colonel . . . but this is the problem with Bird. When you talk about her, you make her too good to be true."[8]

But some Washingtonians thought Mrs. Johnson was on the dull side. "They made me feel," she once confessed, "like putting on red tights and running down Pennsylvania Avenue."[9] Others thought she was absolutely ruled by her husband and had no mind of her own. If LBJ demanded it, Jackie Kennedy once scoffed, Lady Bird would take off all her clothes and run stark naked down the streets of Washington. She couldn't have been more wrong. Mrs. Johnson kept her peace when LBJ stormed around the house (she went about her work at such times quietly whistling to herself as if all was well), but she had her own unobtrusive ways of making her influence felt. She had, a friend observed, "the touch of velvet and the stamina of steel."[10] "If she disapproves of something you say," LBJ admitted, "she never says so, but she somehow makes you want to correct it yourself."[11] Not many women could have handled the impatient, imperious, and at times impetuous Lyndon Johnson with the skill Lady Bird developed after marrying him in 1934.

When LBJ first met Lady Bird she was still a quiet, unprepossessing, and somewhat insecure small-town girl who had studied journalism at the University of Texas in Austin because she thought "people in the press went more places and met more interesting people, and had more exciting things happen to them."[12] Her hometown was Karnack, a little place in East Texas, where she was born in 1912, and her real name was Claudia Alta Taylor. Her nickname (which LBJ soon shortened to "Bird") came from her black nurse who, upon seeing the baby, chortled, "Why, she's as purty as a ladybird!"[13] Lady Bird got her grammar-school education in a one-room schoolhouse in Karnack and then attended high school in somewhat larger towns nearby. LBJ knew that she came from a good home; her father, Thomas Jefferson Taylor, with whom he hit it off upon first meeting, prospered as a farmer and storekeeper ("T. J. Taylor, Dealer in Everything"), owned the only brick house in the region, and was the first to have electricity and install indoor plumbing in his home. When Lady Bird was only five, her mother died after a bad fall and Aunt Effie, a spinster from Alabama, took over her rearing. From her mother (who collected books and phonograph records) and from her aunt, Lady Bird learned to appreciate good books and classical music; neither her father nor, later, LBJ, had time for that sort of thing, but both were indulgent. Of her aunt, Lady Bird once reflected: "She opened my spirit to beauty, but she neglected to give me any insight into the practical matters a girl should know about . . . such as how to dress or choose one's friends or learning how to dance."[14]

Lady Bird's father was generous with his only daughter (he had two sons, both older than Lady Bird). He gave her a car when she was only fourteen so she could drive to school. He also gave her a Buick when she went off to college, and he provided her as well with a checking account of her own and with a charge account in the Nieman-Marcus department store in Dallas (where she attended St. Mary's Episcopal Girls' School before going to Austin). But Lady Bird never splurged. She dressed plainly, ate simply, studied hard, and did first-rate work both in high school and college. In the high school in Marshall she came close to being valedictorian, but mercifully came in third, for she was terrorized by the thought of taking first place and having to give a speech in public. At the University of Texas, where she took her B.A. in 1933, she was among the top ten of her graduating class and stayed on another year to do graduate work. When her father gave her a trip to Washington as a graduation present in June 1934, her friend Eugenia ("Gene") Boehringer told her to look up young Lyndon Johnson, then working as secretary to Congressman Richard Kleburg, but Lady Bird never did. "I would have felt mighty odd calling an absolute stranger," she said; besides she had plenty of Texas-exes (U.T. graduates) to connect with in Washington.[15]

Lady Bird met LBJ soon after. Her friend Gene worked in the State Capitol in Austin and when Lady Bird dropped by to see her one afternoon late in August 1934 she found the twenty-six-year-old Lyndon visiting with her in her office. "Oh, Lady Bird," cried Gene, "here he is at last. Now I'm going to make sure you all get together!"[16] LBJ impressed Lady Bird at once. She found him "terribly, terribly interesting" and "very good-looking," and by all odds "the most outspoken, straight-forward, determined young man" she had ever met. "I knew I had met something remarkable," she said years later, "but I didn't quite know what."[17] LBJ took to her at once and, since he had a date that night, invited her to meet him the following morning for breakfast at the Driskill Hotel. Lady Bird accepted the invitation but had "a queer sort of moth-in-the-flame feeling" about him. She almost stood him up; but when he saw her pass the Driskill on an errand, he "just flagged me down. So I went in."[18] After breakfast they took a drive out into the country and LBJ poured out his soul to her; he talked about his salary, his insurance, his family, and his ambitions, and "all sorts of things" she thought were "extraordinarily direct for the first day."[19] The next day, to her astonishment, when they met again, he asked her to marry him. "I thought it was some kind of joke," she recalled, but it turned out LBJ was quite serious.[20] The following day he took her home to meet his parents and a few days later had her meet Congressman Kleberg at the King Ranch. When the time came for him to drive back to Washington she suggested he go through East Texas en route so he could stop by Karnack to meet her father. LBJ and Tom Taylor hit it off at once. "Daughter," Taylor told Lady Bird after dinner, "you've been bringing home a lot of boys. This time you brought a man."[21]

From Washington LBJ bombarded Lady Bird with daily letters and phone calls and sent her his picture, inscribed, "For Bird, a lovely girl with ideals, principles, intelligence and refinement, from her sincere admirer, Lyndon."[22] Toward the end of October he returned to Texas and stopped in Karnack again. "Let's go out and get married," he told Lady Bird, "not next year . . ., but about two weeks from now, a month from now, or right away."[23] Lady Bird didn't say yes, but she didn't say no either, and the next thing she knew she was on the way to Austin with him to get an engagement ring. Her friend Dorris Powell urged her to wait six months; Aunt Effie also counseled delay. "If he loves you as much as he says he does," her aunt said, "he will wait for you." But her father disagreed. "Lady," he said, "if you wait until Aunt Effie is ready, you will never marry anyone." LBJ finally issued an ultimatum. "We either get married now or we never will," he exclaimed. "And if you say good-bye to me, it just proves to me that you just don't love me enough to dare to. And I just can't bear to go on and keep wondering if it will ever happen."[24]

Lady Bird still hesitated. But she did remark that she had always dreamed of being married in St. Mark's Episcopal Church in San Antonio. LBJ later said he was still twisting her arm as they walked up the church steps the afternoon of November 17. Once inside, though, Lady Bird queried: "You did bring a wedding ring, didn't you?" "I forgot!" cried LBJ and sent a friend over to Sears to find a ring.[25] The friend returned with a tray of inexpensive rings and Lady Bird picked out a temporary one (which cost LBJ $2.50) to use in the wedding ceremony. Sighed the minister as the Johnsons rushed off for a honeymoon in

Mexico: "I hope that marriage lasts."[26] It did. Thirty years later LBJ gave his wife another picture, for her fifty-first birthday, on which he had written, "For Bird, *still* a lovely girl." Sam Rayburn, LBJ's political mentor in Washington, said that marrying Lady Bird was the best thing Johnson had ever done.[27]

For Lady Bird, being married to LBJ was just as hectic as being courted by him. If he was outspoken, straightforward, and determined as a suitor, he was even more so as a husband, and he was peremptory, too, at times. Right after the wedding, when they stopped by to see some friends en route to Mexico, LBJ suddenly exclaimed in front of everybody: "You've got to change your stockings, Bird. You've got a run." When the startled Lady Bird sat tongue-tied for a moment, he repeated the order even more emphatically and she left the room to do his bidding.[28] Later on, when they had settled in Washington, he began nagging her about her conservative clothing. "I don't like muley-lookin' things," he told her, and instructed her in what to wear.[29] To please him she began dressing more fashionably: cut her hair, started wearing spike-heeled pumps, and replaced her tailored suits with dresses in the bold red and yellow colors he liked so much. She learned to serve him his breakfast coffee in bed, too, and brought him the morning paper with it. She also got used to being issued orders in front of other people. "Lady Bird," he might exclaim at a dinner party, "get me another piece of pie." "I will," she would answer, "in just a minute, Lyndon." *"Get me another piece of pie,"* he would repeat impatiently, and his wish became her command.[30]

If Lady Bird was upset by LBJ's imperious manner, she never let on. She had a way of making his demands sound reasonable and his abruptness seem irrelevant. Serving him coffee in bed, she decided, was less trouble than setting the table for breakfast. Putting out his clothes for him in the morning, filling his pen, and even shining his shoes, she reasoned, were simple chores that a good wife could easily perform in order to help a ferociously busy, ambitious, and hard-working husband like LBJ. The bright colors and "dramatic style" he preferred in clothing, she came to acknowledge, "do the most for one's figure," and it was good for her to "learn the art of clothing, because you don't sell for what you're worth unless you look well."[31] Later on, when LBJ became a Congressman and she worked in his office when he was away, she felt that she understood his moods better. "I was more prepared after that," she said, "to understand what sometimes had seemed to be Lyndon's unnecessary irritations."[32] In any case, LBJ had given her fair warning at the outset. "Listen," he told her at one point during his whirlwind courtship, "you're seeing the best side of me. . . . I think you ought to know that."[33]

LBJ certainly had his good side; there were bright moods as well as dark ones. If he was abrupt, he was also affectionate. He gave her orders in public, but he also gave her hugs and kisses, and right in front of everyone, too. He seemed to be surprisingly patient, moreover, as she went about trying, awkwardly at first, to become a competent housewife. She had never learned how to clean a house or cook a meal before she got married, but she bought a cookbook right after the wedding and tried to do her best. Soon after the Johnsons moved into a one-bedroom apartment in Washington, LBJ invited Texas Congressman Maury Maverick and his wife to dinner. When the Mavericks arrived that evening, the first thing Mrs. Maverick noticed was a Fannie Farmer cookbook lying open on the table set in the living room. "Staring at me was a recipe for boiled rice," she recalled. "The menu included baked ham, lemon pie, and, of course, the rice. The ham and pie were very good, but I'll never forget that rice. It tasted like library paste. To this day, I connect boiled rice and library paste." But she and her husband were charmed by the young bride's graciousness and her efforts to cover her nervousness with a smile and try to make them feel at home. Lady Bird's quiet hospitality, even when LBJ brought bunches of people home unexpectedly for dinner, impressed everyone who met her. They always felt she really meant it when she said, as they were leaving, "Y'all come back real soon, hear now!"[34]

In Washington, Lady Bird's quaint way of putting things—a heritage from Alabama (where she lived with her aunt for a time) as well as from Texas—endeared her to LBJ's friends and associates. "I'll see you Saturday night," she would say, "if the Lord be willin' and the creek don't rise." Praising people, she would tell them, "I find myself, as we say down home, in mighty tall cotton." People who talked too loudly were "noisier than a mule in a tin barn" and those she disapproved of were "the kind of people who would charge hell with a bucket of water." She never said, "The fireplace is warm and cozy"; it was, "Doesn't that fire put out a welcoming hand?" Of a friend: "She's a sturdy oak with magnolia blossoms all the year long." Of an unpleasant task: "I look forward to that as much as to a good case of cholera." When strapped for time: "I'm as busy as a man with one hoe and two rattlesnakes."[35]

LBJ, who had just as colorful (but earthier) ways of putting things, felt good about Lady Bird. "She is still the most enjoyable woman I've ever met," he said a few years after marrying her. "As a sweetheart, a swimmer, a rider, and a conversationalist, she is the most interesting woman I know."[36] He may have had his extra-marital affairs, as gossips charged, but there is no question but that Lady Bird was the most important woman in his life. She was absolutely crucial to his happiness and success, and he knew it. When he came home from a busy day's work his first words were, "Where's Bird?" As head of the National Youth Administration in Texas in the mid-thirties, he had to be away from home a great deal and was lonely without her. "It was more difficult to leave you last night than I had anticipated," he once wrote her.

"I have learned to lean on you so much. . . . Never have I been so dependent on anyone—Never shall I expect so much of any other individual. . . . You do every day with your job what I want to do with mine and you know something of my ambitions." He had her handle the family finances, discussed his work with her, and frequently sought her opinions on things. Sometimes, in the midst of discussions with his associates, he would stop and say, "Let's ask Lady Bird."[37]

In 1937, when LBJ decided to run for Congress, he asked his wife for financial help and she telephoned her father to request an advance on her inheritance from her mother's estate. "Daddy," she said, "do you suppose you could put ten thousand dollars in the bank for me? Lyndon wants to run for Congress." "Honey," he said, "couldn't you get by on five?" "No," she said, "we've been told it must be ten." "Well, today's Sunday," he reminded her. "I don't think I could do it before tomorrow morning about nine o'clock."[38] With the loan (later repaid) in hand, LBJ launched his successful race for Congress and Lady Bird was so pleased by the way she had helped him that she carried the canceled check around in her purse for years. "Darlin', I'm for you," she cried after hearing him speak in public. "Lyndon was never so young, never so vigorous, and never so wonderful!" she said of his first Congressional campaign. "My only regret is that I did not have the gumption to share in it, although I suppose at that time I would have been looked upon as absolutely odd if I had gone around with him. In 1937, a wife didn't campaign."[39]

As a Congressman's wife, Lady Bird enjoyed taking her husband's constituents on sight-seeing tours when they visited Washington, but she also helped out with correspondence in his office and made suggestions for speeches. When he ran for re-election in later years, she sat in on discussions of campaign strategy; and when he was on leave for a time to serve in the U.S. Navy during World War II she worked full-time in his Congressional office. She became a businesswoman too. In 1943 she purchased a radio station in Austin with money she inherited from her mother, worked hard learning the business, and supervised its expansion during the next few years into an extremely profitable radio-and-television enterprise. With the birth of Lynda Bird in 1944 and Lucy Baines in 1947, she became a mother, too, and an extremely conscientious one at that. "I don't see how Lady Bird can do all the things she does without ever stubbing her toe," LBJ once exclaimed. "I'll just never know, because I sure stub mine sometimes."[40]

In 1948, when LBJ ran for the U.S. Senate, Lady Bird became an active campaign worker for the first time in her life. She enlisted the services of friends and former high school and college classmates, arranged teas for her husband in the little towns they visited together, went over speeches with him and suggested phrases to use in them. Sometimes, when audiences became restless, she would slip him a note:

"That's enough." Two days before the election, when she was driving to San Antonio to participate in a rally and make her first campaign speech, she had an accident, climbed out of the car, hitchhiked a ride, borrowed a dress from the hostess when she arrived at the reception, shook hands with 200 women, made her speech, and then went to the hospital for X-rays and bandages. "All I could think of as we were turning over," she told LBJ that evening, "was I sure wished I'd voted absentee."[41] On election day she was back in Austin riffling through the phonebook along with LBJ's mother and three sisters to get names to call on behalf of her husband. LBJ was fully aware of the contribution she made to his victory that day in a closely contested election.

In 1954, LBJ won re-election to the Senate and the following year, at forty-six, became the youngest Majority Leader in American history. His new position meant frequent calls on Lady Bird to speak in public, and, still painfully shy about making speeches, she finally decided, at LBJ's prodding, to take speech lessons. "I got real annoyed with myself for being so shy and quiet," she said, "and never having anything to say when asked to speak. I took the course and it turned out to be one of the most delightful, expanding experiences I've ever had."[42] It also turned Lady Bird into a quietly effective speaker and ensured her participation as a speech-maker in LBJ's later campaigns. "Lyndon stretches you," Lady Bird once reflected. "He always expects more of you than you're really mentally or physically capable of putting out. Somehow that makes you try a little bit harder, and makes you produce a little more. It is a very good fertilizer for growth; it's also very tiring."[43] LBJ was as hard on himself as he was on everybody else, and Lady Bird became increasingly concerned about the way he was driving himself. Once, after attending an embassy dinner-dance without him, she said plaintively, "Lyndon, I don't see why you can't take some time off for fun now and then. All the other Senators do. Why, Senator Theodore Francis Green asked me to dance twice tonight." "Senator Green!" boomed LBJ. "Well, what kept me at work so late tonight was passing his pet bill through the Senate."[44]

In July 1955 came a massive heart attack. "Take my hand and stay with me," LBJ told Lady Bird as he as on his way to the hospital. "I want to know you're here while I try to fight this thing." She stayed in the room next to his during the next five weeks in the hospital and she converted it into an office, too, so she could answer the letters pouring in. "Lyndon wanted me around twenty-four hours a day," she said after the crisis had passed. "He wanted me to laugh a lot, and always to have lipstick. During those days we rediscovered the meaning and freshness of life." But to a friend she confessed: "When all this is over, I think I'll go somewhere by myself for about two hours and just cry."[45]

Lady Bird never had time to cry. By August 1955, LBJ was back at the ranch, rarin' to go again, and in January 1956 back in the Senate.

He was no longer a chain-smoker, but he was as hard-driving as ever. When he threw his hat into the presidential ring in 1960, Lady Bird, despite continued misgivings about his tendency to overdo, again went to work for him. She was disappointed when John F. Kennedy nosed him out for the nomination ("Lyndon would have been a noble President") but supported his decision to accept second place and joined him in campaigning for the Kennedy-Johnson ticket with energy and spirit.[46] By this time Democratic professionals were calling her the party's "Secret Weapon." Everywhere she went, with or without her husband, in the North as well as in the South, she created good will for the ticket and admiration for her charm and tact. By election day she had traveled 35,000 miles, by train and plane, spoken briefly at countless breakfasts, luncheons, teas, coffees, dinners, and receptions ("So glad y'all came"), and accumulated so many compliments that she began to feel embarrassed about it: "intelligent and articulate"; "charming, pretty, vivacious, happy"; "a gal with gumption"; "a real love"; "Southern charmer"; "feminine, friendly, and folksy."[47] Her four trips through Texas were particularly effective, and when the Kennedy-Johnson ticket won in November, JFK's brother Robert, who had managed JFK's campaign, exclaimed: "Lady Bird carried Texas for us!"[48]

At the inauguration in January 1961, LBJ impulsively kissed Lady Bird's forehead after taking his oath of office and then went on to do the kind of busy work that goes with being Vice President: filling in for the President on ceremonial occasions, entertaining foreign dignitaries at the LBJ Ranch, and circumnavigating the globe with his wife in an effort to butter up foreign heads of state for the President. Lady Bird set herself three objectives as Second Lady: "To help Lyndon all I can; to lend a hand to Mrs. Kennedy when she needs me; and to be a more alive me."[49] She helped Lyndon by traveling thousands of miles with him to dozens of countries and taking notes in shorthand on the places they visited for use in composing reports for the President. For Mrs. Kennedy she filled in at White House dinners and receptions the First Lady wanted to avoid and came to be known as Washington's number one pinch-hitter. For herself, she took delight in expanding her knowledge and understanding of foreign lands and satisfaction at her increasing skill in public appearances. Once she was scheduled to attend a dinner honoring the Vice President and then received a last-minute call from the White House asking her to substitute for Mrs. Kennedy at a dinner where the First Lady was to receive a TV Emmy Award for Public Service. Lady Bird managed her evening adroitly. She accompanied LBJ to the first dinner, sneaked out when proceedings were under way, took a cab to the second dinner, stepped into a phone booth to compose an acceptance address, then showed up at the TV dinner, accepted the First Lady's award, made her speech, and headed back to the Vice President's dinner afterward.[50] Her husband, famous for the "LBJ trot," couldn't have done better himself.

Mrs. Johnson was in the motorcade with her husband in Dallas on November 22, 1963, when President Kennedy was struck down by an assassin and experienced the shock, incredulity, anger, and then sorrow that other Americans felt over the terrible event. She helped LBJ compose a statement to the nation after becoming President, issued one of her own, and did what she could to help JFK's family in their grief. "Mrs. Kennedy," she told JFK's mother, "we feel as though our heart is cut out, but we must remember how fortunate our country was to have your son as long as we did."[51] The Johnsons invited JFK's widow to remain in the White House as long as she needed to and to continue the nursery school she had set up there for her children and their playmates.

When the Johnsons moved into the White House early in December, Lady Bird was so overawed at first that she found herself tiptoeing around the place. "I feel," she told a close friend, "as if I am suddenly on the stage for a part I never rehearsed."[52] Then she began a series of what she called "educational tours" of the 132-room mansion and began adjusting to her new life there. "My first job," she decided, was "to make this home a place where Lyndon can operate productively, and to add to his operation in every way that I can, because I have never felt so much need on his part, and so much compassion on my part for him."[53] At the outset she gave Chief Usher J. B. West his instructions: "Anything that's done here, or needs to be done, remember this: my husband comes first, the girls second, and I will be satisfied with what's left." She also told him she wanted him to run the place; but before long she was doing it herself, and, he observed, "rather like the chairman of the board of a large corporation."[54]

The Johnsons did a lot of entertaining—there were more than 200,000 guests in their five White House years—and Lady Bird got so she could handle all the luncheons, dinners, and receptions with "a watchmaker's precision."[55] In addition to the customary formal occasions, there were plenty of last-minute informal dinners LBJ thought up on the spur of the moment. "Bird," he said one morning, "let's ask Congress over this afternoon." He did; and she managed the unexpected crowd nicely.[56] Whenever she could, though, she went over guest lists carefully beforehand, learned something about all the guests, and had something personal to say to each of them when they appeared in the receiving line. In preparation for visitors from abroad she went over maps and State Department briefing papers to master information about the countries they represented.

Under the Johnsons there were barbecues, carnivals, and hootenannies as well as more sedate entertainments. There was also a great deal of dancing, for the Johnsons impressed Chief Usher West as "the dancingest First Family" he had ever known. "How funny," Lady Bird once remarked, "that we should do more dancing after fifty than at any other time in our lives."[57] But her favorite occasions were probably the "Women Doers" luncheons she held regularly in the White House. For these she selected the topics to discuss—crime, women's rights, poverty, the arts—and chose the guests—journalists, actresses, singers, public officials, professional women—to lead the discussions. Lady Bird thought that the government paid too little attention to women and made it her business to persuade her husband to see that they got more recognition. When he came home from his office at night, she sometimes said, "Well, what did you do for women today?"[58] She suggested women for top government posts, attended meetings of women's organizations, and made speeches and wrote articles about women's role in American society. "Through the centuries, women have been the prodders," she told the American Association of University Women in February 1966. "Good works go forward in proportion to the number of vital and creative and determined women supporting them. When women get behind a project, things happen."[59]

Lady Bird was herself an indefatigable prodder. The sign she put on her office door—MRS. JOHNSON AT WORK—told all.[60] She was an activistic First Lady, sometimes compared to Eleanor Roosevelt, because of the way she traveled around the country visiting depressed areas and examining the workings of the administration's War-Against-Poverty programs and reporting her findings to her husband. "I like to get out and see the people behind the statistics," she said. "It makes Lyndon's memos and working papers come to life for me."[61] She took part in a "Discover America" program, too, visiting parks, taking raft trips down rivers, and spending time in small towns in the Midwest to encourage people to learn more about their country. "Call it corny if you will," she said, "but I want to boast about America."[62] Her trips ("Lady Bird Safaris") received much publicity.

LBJ was proud of the First Lady. "She's just back from the coal mines," he told some guests with a pleased smile when she walked into the White House one day.[63] He had long since regarded her as his indispensable partner and in making decisions he always turned to her at some point. "She was the last sounding board," according to press Secretary George Christian, "the last reasoner, the voice of calm, and remarkable understanding."[64] LBJ valued her political judgment and respected her candor. Jack Valenti, one of his aides, thought he valued her opinions even more than those of his top advisers "because he knew that, alone among his entourage, she delivered her views without any self-interest or leashed ego that may have been hidden in the breasts of all of the rest of us."[65] But it was more than that; LBJ depended upon her for companionship as well as counseling. As always, he missed her when she was off on a trip; and if she stayed away too long he was like a caged lion, his associates observed, "and his temper grew shorter and his hours at the office longer."[66] One of his aides thought that if anything had happened to Lady Bird, LBJ "would have decayed, quickly, terminally."[67]

When Johnson ran for re-election in 1964, Lady Bird took to the stump, the first President's wife to do so. Her husband asked her to help out in the South, where there was considerable resentment over the civil-rights legislation he was sponsoring, and she agreed to hit the campaign trail for him. "Don't give me the easy towns, Liz," she told her aide, Elizabeth Carpenter, when they were planning the trip. "Anyone can get into Atlanta—it's the new, modern South. Let me take the tough ones." Her strategy in the little towns she visited was to stress her love for the South, resentment over "snide jokes" about it, and faith in the decency and fairness of the Southern people and their willingness to support the Constitutional rights of all Americans. Before commencing her tour she called the Governors of the states she planned to visit to enlist their help. "Guv-nuh," she would say, "this is Lady Bird Johnson." "Howdy, Miss Lady Bird," would come the response, after a startled pause. "How's the President?" "Fine, just fine, Guv'nuh," she would reply. "I'm thinking about coming down to your state. . . . I called to ask your advice." After the Governors reminded her how tough it was for the Democrats that year because of the Civil Rights Act of 1964, she would say, "Well, I know there is a long educational process that is necessary, but I was thinking of coming through on a whistlestop train. You see, I don't want the South to be overlooked in this campaign. And we have lot of good friends and kinfolks there." In the end she persuaded five Governors and four Senators to join her party as it passed through the South.[68]

Lady Bird's whistlestop tour from Washington on October 6 to New Orleans on the 9th—1,682 miles, 47 stops, eight states—was a huge success. Though she encountered heckling in some places, mostly she attracted admiring crowds. "DEAR MRS. JOHNSON," people in one little North Carolina town wired her, "PLEASE STOP IN AHOSKIE. NO IMPORTANT PERSON HAS VISITED HERE SINCE BUFFALO BILL, AND NO PASSENGER TRAIN HAS STOPPED IN 12 YEARS." When she got back to Washington at the end of the tour, LBJ called her "one of the greatest campaigners in America" and exclaimed: "I'm proud to be her husband."[69] On Inauguration Day, January 20, 1965, as he was escorting her to the inaugural luncheon, he suddenly turned, leaned down, and gave her a big kiss that left her glowing.

In her second term as First Lady, Lady Bird sought a special project of her own to which she might devote major energies. She continued to travel around the country inspecting the administration's Great Society projects, but she also began focusing on measures to clean up and beautify the country. Under her prodding, Congress passed the Highway Beautification Act (popularly called the "Lady Bird Act") in October 1965, designed to limit billboards on Federal highways and encourage better planning of the nation's roads. To dramatize her program for preserving and reclaiming the nation's scenic beauty, Lady Bird sponsored a White House Conference on Natural Beauty, gave speeches emphasizing the close connection between ugliness and crime, enlisted the help of architects, conservationists, and philanthropists in the cause, attended ceremonial tree and flower plantings, and dedicated new parks and gardens in various parts of the country. She gave special attention to Washington itself, heading the First Lady's Committee for a More Beautiful Capital and riding around town in an unmarked car, to hunt out places that needed cleaning up. "HER NAME IS CLAUDIA," proclaimed a *Chicago Tribune* headline, "AND BEAUTY IS HER AIM."[70] When she was praised for her beautification efforts, she said modestly, "I only stepped on a moving train." She said she wanted her epitaph to announce: "SHE PLANTED THREE TREES."[71] To her surprise, one day Robert F. Kennedy, no friend of her husband, said to her: "You're doing a wonderful job. Everybody says so." Then, after a pause, he added: "and so is your husband."[72]

Approval of the job LBJ was doing turned out to be short-lived. The Vietnam War soon killed the Johnson administration's plans for social betterment. With increasing involvement in the Vietnam War after the 1964 election came mounting opposition to LBJ's foreign policy in Congress and in the nation at large, and the time came when the First Lady as well as the President was the target of anti-war protesters. When Lady Bird went to Williams College for a speech in October 1967, pickets greeted her with signs reading, "Confront the War Makers in Washington," and some of the students walked out on her as she delivered her speech.[73] The following month, when she persuaded LBJ to attend an Episcopal church in Virginia with her, the minister harshly criticized the administration's Vietnam policy in his sermon, and afterward LBJ told her, with a wry smile, "Greater love hath no man than that he goes to the Episcopal Church with his wife."[74] Like her husband, Lady Bird felt bitter about the "whiners, self-doubters, gloom spreaders," as she called LBJ's critics, and when Arkansas Senator J. William Fulbright, a Johnson friend, joined the anti-war opposition in Congress, she wrote indignantly in her diary: "It will be sheer luxury someday to *talk* instead of to *act*."[75]

On March 31, 1968, LBJ went on television to announce he would not seek another term in the White House and Lady Bird fully supported his decision to retire. "It was a poignant moment, yes," she said of his announcement, "but a relieving one for me."[76] When reporters asked her to sum up the achievements of the Johnson administration afterwards, she said: "We have done a lot; there's a lot to do in the remaining months; maybe this is the only way to get it done."[77] The inauguration of Richard M. Nixon as her husband's successor on January 20, 1969, left her with mixed feelings about leaving the White House. Inauguration evening she went to bed early with a line of poetry—"I seek, to celebrate my glad release, the Tents of Silence and the Camp of Peace"—in mind. "And yet," she wrote in her diary, "it's not quite the right exit for me because I have loved almost every day of these five years."[78]

In retirement, the Johnsons settled down at the LBJ Ranch in Stonewall, Texas, where the former President busied himself working on his memoirs, managing the ranch, entertaining his children and grandchildren, and attending football games. With his death in January 1973, Lady Bird went into a slump and even experienced a decline in health for a time. By the end of the year, however, she had snapped out of it and was following a schedule of work almost as crowded as the one she had observed as First Lady. For the LBJ Library and Museum in Austin she raised funds and sponsored symposia on education, the arts, civil rights, women in public life, and environmental issues. She also served on the Board of Regents of the University of Texas, and, while turning down appointments to the United Nations and to the boards of several big corporations, she took positions with several of her favorite organizations: the National Geographic Society, the National Park Service, and the American Conservation Association. Politics, for the most part, she eschewed; but in 1976 she did take to the hustings once more on behalf of her daughter Lynda's husband, Charles S. Robb, then running for Lieutenant-Governor of Virginia. "When it comes to your daughter and your son-in-law," she explained, "you simply want to do what you can."[79] Lynda insisted her mother was "the best campaigner in the family."[80]

In 1976, Lady Bird did something new for her: she participated in a bicentennial program, "Salute to America," and read excerpts from the Declaration of Independence to crowded audiences in the United States and Canada, while an orchestra played symphonic music in the background. Asked what LBJ would have thought of her performance, she exclaimed: "He would be laughing and he would be proud. He always wanted me to try new things."[81] When asked to sum up her years with the thirty-fifth President, she once declared: "He made me try harder and do more, and for the natural indolence I had, he was its mortal enemy, and I think perhaps sometimes I made him persevere or take a gentler attitude toward people or events or be less impatient."[82] In February 1987, when the executive producer of a TV film about LBJ mustered enough courage to ask her about LBJ's reputation as a womanizer, Lady Bird looked at him for a moment and then said quietly: "You have to understand, my husband loved people. All people. And half the people in the world were women. You don't think I could have kept my husband away from half the people?" Then her eyes moistened and she went on "He loved me. I know he only loved me."[83]

* * * *

Last and Only?

On February 13, 1941, when LBJ was in Congress, the Johnsons were invited to a reception for the Duchess of Luxembourg in the White House. "Tonight," wrote Lady Bird in her diary afterwards, "I went to my first (will it be my last and only!?!) Dinner at the White House!"[84]

Family Work

In 1948, when LBJ was running for the U.S. Senate, he told Lady Bird the night before the election, "Come on, honey, we're going home and spend the night at the ranch." "Oh no we're not," she said, "I'm going back to Austin and I'm going to get your mother, your sister, your aunts, and your uncles, your friends, and your cousins, and I'm going to take the telephone book and I'm going to assign one of them all the A's, one of them all the B's, one of them all the C's, one of them all the D's, right through the Z's, and we're going to call and say, 'Won't you please go to the polls and vote for my husband.' 'Won't you please go to the polls and vote for my son?' 'Won't you please go to the polls and vote for my my brother?'—or 'my cousin?' " And she did just that. By a narrow margin (and not altogether honest election, according to some analysts) LBJ won.[85]

B.C.

When Vice President and Mrs. Johnson were sightseeing in Athens, Mrs. Johnson kept asking, "How old is that?" Finally the guide said: "Mrs. Johnson, everything here is B.C. except you and me!"[86]

Deportment

Once, when Vice President and Mrs. Johnson were in Texas, Mrs. Kennedy invited Lynda and Luci to attend a State dinner at the White House in honor of the President of Sudan. The girls telephoned their mother in Texas to discuss White House deportment. "Read all you can find in the encyclopedia about the Sudan," she told them, "and don't drink any of the wine at dinner."[87]

Impressed

When Lady Bird married LBJ he was Texas Congressman Richard Kleburg's secretary and a hard-working young man who expected much of his wife. One day not long after the wedding he came home with a list of names and handed it to his wife. "I want you to learn the names of all these counties—," he told her. "These are the counties my boss, Congressman Kleburg, represents. These are the county seats. These are the principal communities in each county, and one or two of the leaders in each. Whenever you travel around with me, when we get to this town, you want to know who Mr. Perry is. . . ."

Lady Bird learned her lessons well. Years later, when LBJ was JFK's running mate, Jackie Kennedy was impressed by her work in the 1960 campaign. "She and my sister and I were sitting in one part of the room," she recalled, "and Jack and Vice President-elect Johnson and some men were in the other part of the room. Mrs. Johnson had a little spiral pad, and when she'd hear a name mentioned, she'd jot it down. Sometimes if Mr. Johnson wanted her, he'd say, 'Bird, do you know so-and-so's number?' And she'd always have it down. Yet she would sit talking with us, looking so calm. I was very impressed by that."[88]

Disappointment

Lady Bird's favorite Saturday night indulgence was watching *Gunsmoke* on television. She hated ever to miss it. One Saturday, when there were guests for dinner, she leaned over to the people at her end of the table and said, "I hope you will excuse me in a minute. I have an engagement." "Couldn't you use another word?" cried daughter Luci with a mischievous smile. But Lady Bird was disappointed when she learned that James Arness, who played the part of her hero, Marshal Matt Dillon, was a Republican. "How *could* he?" she wailed.[89]

Lady

Once a gentleman in the Congolese delegation to the United Nations was visiting the LBJ ranch. "Is she a lady?" he asked, puzzled by the First Lady's name. "Heavens, no," said United Press correspondent Helen Thomas. "It's her nickname." "Well," said the Congolese, "she *is* a lady!"[90]

Hat

For a long time Lady Bird preferred not to wear hats and for years did without them. Then one evening, when she and Vice-President Johnson were at a dinner party in their honor, she sat next to Jacob S.

Potofsky, president of the Amalgamated Clothing Workers of America, and he gave her a little lecture. "You ladies in public life don't realize that the way you dress affects the lives and incomes of many people," he told her. "You don't wear hats and you are not doing a thing for the hat industry—it is going to rack and ruin." The very next day Lady Bird went out and bought a new hat. "I am wearing this hat for Mr. Potofsky and the hat makers," she told a friend. Thereafter, whenever she started off for luncheon without her hat and suddenly remembered it, she would return, put it on, and announce: "I must wear my hat for Mr. Potofsky."[91]

Spoon

For years Lady Bird tried hard, but unsuccessfully, to get LBJ to diet. One night she was awakened by a strange clicking noise. She traced it to the kitchen and caught her husband eating one of his favorite desserts—tapioca pudding—with a metal spoon from a bowl. The next day the President instructed an aide to go out and buy him a wooden spoon. "If one thing didn't work," according to White House correspondent Marianne Means, "he'd try another."[92]

Mrs. Johnson

In April 1964, returning to Washington by car from a speech-making engagement in Cleveland, Lady Bird had Liz Carpenter arrange for dinner at a Howard Johnson restaurant near Pittsburgh. "Please reserve a large table for Mrs. Johnson and party," Mrs. Carpenter telephoned ahead. After Mrs. Johnson and party had dined and started on their way again, a newswoman interviewed the waitress. "How did you feel serving Mrs. Johnson?" she wanted to know. "Well," said the young woman, "I was pretty nervous." "Have you ever met a First Lady before?" asked the reporter. "First Lady?" cried the waitress in surprise. "First Lady?" "Yes," said the reporter, "that was Mrs. Lyndon B. Johnson, the First Lady of the Land." "Oh, my God," exclaimed the waitress. "Thank goodness I didn't know it. I would have fainted dead away. I thought it was Mrs. Howard Johnson. That was bad enough."[93]

Dime Box

Lady Bird agreed to deliver the baccalaureate address at Radcliffe College in June 1964, and then became apprehensive as the day approached. "Radcliffe has got the reputation of being one of the most intellectual schools . . .," she reflected, "and I have an education . . . from two quite simple high schools and from a good state university, but I'm far from an intellectual . . . and I was going to appear before people that I wanted to make a good speech for, and I wasn't at all sure I could." But her press secretary Liz Carpenter gave her some advice that quieted her nerves. She reminded her that in Texas's 10th District, which LBJ had represented for twelve years, there was a little town called Dime Box. "Just look out there at them," Mrs. Carpenter told her, "and think they all came from Dime Box." Lady Bird's Radcliffe talk, about the role of educated women in American society, was well received.[94]

White House Worker

Once a White House workman went to Lady Bird's bedroom to do some work, and when LBJ walked in he found the man stretched out on the floor with the First Lady nearby. A look of jealousy and anger swept LBJ's face; then he suddenly realized the man was doing some repair work and relaxed. Lady Bird enjoyed her husband's momentary wrath.[95]

Little Note

In March 1967, LBJ spoke in Asheville, North Carolina, for fifty minutes and the audience became restless. Finally Lady Bird wrote a little note, "close soon," on a piece of paper and slipped it to him. LBJ took it, held it up, and read it aloud. After the audience had stopped laughing he continued his speech.[96]

Yuki

At the wedding of Lynda Bird Johnson and Charles Robb in December 1967, people saw how Lady Bird handled her husband on occasion. As the bride and groom and their relatives lined up in the Yellow Oval Room to be photographed after the ceremony, out wandered Luci Johnson's little white dog, Yuki, dressed in his best red velvet sweater. "We've got to get Yuki in the picture," cried LBJ, reaching for the dog. "We can't have a family portrait without him." "That dog is *not* going to be in the wedding picture," announced Lady Bird firmly. LBJ started to argue, but she refused to be put off. "Mr. Brant," she told a White House aide, "get that dog out of here right now! *He will not be photographed!*" The President retreated, Brant removed the dog, Lynda breathed a sigh of relief, and the cameras started clicking.[97]

Who Would Have Thought?

In the spring of 1968 Mrs. Johnson went up to New York to dedicate the Astor playground at Jacob Riis Plaza in the Hispanic section of the city and then joined Mrs. Vincent Astor for a reception afterward. As they sped down Park Avenue in Mrs. Astor's handsome black limousine, Mrs. Astor giggled, "Who would have ever thought I would be riding down Park Avenue with the First Lady of the land!" "And who," cried Lady Bird, "would have ever thought I would be riding down Park Avenue with Mrs. Astor!"[98]

Little Mouse

When Patty LuPone was preparing to play the part of Lady Bird in NBC's television film, "LBJ: The Early Years," which aired in February 1987, she had a meeting with the former First Lady while making the film. "She was quite charming," LuPone discovered, "and I was completely intimidated." When Lady Bird asked what plays she had done before, the actress mentioned *Evita,* a Broadway musical about Argentina's First Lady, Evita Péron. "It's a far cry from Evita to me," exclaimed Mrs. Johnson. "Evita was a bird of paradise, and I'm just a little mouse."[99]

CHAPTER 35

PAT NIXON

1913–

Thelma Ryan (Pat) Nixon is probably the least known of all our recent First Ladies. She made few speeches, did no writing, avoided big causes, and gave noncommittal answers to reporters seeking her opinions on public issues. David Lester called her "the lonely lady of San Clemente" in his 1978 biography, but other observers, less kind, called her "plastic Pat," "antiseptic Pat," and "Pat the robot."[1] When she appeared in London with her husband, then Vice President, the *Spectator* reported that she "chatters, answers questions, smiles and smiles, all with a doll's terrifying poise."[2] Her daughter Julie, however, once complained that nobody understood her mother. "I feel," she said, "that she kind of lost faith that journalists would interpret things as they really are, and just didn't want to reveal herself at all." Julie insisted that her mother was plucky, not plastic, vivacious, not doll-like, and full of warmth and affection, not robotic.[3]

That Pat Nixon was plucky, vivacious, warm, and outgoing as a young woman is beyond doubt. "She was pretty, well-groomed, and had a radiant smile," remembered one of the students in a typing class Mrs. Nixon taught when she was only twenty-four.[4] "She was a different person entirely from the woman who was the wife of the President," insisted another former pupil. "She was approachable, friendly, and outgoing. She was happy, enthusiastic, sprightly. Her disposition was sunny, not intermittently, but all the time. She was a happy young woman. We liked her enormously because she never talked down to the students, always meeting them on an adult level, never intimidating them. . . . She enjoyed her life and her work."[5]

Work was something Pat Nixon knew in abundance in her early years. Born Thelma Catherine Ryan on March 16, the eve of St. Patrick's Day, in 1912, she grew up on a ten-acre truck farm in Artesia (later Cerritos), California, about twenty miles from Los Angeles, and did her share of the farm chores as a young woman. "It was very primitive," she recalled. "It was a hard life . . . I didn't know what it was not to work hard."[6] She seems, though, to have had a sunny disposition like her mother's and to have made fun out of work. "It was a good kind of life when you look back on it," she once said. "I worked right along with my brothers in the fields, really, which was lots of fun. We picked potatoes; we picked tomatoes; we picked peppers and cauliflowers. When I was real tiny I just tagged along. But when I got older I drove the team of horses. . . ."[7] Sometimes she went to town with her father to buy the weekly supplies. "I would never ask for anything, but how I hoped! I'd watch the corner to see if he came back carrying a strawberry cone. That would be the big treat."[8]

When Pat was only fourteen her mother died of cancer and she took over as housekeeper for her father and two older brothers. "As a youngster life was sort of sad," she admitted, "so I had to cheer everybody up. I learned to be that kind of person."[9] She "had a big heart," like her mother, according to her brother Tom, and she "sacrificed and did things without complaining."[10] Despite long hours in the kitchen and in the fields, moreover, she did splendid work in school and even found time to squeeze in some amateur theatricals. When she was a senior in high school, her father became seriously ill and she added nursing to her other skills. But in 1930, when she was seventeen, he died of silicosis (he had once worked as a miner in Ely, Nevada, Pat's birthplace), and since he had always enjoyed calling her his "St. Patrick's babe in the morn," she decided to change her name from Thelma to Patricia. The loss of her father, stern and reserved, but occasionally affectionate, was devastating. "I don't like to think back to that time," she told Jessamyn West years later.[11] When Gloria Steinem asked her about her childhood in an interview in 1974, Pat Nixon turned suddenly resentful. As a girl, she told Steinem, she worked too hard to have any childhood dreams. "I've never had it easy," she said, "I'm not like all you . . . all the people who had it easy."[12]

But young Pat did in fact have her dreams. She longed to travel to faraway countries the way her father did before his marriage. She also dreamed of getting a college education and broadening her horizons. "I always wanted to do something else besides be buried in a small town . . . ," she once reflected. "I wanted to start with an education."[13] After graduating from Excelsior High School, where she came close to being valedictorian, she enrolled in Fullerton Junior College, and to support herself, worked as a cleaning woman and then as a clerk in the First National Bank of Artesia. In 1931, though, she interrupted her college education to go to New York. To pay her way she drove an elderly (and ailing) couple across the country in "a huge and ancient Packard" in return for bus fare back.[14] But she stayed two years in New York, working first as a stenographer, and then, after a course in radiology at Columbia University, as an X-ray technician in a Catholic hospital in the Bronx. Working with lung patients like her father, she said later, was a "haunting" experience. "I wanted to reach out and help them," she said. "That is what gives one the deepest pleasure in the world—helping someone."[15]

In New York young Pat Ryan had a busy social life. She dated the young doctors and interns at the hospital and occasionally sneaked off to go bobsledding with the patients. But for the most part she lived austerely, for she was saving up money to go back to college. "The world is just what we make it—," she wrote her brother Tom in February 1933, "so let's make ours a grand one. Too, it's fun to work and then enjoy the fruits of the success. I love to learn new things, no matter now difficult—also go to new places."[16] Later that year she returned to Los Angeles, moved into an apartment with brothers Bill and Tom, and in the fall of 1934 entered the University of Southern California.

To put herself through USC, Pat Ryan held a variety of jobs: as a bit player in movies (in which she showed real promise but was uninterested), as a store clerk, as a dental assistant, and as a telephone operator. She also took numerous part-time positions at USC. Frank Baxter, who taught her Shakespeare, remembered that she often looked weary when she came to class. "There seemed to be plenty of reason for it," he later said. "As I recall it, if you went into the cafeteria, there was Pat Nixon at the serving counter. An hour later, if you went to the library, there was Pat Nixon checking out books. And if you came back to the campus that evening, there was Pat Nixon working on some student research program. Yet with it all, she was a good student, alert and interested. She stood out from the empty-headed overdressed little sorority girls of that era like a good piece of literature on a shelf of cheap paperbacks."[17] In 1937 she graduated from USC, cum laude, with a teacher's certificate as well as a degree in merchandising, and hoped for a position as a buyer in a department store. But when she received a teaching offer from the high school in Whittier, she decided to accept. At Whittier High School she taught typing and shorthand and did extracurricular work as well; she put on plays, coached cheerleaders, and served as faculty adviser to the student Pep Committee. "She looked so young to us," said one of the girls she taught. "She was very attractive, red hair, a very slim face. We were fascinated with her. She was soft-spoken, firm, and quite a good teacher."[18] Another former student remembered her as gentle but firm. "She treated us warmly, but she insisted on results. By the same token she expected clockwork punctuality from us and we absorbed the gentle hint that questions to her should be prefaced by her name. . . . Miss Ryan followed the book. She allowed no compromises, no errors, no second-rate job. Perfection and high standards were the only things she accepted."[19]

One evening in the fall of 1938 Pat Ryan decided to try out for a part in a play being put on by the Whittier Community Players. She got the part—as Daphne in the Alexander Woollcott-George S. Kaufman mystery melodrama, *The Dark Tower*—and also met Richard M. Nixon, ambitious young lawyer who also tried out for and won a part in the play that evening. Nixon was dazzled by the "beautiful and vivacious young woman with titian hair" he met for the first time at the rehearsal; for him "it was a case of love at first sight."[20] Pat was less impressed. After the tryouts Nixon told her: "I'd like to have a date with you." "Oh," she said, "I'm too busy." "You shouldn't say that," said Nixon boldly, "because someday I am going to marry you." Pat was not amused. "I thought he was nuts or something," she recalled. "I guess I just looked at him. I couldn't imagine anyone ever saying anything like that so suddenly." When she got home she told a friend: "I met this guy tonight who says he is going to marry me."[21]

Nixon's courtship was long (two and a half years) and at times discouraging. It wasn't until a few weeks after he first met Pat that she began accepting occasional dates: movies together, trips to the beach on Sundays, meetings at ice-cream parlors. For Pat's sake Nixon tried ice-skating and almost broke his neck; to please her he took up dancing, too, with not much more success. He was better at helping her grade papers; he also sent her poems and flowers and slipped tender notes under the door when she pretended not to be at home. So devoted a suitor was he that when Pat had dates with other fellows in Los Angeles he insisted on driving her to the city and then killing time until the evening was over and she was ready to return home. "He chased her," said one of Pat's girl friends, "but she was a little rat. She had Dick dating her roommate and all he did when he took the roommate out was talk about Pat." Once Nixon told her she seemed to like her Irish setter better than she did him.[22]

But Pat was not entirely uninterested. "I admired Dick Nixon from the very beginning," she later declared. "But I was having a very good time and wasn't anxious to settle down."[23] Gradually, however, she found herself strongly drawn to the persistent young lawyer, with his enormous drive and energy, enjoying the walks and talks together, and beginning to take him seriously. "He's going to be President someday," she told her friends.[24] To Nixon's delight she began showing up at the Nixon country store in Whittier now and then to help out behind the counter; sometimes she arrived there before school-time to help Nixon's mother bake pies and cakes for the day. By the time Nixon proposed, in March 1940, she was deeply in love, and readily agreed to an engagement. Soon after, he jubilantly sent her the engagement ring, which he had picked carefully, in a basket of flowers. The wedding—a Quaker service, for Pat adopted Nixon's faith—took place on June 21, 1940, at the Mission Inn in Riverside. The honeymoon was a two-week trip to Mexico in an automobile (filled with canned food) Pat helped buy for the occasion.

After the honeymoon the Nixons moved into a little apartment over a garage near Nixon's law office and Pat helped make ends meet by continuing to teach as well as keeping house. World War II took the young attorney to Washington for a time, working for the Office of Price Administration (OPA), and then into the U.S. Navy, with assignments in the Pacific as well as at various naval posts in the United States. Pat worked for the OPA in Washington, too, for a time, and then, when Nixon went overseas, got a position as a price analyst for the OPA branch in San Francisco. "Your job is far more important than mine was at OPA," Nixon wrote her. "I'm really *very* proud. I like to tell the gang how smart you are as well as being the most attractive person they'll ever see."[25] The two exchanged letters every day when he was overseas.

After the war Nixon entered politics instead of resuming law practice in Whittier. Late in 1945, when he was still in the service, Republican leaders invited him to run for Congress the following year, and, after talking things over with Pat, he decided to make the bid. Pat wondered where the money to run the campaign was coming from, and when he suggested biting into funds they had accumulated during the war she made no objection. She wasn't particularly interested in politics, she admitted, but "I could see it was the life he wanted," she said later, "so I told him that it was his decision and I would do what he liked."[26] But she laid down two conditions. One was that her home would always be a quiet refuge from politics where she could give the children (Tricia was born in February 1946 and Julie in March 1948) a normal life; the other was that she would never be called on to make any political speeches. Nixon readily agreed to both conditions, but, as it turned out, both were eventually forgotten.

In the spring of 1946 the so-called "Dick and Pat team" was born. While Nixon stumped the Twelfth Congressional District, Pat worked at campaign headquarters typing up campaign literature, answering letters, and soliciting contributions. Sometimes she accompanied her husband on speech-making tours and handed out leaflets to crowds on streetcorners and in meeting halls. She didn't especially enjoy any of this. "She gritted her teeth," observed a friend, "and did it because she felt it her duty to do it."[27] At teas, coffees, and receptions, she was "nervous, uptight and tense," according to Roy Day, Nixon's campaign manager. "It was all so new to her." Still, she made a good impression in public and Day thought she was "a hell of an asset. She won a lot of Brownie points for Dick with those appearances."[28] In 1946, Mrs. Nixon recalled, "not many women were active in politics. But we were so anxious to win I just thought of ways I could be helpful."[29]

Some people thought Mrs. Nixon was shocked by her husband's slashing attacks on his Democratic opponent, Jerry Voorhis, during the campaign. But those who knew her best were convinced she loyally supported her husband in whatever he said or did and never questioned his methods. In time, though, she came to dislike politics because of the vicious give-and-take that seemed always to characterize the Nixon campaigns. "She didn't want politics ever," friend Earl Mazo insisted. "Her friends were never political friends. She hated the idea of ever facing another campaign. Every time Nixon entered one she was in deep despair."[30] But when he ran for re-election to the House of Representatives in 1948 and made his successful race for the U.S. Senate in 1950, she helped out even more than she had in 1946. She took notes in shorthand of the speeches of her husband's opponents so he could use them in his responses; she also made critiques of his own speeches which he found "thoughtful, and sometimes even persistent."[31]

In Washington, where the Nixons settled in 1947, Mrs. Nixon tried to keep out of the limelight as much as possible, but at the same time she tried to be a good politician's wife. "She is basically very, very jealous of her privacy," Nixon told a reporter. "While she does the protocol things with a great grace, skill and charm, she's not the one who seeks it—none of that showboat business."[32] Mrs. Nixon took over all the household chores, at her husband's request, but she also showed up in his office when the mail was especially heavy. "Dick had the capacity to do great things," she explained. "I wanted to save his time. So I did all the chores so that he could use his energy for the problems at hand."[33]

But Mrs. Nixon balked at first when the chance came for her husband to become Dwight D. Eisenhower's running mate in 1952. She was reluctant to face the ordeal of another campaign and hated the thought of neglecting the two children to accompany her husband on speaking tours. In the end, however, she yielded. "I guess I can make it through another campaign," she sighed after talking it over at length with her husband.[34] She was eating lunch with a friend in a Chicago restaurant when she heard the news that her husband had received the vice-presidential nomination by acclamation. "The bite fell out of my mouth," she recalled, "and we rushed back to the convention hall."[35] As she joined her husband and headed for the platform she felt suddenly elated and kissed him twice on the cheek. "We work as a team," she told reporters afterwards and listed the contributions she expected to make to the campaign: handling mail and news releases, attending women's meetings, and speaking informally, though not about politics.[36] Republican officials adored her. "She's the best one on the whole ticket," one of them exclaimed.[37] "She had a graciousness about her and yet she was spontaneous," said Herbert Brownell admiringly. "It's an unusual combination and hard to define. The crowds just loved her. She was *simpático*."[38]

If Mrs. Nixon took pleasure in helping out in 1952, it was short-lived. Early in September came the revelation that Nixon had been receiving secret contributions from California businessmen for his political expenses, and there were immediate demands by some Republicans that he leave the ticket. Plunged into despair, Nixon began thinking of dropping out of the race. But his wife stood firm. "You can't think of resigning," she exclaimed. "If you do Eisenhower will lose. He can put you off the ticket if he wants to but if you, in the face of attack, do not fight back but simply crawl away, you will destroy yourself. Your life will be marred forever and the same will be true of your family, particularly your daughters."[39] Her reasoning convinced Nixon, and he resolved to "fight the thing through to the finish, win or lose."[40] At Eisenhower's suggestion he decided to go on the air and bare his private finances in order to show that he hadn't profited personally from the secret fund. His wife was bothered by the thought of abandoning their privacy. "But why," she cried, "do we have to parade how little we have and how much we owe in front of all those millions of people?" "All candidates' lives are public," explained Nixon. "It seems to me," she said, "that we are entitled to at least some privacy." "Right now we're living in a fishbowl," Nixon told her. "If we don't itemize everything we've earned and everything we've spent the broadcast won't convince the public. I just don't have a choice."[41]

Nixon's *apologia pro sua vita* on September 23 was warmly received by the public and ensured his place on the Eisenhower ticket. But just before going on the air he had sudden misgivings. "I just don't think I can go through with this one," he told his wife. "Of course you can," she said firmly and took his hand as they walked onto the stage at the NBC studio in Los Angeles.[42] In his speech, which he delivered from behind a desk, with Mrs. Nixon sitting impassively in an armchair at his side, Nixon mentioned his wife three times. "I am not a quitter," he announced at one point, "and Pat's not a quitter." Then, rearranging dates a bit, he added: "After all, her name was Patricia Ryan, and she was born on St. Patrick's Day, and you know the Irish never quit." He referred to his wife again in an oblique reference to Senator John Sparkman, Democratic vice-presidential nominee, whose wife worked in his office. "I'll tell you what some of them do," said Nixon. "They put their wives on the payroll, taking your money and using it for that purpose. And Pat has worked in my office night after night after night, and I can say this, and I say it proudly, she has never been on the government payroll since I have been in Washington, D.C." And, finally, with revelations in mind of influence-peddling among Democratic officials, involving refrigerators and mink coats, he declared, after detailing his personal finances, "It isn't very much, but Pat and I have the satisfaction that every dime that we've got is honestly ours. I should say this—that Pat doesn't have a mink coat. But she does have a respectable Republican cloth coat. And I always tell her that she'd look good in anything."[43]

After the speech Nixon's misgivings returned. "Pat," he said gloomily as they headed back to the hotel, "I was a failure." "Dick," she said sharply, "I thought it was great."[44] She expected Eisenhower to back her husband at once and was baffled by his continued uncertainty. "What more does that man want?" she exploded. To a close friend she exclaimed, with tears in her eyes, "Why, why should we have to keep taking this?"[45] Eisenhower finally came around and kept Nixon on the ticket, but Pat remained resentful. Later on, when Mrs. Eisenhower remarked that the Nixon fund crisis had hurt the campaign, Mrs. Nixon cried: "But you just don't realize what *we've* been through!"[46] She never really enjoyed her husband's political involvement after 1952. From that time on, Nixon wrote later, he knew that "although she would do everything she could to help me and help my career, she would hate politics and dream of the day when I would leave it behind."[47]

As Vice-President, Nixon settled his family in a big fieldstone house in Washington and began employing housekeepers, baby-sitters, and servants to manage the chores that Pat had once handled. At first she tried to stay on top of things; she planned meals, made supermarket lists, did some cooking, and even continued pressing her husband's pants, though he told her to send them to the dry cleaners. "He scolds me," she admitted, "but I've always done it, and I like to."[48] Less and less, though, did the image of the busy housewife she projected accord with reality. With continual calls on her time as Second Lady to be with her husband on ceremonial occasions and to make goodwill tours with him abroad, she soon found herself turning the household chores over to the servants, enlarging her wardrobe, joining country clubs, and, in general, living the life of luxury and privilege that she and her husband had once envied. Socially, the Nixons were considered a bit "wooden and stiff," but the Second Lady received high praise for her discreet behavior in public and for her scrupulous avoidance of any kind of political comments that might embarrass the administration.[49] "Pat's smart enough to say the right things," said Nixon proudly, "and avoid the sticky questions."[50] One of the writers for *The New Republic* was struck by her evasiveness in interviews. "She knows when to be the silent, demure partner of the great man who is only a heartbeat away from the White House . . .," he wrote. "She always has the right reply, the right greeting, the gracious smile."[51]

And what was behind the gracious public smile? Was Mrs. Nixon imperturbable at home too? Her closest friends insisted she let her hair down when she was with her best friends and that she laughed, joked, and even teased her solemn husband when they weren't in the public eye. One evening, when Nixon was pontificating about something, they reported, Mrs. Nixon walked into the room with a tray loaded with goodies and cried merrily: "Try some of these. They're better than that baloney he's handing out!"[52] There were rumors of quarrels, too, for Pat had a temper, mainly about the long hours Nixon spent away from home on the job, and about the way his political ambitions intruded upon their family life. "We don't have as many good times as we used to," she told a friend candidly not long after her husband became Vice President.[53] When the quarrels came, Nixon frequently got his mother on the phone to pacify his wife. Once, though, the quarrel was so protracted that he finally brought his mother from Whittier to Washington for a few days to thaw things out between the two of them.

But Mrs. Nixon's temper never flared in public. On the goodwill tours she made with the Vice President—to the Far East, Africa, England, Russia—State Department officials, American and foreign journalists, and foreign dignitaries had only praise for her. She studied up on the countries she was to visit, avoided conventional routines, visited schools, hospitals, and orphanages instead, and learned a great deal about local customs, and tried to follow them herself, too, whenever she could. Above all, she insisted on meeting women's groups in every country she visited, for she was anxious to dramatize the achievements of women in the world, and bring their problems before the public, especially in countries where their status was low. "Everywhere I went," she said after one of her trips, "it helped women."[54] Eisenhower was delighted by the splendid showing she made abroad. "Dick," he said after one such trip, "I've heard some pretty good reports on you." Then he turned to Mrs. Nixon. "But the reports on you, Pat," he added, with a grin, "have been wonderful."[55]

When the Nixons visited Latin America in May 1958, however, they were the target of virulent anti-American demonstrations. In Lima, Peru, an angry crowd followed the Nixon motorcade to the hotel where they were staying and hurled rocks and imprecations all the way. Five days later a mob in Caracas, Venezula, came close to killing the Vice-President and his wife. So hostile was the crowd lining the streets that the Nixons abandoned their planned itinerary and drove straight to the American Embassy from the airport instead. During the twelve-mile drive hostile demonstrators spit at them, threw garbage and rocks, and even attacked their cars with baseball bats. But Mrs. Nixon kept calm; she was, Nixon later wrote, "probably the coolest person in the whole party."[56] At one point she even leaned across the barricade to pat the shoulder of a young girl who had been cursing and spitting and saw the latter turn her head away in embarrassment. One of the Secret Service men thought the Vice President's wife displayed "more guts than any man I've ever seen."[57] Mrs. Nixon later confessed she was more angry than frightened during the ordeal. "On occasions like that," she said later, "it wasn't your own personal fear that you think of particularly. You just feel sick that anything like this could happen to what was meant to be a goodwill trip."[58] For the *Los Angeles Times* Robert Hartman, who was in the press truck accompanying the motorcade, filed a report: "Pat was magnificent today. . . ." He also reported that when newsmen cheered upon her arrival at the embassy, her eyes filled with tears.[59]

Fortunately, the Vice Presidency was mostly pleasant for the Nixons. And Mrs. Nixon hoped it meant the end of the political road for her husband. She was quite upset when he began toying with the idea of running for President in 1960. But as usual, once he made up his mind to run, she threw all of her energies into the campaign, accompanied him once more on speech-making tours, and attended the customary vote-seeking meetings on his behalf. During the 1960 campaign, however, Republican officials, with Nixon's full approval, decided to make her the center of things, too. "When you elect a President, you are also electing a First Lady whose job is more than glamour," they announced. "The First Lady has a working assignment. She represents America to all the world. Pat Nixon is part of the experienced Nixon team. She's uniquely qualified for the position of First Lady."[60] In October, the women's division of the Republican National Committee proclaimed a Pat Week, distributed Pat-Nixon-for-First-Lady buttons, held a series of coffees and rallies in her honor, and encouraged precinct workers to canvass their neighborhoods in her name that week. Some people thought the Vice-President's wife was getting too involved in politics. But she told her critics that her role as a campaigner was "reflective of women all over America taking an active part, not only in political life, but in all activities. There was a day when they stayed at home . . . but they have emerged as volunteers for a cause they believe in."[61] With Jackie Kennedy playing some part in Democratic candidate John F. Kennedy's campaign in 1960, syndicated columnist Ruth Montgomery concluded that "for the first time in American history, one woman could conceivably swing a presidential election."[62] In the end, Jackie seems to have outshone Pat, partly because of her youth, but there is no way of course, of measuring the impact of either woman on the outcome of the contest. Kennedy's narrow victory over Nixon in November came as a severe blow to Mrs. Nixon and she was convinced that ballot-box frauds in Texas and Illinois had cheated her husband out of victory. She felt so bad about it, in fact, that she snubbed Mrs. Kennedy at a pre-inaugural meeting in January 1961. "I guess I'll have to get a job teaching again," she quipped to friends, but she wasn't really laughing it off.[63]

For all her disappointment, Mrs. Nixon found that the 1960 defeat had its compensations. It meant a return to the private life she cherished and a resumption of old friendships and associations in California. In 1961, Nixon joined a law firm in Los Angeles, built a home with a swimming pool in a fashionable part of town, and pleased his wife by giving her the impression he had left politics forever. "You think people in the movie business are competitive," she told producer Samuel Goldwyn, Jr., at a dinner party. "They may be competitive, but they are not mean. In politics they are the most vicious people in the world."[64] Her loathing of politics was so strong by this time that when Nixon changed his mind and decided to make a bid for the California governorship in 1962, he postponed talking it over with his wife until the last possible moment. When he finally broke the news to her at a small restaurant party, she retired to the restroom in tears. "If you run this time," she told him, "I'm not going to be out campaigning with you as I have in the past." Her obduracy led Nixon to reconsider his plans. One day, however, as he sat down to draw up a statement announcing his intention to stay out of the contest, his wife came into the room, and, fighting back her disappointment, declared: "I have thought about it some more, and am more convinced than ever that if you run it will be a terrible mistake. But if you weigh everything and still decide to run, I will support your decision. I'll be there campaigning with you just as I always have." "I'm making some notes to announce I won't be running," Nixon told her. "No," she said, "You must do whatever you think is right." Then she put her hand on his shoulder, kissed him, and left the room. When she was gone Nixon threw the paper he had been writing on into the wastebasket, took out some fresh sheets, and began preparing an announcement that he was entering the race.[65] For all her reservations, Mrs. Nixon helped him campaign again in 1962 (he called her his "secret weapon") and wept at his defeat. But she told friends afterward she was glad to be "out of the rat race" and rejoiced when her husband became a partner in a Wall Street law firm in 1963. She called the next period in her life in New York a "six-month vacation."[66]

The vacation ended in January 1968 when Nixon entered the New Hampshire presidential primary with his wife's approval. She had initially opposed the return to politics, of course, but agreed to help when she realized how important trying again for the Presidency was to her husband. She was convinced, too, that "he alone was capable of solving some of the problems we were facing in the country then" and she was still rankling over the 1960 defeat and eager to see him vindicate himself this time around.[67] During the 1968 campaign she strove mightily to counteract the popular image of her husband as a solemn and at times mean-spirited gut-fighter. In press conferences and TV interviews she portrayed him as a warm, affectionate family man with a nice sense of humor, as well as a highly informed and experienced statesman who would make an outstanding Chief Executive. But as always she avoided politics. "I don't think one person can speak for another," she declared, when pressed to speak out on the issues. "The candidate should speak for himself."[68] Asked by one reporter what her greatest contribution to her husband was, she said simply: "I don't nag him. The best I can [do] is cheer him up." She did, though, admit to giving him advice on occasion. "I fill him in on what women think," she told one reporter. "They are thinking peace at home and peace abroad."[69] On election night she wept with joy when her husband defeated Hubert H. Humphrey at the polls. "I felt at last," she told daughter Julie, "that Daddy was where he could really be of value to the country and to the world."[70]

To her surprise Mrs. Nixon enjoyed being First Lady. Once she even called her White House years "glorious."[71] She liked being hostess, not only for foreign dignitaries and domestic celebrities, but also for humble people from nursing homes, orphanages, and homes for the aged. She continued, too, to keep in close touch with women's organizations, frequently inviting members to the White House for luncheons and teas. The women of the press corps found her more relaxed and at ease with them than she had been as the wife of a Vice-President. "She is warm and kind and she goes the extra mile to shake a hand and greet a stranger," UPI reporter Helen Thomas observed. "She is concerned about people's feelings. . . . As a hostess, she has kept her promise of not entertaining just the big shots. She never forgets her days of poverty when she was growing up. . . ."[72] The Nixon dinner parties were more formal than those of the Kennedys and the Johnsons—State Dinners were formal white tie-and-tails affairs—but Mrs. Nixon's quiet friendliness kept them from being stuffy events. "She feels that person-to-person contact is her strongest asset," noted one observer. "People respond so warmly to her and she is so spontaneous about it. For instance, she will be in a crowd and suddenly she will put her arm around someone or smile and make some personal comment. She is so gracious and warm and is absolutely unflagging. She wants to shake the hands of all who are waiting to see her, and will continue to do so even though she becomes very tired sometimes doing it."[73] Some people thought there was a "new Mrs. Nixon" in the White House. "Pat Nixon," noted *Newsday* in July 1969, "has suddenly emerged from an icy cocoon of literal anonymity and proven herself a living, breathing, thinking, loving woman."[74]

Going over the mail was for a time almost as exciting for the new First Lady as presiding over White House receptions. Huge bags of mail arrived for her every day—more than previous First Ladies had received—and she spent hours at first trying to read every letter and making personal responses to as many as she could. When reporters wanted to know why she spent so much time on her mail, she explained: "When a letter from the White House arrives in a small town, it's shown to all the neighbors, and often published in the local paper. It's very important to the people who receive it."[75] In time, though, she entrusted her mail for the most part to members of her staff and began focusing her energies on another labor of love: redecorating the White House. Critics accused her of trying to eliminate the Kennedy influence from the White House, but Mrs. Nixon denied the charge. The White House was, in fact, run down when the Nixons moved in, from constant use since the Kennedy days, and what Mrs. Nixon did to restore it—raising funds for redecorating the rooms and acquiring paintings and furniture of historic interest—was very much in line with Jackie Kennedy's restoration project eight years earlier. The latter visited the White House in February 1972—her first visit since her husband's funeral in 1963—and expressed pleasure with what Mrs. Nixon had accomplished. "I think it looks lovely," she said. "I never intended Boudin's works to remain in the White House forever. Every family that lives there should put its own imprint there."[76] White House curator Clement E. Conger had high praise for Mrs. Nixon's imprint and thought she was "always regrettably modest" about the way she supervised the assembling of the best collection of "Americana furnishings" in the country.[77]

Goodwill tours had been the highlights of Mrs. Nixon's public life when her husband was Vice President, and she continued to derive great pleasure from the trips—more of them than ever before—she made abroad as First Lady, sometimes with the President and sometimes on her own. As in the vice-presidential days she did her homework before embarking on her journeys and created good feelings wherever she went—Russia, China, Africa, Europe—by her knowledge of the countries she visited and curiosity about the people there. "I have known the wives of several American Presidents," said Soviet Foreign Minister Andrei Gromyko's wife, "but Mrs. Nixon is the nicest."[78] After a terrible earthquake hit Peru in May 1970, Mrs. Nixon insisted on flying to Lima with tons of relief supplies, and touring the areas of devastation with President Velasco's wife, and the result was an improvement in relations between the United States and Peru. No other First Lady, observed AP correspondent Fran Lewin, not even Eleanor Roosevelt, had "ever tried such a person-to-person mercy mission of human concern and diplomatic side effects."[79] And in 1972, after she represented the United States at the inauguration of a new President in Liberia, even the West Wingers (the President's aides, housed in the West Wing of the White House, who tended to be hostile to the East Wingers, members of the First Lady's staff) finally came around. "Mrs. Nixon has now broken through where we have failed," they acknowledged in a memorandum to the President. "She has come across as a warm, charming, graceful, concerned, articulate, and most importantly, a very human person. People, men and women—identify with her—and in return with you."[80]

If Mrs. Nixon had any special project as First Lady—and she denied wanting to have one—-it was encouraging what she called "volunteerism," that is, programs in which people offered their services to help the needy without pay. "Government is impersonal," she declared, "and to really get our problems solved we have to have people too. We need the personal touch."[81] In March 1969 she made a long trip across the country to examine various community projects, organized by volunteers, to help the poor, the blind, the aged, and the handicapped. "Volunteerism comes from the heart," she told people. "When you're paid, it's too commercial. You volunteer because you love your country, your people, and because it makes you feel good."[82] She was especially anxious for women to participate in volunteer work. "So many women can't work full-time; volunteerism gives them a chance to be useful with the hours they can afford. In volunteering, it is your heart that is speaking, and it's listened to. It's working with people on a one-to-one basis."[83] When she visited college campuses on behalf of volunteerism her sincerity and devotion even impressed students opposed to her husband's Vietnam policy. "She wanted to listen," said one student after she visited his campus. "I felt like this is a woman who really cares about what we are doing. I was surprised. I didn't expect her to be like that."[84]

Some observers thought the White House was transforming Mrs. Nixon. She seemed increasingly willing to express her opinions, even if mildly, in public, and less reluctant to go out on a limb. She told reporters she thought abortion "should be a personal decision" (though her husband called it "an unacceptable method of population control"), came out for the appointment of a woman to the Supreme Court, began wearing pants suits in public (though Nixon had once deplored them), and encouraged women to get involved in politics. "It is important for wives to campaign," she said. "Unless you are willing to work for good government, then you won't have it."[85] When Nixon ran for a second term in 1972 she did something new: went on a campaign tour by herself through seven midwestern and western states. She still found it a trial to speak in public, but she came through nicely and, according to a close friend, seemed to have "gotten the feel, the instinct if you will, that all political people need to possess if they want to survive."[86]

The Watergate crisis soon destroyed the pleasure Mrs. Nixon took in her husband's triumphant re-election as President in November 1972. At first she thought revelations of the attempted burglary of Democratic headquarters in Washington during the campaign were "blown all out of proportion."[87] Then, when it became clear that her husband had secretly tried to block an investigation of the Watergate break-in, she became convinced the Democrats were out to get her husband. "It's right out of *The Merchant of Venice,*" she told a friend. "They're after their last pound of flesh."[88] There were stories that the Watergate ordeal was driving her to drink and even leading her to contemplate divorce, but they were all false. Even after the House Judiciary Committee voted articles of impeachment, she remained unswervingly loyal to her husband and believed, almost to the end, that somehow he would win out. "You know I have great faith in my husband," she told UPI's Helen Thomas, clenching her fists. "I happen to love him."[89] Like her daughters, Tricia and Julie, she wanted him to fight on, but when he decided on resignation, she set to work at once planning the move out of the White House. "We're all very proud of you, Daddy," she said when the family gathered for their last dinner in the White House on August 7, 1974.[90]

On the day the Nixons left the White House, Mrs. Nixon was too distraught to say farewell to her own personal staff. And when her husband spoke to his own staff just before leaving for California, she stood at his side without saying a word. Some of those present were surprised when Nixon mentioned his mother ("she was a saint") in his mournful valedictory, but not his wife. But if Mrs. Nixon shared their chagrin, she gave no sign of it. "Her whole life was in her face," according to one of those present. "It just said everything. There are no words that could ever describe that look. I tried to, myself. I even tried to write it down, but I couldn't. It wasn't even just her face. It was her whole self, the way she walked and moved. You could see that she was steeling herself."[91] In an entry in his diary during the Watergate crisis Nixon wrote: "She has always conducted herself with masterful poise and dignity. But, God, how she could have gone through what she does, I simply don't know."[92]

In San Clemente, where the Nixons settled after leaving the White House, Mrs. Nixon went into seclusion. "It's as if she went underground," said a friend, "or vanished into the sea."[93] Her despair was so profound for a time that she stopped communicating with her closest friends. "She's a recluse," lamented Roy O. Day, Nixon's first campaign manager. "It's a damned shame. She's a great lady."[94] When Helen McCain Smith told her several people wanted to interview her, she exclaimed: "Oh, Helen, we're out of it now."[95] But the Watergate disgrace seems to have brought her and her husband closer together than they had been in a long time. During the White House years Nixon had treated his wife at times with such indifference and lack of consideration that members of her personal staff and the women's press corps were practically up in arms about it. On the San Clemente estate (where Mrs. Nixon arranged things in the bedroom the first day there to please her husband) the Nixons began finding solace again in each other's company. "At least we have the chance to spend a lot of time together . . . ," Nixon told a British journalist. "We've discovered, in this time of crisis, that we need each other. We've grown closer than ever before. . . . I don't know what history will say about me, but I know it will say that Pat Nixon was truly a wonderful woman."[96] When a severe attack of phlebitis sent Nixon to the hospital in Long Beach for surgery in October 1975, Mrs. Nixon stayed by his side until the crisis was over. According to a *Ladies' Home Journal* poll, she continued to be the most admired woman in America even after Watergate.

In July 1976 Mrs. Nixon suffered a stroke that left her partially paralyzed on the left side and slightly affected her speech. But her spirits remained high—*McCall's* called her the "unsinkable Pat Nixon"—and she started doing exercises right after leaving the hospital to regain use of her left side.[97] "I do or die," she once said. "I never cancel out."[98] But she kept to the San Clemente compound most of the time and refused all invitations to do interviews or preside at community charity functions. Her health continued to be frail, but she spent as much time as her strength allowed in her beloved garden; she also read a lot (biographies and historical novels), listened to operettas and musical comedies on records, and took occasional swims with her husband in their outdoor pool. In February 1980, the Nixons moved to New York City, partly to be near their children and grandchildren, and then, in October 1981, to Saddle River, New Jersey, where they would be within easy commuting distance of Manhattan, where Nixon had his office.

"Watergate," Mrs. Nixon once told her daughter Julie, "is the only crisis that ever got me down."[99] She tended to blame White House aide Robert Haldeman for her husband's downfall and to hope for eventual vindication. She was bitter about the fact that Watergate seemed to be about the only thing most people remembered about her husband's Presidency. When Julie began working on *Pat Nixon: The Untold Story* (published in 1986), her mother sighed: "I hate when it is published the whole focus will be on Watergate."[100]

* * * *

Staunch Democrat

When the Nixon girls, Tricia and Julie, then teenagers, visited the Mission Inn in Riverside, California, in 1963, they were anxious to see the Presidential Suite (so named because several Presidents had stayed there), where their parents had been married, and joined a group of tourists who were being shown through the famous old Spanish-style hotel. When the touring group reached the Presidential Suite the woman who was guide announced: "And here is where Richard and Pat Nixon were married." As the Nixon girls looked in surprise at the long bar, banquettes, and cocktail waitresses busy serving drinks, Janet Goeske, a Republican leader who was with them, exclaimed: "I beg to differ with you; they were married in here, but then it was a *Presidential* suite."

Then she turned to the two girls: "Your father and mother," she said, "were not married in *this* place." Afterwards Mrs. Goeske insisted that the "lady guide was a staunch Democrat."[101]

Wifely Secret

During the 1956 campaign, Pat Nixon always looked so well-groomed when she was on the road with her husband that reporters finally asked her how she managed it. She told them she brought five dresses, four suits, two pairs of shoes, and eight hats on the trip. "In one bag?" cried a newspaperwoman. "All the rest of us were told we're limited to one bag!" "And a briefcase," added another newshen. Mrs. Nixon explained. "I don't have a briefcase," she said, "so I felt I was entitled to substitute a hatbox." "But one bag?" persisted the reporter. "Please—please, Mrs. Nixon. Tell us your secret!" Mrs. Nixon finally yielded; she agreed to tell them a secret which, she said, "every wife knows." Grinning she said: "I do the packing for these trips, and I pack at least two of my suits in Dick's bag when he isn't looking."[102]

Haimisheh

Pat Nixon seemed distant and aloof in public, but in private her friends found her warm and fun-loving. Journalist Earl Mazo's wife Rita, who knew her well, used the Yiddish word *haimisheh* (meaning warm, cozy, informal, unpretentious) to describe her. Once Mrs. Mazo ran into Mrs. Nixon, when she was Second Lady, in a large Washington department store with her two daughters. "Oh, Rita," exclaimed Mrs. Nixon, "I'm go glad to see you. We've done some shopping, been to the dentist, and now we're going to lunch. Come join us." "I'm sorry," said Mrs. Mazo, "but I can't. I must get home to meet the children when they get home from school." Afterward, she couldn't help thinking: "Here was the wife of the Vice-President of the United States asking me to lunch, and here I was saying no to her. It just didn't occur to me that she *was* the wife of the second man in the country. She was just a mother taking her two children shopping and to the dentist and I was another mother on her way home to meet her children when they came from school."[103]

Mrs. Nixon's Tureen

One afternoon Mrs. Nixon attended an auction in Virginia with some friends, and when the auctioneer brought out a Rosenthal china tureen, she thought it would make a perfect centerpiece for flowers and began bidding for it. But someone on the other side of the room made a higher bid and Mrs. Nixon quickly went up too. After a few more challenges she finally won, though she ended by paying more than she thought the tureen was worth. Afterward, to her surprise, she learned that the other bidder was a friend who knew Mrs. Nixon was looking for just such a tureen and didn't realize she was bidding against the Second Lady. There was another surprise when Mrs. Nixon got home. When she unwrapped the package she found the tureen contained only one handle, which she hadn't noticed from a distance. And then one more surprise: it wasn't a tureen at all, but an elegant chamber pot.[104]

Can't Stand Pat

In his fourth debate with John F. Kennedy in 1960, some people thought Nixon made a real gaffe at one point and made it three times. "America can't stand pat," said Nixon. "We can't stand pat for the reason that we're in a race, as I have indicated. We can't stand pat because it is essential with the conflict we have around the world, that we not just hold our own. . . ."[105]

Louise Bird

During the Johnson-Goldwater campaign of 1964, Nixon went on the road seeking votes for Goldwater while his wife went to Europe with her daughters and her friend Louise Johnson for a little vacation. But Europeans who knew something about the campaign were surprised when they came across the names, Mrs. R. Nixon and Mrs. L. Johnson, in hotel registers and ticket reservation offices. When the two women visited Edinburgh, the American Foreign Service officer there, having heard that Mrs. Richard Nixon and Mrs. Lyndon Johnson were arriving, rushed to meet them at the airport with a military aide and a plainclothes detective with him. But after seeing Louise Johnson, he whispered to Mrs. Nixon: "That isn't really Mrs. Lyndon Johnson, is it?" "No," said Mrs. Nixon, and then, deciding to tease him, added: "It's the President's sister." The officer had lost interest by then and was anxious to get off the hook. When he asked lamely whether there was anything he could do for them, Mrs. Nixon, still in a bantering mood, said, "Why, yes. We would love to have tea with the queen." In great embarrassment, the officer gave her a long list of reasons why he couldn't possibly arrange a meeting with Queen Elizabeth. He was happy to leave them when they got into the city. Afterward, for the rest of the trip, Mrs. Nixon called her friend "Louise Bird."[106]

Cottage Cheese

The first day the Nixons were in the White House, the chef and his staff awaited their first order for dinner with some trepidation. Finally it came. "Tricia, Julie, David [Eisenhower, Julie's husband] and the President would like steak for dinner in the upstairs dining room," Mrs. Nixon told the chef. "I'd just like a bowl of cottage cheese in my bedroom." There were plenty of juicy steaks in the White House kitchen, it turned out, but not a smidgen of cottage cheese. The chef hastily requested a White House limousine. "For two weeks we've laid in supplies in the kitchen," he moaned. "I think we could open a grocery store in the pantry. We've tried to find out everything they like . . . But we don't have a spoonful of cottage cheese in the house. And what in the world would be open this time of night—and Inauguration night to boot?" But the head butler got in the limousine and prowled around Washington until he finally located a delicatessen that was still open and had plenty of cottage cheese. After that the White House chef kept the White House kitchen stocked with Mrs. Nixon's food. "I'd worried that Mrs. Nixon was so thin," Chief Usher J. B. West recalled. "Now, I realized, she intended to stay that way."[107]

No Slobbering

On March 16, 1974, Mrs. Nixon returned from a six-day visit to South America and met her husband in Nashville to celebrate her sixty-second birthday at the opening of the new Grand Old Opry there. After the audience sang Happy Birthday, with the President at the piano, Mrs. Nixon got up, went over to her husband, arms extended, to hug him, and he turned away. Some of the women reporters present were shocked. "He absolutely, in front of thousands of people, turned his back on her outstretched arms—totally ignored her," exclaimed one reporter afterward. "He turned his back! It was simply incredible." But Julie Eisenhower later explained that her father was turning to the center of the stage when her mother came over, so he could indicate to the master of ceremonies that he had interrupted the program long enough and that it could now proceed. She also emphasized the fact that her parents were very private people who refrained from expressing their affection for each other in public. Still, some people couldn't help noticing that while Nixon often referred to his wife as a wonderful person in speeches, he often ignored her when they were together in public. In his farewell remarks upon leaving the White House on August 9, 1974, he mentioned his beloved mother but not his wife.

Not surprisingly, Nixon deeply resented speculations about his relations with his wife. In a CBS interview in 1984, when Frank Gannon asked why he had never said publicly that he loved his wife, he exclaimed impatiently: "When I hear people slobbering around . . . 'I love her' . . . that raises a question in my mind as to how much of it is real. We just don't go in for those public declarations of—of love. . . . We never held hands in public. She isn't a public kisser; I am not either.

Sometimes love . . . is much greater when you don't make a big point of showing it off."

This much seems clear: Mrs. Nixon stuck loyally by her husband through all of his crises. She thought he should have burned the incriminating tapes during the Watergate crisis and wanted to fight to the end against removal from office. "She loved him very much," declared UPI's Helen Thomas. "That was genuine."[108]

CHAPTER 36

BETTY FORD

1918–

Betty Ford was fetchingly forthright. "She's the most up-front person I ever knew," exclaimed White House press secretary Sheila Weidenfeld.[1] When *Newsweek* chose her as "Woman of the Year" in 1975, the editors singled out plain-spokenness as one of her chief traits. "I don't like to dodge a question," she admitted, "and I guess I'm not astute enough to walk around it."[2] She wasn't First Lady long before she made it clear she favored the Equal Rights Amendment, approved the Supreme Court's recent decision legalizing abortion, and, while not approving premarital sex and the use of marijuana among young people, refused to get hysterical about it. She was the first President's wife to be picketed—by antifeminists—and rather proud of it. "She's the best kind of liberated woman," said feminist Betty Friedan.[3] Even people who thought she talked too much were impressed when she went public with her mastectomy in 1974. Said one of her doctors admiringly: "She's a gutsy lady."[4]

Some of Mrs. Ford's opinions upset people and bothered the President's staff. But her habit of speaking her mind in public probably helped rather than hurt her husband during his two and a half years in the White House. In the wake of Watergate the American people yearned for honesty and decency in government, and the First Lady's outspokenness had the effect of reinforcing her husband's efforts to make his administration as open and accessible as he could. "I don't believe that being First Lady should prevent me from expressing my ideas," Mrs. Ford insisted. "Why," she asked a women's group, "should my husband's ideas, or your husbands' prevent us from being ourselves?"[5] Once, after President Ford had outlined his conservative views on abortion in a television interview, she sent him a message on a piece of paper: "Baloney! This is not going to do you a bit of good."[6]

Gerald Ford seems to have found his wife's independence refreshing. At any rate he always reacted good-humoredly whenever her public statements threatened to embarrass him politically. After a teapot tempest stirred up by some things she said in an interview, his press secretary, Ron Nessen, announced that the President had "long ceased to be perturbed by his wife's remarks."[7] Mrs. Ford appreciated his forbearance. "I've been told that I didn't play it safe enough," she once said; "but my husband has always been totally supportive."[8] In January 1975, just before putting his name to an executive order establishing a National Commission on the Observance of International Women's Year, President Ford turned to the First Lady with a smile. "Before I sign this, Betty," he said, "if you have any words of wisdom or encouragement, you are welcome to speak." "I just want to congratulate you, Mr. President," said Mrs. Ford, as the audience tittered. "I am glad to see you have come a long, long way."[9]

Some people thought Mrs. Ford had come a long way herself. Once a "silent, smiling plastic politician's wife," according to Ron Nessen, she "blossomed" in the White House into "an outgoing, witty, and warm public personality with strong and independent views."[10] Betty Ford herself denied the transmogrification. The White House hadn't really changed her, she insisted, and her views on women's rights weren't new. It was just that when she became First Lady people began listening to her for the first time. The White House gave her a forum; it was now possible for her to say something about issues that had long been important to her with some expectation of getting a respectful hearing. It "wasn't so much that the White House altered me in any essential way," she wrote in her autobiography, "as that I found the resources with which to respond to a series of challenges."[11]

Still, there was a big change in Mrs. Ford's life. As First Lady, she found that for the first time since marrying Jerry Ford she could be more than a conscientious wife and mother if she so chose. She could, she discovered, do things on her own that gave her personal satisfaction as well as contributed to her husband's work. It was even possible, she found to her delight, to influence her husband's policies, especially when it came to women's rights. In the White House, she told a reporter years later, she learned she could say: "Look, I'm important too!"[12]

Betty Ford was never a shrinking violet. A dancer and a model as a young woman, she was used to asserting herself long before becoming First Lady. "As I look back," she once mused, "for everything I did there was an audience."[13] Her mother liked to say her daughter had "popped out of a bottle of champagne," and Betty, born in April 1918, and named Elizabeth Ann, liked the idea.[14] Her parents named her Elizabeth Ann, but from the beginning always called her Betty. Growing up in Grand Rapids, Michigan, she was a "terrible tomboy," she recalled, disgruntling her two older brothers by trailing them around and horning in on their football and ice hockey games.[15] When she was eight, however, her mother enrolled her in a dancing class and she quickly discovered the love of her life. "Dancing," she decided, "was my happiness."[16] When she was in high school, she did good work in all of her classes, but reveled in the dancing sessions at the Calla Travis Dance Studio after school every afternoon. When she was only fourteen she began giving dancing lessons herself. Her great dream was to become a professional dancer.

In her autobiography Mrs. Ford insists she had "a sunny childhood" and "a wonderful girlhood." Elsewhere, however, she admits that her mother's drive for perfection was hard on her.[17] "My mother was a very strong woman," she recalled. "I wanted to be as strong as she was." Still, she worshipped her and felt "she was always there when I needed her. . . ."[18] One day the children at school ridiculed her for having a birthmark on her left arm and she went home in tears. But when her mother told her she was the only girl in the world with a birthmark like that and therefore "a very special child," she returned to school, head held high.[19] But later on when she did a sloppy job in one of the shows put on in school her mother told her: "If you don't do it well, don't do it at all."[20]

Betty's father, William Bloomer, a machinery salesman, seems to have been a minor figure in her life. In her autobiography she says little about him except that he was away from home a great deal and that he died when she was only sixteen. Years later she learned he had been an alcoholic. To help out with the family finances right after his death (her mother later remarried), Betty worked as a teen-age model in the local department store and held dancing classes as well. But when she graduated from high school, where she did good work, there was money enough to enable her to spend two summers at the Bennington School of the Dance at Bennington College in Vermont, where she became acquainted with Martha Graham and experienced "the ecstasy of being able to dance eight hours a day."[21] She was now sure she wanted to make dancing her life work. In 1939 she went to New York City to study with Martha Graham, and though she didn't make it into the main dancing group, she did well enough to get into the auxiliary group and performed in public whenever the Graham troupe played a New York concert. To help pay her way in New York she did some modeling for the John Powers Agency and appeared in fashion shows. Attractive, energetic, and vivacious, she also did a lot of socializing. "You can't carouse," Graham once lectured her, "and be a dancer too." Graham thought Betty had "a nice animal-like movement that was appealing," but made it clear she had a future in dancing only if she gave up everything else. Betty knuckled down to harder work after the reprimand, but began wondering whether she had the kind of dedication that professionalism requires.[22] When her mother suggested returning home for six months and then rejoining the Graham troupe if she still wanted to make a career out of dancing, she readily assented. "I think it's a wise thing to do," Graham told her.[23]

Betty never returned to the Graham troupe. Instead, she became the "Martha Graham of Grand Rapids." She started her own dancing troupe, introduced religious dance to the city, and taught modern dance at Calla Travis's studio. She also became fashion co-ordinator for Herpolscheimer's Department Store and in that capacity arranged fashion shows, supervised window displays and advertising, trained models, and took trips to New York to place orders for Herp's. In her spare time she held dancing classes for children and, like her mother, took an interest in handicapped youngsters and made use of dancing as therapy for them. In an era in which motherhood was considered a woman's highest goal she seemed to be an exception. "I deviated from that," she said years later, "but not to a great extent," pointing out that being a teacher and a fashion director were both acceptable "women's careers" in those days.[24] In her maturity she came to feel she hadn't lived up to "her mental capabilities" as a young woman. "I wasted a lot of time," she said ruefully.[25]

In 1942, when she was twenty-four, Betty married William C. Warren, a Grand Rapids insurance salesman, but the marriage was not a success. Shortly after the wedding, Warren gave up insurance and became a furniture dealer, but he was never able to hold a job for long. He took Betty from town to town in a quest for work and even when he settled down at a new job for a time, he spent more time with the boys at the bar than he did with Betty, and she finally decided to seek a divorce. But when he suddenly became seriously ill (he was a diabetic), she changed her mind and spent the next two years taking care of him. Only when he was fully recovered did she start divorce proceedings, on the grounds of incompatibility, and her husband did not contest the suit. In 1947, then, at twenty-nine, Betty was a grass widow, back at work at Herp's, and not eager to try marriage again.

In August 1947, shortly before the divorce decree became final, Betty had her first date with Gerald Ford, former football star and Grand Rapids lawyer who was just getting over what he called "a torrid four-year love affair" with a New York model.[26] One evening Betty's friend Peg Newman called her at Herp's to say that her friend Jerry, looking for a date, wondered whether she would join him for a drink. Betty said she was too busy, but Jerry got on the line and told her she needed a break. "But I can only be gone twenty minutes," she said, after reluctantly agreeing to meet him. In the end they spent an hour or two together; they had met casually before, but this time they found they got along famously. Jerry later said he "had no idea that someone special had just come into my life" and Betty felt the same way. But they began seeing each other from time to time—for dances, football games, bridge parties—and although they found their values and goals were "almost identical," they both agreed not to take each other seriously.[27]

The relationship soon turned serious. Right after Christmas, when Jerry went to Idaho for a ski vacation and Betty was in New York attending fashion shows for Herp's, they suddenly discovered they missed each other very much. Jerry wrote her every day; he also bought her a hand-tooled leather belt with a silver buckle. "What?" cried one of his friends when she heard about the gift. "Do you mean Jerry Ford actually gave a present to a *girl?* This must be serious."[28] In February 1948 Jerry proposed and Betty accepted. "I'd like to marry you," was his way of putting it, "but we can't get married until next fall and I can't tell you why."[29] The reason was, she soon learned, that he was planning to run for Congress and had promised to keep his candidacy a secret until the time came for him to declare.

Ford declared his candidacy in June and his love for Betty (which he neglected to mention when proposing) about the same time. Betty offered to help in the campaign in her spare time and was soon putting stamps on envelopes at his Quonset hut headquarters, recruiting friends and associates to work for him, and seeing that all the stores in town displayed Ford-for-Congress posters. To her dismay, though, her fiancé spent more time campaigning than courting. "You won't have to worry about other women," Jerry's sister-in-law told her. "Jerry's work will be the other woman." At the wedding rehearsal Ford had to slip out to give a speech.[30] And when the wedding day came—an Episcopalian service on October 15—he was almost late; "all of a sudden he came flying in," Betty recalled, and there was mud on his shoes when they walked to the altar.[31] During the honeymoon the newly-weds took time out to hear Thomas E. Dewey, Republican candidate for President, give a speech; and they went separate ways a couple of times when Jerry had to attend political meetings. A day or two after they settled down to married life in an apartment in Grand Rapids, Ford called from his office to say he would have to miss dinner. "Can you make a sandwich?" he asked. "There's a meeting tonight that I have to attend."[32]

Betty soon learned that being a politician's wife meant doing without a husband much of the time. At first it wasn't so bad. When Ford took her to Washington to live after his election, she enjoyed life in the capital for a time. She cooked, cleaned, paid the bills, took care of the children (eventually three boys and a girl), and, like other Congressional wives, helped out in her husband's office when he needed her and showed his constituents around town when they arrived for sight-seeing. She also taught Sunday School, became den mother for the Cub Scouts, and was active in the Congressional Club, a social organization made up of the wives of Congressmen and other public officials. But life changed when her husband became Minority Leader in the House of Representatives in January 1965. Betty was pleased by the honor but appalled by the enormous drain on her husband's time and energy that the position demanded. "The Congress got a new Minority Leader," she said later, "and I lost a husband. There followed a long stretch of time when Jerry was away from home 258 days a year. I had to bring four kids up by myself."[33] Jerry later admitted that his frequent absences put a strain on the marriage. Though he called Betty every night when away and tried as often as he could to join the family for Sundays, no matter where he was, it is clear that Betty felt increasingly overlooked and overworked. "I was resentful of Jerry's being gone so much," she told an interviewer years later. "I was feeling terribly neglected." One night she rolled over in bed and saw Jerry beside her. "What are *you* doing here?" she cried in mock-surprise.[34]

Illness exacerbated the emotional stresses and strains of Mrs. Ford's life. One night in August 1964, when she reached over the kitchen sink to raise a window, she strained her neck and a few hours later started experiencing excruciating pain. "I was scared," she said later, "but I didn't want to disturb my husband so I climbed out of bed and went downstairs. I tried, without any success, to sleep on the couch. Early in the morning, when Jerry found me, my left arm had gone totally numb." Jerry rushed her to the hospital, where the doctors found she had a pinched nerve and put her in traction for a couple of weeks. But she never fully recovered. The pinched nerve continued to give trouble and she also developed arthritis. Soon she was depending on pain-killing drugs to alleviate her misery. Dancing—"my purest pleasure all my life"—came abruptly to an end.[35] By 1970 the combination of physical pain and emotional distress led her to the verge of a nervous breakdown and she sought psychiatric help. "I don't believe in spilling your guts all over the place," she said, looking back on those years, "but I no longer believe in suffering in silence over something that's really bothering you."[36] She found her psychiatrist immensely helpful, and in eighteen months of therapy developed more self-confidence and a stronger sense of self-worth than she had had in some time. She also learned to reserve some "private space" of her own, in which she could find an outlet for her own feelings and interests apart from her responsibilities to her husband and children.[37] She convinced her husband, too, that it would be better for both of them if he retired from office after running once more in 1974.

The resignation of Vice-President Spiro Agnew in October 1973 changed the Fords' plans. President Nixon quickly named Ford as his

Vice President and after his confirmation by Congress in December Betty suddenly plunged into a series of new activities as Second Lady which she found she rather enjoyed. "When Jerry was selected as Vice President, it gave me a challenge I needed," she admitted. "I'm not a bridge-player or clubwoman. I can enjoy something only if I feel I'm working at it and contributing something."[38] Eight months later, the resignation of President Nixon in the midst of the Watergate scandal took the Fords by surprise, for they both believed in his innocence and felt betrayed. But they joined the Nixons for the President's resignation address on television on August 9, 1974, and afterwards accompanied them to the helicopter that was to take them away. "My heavens, they've even rolled out the red carpet for us, isn't that something?" exclaimed Mrs. Nixon. "Well, Betty," she added dolefully, "you'll see many of these red carpets, and you'll get so you hate 'em."[39] But Betty never got to hate them. She was thrilled when her husband took the oath of office as President. "I really felt like I was taking that oath too," she said. And she was deeply moved when her husband declared: "I am indebted to no man and only to one woman—my dear wife." But after the ceremony she thought: "My God, what a job I have to do!"[40]

Betty Ford did a splendid job. With the help of her husband she succeeded in developing "an open, friendly White House" which people enjoyed visiting. "This home has been a grave," she remarked shortly after taking up residence in the Executive Mansion. "I want it to sing!"[41] She made it sing, and dance, too, as well as reverberate with the sound of good talk and pleasant laughter. "Boy, what a change!" exclaimed one couple as they left the Fords' first State Dinner.[42] "I feel good vibes about this White House," said ex-Beatle George Harrison after having lunch there.[43] The Fords' natural warmth extended to the White House help too. When Mrs. Ford greeted the guards during her first days as First Lady she was puzzled when they remained silent. Later, when she learned that during the Nixon years they were instructed to refrain from talking to the President and his wife, she quickly changed the rules and made the White House atmosphere friendly and cheerful again. "It's time now that we get back to simplicity,'" she said. "Jerry and I are very ordinary people who enjoy life and aren't overly impressed with ourselves."[44]

When Betty Ford moved into the White House, the first question reporters asked her was, "What is your program going to be?" She had no program, she told them, but she did have several interests she intended to pursue. As a former dancer, she was interested in ballet, opera, music, writing, and photography, she said, and she hoped to support the creative arts while she was First Lady. But she wanted to work with handicapped children, too, she added, and she also wanted to encourage better treatment of poor and ailing elderly Americans whom she thought were being shamefully neglected.[45] She ended by doing all of these things. She included people from the arts and humanities in White House guest lists and persuaded her husband to give her old dance teacher Martha Graham a Presidential Medal of Freedom. She spent many hours in the Washington Hospital for Sick Children, too, befriending the retarded youngsters there, and encouraging others to do the same. She also supported efforts to eliminate abuses of older people and improve conditions in nursing homes for the elderly. But it was as a champion of women's rights that she achieved her greatest fame. Within a few weeks after her husband became President she had emerged as one of the most articulate women's-rights advocates ever to reside in the White House.

Mrs. Ford's feminism took several forms. First and foremost, it led her to campaign for women in high office. She encouraged her husband to appoint women to the Cabinet (he made Carla Hill his Health and Education Department Secretary) and to the foreign service (he selected Anne Armstrong as ambassador to Britain). She also wanted him to add a woman to the Supreme Court. Right after William O. Douglas stepped down from the Court she dashed into her husband's office and he cried, good-naturedly, "Now just hold your water!" "Well," she said, "if you aren't going to let me have my say, then why don't you just make a deal?" "Deal?" said Ford puzzledly. "What deal?" If he put a woman on the Court, she said, she would "put in a good word for him with my friends in New York" (where he was unpopular for opposing federal aid in the city's financial crisis). Ford assured her he had already directed his staff to make a list of qualified women for the vacancy and she left his office pleased by what he told her.[46] But in the end John Paul Stevens got the nod and though she acknowledged the appointment was a good one, she regretted not having pushed harder for a woman. She was happy, though, to see her husband become increasingly receptive to the idea of women in politics. "I've worked hard on my husband," she wrote in her memoirs, and she said she had a feeling of triumph when he remarked one day that he thought there would be a woman Vice-President in the next ten years and perhaps a woman President in the next twenty.[47]

As a feminist Mrs. Ford was also a vigorous champion of the Equal Rights Amendment (ERA). Not only did she express warm support for ERA in speeches and interviews; she also got in touch with state legislators around the country encouraging them to support the Amendment or at least allow it to come to a vote. Whether her calls to members of state legislatures—in Illinois, Missouri, Georgia, Nevada, Arizona—had an influence it is hard to say. Certainly they aroused no antagonism, for she was discreet and disarming in her approach. "I know you come from a pioneering family," she would say, "and perhaps in that spirit you can look at ERA as giving women at least a choice. . . ." Or: "I know you don't feel the way I do on this, but I admire the fairness with which you have approached it." Or: "We like to think of the Republican Party as being a leader in supporting the cause of basic human rights." Her ERA stand provoked considerable opposition. "Betty Ford is trying to press a second-rate manhood on American women!" charged her critics. But Mrs. Ford never wavered. "The Equal Rights Amendment is a must," she insisted.[48] She was elated when in February 1975 the White House began receiving more mail supporting her position than opposing it.

Though Mrs. Ford called for equal rights and opportunities for women, she denied wanting to denigrate women who devoted their lives to being good wives and mothers. The housewife deserved to be honored, she said, just as much as the woman who earned her living in the marketplace. "In fact," she once declared, "being a good housewife seems to me a much tougher job than going to the office and getting paid for it. What man could afford to pay for all the things a wife does, when she's a cook, a mistress, a chauffeur, a nurse, a baby-sitter? But *because* of this, I feel women ought to have equal rights, equal social security, equal opportunities for education, an equal chance to establish credit."[49] When she heard that anti-feminist leader Phyllis Schlafly was denouncing her in the name of motherhood, she exclaimed: "Well, I was a mother. I thought motherhood was swell. But I [am not] so sure mothers shouldn't have *rights*."[50]

Mrs. Ford's willingness to go out on a limb made headlines when she appeared on CBS-TV's "60 Minutes" in August 1975 and gave frank answers to interviewer Morley Safer's frank questions. She called the Supreme Court's ruling allowing abortion "a great, great decision," suggested that although she didn't favor premarital sex, it "might lower the divorce rate," and acknowledged that, like most young people in the 1970s, she might have tried marijuana if she had belonged to their generation. And asked what she would do if her daughter told her she was having an affair, Mrs. Ford replied: "Well, I wouldn't be surprised. I think she's a perfectly normal human being, like all girls. If she wanted to continue it, I would certainly counsel and advise her on the subject. And I'd want to know pretty much about the young man." Afterwards, when her remarks touched off a storm, her daughter Susan told reporters she had no affair to discuss, and Mrs. Ford insisted that all she had meant to say was that "while I couldn't condone an affair, I wouldn't kick my daughter out of the house for having one."[51]

Response to the First Lady's appearance on "60 Minutes" was at first hostile, and a flood of angry letters and telegrams poured into the White House. "I cannot think," cried a prominent Dallas preacher, "that the First Lady of this land would descend to such a gutter type of mentality!" The immorality of her remarks, raged a New Hampshire newspaper, is "almost exceeded by their utter stupidity. Involving any prominent individual, this would be a disgusting spectacle. Coming from the First Lady in the White House, it disgraces the nation itself." Queried about the interview, President Ford said that when he first heard what his wife said he thought he had lost ten million votes. "Then, when I read about it," he added, "I raised that to twenty million." That night he playfully threw a pillow at his wife.[52]

But there was a gradual turnabout during the next week or so. Soon there was more friendly than hostile mail coming into the White House. "AT LAST," came one telegram, "A REAL FIRST LADY." Betty Ford "should be banned from television," wrote *Washington Post* columnist Sander Vanocur playfully. "She is too honest. Mrs. Ford wears her defect like diamonds. And they dazzle." The Harris Poll, taken in the wake of the TV interview, showed her rating higher with the public than her husband's. And a poll taken by the *New York Daily News* revealed that a majority of New Yorkers approved of what she had said. "Betty Ford has now become one of the most popular wives of a President to occupy the White House," declared pollster Louis Harris. "Mrs. Ford's outspoken statements," he went on to say, "have won support from those younger and more independent elements in the electorate who are indispensable to her husband in a contest for the White House next fall. Betty Ford has a wide and deep following in the mainstream of American life, and surely must be judged a solid asset to her husband in the White House." As her name soared in public-opinion polls, Ford began joking in speeches: "If I could just get my rating up to hers!"[53]

Mrs. Ford's candor extended to the radical mastectomy it was necessary for her to have in September 1974 when her doctors discovered a small lump in her right breast was malignant. After she left the hospital and had recovered from the operation, she insisted on talking frankly about her illness before various groups in order to dramatize the importance of regular physical examinations for the early detection of breast cancer. She was also anxious to reassure women about the operation. "It isn't vanity to worry about disfigurement," she said in a speech in New York in November 1975. "It is an honest concern. I started wearing low-cut dresses as soon as the scar healed and my worries about my appearance are now just the normal ones of staying slim and keeping my hair and makeup in order. When I asked myself whether I would rather lose a right arm or a breast, I decided I would rather have lost a breast."[54] Thousands of letters commending her behavior poured into the White House, and many donations (which she turned over to the American Cancer Society) came in too. But she was especially pleased to learn that hundreds of women throughout the country lined up for breast examinations following news of her own mastectomy.

Mrs. Ford loved being First Lady. She saw more of her husband than ever before, for one thing, and, for another, she came into her own, for the first time, as a person of some importance in her own right. In the White House she gradually "began to realize that she was special," according to Sheila Weidenfeld, "not because she was Mrs. Gerald Ford, but because she was Betty Ford, a woman with some special personality traits. People liked her honesty, her sparkle, her frankness. What's more, as First Lady, she had some national influence. She could push causes and areas that she believed in. For the first time in her life she was in the spotlight because of her *own* characteristics."[55] Mrs. Ford enjoyed the White House entertaining and the trips abroad with her husband as well as the chance to publicize issues dear to her heart. She was also warmed by the friendly headlines: BETTY FORD RANKS HIGH; FIRST LADY OUTSPOKEN BUT LISTENED TO; BETTY A BIG PLUS; A SOLID ASSET. When she learned she had become one of the most popular women in the world, she was astonished but pleased. "I loved it," she admitted. "I'd be dishonest if I said it didn't please me. I hadn't expected it, but so long as it was forthcoming, I enjoyed it. Though it bothered me that while I was getting so much praise Jerry was getting criticism. He was a good sport. He was proud of me and even in cases where he didn't agree with my views, he was all for my spouting them." Her husband accepted her higher ratings in the polls with good humor. "Frankly," he said, "I think we're going to have to run Betty for President."[56]

In July 1975 Ford announced his intention of running for President the following year. Though Mrs. Ford had been looking forward to retirement, she released him from his promise to leave office in 1977 and volunteered to help out in the campaign. She knew he was anxious to win the Presidency on his own, felt he had done a good job as Nixon's successor, and thought he was by far the best man for the job. At first she wanted to confine her work to joint appearances with him on speech-making tours, but she was soon drawn into independent activities of her own as well. Republican officials were well aware of her immense drawing power and gave her "top priority" in planning their strategy. "Our people very much want her," they declared, "and consider her one of the most valuable assets for building momentum in the campaign." There were Betty as well as Jerry buttons in 1976: ELECT BETTY'S HUSBAND; KEEP BETTY IN THE WHITE HOUSE; I LOVE BETTY; BETTY'S HUSBAND FOR PRESIDENT. Betty's husband didn't seem to mind a bit.[57]

Mrs. Ford's role, as she saw it, was to "see as many people as possible and tell them about the integrity, leadership and honesty of the President."[58] To that end she gave interviews, made short speeches in scores of towns and cities around the country, was on hand for numerous parades, parties, luncheons, and dedications, got on the telephone to line up support for her husband, and even participated in a door-to-door walk in Pittsburgh to get out the vote, and talked on citizen's-band radio (her call word was "First Mama") to encourage Ford workers in Wisconsin. Despite the rigorous schedule assigned her, she retained her sparkle and good humor throughout. In Sioux Falls, South Dakota, she asked the crowd: "Do you mind if I turn around and say hello to my husband first?" Then, as people cheered, she planted a big kiss on a poster of the President on the wall behind the platform.[59] In the reviewing stand with Democratic Vice-Presidential nominee Walter Mondale for the Pulaski Day Parade in Buffalo, New York, she mischievously pinned a "KEEP BETTY IN THE WHITE HOUSE" button on Mondale's coat when he wasn't looking.[60] And just before her husband's last television debate with Democratic Presidential candidate Jimmy Carter, she sent the latter a little note: "Dear Mr. Carter—May I wish you the best tonight. I'm sure the best man will win. I happen to have a favorite candidate, my husband, President Ford. Best of luck, Betty Ford."[61]

President Ford's defeat by Carter by a close margin on election day in November was a hard blow for Mrs. Ford and she was convinced that if the campaign had lasted another week her husband might have won. But she put on a brave front. "Mom," her daughter Susan asked her the following morning, "did we win?" "No," she replied, "you kids got a father back, and I got a husband back." She also told her husband: "I think we won! There's so much to look forward to. I think it will be another new adventure."[62] But her words belied her feelings. She was crushed by her husband's rejection by the voters and felt as though all of his services to the nation "had just gone down the drain." Though she looked forward to seeing more of him now that he was out of politics, she knew she would miss being First Lady. "The truth is," she said years later, "that I was at the end of my rope."[63] In Palm Springs, California, where she and her husband took up residence in a large ranch-style house, she began losing her sense of purpose. She also became dangerously dependent on pain-killing pills for her osteoarthritis as well as tranquilizers and alcohol for peace of mind, with the combination putting her into a groggy haze much of the time. By the end of 1977 her family was becoming alarmed at her condition. "I found myself almost in a position of baby-sitting for her," Susan recalled. "She had no friends. You couldn't trust her. She wouldn't show up for appointments. I feared she'd fall and crack her head open. She was walking into a dead-end street."[64]

In the spring of 1978 Mrs. Ford's family finally decided on an "intervention." Late in March they called a meeting, confronted her with instances of her erratic behavior, and urged that she seek help. Mrs. Ford was at first bewildered and then deeply hurt by what they told her. "I was completely turned off," she recalled. "I got very mad, and was so upset that, after everyone left, I called a friend and complained about the terrible invasion of my privacy."[65] But a second "intervention" on April 1, with a doctor and a nurse present, was more successful. As she talked things over with her husband and children, Mrs. Ford came to realize that they "were there because they loved me and wanted to help," and by the end of the conference she had agreed to check into the Alcohol and Drug Rehabilitation Service of the Long Beach Naval Hospital.[66] When she first began treatment, she was willing to admit only to addiction to medication, but after about ten days she found she was able to confess to alcoholism as well. But when her doctors urged her to make a public statement about her alcoholism, she told them, "I don't want to embarrass my husband." "You're trying to hide behind your husband," they told her. "Why don't you ask him if it would embarrass him if you say you are an alcoholic?" "There will be no embarrassment to me," Ford assured her. "You go ahead and say what should be said." To help herself as well as other people with similar problems she finally decided to make a public statement. "I have learned," she announced, "that I am not only addicted to the medications I have been taking for my arthritis, but also to alcohol." She went on to urge people with problems like hers to seek professional help.[67]

By going public with her alcoholism, as with her mastectomy four years before, Mrs. Ford hoped to encourage people needing help to be frank and open about it. "I draw strength from that," she said in an interview in 1981. "I think that doing constructive things and helping people is probably the best cure in the world for your personal problems."[68] Once out of the hospital and feeling relaxed and at ease with herself for the first time in years, she was soon almost as busy as she had been in the White House: answering letters from people seeking her aid and advice, appearing at fund-raising events for arthritis research, giving speeches for cancer organizations, and doing television spots for the National Arthritis Association. She also helped found and raise money for the Betty Ford Center for Drug and Alcohol Rehabilitation which opened in 1982 in Rancho Mirage, California, and was soon attracting celebrities from the entertainment world as well as ordinary folk with drug and alcohol problems.

Early in 1987 Mrs. Ford published a book, *Betty: A Glad Awakening*, discussing her drug and alcohol dependency in some detail and presenting the case histories of some of the patients who received successful treatment at the Betty Ford Center. About this time ABC released a television movie, "The Betty Ford Story," dramatizing her problems. For years Mrs. Ford had withheld her approval of the production, and when she finally gave her consent, she insisted it be "absolutely accurate." When the film was aired in March, she found parts of it "painful for me and my family to watch, but I was convinced it would help thousands of people."[69] Afterward, a reporter asked Jerry Ford why his wife's and not his own life story had been made into a movie. Said he with a big smile: "My wife is much more interesting!"[70]

* * * *

Nobody Asked

Not long after President Nixon made Jerry Ford his Vice-President, the news came out that Mrs. Ford was a divorced woman. Bonnie Angelo, a correspondent for *People* magazine asked her why she had never told anybody about it. "Well," said Mrs. Ford, "nobody ever asked me."[71]

Objection

"Something else no one seems to expect is for a First Lady to sleep with the President," said Mrs. Ford as she was showing Sheila Weidenfeld around the White House for the first time. "This," she said, gesturing, "is *our* bedroom, and that is *our* bed. We are the first President and First Lady to share a bedroom in an awfully long time. To my great surprise, though, people have written me objecting to the idea of a President of the United States sleeping with his wife."[72]

As Often as Possible

Soon after becoming First Lady Mrs. Ford told the writer Myra MacPherson that she had been asked everything except how often she slept with her husband. "And if they'd asked me that," she added, "I would have told them." "What would you have said?" MacPherson wanted to know. "As often as possible!" exclaimed Mrs. Ford.[73]

First Lady's Flag

"I have a nice car," Mrs. Ford told Rick Sardo, the President's military aide, one day, "but no flag. If the President gets flags, why shouldn't the First Lady?" A week later Sardo came in with a flag to fit on one of the fender poles of her car. It was blue satin, trimmed with lace and braid, and decorated with red, white, and blue stars. In the middle was a pair of red and white calico bloomers, in honor of her maiden name, with the legend, "Don't tread on me," above, and the initials, ERA, below.[74]

Back in the Oval Office

In November 1974, when Ford was in Vladivostok conferring with Soviet Premier Leonid Brezhnev, his wife was going through the big warehouse the White House maintains for furniture and art objects to use in the Executive Mansion. When she ran across an elegant bust of Harry Truman (whom the Fords admired), she took it gleefully back to the White House with her. That night Ford phoned from Vladivostok and she told him to hurry home. "You'd better come quick, honey," she exclaimed, "because Harry Truman's back in your Oval Office!"[75]

Mexican Dish

At a dinner in November 1974 honoring Supreme Court Justice William O. Douglas, the Mexican-American singer Vicki Carr performed. When President Ford congratulated her afterward she offered to invite him to her Los Angeles home for dinner. "What Mexican dish do you like?" she asked. "I like *you,*" he said playfully. "That woman," cried Mrs. Ford, who overheard the exchange, "will never get into the White House again!"[76]

Gilded Grasp

When Mrs. Ford was showing actress Candice Bergen around the White House and took her into the Oval Room, she paused before a winged gold figure of a woman with a bowl resting on her head. "She used to be holding a scrowl, but it disappeared," Mrs. Ford said, "so whenever I pass her I try to give her something to hold." At this point Mrs. Ford slipped a cigarette into the gilded grasp and smiled. "Then I wait to see how long before someone notices," she laughed. "I wonder if they know it's me." The White House help always removed the cigarette.[77]

Remarkable Woman

In June 1976, Mrs. Ford went to New York to attend a dinner at which Rabbi Maurice S. Sage, director of the Jewish National Fund of America, was to present her with a silver Bible from Jerusalem. When it came time for the presentation, Dr. Sage went to the microphone holding the Bible and began to introduce the First Lady. Suddenly he stopped, clutched his chest, sat down on the steps behind the podium, and slumped over from a heart attack. "Is there a doctor?" cried one of the men on the podium. "Please come forward. We need a doctor." There were gasps from the audience and consternation reigned in the ballroom. As Secret Service men tried to revive Dr. Sage, Mrs. Ford went over to the microphone, pale and shaken, and said quietly: "Can we all bow our heads for a moment for Dr. Sage?" When the crowd was still, she asked for God's blessing on Dr. Sage and then suggested: "Let's all join together in silent prayer for Dr. Sage." By the time she finished an ambulance crew had arrived and was taking Dr. Sage to the hospital where, an hour later, he died. On the plane back to Washington Mrs. Ford's assistant, Sheila Weidenfeld, couldn't help thinking of how skillfully Mrs. Ford had kept the situation under control by her words. "It was the sound of strength, inner strength, the sound of a truly remarkable woman," she mused. "Few of us are ever put to that kind of test, but how many of us could rise to the occasion so splendidly?"[78]

Breaking the Tension

On the Fords' last day in the White House they had breakfast together and then said farewell to the household staff in the State Dining Room. After that they went to the Blue Room with members of their immediate staff to await the arrival of the Carters. As everyone waited solemnly for the new President-elect and his wife, Mrs. Ford decided to break the tension. She looked over at Sheila Weidenfeld with a mischievous gleam in her eye and cried: "I've got some unfinished business to take care of!" Then she grabbed the press secretary, put her over her knee, and started spanking her. When she finished, the two embraced, and everyone, the out-going President especially, was smiling.[79]

Nightcap

About a year after Mrs. Ford stopped drinking, her husband came home from a plane trip and she said, "A nightcap will relax you. Let me make you one." "No, thanks," he said. "Come on," said Mrs. Ford. "It will make you feel so much better, and I don't really mind. "I don't want one," he repeated. Mrs. Ford was puzzled. "We always used to have a nightcap before we went to bed," she reminded him. "Yeah," he said. "And I never enjoyed it." "Then, why did you do it?" "Because," he said, "I didn't want you to drink alone." She was moved by his thoughtfulness.[80]

Making Jerry Happy

At the Washington Press Club in 1979, Mrs. Ford handled questions with wit and charm. Her timing was also good. "Of course," she said in answer to a question about her marriage, "you've heard me say many times whatever makes Jerry happy makes me happy." She paused for a moment. "If you all *believe* that," she went on, and finished amid laughter, "you're indeed unworthy of your profession."[81]

CHAPTER 37

ROSALYNN CARTER

1927–

In the spring of 1977, a few months after he became President, Jimmy Carter asked his wife Rosalynn whether she would be willing to go on a special mission to Central and South America to promote human rights and democracy and convey his hopes for a "nuclear free zone" in the southern hemisphere. "Yes, I'll go!" she told him excitedly. Before leaving Washington on May 30, she studied up on the history and culture of the seven countries she was to visit, made a summary of her husband's foreign policy to present to the heads of state she was to meet, and took time out to brush up on her Spanish. "I was *determined,*" she said, "to be taken seriously."[1]

The new First Lady had her wish; she was taken quite seriously. Her venture into diplomacy, in fact, provoked a great deal of criticism at first in both the United States and Latin America. In "electing her husband to office," growled the editor of *America,* "we did not elect her ambassador extraordinary." Wheezed a prominent Brazilian official: "Sending her down here to talk about atomic bombs and human rights is the most ridiculous thing I ever heard of." An American reporter confronted her in Ecuador. "You have neither been elected by the American people nor confirmed by the Senate to discuss foreign policy with foreign heads of state," he told her. "Do you consider this trip an appropriate exercise of your position?" "I am the person closest to the President of the United States," replied Mrs. Carter, "and if I can explain his policies and let the people know of his great interest and friendship, I intend to do so."[2] As it turned out, her charm, tact, and businesslike efficiency won over all the foreign dignitaries she encountered on her 12,000-mile tour and she created a great deal of good-will for her country—and her husband—in lands to the south. By the time she returned to Washington on June 10, the American people had concluded that Rosalynn Carter was going to be the most activistic First Lady since Eleanor Roosevelt.

Mrs. Carter vigorously denied trying to emulate Mrs. Roosevelt. "I've read that I've tried to be like Eleanor Roosevelt but that's absolutely false!" she declared. "She was a great person, but I've never ever tried to pattern what I do after what she did. I've been doing the things I think are important, and I've not tried to copy anybody."[3] There were, however, similarities between the two. Like Mrs. Roosevelt, Rosalynn Carter talked over policies with her husband, presented him with information she gathered on her travels that he might otherwise not have received, and sometimes played the role of devil's advocate that presidential advisers were reluctant to assume. But the differences between the two First Ladies were also striking. Mrs. Roosevelt developed an identity of her own that enabled her to continue playing a prominent role in public life long after her husband's death. Mrs. Carter, for all her activism, remained what her husband called "almost a perfect extension of himself."[4] She was, reporters observed, a woman who had a clear sense of mission but who put her husband first.[5]

The Carters saw eye to eye on most matters. "You can't be around them and not know how attuned they are," observed one White House assistant. "They have antennae-like ability to know what the other is feeling."[6] Mrs. Carter once summed it all up: "We co-ordinate."[7] An interviewer who talked to them separately exclaimed afterward that he had just met "two Jimmy Carters!"[8] When the Carters came to write their memoirs after leaving the White House (Carter's appeared in 1982, his wife's in 1984), they both used the collective pronoun, we, throughout, no matter what they were discussing. "We've always worked together on everything," Mrs. Carter declared. Carter agreed. "I share almost everything with her . . .," he told a reporter in 1979. "There is very seldom a decision that I make that I don't discuss with her, tell her my opinion and seek her advice . . . On matters where her knowledge is equal to mine, she prevails most of the time."[9] While he was in the White House, he invited her to sit in on Cabinet meetings, quoted her frequently in discussions with his advisers, and bombarded her with memos on which he had scribbled, "Ros, what think?"[10] His press secretary, Jody Powell, who knew them well, saw no wifely subservience in the relationship. "Everything they've done has not been the case of Jimmy Carter doing it with a *supportive* wife," he said. "It has been Jimmy Carter and Rosalynn Carter doing it together as a team."[11] The Carters regarded themselves as "full and equal partners," but it was of course

Jimmy Carter's career which both of them were advancing; his wife was junior partner.[12]

The Carters shared the same values from the outset. They both came from Plains, a little town in southwest Georgia, centered their lives as youngsters on family, church, and school, and, through their childhood reading, grew up eager to transcend the boundaries of small-town life and willing to work hard to prove themselves. Three years younger than Jimmy, Eleanor Rosalynn Smith was born on August 18, 1927, on the family farm a few miles outside of Plains, the first of four children, and grew up, she said, "secure and isolated from the outside world."[13] Her father, whom she worshipped, drove the school bus, ran an automobile repair shop, and took care of his farm on the outskirts of town; and her mother, who had gone to college, ran the household and brought her daughter up to make beds, sweep the porch, and do the dishes. Her parents were strict but affectionate and she remembered "many warm and wonderful times," especially with her father, when growing up.[14] Quiet and shy as a child, she was so well-behaved she could wear a white dress all day without getting it dirty. Years later, when someone asked her mother whether she could recall any instances of her daughter's misbehavior, she said, "Well, I can't think of any." Then she remembered: "She used to run away. Well, I don't really mean run away, she would just go off and not tell us."[15] Her father once switched her little legs for going off without telling anybody, and her mother thought he was being unduly hard on the little girl. "I just wanted her to live the good life," she said, "and be a good person."[16]

Rosalynn was good, without being a goody-goody, for she roughhoused with her younger brothers when she was little. But she took pride in pleasing her parents and, since her father thought she was bright and could do everything just right, she was anxious never to let him down. She studied hard in school, did splendid work in all her classes, and delighted her parents when she won a five-dollar prize for the highest yearly average in school when she was only twelve. In 1939, when the war in Europe commenced, a young teacher brought a map to class, talked about the war, and encouraged the youngsters to read newspapers and listen to the radio in order to keep up with things. Rosalynn dutifully started reading the local papers, was soon fascinated by the "world of interesting persons" she discovered in them, and began dreaming of visiting faraway places when she was older.[17]

With Rosalynn, as with Jimmy, the church was a major factor in her life as a child. Reared a Methodist (she adopted Jimmy's Baptist faith when she married), she went to Sunday School, attended Sunday morning church services, had Bible classes, took in revivals, and, like Jimmy, became a "born-again Christian" (that is, renewed her faith) when she became an adult. Rosalynn's God, like Jimmy's, was a well-meaning Deity whom she was taught to love. "But we were also taught to fear God," she recalled, "and though I loved Him, I was afraid of displeasing Him all my young life. I didn't think about Him as a forgiving but as a punishing God, and I was afraid ever to have a bad thought."[18] When her father developed leukemia in 1940, while she was in high school, she was overwhelmed with guilt. "I thought he was suffering because of the mean thoughts I had had about him in the past, that somehow I was part of the cause of his illness. I felt so guilty that I tried to do everything I possibly could to erase those bad thoughts, and let him know how much I loved him. I waited on him hand and foot, brushed his hair for him, and read to him for hours. . . ."[19]

One Sunday morning Rosalynn's father called the family together in the bedroom. "I want you all to listen very carefully to what I have to say and be very brave," he said. "The time has come to tell you that I can't get well and you're going to have to look after Mother for me. You are good children and I'm depending on you to be strong." He then told them he regretted not having gone to college and wanted his children to have college educations so they could "have a better position in life."[20] He died shortly afterwards, and Rosalynn, then thirteen, felt guiltier than ever and even had a crisis of faith. "I could do anything except the most important thing: I couldn't keep my father alive," she recalled. "I had prayed and prayed for him to get better, and because of those prayers, I'd expected him to get better. But he hadn't, . . . I felt very sorry for myself and didn't understand why this had to happen to me. Had I been so bad? Didn't God love me anymore? I had doubts about God, and I was afraid because I doubted. . . ."[21] Later, she learned, with Jimmy's help, to leave something to Providence, after she had done her best, with the assurance that the Lord helps those who help themselves. But her father's influence continued to be powerful. "In a curious way," she wrote later, "my father affected me even more after he died than while he was alive. It seemed more important than ever to do what he had expected me to do. Whenever I was faced with a decision or even a temptation, I would think about whether Daddy would like it or not."[22]

Rosalynn's childhood, she later wrote, ended with her father's death. To make ends meet, her mother "took in" sewing, worked in a grocery store, and eventually got a job in the Plains post-office, while Rosalynn took over many of the household chores and, as eldest child, looked after her younger sister and two brothers. She also helped with the sewing and, to earn spending money, shampooed hair in the local beauty parlor. Despite her youth, she became her mother's chief counselor, too, trying as best she could to give the advice her mother sought on household expenses, clothes, jobs, and child-rearing. "She expected me to be responsible," Rosalynn recalled. "I felt very sorry for myself at times, always having to be so grown up."[23] But she played basketball in high school, had boy friends, and went to picnics and parties, and in retrospect felt she hadn't done enough for her mother in those days. Her mother was grateful for her help, though, and saw to it there was money enough for college when her daughter graduated from high school, as valedictorian, in 1944. That fall the 17-year-old Rosalynn enrolled in Georgia Southwestern, a junior college in nearby Americus, with the vague idea of studying interior decorating. It was while she was in college that she began dating Jimmy Carter.

Rosalynn had known Jimmy casually for years, for the Smiths and the Carters were on good terms, and Ruth Carter, Jimmy's younger sister, was her best friend. It wasn't until she was a college sophomore, however, that she became seriously interested in him. It was a picture of Jimmy (then a midshipman in the U.S. Naval Academy in Annapolis), which she saw in Ruth's bedroom, that suddenly caught her fancy. "I couldn't keep my eyes off the photograph . . . ," she later wrote. "I thought he was the most handsome young man I had ever seen. I had known him as long as I could remember, the way everyone in a small town knows everyone else, but he was three years older than I and had been away at school for four years. . . . He seemed so glamorous and out of reach."[24] She thought she would be tongue-tied if she saw him in person, but when Ruth arranged a little picnic for the three of them one afternoon in June 1945, while Jimmy was home on leave, she discovered she "could talk, actually talk to him."[25] Not only that; he took to her at once, invited her to a movie that evening, and kissed her—she had never let a boy do that before on a first date—on the way home. Later, she learned, he told his mother when he got home: "She's the girl I want to marry."[26] He had another date the following evening, however, and had to leave for Annapolis right afterward. But Rosalynn joined the Carters to see him off at the station late that night and just before he left he took her aside and said: "I'm sorry about tonight. I would much rather have been with you. Will you write me?" She readily agreed and he kissed her goodbye.[27]

The courtship proceeded apace by mail. Jimmy teased Rosalynn a great deal, told her to see other boys and then became furious when she pretended she did, wrote about a beautiful girl he was seeing (the eight-year-old daughter of the Commandant), and taught her to sign letters with the initials, I.L.Y.T.G. (I Love You the Goodest). When Christmas rolled around and he was in Plains again on vacation he proposed and she turned him down. "It was all too quick," she thought. "I wasn't ready to get married yet."[28] But they saw each other every day for movies, parties, and long drives, and on his return to Annapolis continued to exchange letters every day. And "just as I had fallen in love with his photograph, I now fell in love with his letters," Rosalynn recalled, "and my indecision slowly faded."[29] In February 1946 she joined the Carters for a weekend visit to Annapolis and when Jimmy proposed again she accepted and they decided to get married in July after he graduated from the Academy and she finished at George Southwestern. Soon after she got back home he sent her a copy of *The Navy Wife* and she plunged into the guidebook with mounting eagerness about the life ahead of her away from Plains. The wedding was in the Plains Methodist church and the honeymoon in a friend's summer place in Chimney Rock, North Carolina.

Rosalynn Carter's next seven years as a Navy wife were both frustrating and fulfilling. Navy life, she and Jimmy discovered, was one of "constant separations interspersed with ecstatic reunions."[30] The separations, especially at first, were hard on Rosalynn. She was only nineteen, had never ventured far from home before, and found herself suddenly in charge of running a household by herself. But she learned fast and was soon handling the bills, fixing up their quarters, and dealing with landlords, plumbers, and electricians, with an efficiency that brought compliments from her well-organized husband. And when the babies came (eventually three boys) she found she could be a good mother too. Her responsibilities as a young wife and mother were enormously stimulating; she took pride in being able to manage so many things on her own for the first time in her life and she was also extremely pleased by her husband's pride in her performance. Seeing the world—Norfolk, New London, Honolulu—as her husband moved from post to post, was an added bonus; and when her husband's duties permitted him to be at home, Navy life seemed almost idyllic. "I loved it . . . ," she recalled. "I really loved it. I would wake up all excited about the day ahead."[31] Whenever she and Jimmy were reunited after he had been at sea for a time, they enjoyed doing things together: cooking, studying art, reading to each other, listening to classical music, and memorizing Shakespeare. There was discord, to be sure, in the early years together. "I was very domineering and demanding," Carter once confessed to a reporter. "That means he made all the decisions," his wife explained. "I never thought of asserting myself or trying to do anything different."[32] But as she gained in self-confidence she began to insist on playing some part in family decisions and this produced arguments; she raised her voice and he retreated into silence. In the end, though, Carter came down off his high horse and began listening to her and yielding ground. Gradually, he recalled, "we each grew and we learned to understand one another and ourselves."[33] On one major point, however, he absolutely refused to back down: his decision to abandon his Navy career. When his father died of cancer in 1953, he decided to resign from the Navy and return to Plains to assume management of the family's wholesale peanut-farming business. Rosalynn was horrified at the thought of returning to small-town life again and jeopardizing the hard-won freedom she had achieved away from home. "I didn't want to go back," she recalled. "I had been so independent. I didn't want my mother and Jimmy's mother to tell me what I was doing wrong. I wanted to keep our life the way it was. We had a battle that lasted for days. I screamed and yelled and did everything to make him change his mind, but I couldn't do it."[34] In the fall of 1954, as they drove into Plains, their car loaded with possessions, Jimmy had a big smile on his face but his wife stared straight ahead in sullen silence.

For months afterward Rosalynn was miserable. She concentrated on housework, turned down invitations to socialize with neighbors, and whenever Jimmy said that some day she would be glad they had returned to Plains, she exclaimed: "I will not ever be glad! Don't say that any more."[35] But after Carter managed to put his father's business back on its feet and started expanding operations, he decided to turn to his wife for help. One afternoon in the spring of 1955 he asked her if she would be willing to handle the office calls while he was out with some customers, and she said she would and went over to his office at once with her children. The next thing she knew she was putting in one day a week there, then two days a week, and finally working full-time, answering the phone, making out bills, and doing office chores. "It was a pleasant change for me," she recalled. "I felt I was doing something more important than cooking and washing dirty blue jeans. . . ."[36] Eventually she took a course in accounting, fell in love with figures, and began keeping the books for the Carter enterprise. "I loved it," she declared, "To make all those books balance? I liked it better than anything I've ever done."[37]

Within a few years Rosalynn knew so much about the business that her husband was looking to her for advice. "Does this work?" he would ask her. "Should we continue to do this in the business?" To her delight she found herself explaining things to him. "I knew more about the books and more about the business on paper than Jimmy did," she recalled. "I knew which parts of the business were profitable, which were not, how much money we had, how much credit, and how much we owed on our debts." Being a businesswoman, she decided, was more exciting than being a Navy wife.[38] "We grew together," she said later, "as full partners."[39] For recreation the Carters continued to do things together: eat out, see movies, take in car races in Atlanta, go bowling. They also went in for self-improvement; they took ballroom dancing lessons and enrolled in courses at Georgia Southwestern centered on "great issues" and "great books." They were active in the Baptist church, too, and in civic organizations, and, despite their advanced views on race, came to be accepted in respectable upper-middle class circles in the region. "Now aren't you glad we're home," said Carter one day, "and aren't you glad I'm my own boss?" Rosalynn had to agree.[40]

In 1962 Carter ran for and was elected to the Georgia State Senate. His wife warmly supported his decision to enter politics, and in her spare time even helped out on the campaign: addressed letters, telephoned voters, kept records. She was pleased, too, that she could be in full charge of the business whenever her husband was in Atlanta for sessions of the legislature. "I liked the feeling that I was contributing to our life," she said, "and making it possible for him to pursue a political course."[41] She liked being a "political wife," moreover, and when he ran for Governor in 1966, she traveled around the state for him with their sons, talking to voters, handing out brochures, and doing interviews on radio and television.[42] She took his defeat hard, and when he ran again she redoubled her efforts on his behalf. Leaving little Amy (born in 1967) with Jimmy's mother, "Miz" Lillian, she toured the state, talked to thousands of people at shopping centers, factories, county fairs, and sporting events, and even made a few brief speeches (though she found public speaking an ordeal) for him. Carter won this time and his wife regarded his victory as a major turning point in her own life. She had a new challenge: learning to be First Lady of Georgia.

At first Mrs. Carter was "scared to death" of her new responsibilities.[43] But after resolving to do the best she could and then trusting to the Lord to take up the slack, she plunged into her duties as Governor's wife with her usual zeal and energy. With the help of Ruth Gottlieb, the German consul-general's wife, author of a book entitled *Gracious Entertaining,* she learned how to entertain Very Important People, domestic and foreign, with tables set properly, flowers arranged nicely, and seating arrangements prepared precisely according to protocol. She was soon having open house, too, for people in whom she took a special interest: senior citizens, school children, and, above all, mentally retarded youngsters. Queried years later about "the most rewarding things" she had done as Governor's wife, she said it was her work with mentally retarded children.[44] While First Lady of Georgia, she worked hard to improve the state's services for them; she visited hospitals, gathered data, and helped formulate recommendations that led to the development of new and better facilities for handicapped young people. She worked for prison reform, too, as well as for highway beautification, and at times became so overwhelmed with her varied commitments that she had to remind herself: "When problems come, when you are burdened down, when you have tried everything and nothing works, release it to God. That was the key. Release! Release!"[45]

One day Carter asked his wife to outline his plans for mental-health programs in a speech to the Georgia Association for Retarded Children. At first she refused. Though she had made brief talks during the 1970 campaign she still found public speaking intimidating and was unwilling to go out on a limb even for a cause in which she deeply believed. But when Carter offered to help her with the speech, she reluctantly agreed to go through with it. Afterward, though, she told him that giving the speech had been a horrible experience that she never intended to repeat. "That's it," she said with finality. "I just can't do it." "Why don't you do what I do?" said Carter. "Write down a few words that will remind you of the things you want to say and then just get up and talk about them." A little later, when the Atlanta Women's Chamber of Commerce asked her to give a talk about the Governor's Mansion (through which she conducted scores of tourists every week), she decided to try out her husband's method. She went to the luncheon with a few key words jotted down on a card and when she got up to speak pretended the people in the audience were simply tourists she was showing around. The next thing she knew she was explaining things with ease and even enthusiasm. "I did it!" she exulted to her husband afterward. "I did it!"[46] After that she took to speaking in public with zest and soon became better at it, in the opinion of some observers, than her husband.

With her newly acquired flair for making public appearances, Mrs. Carter was able to play a major role in her husband's quest for the Presidency in 1976. Carter was relatively unknown nationally when the campaign began, and, to get his name before the public, the Carters decided to stump the country separately most of the time. "It was like having two candidates," said their oldest son Jack, who hit the campaign trail himself. "It meant we could travel twice as far and meet twice the number of people. I think that won it for us."[47] Mrs. Carter put in eighteen-hour days ("labors of love") campaigning for her husband in more than thirty primary states, and then, after he won enough delegates to secure the Democratic nomination in July, toured the country again on his behalf during the general election.[48] Traveling in her own chartered Lear-jet, she visited close to a hundred cities, outlining her husband's views in countless speeches, seeking newspaper, radio, and television interviews, mustering support for him among local politicians, and talking to thousands of people in schools, factories, hospitals, nursing homes, senior citizen centers, courthouses, and city malls. She canvassed black neighborhoods, too, speaking with great effectiveness in black churches.

As Mrs. Carter made the rounds, reporters were struck by the way she combined soft-spoken Southern charm with shrewdness and determination. She was "like a Sherman tank in a field of clover," they said; she was "a Steel Magnolia." Their comments didn't offend her a bit. "I don't mind being called 'tough,' " she said. "I *am* strong. I do have definite ideas and opinions. In the sense that 'tough' means I can take a lot, stand up to a lot, it's a fair description."[49] She was skillful at fielding questions, especially hostile ones. "Mrs. Carter," someone would ask her, "do you like to cook?" "Yes, I like to cook," she would say, "but I'm not doing much of it this year. I'm trying to get Jimmy Carter elected President. As a housewife I know our country needs him." Then she would launch into a summary of his qualifications for the Presidency. When asked about the famous *Playboy* interview in which her husband, talking about his religion, admitted to "lusting in his heart" like any other "sinner," she said: "Jimmy talks too much, but at least people know he's honest and doesn't mind answering questions." Asked if her life was submerged in her husband's, she declared: "I have never felt submerged. Jimmy has never ever considered me submerged. . . . I've always felt that I had my own identity."[50] Once she had to introduce her husband at a rally and as she spoke movingly about their life and work together, his eyes moistened. Suddenly, just as she was finishing, he jumped up, grabbed her, kissed her warmly, and asked the crowd: "How many of you would like to have this woman as First Lady?"[51] A few weeks later, when he bested Jerry Ford at the polls, he gave his wife major credit for helping him achieve victory.

At the inauguration in January 1977, Mrs. Carter held the Bible for her husband as he took the oath of office, listened to his inaugural address (for which she had made suggestions) with pride, and walked hand in hand with him afterward from the Capitol to the White House. "I believe we're going to be happy in the White House," she murmured as they approached the Executive Mansion.[52] They were happy, for the most part, as they plunged into the new tasks facing them with their customary diligence, until double-digit inflation and the seizing of American hostages by Iranian terrorists in 1979 destroyed the Carter Presidency. But through thick and thin the Carter partnership continued undiminished in the White House. Mrs. Carter sat in on Cabinet meetings to keep abreast of things, met her husband for lunch once a week to co-ordinate their work, and talked over policies, domestic and foreign, with him whenever they were together at the end of the day. White House staffers as well as reporters soon realized that the First Lady could speak for the President with authority. "She and I continued to discuss a full range of important issues," wrote Carter in his memoirs, "and, aside from a few highly secret and sensitive security matters, she knew all that was going on. When necessary, she received detailed briefings from members of my domestic and national security staff; it was most helpful for me to be able to discuss questions of importance with her as I formed my opinions."[53] He talked over the Iranian hostage crisis with her, kept her informed of developments during the Camp David meetings in 1978 to bring about an accord between Egypt and Israel, and sent her around the country and the world from time to time as his "roving ambassador." "Her influence on her husband was considerable," noted Carter's national security adviser, Zbigniew Brzezinski, "and was exercised almost openly. Carter—at least to me—was not embarrassed to admit it."[54] Some people groused about it. "After all, who elected her," they grumbled, "and to what?" But the Harris polls revealed in the summer of 1977 that she was one of the most popular First Ladies in recent years.[55]

While helping her husband in his work, Mrs. Carter had her own agenda too. At the top of the list was mental health. Though a friendly reporter warned her that "mental health isn't sexy," she made the study of the nation's mental-health needs the major focus of her energies during her early months in the White House.[56] After persuading her husband to appoint a Commission on Mental Health, she worked closely, as honorary chairman, with the professionals on the Commission gathering data, making recommendations, and devising legislation (for which she lobbied) to improve mental-health programs at the local, state, and national levels. She took an interest in the needs of the elderly, too, visiting nursing homes and senior-citizen centers around the country, holding round-table discussions in the White House on the problems of aging, and championing legislation to reform Social Security, extend the mandatory retirement age for federal workers, and improve medical services for older Americans in rural areas. She supported women's causes, too, providing her husband with lists of women qualified for positions in the Federal Government and making dozens of calls to state legislators urging ratification of the Equal Rights Amendment. She was proud of the fact that her husband appointed more women to high office (Cabinet, Federal courts, the foreign service) than any other President, but was disappointed, as he was, when ERA failed of ratification while they were in the White House.

"Ye gawds," cried a reporter when Mrs. Carter announced her concern for conditions in American cities, "she's trying to take on all the problems we have. . . ." "And what's wrong with that?" retorted the First Lady's press secretary.[57] Mrs. Carter did not of course take on all of America's problems, but she maintained a frantic pace all the time. At one point the *Washington Star* summarized her activities for her first fourteen months in the White House: visited 18 nations and 27 American cities; held 259 private and 50 public meetings; made 15 major speeches; held 22 press conferences; gave 32 interviews; attended 83 official receptions; held 25 meetings with special groups in the White House. But while she worked hard to develop what she called "a caring society," Mrs. Carter also took seriously her duties as White House hostess.[58] She planned her entertainments carefully, laboring to tailor them to the needs and interests of the guests and ascertaining in advance precisely what topics to discuss with them and what topics to avoid. But she soon decided that "real protocol" was "warmth and putting your guests at ease, an important part of traditional southern hospitality, not just rules."[59]

By 1980 the First Lady was involved in a new activity: campaigning for her husband's re-election. Since Carter himself insisted on staying in Washington because of the Iranian hostage crisis, Mrs. Carter felt that her responsibilities were even greater than they had been in 1976 and poured all her energy into the contest. She made so many speeches that her mouth became sore. And while stumping in Illinois she developed a big welt under one of her eyes. The doctor blamed it on something she ate, but Mrs. Carter decided she had become "allergic to campaigning!"[60] Still, she had high hopes for her husband until almost the end, despite the gloomy forecasts of her husband's polltakers. About five hours before the voting booths closed, however, she finally faced the truth. "We're going to lose," she told her son Jack resignedly. "I know, Mom," he sighed. When she got back to Plains on election day and her husband arrived from Washington, she asked: "How bad is it?" "It's gone . . . ," he said, "it's all over."[61] A little later, when it was clear that Ronald Reagan had won overwhelmingly at the polls, someone exclaimed: "Mr. President, you're a great example. You don't seem bitter at all." Interposed Mrs. Carter: "I'm bitter enough for both of us."[62]

Mrs. Carter never really reconciled herself to the 1980 defeat. Looking over her speech cards after the election, she still found them so convincing that she "couldn't see how anybody could have looked at the facts and voted for Ronald Reagan."[63] She also said she would willingly take to the stump if her husband decided to run again and confessed she missed the world of politics. "Nothing is more thrilling than the urgency of a campaign," she declared, "the planning, the strategy sessions, getting out among people you'd never otherwise meet—and the tremendous energy it takes that makes a victory so sweet and a loss so devastating."[64]

For a while, after the return to Plains following Reagan's inauguration, Mrs. Carter was so depressed that her husband had to work hard to cheer her up. Then, as she became immersed in new activities—paying off campaign debts, planning the Carter Library in Atlanta, helping her husband write his memoirs and beginning work on her own autobiography—she began enjoying life again.

After disposing of the peanut warehouse, which had fallen deeply into debt during their White House years, the Carters resumed the cooperative activities they had commenced when Mrs. Carter began managing the business' finances years before. They were partners, as of old, at work and at play, in retirement. They cooked together; fished, walked, and biked (sometimes on a bicycle built for two) together; and worked together raising money for Habitat for Humanity, a non-profit Christian organization involved in building low-cost housing for the poor. They spent summers together in New York, too, helping renovate an abandoned tenement on the Lower East Side under the auspices of Habitat for Humanity. They also planned a book together, with the working title, *More Years, More Life,* on how to make the most of the second part of one's life. When the book, essentially a self-help manual, appeared in June 1987, it had been retitled, *Everything to Gain: Making the Most of the Rest of Your Life* for publication. But the collaboration had been stormy. The Carters agreed never to write a book together again.

Though the Carters kept busy in retirement, Mrs. Carter continued to look back on the White House experience with nostalgia. "I won't say it's a relief not to be First Lady," she told an interviewer in 1984, "because I enjoyed every minute of it."[65] Once a reporter asked her, "How will history books describe you?" She laughed, blushed, looked out of the window, and then said quietly: "As an ordinary woman, who never for one moment doubted she was anything but an ordinary woman, who did the very best she could, and took advantage of every opportunity that came her way."[66]

* * * *

Put in Charge

Mrs. Carter campaigned energetically for her husband when he ran for Governor of Georgia in 1970. In one town she met people as they changed shifts at a cotton mill, and one woman whom she engaged in conversation told her how hard it was trying to take care of her mentally retarded child and work to support him at the same time. Afterwards Mrs. Carter decided to go to a meeting in the same town where her husband was speaking. That night she sneaked into the back of the audience, listened to his speech, and then joined the receiving line with everybody else. Carter actually shook her hand before he realized who she was. "What are *you* doing here?" he exclaimed when he saw her. "I've come," she said, "to find out what you are going to do about mental health if you're elected Governor." "Well, Rosalynn," he replied, "we are going to have the best mental health program in the nation—and I am going to put you in charge of it!" After the election he did just that.[67]

Love Tap

One day, when Mrs. Carter was First Lady of Georgia, she gave a group of mentally retarded youngsters a tour of the Governor's Mansion. One little boy became so excited that he suddenly swatted Mrs. Carter hard with the palm of his hand. The teacher began reprimanding him, but Mrs. Carter interposed. "Oh, that's all right," she said. "He was just giving me a little love tap. I love that child!"[68]

For ERA

In January 1974, when the Equal Rights Amendment came to a vote in the Georgia House of Representatives and there was talk of inviting feminist leader Gloria Steinem to lead an ERA march in Atlanta, Mrs. Carter told her husband that such a demonstration would hurt the ERA cause. A little later Carter met an anti-ERA group in his office. "I respect your right to oppose the ERA," he told them, "but my mind is made up. I am for it—but my wife is against it!" That evening, at the Carters', one of Mrs. Carter's friends exclaimed: "Rosalynn, I'm surprised that you're against the ERA." "I'm not against it," cried Mrs. Carter shaking her head in surprise. "That's not what your husband said this morning," the friend told her. Mrs. Carter turned reproachfully to her husband and cried: "How could you?" Apparently he had misinterpreted her remark about Gloria Steinem. The next day, when she went to the capitol with him for lunch, she wore a big I'M FOR ERA button and was booed by the anti-ERA demonstrators. The amendment, she told reporters, "talks about rights, equal rights and I am for that. Jimmy was just mistaken, that's all." "I thought I knew what Rosalynn thought," murmured Carter. "But I was wrong."[69]

Locked In

When Rosalynn Carter was First Lady of Georgia she went to a high school to discuss, "Life in the Governor's Mansion." Her hosts met her at the front door and presented her with a large corsage which she pinned on her suit. Then she excused herself to go the ladies' room. When she was ready to return, the door of the bathroom stall wouldn't unlock, though she shook it repeatedly. She thought about screaming, but couldn't bring herself to do that, and finally decided to exit over the top. So she put one foot on the commode and the other on the toilet tissue rack, and hoisted herself up and over the top, corsage and all, worried all the time lest someone come in and find her suspended five feet off the floor. But no one came in, and a couple of minutes later she walked calmly out to deliver her speech as though nothing had happened.[70]

Cahtuh

When campaigning for her husband in New Hampshire during the 1976 campaign, Mrs. Carter would introduce herself: "I'm Mrs. Jimmy Carter." "Who?" "Mrs. Jimmy Carter." "Who?" "My husband's running for President." "Oh, you're Mrs. Jimmy Cahtuh!" She soon learned to overcome her southern accent and say: "I'm Mrs. Jimmy Cahtuh." Then, to her delight, people would say, "Oh, your husband's running for President."[71]

Two-fifty

In the summer of 1975 Mrs. Carter spent long hours on the telephone trying to raise money for her husband's presidential campaign. The Carters were anxious to qualify for matching government funds. According to the Federal Elections Act of 1974, if they raised $5,000 in each of twenty states in contributions of $250 or less, they would qualify for federal money. Some of her calls produced promises of contributions that never came; others ended in misunderstanding. One day she called a wealthy Texan who was originally from Georgia, explained the "two-fifty" limit to him, mentioned some mutual friends in Atlanta, and asked if he would be willing to help. To her delight he said he would. Soon after came his contribution: $2.50. (Later on he sent more.)[72]

Cheerleader

Campaigning for Jimmy Carter in 1976 produced a series of crises and frustrations for Mrs. Carter. When she made her second trip to a Tennessee town and checked the press clippings, she found she was wearing the same dress she wore the first time. When she flew to Boston to begin a twelve-day campaign tour, the seat belt in the car taking her from the airport into town jammed, and she had to borrow scissors to cut her way out. Then, after attending a reception in Boston, she found all her luggage had been stolen out of the car. She especially lamented the loss of her wig. "My wig was my security," she recalled, "against wind, against rain, and if I was just absolutely too tired at night to wash my hair, my security against worry. It was the one and only wig I ever had that was perfect. Even my mother didn't know when I had it on, and now it was gone."

In Pittsburgh the following day she made do with the same clothes. But late in the afternoon, on the way to the airport to catch a plane to Los Angeles, she spotted a shopping mall. With only a few minutes to spare, she rushed into a dress shop only to discover that it carried only junior clothes. But there was one longer skirt hanging on the rack among all the short skirts, so she grabbed it, along with a short-sleeved blue sweater, a scarf, and a half-slip. That evening she arrived in Hollywood in her new outfit, looking, she sighed, "like a cheerleader."[73]

Wouldn't Tell

After Carter's remark, in a *Playboy* interview, about "lusting in his heart," a television reporter in Shreveport asked Mrs. Carter on camera whether she ever committed adultery. "If I had," she exclaimed indignantly, "I wouldn't tell you!"[74]

Going Home

When Mrs. Carter was in Canada in August 1977, two local interviewers cornered her and asked if she saw herself as a potential woman

President. "I've never said I could do *any*thing." When the question came up a second time, she said firmly: "When the Carter administration ends, I'm going home."[75]

Too Kind

During the 1980 campaign, when Carter made his unsuccessful bid for a second term, Mrs. Carter received both brickbats and bouquets while on the campaign trail for her husband. She also received a few left-handed compliments that amused her. "You're more beautiful than in person!" one admirer exclaimed. "Me and my husband have your full support," one couple assured her. "The President has aged so much . . . ," one young man told her. "His hair has turned gray; his face looks so wrinkled; the four years as President have taken such a toll—and I'm too kind to say what has happened to you."[76]

Never Again

In June 1987 the Carters published *Everything to Gain,* a kind of personal-advice book for retired people, which presented nine ways in which "most of us can put ourselves in control of our life span." It was the first book they wrote together and, they said, the last. "We've been married 40 years, and this is the worst thing we have ever been through," said Carter frankly while on a promotion tour for the book. "It was terrible," agreed his wife. "We had different styles of writing and different work habits. Jimmy writes early and fast. I like to write into the night, and it takes me a long time to write a chapter." "To write a paragraph," interjected Carter.

The Carters used word-processors in their home offices for writing the book. And when they showed each other what they had written each day there were usually fireworks. "We are both very strong-willed," said Carter. "If Rosalynn wrote something, it was sacred. It was like she just came down off Mount Sinai with it. It was painful to her if I suggested that we change a few words." His wife, for her part, dismissed Carter's speedily written chapters as merely first drafts. The situation grew so tense at times that the two began communicating with each other through their word-processors. "That is, we weren't talking to each other about the book," said Mrs. Carter, "although we were talking to each other about everything else. We would leave nasty messages for each other on our word-processors. I thought, 'Well, maybe we can have a sensational final chapter in which we announce the end of our 40-odd year marriage.' " In the end they reached a firm agreement. "We'll never write a book together again," said Mrs. Carter. "We promised, promised," cried Carter. The two looked at each other, smiled, and nodded emphatically.[77]

CHAPTER 38

NANCY REAGAN

1921–

After Ronald Reagan's inauguration in January 1981, his wife Nancy felt suddenly overwhelmed. She had enjoyed enormously the inaugural festivities, the most lavish in American history, but she wasn't sure what came next. For reassurance she decided to call her old friend, silent film star Colleen Moore, godmother of daughter Patti, in Paso Robles, California. "Nancy *who?*" exclaimed Moore, who couldn't quite believe that the First Lady was on the line. "It's me," said Mrs. Reagan "Nancy." "Nancy!" cried Moore. "Why, it's you! Why. . . ." "Colleen, I'm so scared," sighed Mrs. Reagan, "so scared and lonely." "Oh, Nancy, you aren't a movie star now, not the biggest movie star," Moore told her. "You're the star of the whole world. The biggest star of all." "Yes, I know," said Mrs. Reagan "and it scares me to death."[1]

Mrs. Reagan soon conquered her fears. Ronnie, as she called her husband, was obviously having a good time being President; why shouldn't she enjoy herself as First Lady? "You can only be yourself," she decided, "and if you try to do anything else, it's phony."[2] She plunged into the work of redecorating and refurbishing the family quarters of the White House with funds—$700,000—provided by private donors. She acquired a new set of china costing over $200,000—also private funds—for use in the State dining room. She appeared in public—at White House teas, luncheons, dinners, and receptions, and on trips around the country and abroad—in some of the most striking creations America's leading fashion designers could come up with. She saw her wealthy old California friends as much as she could and hobnobbed with celebrities from the entertainment world like Frank Sinatra as well. And she sought the acquaintance of the BOTs ("Bright Old Things"), that is, socially established people who had lived in Washington and Georgetown for years.

In "Nancy's White House," it was fashionable to be fashionable again. White ties, good jewelry, floor-length formal wear were in; short evening gowns, pants suits, informal attire were out. "People feel they can dress up again," remarked First Hairdresser Monsieur Marc contentedly.[3] "I don't see spangled jeans anymore," said society orchestra leader Howard Devron. "Now there's flowing chiffon and silks. It's gracious and dignified."[4] Mrs. Reagan's friends praised her for bringing glamour, polish, and sophistication to the White House after the "pinchpenny style" of Jimmy and Rosalynn Carter.[5] "I believe very strongly that the White House is a special place," declared Mrs. Reagan, "and should have the best of everything. I think people want it that way."[6] Well, not all the people, it turned out. To the new First Lady's surprise and grief there seemed to be more brickbats than bouquets during her first year in the White House.

Mrs. Reagan's first meeting with newswomen soon after the "Big Bucks Inaugural," as carpers called it, was awkward and strained.[7] She didn't even have a honeymoon with the press the way her husband did. Editorial writers and columnists seemed out to get her from the start. They hated to criticize the President, it seemed, for he was so likable, so they took it out on his wife. They complained about lavish expenditures on the White House's family quarters and on fancy china at a time of recession and unemployment and just when the President was calling for reduced government spending, especially on social programs for the needy. They groused about the First Lady's expensive wardrobe, too, and about the way she accepted generous gifts from famous designers and jewelers and strove to outshine British royalty when she attended the wedding of Prince Charles and Lady Diana in July 1981. White House spokesmen finally announced that the designers' gowns were on loan, that they were being given to museums, and that Mrs. Reagan would accept no more such loans. They also insisted that much of what the First Lady wore she had acquired before coming to the White House. But Mrs. Reagan's press secretary, Sheila Tate, was continually on the defensive. "It's a seven-year-old Adolfo," she said, when reporters asked her about Mrs. Reagan's clothes. "It's a seven-year-old Galanos," she would say, or, "It's a seven-year-old Blass." "My, Sheila," one reporter exclaimed after a few months of this, "Nancy must have done an awful lot of shopping seven years ago!"[8]

In a cover story about Mrs. Reagan in December 1981, *Newsweek* reported that to many people the First Lady appeared to be an "idle rich, queen-bee figure" who was "obsessed with fashion and society." According to a *Newsweek* poll, 62 percent of the American people thought she "puts too much emphasis on style and elegance" at a time of economic recession.[9] *Time* also reported considerable criticism of the "extravagant new Tory chic" the First Lady seemed to be sponsoring.[10] "There is a little element here of Louis XIV's French court and *les precieuses*—the affected ladies," admitted one of Mrs. Reagan's former aides. "She had a liking for witty, amusing, well-dressed men who were willing to walk three paces behind and carry the purse."[11] Even friends like Bob Hope were joking about it. Instead of saying, "Coochie, coochie, coo," said the popular comedian, Nancy's nursemaid said, "Gucci, Gucci, goo." Johnny Carson announced that Mrs. Reagan's "favorite junk food" was caviar.[12] "What shall I do about these press attacks?" wailed Mrs. Reagan. "Forget them!" her husband advised her.[13] But he told reporters he thought his wife was getting "a bum rap" and that the attacks were "absolutely uncalled for."[14]

But the "Nancy problem," as the President's advisers called it, persisted. "It's been very hurtful to her," said one of her aides, "to be transformed into this all-consuming giant Adolfo mannequin."[15] Reagan thought his wife was "warm and caring" and so did many of his associates.[16] "If more people could see her on a one-to-one basis, they'd love her," insisted the wife of a Washington official who met her at a luncheon. "She's funny and she laughs so easily. I can see why the president loves to go home to her. She has a very cozy manner. We were all perfectly charmed by her."[17] Bill Libby, who saw much of Mrs. Reagan while writing up her reminiscences for the book, *Nancy* (1980), was favorably impressed and saw none of the haughtiness some critics discerned in her. "She is relaxed and laughs easily," he reported. "She is sentimental and tears come fast to her eyes. . . . She is highly intelligent and articulate and speaks softly with a warm voice, but she speaks firmly and is careful to say what she wants to say."[18] But *Nancy,* the first autobiography of a First Lady to appear in print before she got into the White House, makes it clear that Mrs. Reagan grew up in comfortable circumstances as the stepdaughter of a prominent and well-to-do Chicago surgeon, that she took for granted an elegant mode of living that most Americans considered unusual, and that she was simply doing what came naturally when she stressed style as First Lady.

The name, Nancy, was her mother's nickname for her and it stuck. She was born Anne Frances in New York City on July 6, 1921 (in Hollywood she later subtracted two years from her age), daughter of Kenneth Robbins, a New Jersey automobile dealer, and Edith Luckett, a professional actress who worked with such celebrities as Alla Nazimova, George M. Cohan, and Walter Huston. When she was only two, her parents separated, and her mother, who returned to the stage, placed her with an aunt and uncle in Bethesda, Maryland. The latter treated her as if she were their own daughter and she had "wonderful memories" of her five years with them.[19] She never saw much of her father after her parents' divorce and he came to seem like a stranger to her. But she was lonesome for her mother, reveled in her periodic visits, and saw her shows whenever they opened on Broadway. She also enjoyed dressing up in her mother's stage clothes, putting on her makeup, and pretending she was an actress herself.

In May 1929, when Nancy was seven, her mother married Dr. Loyal Davis, eminent Chicago neurosurgeon and chairman of the Department of Surgery at Northwestern University, and Nancy went to live with them in a commodious apartment on Chicago's Lake Shore Drive. She was overjoyed to be living again with her mother (who abandoned her career) and soon became deeply attached to her stepfather. "He was . . . a man of more strength and integrity than any I have known other than Ronnie," she wrote in *Nancy,* "and as I grew up I came to understand this and to love and respect him."[20] When she was fourteen, she readily agreed to adoption, journeyed to New Jersey to ask her father to sign the necessary papers, and was delighted when he consented, "although I'm sure it hurt him to do so."[21] She was now officially Nancy Davis; and by this time some observers thought she was more like the earnest, disciplined, perfectionist "Dr. Loyal" (as she called him) than like her more outgoing and vivacious mother. Davis's ultraconservative political views (he loathed Franklin Roosevelt and the New Deal) may well have shaped her thinking, too, though she was uninterested in politics as a young woman.

The Davises made Nancy the center of things; they gave her fine clothes, took her to fine restaurants, sent her to fine schools (Girls' Latin School in Chicago and Smith College in Northampton, Massachusetts), and accustomed her to fine company (Chicago socialites and luminaries from the theatrical world). "Nancy's social perfection is a constant source of amazement," announced the Girls' Latin yearbook in 1939. "She is invariably becomingly and suitably dressed. She can talk, and even better listen intelligently to anyone from her little kindergarten partner of the Halloween party, to the grandmother of one of her friends."[22] At Girls' Latin she was president of the Dramatic Club and acted in school plays, and at Smith (which she entered in 1939) she majored in drama. She made her debut at an afternoon tea dance in Chicago's Casino Club in December 1939, was noticed on the *Chicago Herald-American's* society page ("cute little Nancy Davis"), and dated boys from good families going to good colleges.[23] During World War II, Dr. Davis joined the Army Medical Corps, his wife put in time at a servicemen's canteen in Chicago, and Nancy became a nurse's aide after graduating from Smith in 1943.

Nancy wasn't a nurse's aide for long. Late in 1943 comedienne ZaSu Pitts, her mother's long-time friend, offered her a small part in a play in which she was starring and she eagerly accepted. The play, *Ramshackle*

Inn, a comedy-melodrama, in which Nancy played a girl being held captive in an upstairs room, opened on Broadway in January 1944 and ran six months. "At one point," she recalled, "I came downstairs, spoke my three lines, and was returned to my room. It wasn't much, but it was a start, and I was out on my own with the best wishes of my parents."[24] When the play closed, she stayed on in New York, seeking other parts, working as a model between engagements, and socializing with big shots in the show world (including film star Clark Gable when he was in town) whom she met through her mother. She usually played ingenue roles, and her notices, when they came, were generally friendly ("unusually attractive and talented").[25] When M.G.M. invited her to Hollywood for a screen test in the spring of 1949, her mother helped out again. She called her actor-friend Spencer Tracy and had him arrange for the distinguished director, George Cukor, to supervise the test. Nancy tested successfully, received the standard seven-year contract for M.G.M. starlets, and began making her first film, a B-picture, soon afterwards. Her early films, all forgettable, brought good reviews and even fan letters. "She was a damned good workman," said Ray Milland, who played with her in one film.[26] In the questionnaire she filled out for Metro, however, she wrote that although her childhood ambition was to be an actress, her "even greater ambition" was "to have a successful, happy marriage."[27]

In the fall of 1949 Nancy Davis met Ronald Reagan. It wasn't fortuitous. She had seen his movies, liked the cut of his jib, knew he had broken up with his wife, actress Jane Wyman, and was eager to meet him. At her prompting, producer Dore Schary and his wife Miriam arranged a little dinner party for the two of them. Soon afterward they struck up a friendship that gradually turned into a courtship.

In her autobiographical book, *Nancy,* Mrs. Reagan gives a more dramatic account of her first encounter with her husband. Her name, she says, turned up on a list of show people suspected of Communist sympathies, and, fearing it might hurt her career, she talked to director Mervyn LeRoy, then directing one of her pictures, and he promised to talk to Reagan, then president of the Screen Actors Guild, about it. That evening, she says, she stayed in her apartment awaiting a call from Reagan, but it never came. On the set the following day she asked LeRoy what had happened and he told her he had called Reagan and the latter had cleared everything up. Reagan checked the Guild files, according to LeRoy, found there were at least four other Nancy Davises connected with show business, and called back to say the Guild would defend Nancy if there was any trouble. "Fine, fine," said Nancy. But she added: "I'd feel better if the Guild president would call me and explain it all to me."[28] At this point LeRoy realized she wanted to meet Reagan, and, thinking they would make a good pair, quickly made the arrangements.

That evening, according to *Nancy,* Reagan did call. He invited her to dinner and she accepted. But they were both coy about it; they pretended they had early morning calls and couldn't stay out late. That way, Nancy upheld her pride about a last-minute invitation and Ronnie could cut out early if he found her a dud. But the night, she said later, turned out to be a long one. Reagan showed up on crutches (he had injured his leg in a charity baseball game), took her to LaRue's on the Sunset Strip for dinner, and then to Ciro's to see Sophie Tucker perform. It was the beginning of a beautiful romance.

There is fancy as well as fact in Nancy's story. She seems to have forgotten the meeting at the Scharys as well as the fact that the Nancy Davis mix-up didn't occur until a couple of years later. Still, there is no question but that she and Reagan hit it off well from the outset and were soon seeing each other regularly. Naturally garrulous, Reagan poured out a steady stream of stories, jokes, one-liners, and strong opinions that held Nancy spellbound. Jane Wyman had become bored with Reagan's endless talk ("If you ask Ronnie what time it is, he tells you how to make the watch"); Nancy was fascinated by everything he said.[29] "I loved listening to him," she wrote in *Nancy,* "and still do!"[30] And Reagan, for his part, enjoyed having a good listener with him for a change. "I don't know if it was love at first sight," Mrs. Reagan mused about their first date, "but it was something close to it. We were taken with one another and wanted to see more of each other. We had dinner the next night and the night after that and the night after that. We took in all the shows at all the clubs, and there were a lot of clubs in those days. . . . As soon as we realized that a steady diet of night life wasn't what we really wanted, we started to have quiet evenings, often at the home of Bill and Ardis Holden . . . Ronnie's good friends. . . ."[31]

The courtship moved more slowly than Nancy remembered; it took Reagan close to two and a half years to propose. He liked Nancy from the beginning, to be sure, but he was in no hurry to get married again. The breakup of his marriage to Jane Wyman (considered "ideal" when it took place in 1940) had upset him deeply and for a long time he hoped for a reconciliation. When it became clear that the break was final (Wyman filed for divorce in 1948 on grounds of "extreme mental cruelty"), he began squiring a lot of women around town and Nancy was only one of many. Gradually, though, he realized he was happiest when he was with her, and he finally invited her to visit his ranch, meet his children, Maureen (born in 1941) and Michael (adopted in 1945), and help him out with chores around his place. In February 1952 he and Nancy became engaged, but when a film came along for him to do, they postponed the wedding. Then another film interceded and then a third one. Finally, one afternoon, when Reagan was at a meeting of the Motion Picture Industry Council with William Holden, he suddenly scribbled a note for Holden on a piece of paper: "To hell with this, how would you like to be my best man when I marry Nancy?" "It's about time," cried Holden.[32] The marriage took place on March 4 at the Little Brown Church in San Fernando Valley with Holden as best man and his wife as matron of honor. Nancy was in a daze throughout. When Holden leaned over and said, "Let me be the first to kiss the bride," she cried: "You're jumping the gun." "I am not," he said; and then Nancy realized the ceremony was over.[33] In Phoenix, Arizona, where the newlyweds spent their honeymoon, Reagan met Nancy's mother and stepfather for the first time and got along famously with them.

Mrs. Reagan's life, she said many times, "really began with Ronnie." Her major aim after her marriage was "to be a wife to the man I loved and mother to our children."[34] Being a wife didn't mean cooking and cleaning the house; there was a housekeeper to do the chores. "We all have our talents but cooking isn't one of mine," she admitted in *Nancy*. Once, she revealed, when she wanted to make coffee for her father and put the pot on the stove, she couldn't understand why nothing happened, until her father told her the pilot light didn't provide enough heat.[35] But she enjoyed shopping for groceries, studying cookbooks to locate interesting recipes for the cook to use, and, above all, as a "frustrated interior decorator," fixing up the houses in which she and Ronnie resided.[36]

For a few years after her marriage Mrs. Reagan continued her film-making; in all, between 1949 and 1957, she made eleven films, including one, *Hellcats of the Navy* (1957), in which she played opposite her husband and found the love scenes a sheer delight to perform. She also did a little television and co-starred with her husband again a couple of times. But in the end she bowed out of show business completely and was happy to center her life thereafter on her husband, and, when they came, her children (Patricia Ann, later called Patti, born in October 1952, and Ronald Prescott, born in May 1958). She was an affectionate, but demanding and at times over-protective mother, for she wanted her children to be as neat and orderly as she and Dr. Loyal were, and it caused her considerable grief when, for a time, both Patti and Ron seemed to be succumbing to the blandishments of the counterculture.

Gradually the Reagans moved up in the world. Reagan's film career was waning when Nancy married him, but in 1954 he landed a generous contract with General Electric to serve as host for the *General Electric Theater,* popular Sunday night television series, and was soon doing better in the TV world than he had done in films. Before long the Reagans acquired a fine home in the Pacific Palisades (Nancy's "dream-home"), as well as a ranch within easy driving distance, and began mingling with southern Californians prominent in the social world. Mrs. Reagan joined the "Colleagues," a group of wealthy Los Angeles women who met regularly for lunch and sponsored charity fund-raising events, and was soon spending a lot of time with Betsy Bloomingdale (whose husband founded the Diners Club), Marion Jorgensen (whose husband was a steel-company executive), and Mary Jane Wick (whose husband founded the Wick Financial Corporation and Mapleton Enterprises). Reagan's work for General Electric also took him into the business world and made him, like his wife, increasingly familiar and at home with its folkways and mores. His contract with General Electric called for spending ten weeks each year visiting General Electric plants around the country, lunching with GE executives, boosting GE products, and meeting GE employees. Reagan quickly found he was good at public relations. He also found himself good at making speeches to GE audiences about the virtues of big business and the sins of big government. Eventually he became so vehement about the iniquities of government (he even denounced the Tennessee Valley Authority, a GE client with millions to spend) that GE executives had to remonstrate with him.

Reagan had once been a fervent New Deal Democrat. For years he had argued politics with his Republican brother Neil, an advertising man, and with his friend, film star Dick Powell, also a Republican, and for a long time the latter had practically written him off as a hopeless knee-jerk liberal. Reagan's shift to the right began about the time he married Nancy. He supported Eisenhower in 1952 and 1956 and Richard Nixon in 1960, and in 1962 he finally registered as a Republican. He wasn't the only liberal to turn conservative as he moved up in life and as the United States moved out of the Great Depression during World War II and entered a period of seemingly endless prosperity in the fifties. But his entry into politics in the sixties as a right-wing Republican touched off questions about his pronounced change in outlook. Some observers thought Nancy was primarily responsible, though she had never taken much interest in politics; others attributed the change to Nancy's stepfather, Dr. Davis, a militant reactionary even in the New Deal days. Brother Neil liked to take credit for the change; and so did Dick Powell. "You finally heard what I've been telling you!" exalted Powell when he learned that Reagan had changed parties. "It wasn't just you, Dick," said Reagan. "I began looking at my friends. I began counting heads and I found out that most of them were Republicans. And so I finally decided I must be a Republican, too."[37]

In 1966, when Reagan made his successful race for Governor of California, with the backing of his rich friends in the state, his wife joined the campaign trail. Timorous at first about getting involved in campaigning, she soon found it exhilarating. "Yes, it's tiring," she admitted, "but I wouldn't have traded it for anything."[38] She was a real help to her husband: gave him shrewd advice about the people with whom he was working, saw to it that he was not overscheduled, sat adoringly at his side when he gave his speeches, and though making no speeches herself, conducted question-and-answer sessions about his views that went over well with audiences.

To Mrs. Reagan's dismay, however, the campaign was still young when reporters covering the race began making fun of the worshipful way she watched her husband whenever he was making a speech. They called it, "the Look," "the Gaze," and "the Stare." A *Washington Post* correspondent described it as "a kind of transfixed adoration more appropriate to a witness of the Virgin Birth." "Her eyes sparkle as if she were in some kind of trance," according to a *Chicago Tribune* writer. "Hearing a one-liner for the one-hundreth time, she laughs on cue and then resumes the adoring gaze." Even her friends were bothered by it. "Although I truly love that Nancy, I simply can't stand watching her watch Ronnie when he's making a speech," said one woman. "Her face freezes into this adoring expression, and not an eyelash moves from the time he starts until the time he finishes. Oh, does it make me nervous!" Another friend finally confronted Mrs. Reagan. "Nancy," he cried, "people just don't *believe* it when you look at Ronnie that way—as though you are saying, 'He's my hero'." "But he *is* my hero!" insisted Mrs. Reagan.[39] And though she gradually modified "the Look" in later years, she was still defensive about it when she prepared her autobiography. "Well, first of all, I think it is only polite to look at the person who is speaking," she declared. "Of course, I'll admit I don't look at other speakers the way I look at Ronnie, but I really can't digest what someone is saying unless I look at him or her. Besides, I like to hear Ronnie speak even when I've heard the speech before. I think he's great. I've often wondered what the remarks would have been if I'd looked at my plate or counted the house while he was speaking."[40]

At Reagan's inauguration as Governor in January 1967, Mrs. Reagan's eyes moistened while she listened to him outline his plans for the state. But she was also reduced to tears when she saw the ugly old Governor's Mansion, located in a noisy part of Sacramento, in which she and her family were expected to live. They didn't spend much time there. When she discovered the place was a firetrap, she insisted on moving to a nicer part of town and renting a pleasanter place at their own expense. The press complained about the move, but she stilled the criticism by taking the wives of state legislatures on tours of the Mansion so they could see firsthand how bad it was. For her husband's birthday in February she had his office in the State Capitol redecorated; and then, pleased by all the compliments she received, she went on to renovate other rooms in the building.

As First Lady of California Mrs. Reagan was a glamorous figure. Her style and taste and clothes (Galanos) elicited a great deal of favorable comment and delighted the fashion industry. "At 44, Nancy Reagan looks a little like a Republican version of Jacqueline Kennedy," observed a writer for *Look*. "She has the same spare figure, the same air of immaculate chic. I had the impression that even in the high wind, her short, reddish-brownish-goldish hair would stay in precise order. . . . Her handsome face, her large eyes, and full mouth give away whatever she is feeling at all times."[41] Still, Joan Didion, who interviewed her for the *Saturday Evening Post* in 1968, found it hard to warm to her. "Nancy Reagan has an interested smile," she wrote, "the smile of a good wife, a good mother, a good hostess, the smile of someone who grew up in comfort and went to Smith College and has a father who is a distinguished surgeon . . . and a husband who is the definition of a Nice Guy, not to mention governor of California, the smile of a woman who seems to be playing out some middle-class American woman's daydream circa 1948. The set for the daydream is perfectly dressed, every detail correct."[42] Mrs. Reagan was shocked when she read Didion's article. "I thought we were getting along well," she said puzzledly. Then she became angry. "Do you think she might have liked it better if I had snarled?" she exclaimed.[43] But Mrs. Reagan never snarled in public. She took out her frustrations by getting into a warm bath and, while soaking, carrying on imaginary conversations with journalists and politicians who had criticized her and her husband and telling them off at length the way she wished she could do in person.

Like many other Governors' wives, Mrs. Reagan showed her social concern by making the rounds of state hospitals. She visited hospitals for the young, for the elderly, for the mentally retarded, and, of course, veterans hospitals. She spent a great deal of time with servicemen who had been wounded in Vietnam, for she was anxious to show that, like her husband, she regarded the Vietnam War as a noble endeavor. But she also became interested in the Pacific State Hospital's "Foster Grandparent Program"—a program which brought older people who wanted to be of service into touch with handicapped and mentally retarded children who needed affection and attention—and made it into her "pet project" during her eight years in Sacramento. She persuaded her husband to expand the program to other hospitals in the state and to include juvenile delinquents in it too.

Reagan wasn't Governor of California long before Republican conservatives began talking of him for the Presidency. Mrs. Reagan wasn't particularly eager for him to run; she disliked the invasion of privacy and the political sniping that went with public office. Still, she thought him superbly qualified for moral leadership of the nation, and when he tried to take the nomination away from Gerald Ford in 1976 she was in there pitching, and wept when he went down in defeat. In 1980, when he finally won his party's nomination and took on Jimmy Carter, seeking a second term, she was at his side cheering throughout the campaign. But she did more than cheer him on. She smiled pleasantly through countless receiving lines, luncheons, and dinners, talked campaign strategy over with her husband, gave him advice on the people he picked as advisers, mediated disputes among members of his staff, and sometimes gave little speeches of her own expounding his views at rallies he couldn't attend. Above all, she boosted his morale and self-confidence by her adoring presence when he made campaign speeches. Whenever she was there, according to Reagan's aides, he always spoke more effectively. "She charges his batteries," they insisted.[44] On occasion, too, she helped out at press conferences. Once, when a reporter challenged Reagan's knowledge of marijuana-smoking, she whispered, "Tell him you wouldn't know," and he exclaimed: "I wouldn't know."[45] Reagan's staff considered her a "key element" in the 1980 victory.[46]

Soon after the election, UPI correspondent Helen Thomas asked Mrs. Reagan what her project in the White House would be. "I don't have one," Mrs. Reagan told her. "My husband is most important." She went on to say that she would concentrate on her husband and her home. "You will make a big mistake if you don't pursue some great goal," Thomas warned her. "You will have the help of a big staff so don't miss the chance to do some great good that will go down in the history books."[47] When Mrs. Reagan ignored her advice and spent her initial energies on redecorating the White House's private quarters, acquiring new china for the White house, and planning elegant White House entertainments, the dean of the White House press corps was not surprised to see her running into trouble with the media.

Mrs. Reagan's "image," so crucial in the Age of Television, was, unlike the President's, extremely negative at first. She was "a frivolous clotheshorse and chum of the rich" to many Americans, "the marzipan wife," a let-'em-eat-cake Marie Antoinette, a "Fancy Nancy."[48] Not even her courageous behavior at the time of the assassination attempt on her husband (she concealed her fears for his life to bolster his spirits at the hospital) at the end of March 1981 stilled the criticism. Reagan's advisers began regarding her as a political liability, and Mrs. Reagan herself became acutely aware of the fact that "everybody was not just cuckoo about me."[49] After several emergency meetings to discuss the crisis, White House officials finally launched what has been called "Project Nancy-Has-a-Heart." Their scenario called for reducing her associations with the fashion world and fancy friends and portraying her as a serious person doing unexpected things. "Whenever people see her doing something they didn't expect," explained James Rosebush, her chief of staff, "they say, 'Hey, there's more to this woman than we thought.' "[50]

First came the unexpected humor. At the annual Alfred E. Smith Memorial Dinner held in New York late in October 1981, Mrs. Reagan, an honored guest, didn't simply express pleasure at being there when the master of ceremonies introduced her; she remained standing and made a few remarks of her own. She reminded the guests of a postcard then making the rounds, picturing her as Queen Nancy with a crown on her head, and declared: "Now that's silly, I'd never wear a crown.

It musses up your hair." Then, as the crowd roared, she went on to announce her new pet charity: "The Nancy Reagan Home for Wayward China." When the laughter subsided, she commented on criticisms she was receiving for accepting money from friends for redecorating some of the White House rooms. "I'm glad I raised as much as I did for the White House," she said. "Ronnie thinks I did such a good job, he wants me to help work on the deficit." She received a big ovation when she sat down.[51]

At the annual dinner of the Gridiron Club in Washington in March 1982, Mrs. Reagan put on another unexpected performance. One of the highlights of the evening was a satirical skit in which a singer teased the First Lady about her love for high fashion by doing a takeoff on Fanny Brice's hit song of the 1920s, "Secondhand Rose from Second Avenue," about "Secondhand Clothes." As the song ended and the crowed applauded, Mrs. Reagan, who was sitting next to her husband at the head table, jumped up and hurried out of the room. Reagan though she had gone to the ladies' room, but one of the publishers nearby murmured: "Oh, oh, Nancy Reagan is leaving. I'll bet she's really ticked off." Then, to the surprise of everyone, including the President, Mrs. Reagan suddenly appeared on the stage, dressed like a bag lady, with raggedy dress and hillbilly boots, and, as the pianist played "Secondhand Rose," again, began capering around singing some silly lines about herself as a clotheshorse. Reagan, according to one observer, "looked stunned." But he loved her act; and so did most of the people there, though a few reporters thought it unintentionally callous about the nation's poor.[52]

But the new First Lady showed that she had a sense of high purpose as well as a sense of humor. She renewed the activities she had begun as Governor's wife on behalf of the Foster Grandparents Program (FGP) and encouraged its extension to the nation at large. In the fall of 1981 she signed a contract to help prepare a book, *To Love a Child,* containing accounts of senior citizens and children who had participated in the program, persuaded her friend Frank Sinatra to record a song, "To Love a Child," to popularize the program, and arranged for proceeds from the sale of both the book and the record to go to FGP. She also took on a new project: fighting drug abuse among young people. She began traveling around the country visiting drug rehabilitation centers to give her moral support to youngsters fighting addiction, gave numerous speeches on drug abuse, attended countless anti-drug conferences, appeared on a two-hour television talk show discussing drug use, narrated an anti-drug documentary, "The Chemical People," for the Public Broadcasting Service, and even appeared in a cameo role for one episode of the sitcom, "Different Strokes," discouraging drug use among young people. "What should I say if someone offers me drugs," one child asked her, "if someone wants to give them to me?" "Just say

NO!" exclaimed Mrs. Reagan. "That's all you have to do. Just say no and walk away." To her delight children began forming "Just Say No to Drugs" clubs when they heard about her advice; and the next thing she knew she was raising funds for the Just Say No movement, attending a national conference of club members in Granville, Ohio, and, on junkets abroad with her husband, discussing the drug-abuse problems with the wives of foreign leaders.[53] In September 1986 the President himself joined the anti-drug campaign. As the TV cameras rolled, he praised his wife's anti-drug work and called on Congress to appropriate funds to launch a "National Crusade for a Drug-Free America."

Project Nancy-Has-a-Heart was a success. There was a gradual turnabout in the nation's perception of the First Lady. In January 1985 a *New York Times*-CBS poll revealed that she was even more popular than the President: 71 percent to 62 percent. "In fewer than three years," noted *New Republic* writer Fred Barnes in September 1985, "the least popular president's wife in decades" had become "the most popular First Lady of the last seven, topping even Jacqueline Kennedy." Once a political liability, Mrs. Reagan had become, in Barnes's opinion, a distinct asset to the President.[54] "I don't think I've changed in what I believe in or in my fundamental values," Mrs. Reagan told an interviewer. "I think I've grown. I hope I've grown. I don't know how you could be in this position and not grow. You're in the middle of history. How could you not grow?"[55]

But some people remained skeptical. Feminists continued to be rankled by her opposition to the Equal Rights Amendment and by what they considered her essentially old-fashioned view of women's role in society. Experts in the drug field wondered whether she had a serious grasp of the problem, and they pointed out that while she was crusading against drug abuse, her husband was calling for drastic cuts in Federal appropriations for drug prevention and education programs and that the latter were being seriously curtailed as a result. But Mrs. Reagan insisted that the solution lay in the realm of private morality. "I don't think throwing a lot of money into this problem is going to solve it," she declared. "It's going to be solved by people standing up and taking a position that it is wrong and they won't put up with it. It's morally wrong."[56] She called on film-makers and television producers to stop portraying the casual taking of drugs in their productions and to create characters who firmly rejected narcotics. She also wanted them to produce movies glorifying the American way of life, exalting the family, and presenting heroes for young people to emulate. That a pecuniary culture centered on consumer goods might produce hedonistic rather than home-spun values seems not to have occurred to her. "Goodness, God, home and country," according to Bill Libby, "are what Nancy believes in."[57] It was what her husband believed in too.

Some people thought Mrs. Reagan was a major influence on her husband. "They are joined at the hip," one of the President's friends observed.[58] But when Mrs. Reagan read that the press corps regarded her as a "controlling influence" over the President, she was indignant. "That's just all wrong," she exclaimed. "He makes the decisions. We can get into a discussion of how to approach an issue, but the final decision is his. Of course, after twenty-eight years of marriage, he influences me, and I influence him, to an extent. But that doesn't mean that you go in there and say, 'Do thus and so.' That just isn't so."[59] She did, to be sure, take some credit for his shift from a pro-ERA to an anti-ERA position; she also thought she may have strengthened his opposition to abortion (he had refrained from vetoing a bill legalizing abortion when he was Governor of California and had come to regret it). When it came to major domestic and foreign policies, however, she insisted she lacked the necessary expertise to be of any help to him. "I figure my husband is well equipped to handle all that," she declared. Asked once whether she was "the brains behind Ronald Reagan," she laughed and exclaimed: "Ronnie has enough brains of his own and he doesn't need mine." The most she would admit to was acting as a "sounding board" for him.[60] "Sure, she's influential," said longtime friend Justin Dart shortly before his death. "I don't know of a good marriage where a wife is not influential. But is she domineering? No. . . . There's no ring in [Reagan's] nose. . . . "[61] Frank Donatelli, a Reagan campaign worker, thought Mrs. Reagan's biggest contribution to her husband's political career was in encouraging him to trust his own natural instincts. "You have to listen to advice, but you must follow your instincts to make a decision," he said. "And she was very good at leading him back to his natural instincts."[62]

But Mrs. Reagan did more than simply encourage her husband to be true to himself, particularly after his run-away re-election in 1984, to which she contributed much by the way she counseled him. During Reagan's second administration she began taking the lead in White House matters in ways that offended some of her husband's associates and gave them the impression she was beginning to think of herself as a kind of Associate President. In February 1986 *Parade* magazine called her the second most powerful person in the White House, with Donald T. Regan, the President's chief of staff, coming in third.[63] *New York Times* columnist William Safire thought her activities were hurting the President; "at a time when he most needs to appear strong," he wrote, "President Reagan is being weakened and made to appear wimpish and helpless by the political interference of his wife."[64] But some people welcomed her activism. "Thank God he's got her," exclaimed one Reagan intimate. "She's the engine that makes damn sure things get done."[65]

In taking a more active role in her husband's administration after 1984, Mrs. Reagan had no particular desire to wield power for its own sake or to seek glory for herself. Her sole concern, as always, was the welfare of her husband. "Her agenda is Ronald Reagan," insisted Sheila Tate, "and that's it."[66] She regarded herself as the President's chief protector, and during his second term a series of crises seemed to her to cry out for her intervention: his colon cancer operation in July 1985; the outbreak of the Iran-contra scandal, putting him on the defensive, in November 1986; and his prostate surgery in January 1987. Her deep concern for her husband's health, honor, and happiness during these crises produced a bitter confrontation with White House chief of staff Donald Regan that kept Washington agog for months.

Mrs. Reagan hadn't liked Regan much from the outset. He didn't seem to take her seriously, for one thing, and sometimes he even refused to take her calls. He seemed to be blowing his own horn, for another, and talked to reporters too much, she thought, about how he had run the government when Reagan was in the hospital for colon surgery in July 1985. And when the President left the hospital, and she wanted to oversee his schedule to keep him from overdoing, she ran at once into a Regan roadblock. "That's my turf," he told her, and a long argument followed about how much the President was to do. "I have never seen a wife who gets into her husband's affairs so much," Regan grumbled to an acquaintance of his. "It's unfortunate that she doesn't realize how damaging that is to him."[67]

But Mrs. Reagan thought Regan was doing all the damage. She held him responsible for her husband's poor performance at a press conference after the Iran-contra affair hit the headlines in November 1986. She placed major blame on Regan, too, for her husband's involvement in the secret scheme to sell arms to Iran in exchange for hostages (violating his pledge never to deal with terrorist states) and to divert profits from the sales to U.S.-backed contra rebels in Nicaragua (thus violating Congressional legislation). And when, in the midst of Iran-contra headlines, Regan told reporters he was a "shovel brigade" cleaning up after the Reagan administration, she was outraged. "Regan could never recover with her after a remark like that," said a White House insider.[68] In January 1987, moreover, there was more tension; when the President returned to the White House after prostrate surgery, his wife wanted to emphasize rest and recovery again, while Regan insisted the President resume his normal activities as soon as possible to show people he was in good shape and in no way upset by the agitation over the Iran-contra affair. When Regan finally hung up on her a couple of times after she called him, it was apparently the last straw. "That's the kiss of death for Don Regan," said a former White house aide. "You don't do that to Nancy Reagan and get away with it."[69]

In December 1986, Mrs. Reagan began pushing vigorously for Regan's replacement. Her husband at first resisted; he hated to fire people, and, besides, he got along well with Regan. But Mrs. Reagan persisted. The *Washington Post* reported that she pushed him so hard that at one point he finally yelled, "Get off my goddamned back!" Mrs. Reagan flatly denied the story. "He never uses that word—to anybody," she declared. "They just happened to pick a word that he never uses. He'll say damn, but he'll never put the two together. He feels very strongly about that—he'll stop people; he'll say, 'Please don't say that'."[70] In the end, though, she had her way. In February 1987 Donald Regan resigned his post and former Tennessee Senator Howard Baker became the President's new chief of staff.

No one doubted that Mrs. Reagan was primarily responsible for the sacking of Regan. When Mrs. Reagan "gets her hackles up," remarked the new chief of staff Baker to reporters, "she can be a dragon." The *New Republic*'s Fred Barnes thought Regan's departure proved that Mrs. Reagan was now the White House's "chief honcho," and the *New York Times's* William Safire charged that she was making the President look weak and helpless by her efforts to be another Edith Wilson.[71] The *New York Daily News* ran a cartoon showing Mrs. Reagan sitting on her husband's shoulders as he tells the nation in a television address, "Regardless of reports to the contrary, I did NOT tell Nancy to get off my damned back!"[72] The President was understandably irked by stories of his wife's power in his administration. When a reporter asked him about Mrs. Reagan's role in running the government, he exploded. "That is fiction," he exclaimed, "and I think it is despicable fiction. And I think a lot of people ought to be ashamed of themselves." He went on to say that "the idea that she's involved in governmental decisions and so forth and all of this and being any kind of dragon lady, there is nothing to that."[73] But a Gallup Poll in the spring of 1987 reported that 62 percent of the people polled believed Mrs. Reagan had more influence on the President than any other First Lady had had, and 59 percent approved of the way she handled her role as First Lady.[74]

In a speech to the American Newspaper Publishers Association on May 4, 1987, Mrs. Reagan discussed her behavior as First lady in public. First she joked about it. She said she almost had to cancel her appearance. "You know how busy I am," she said. "I'm staffing the White House and overseeing the arms talks. I'm writing speeches." Then she turned serious. "Although I don't get involved in policy," she declared, "it's silly to suggest that my opinion should not carry some weight with a man I've been married to for 35 years. I'm a woman who loves her husband and I make no apologies for looking out for his personal and political welfare. We have a genuine, sharing marriage. I go to his aid. He comes to mine. . . . I have opinions; he has opinions. We don't always agree. But neither marriage nor politics denies a spouse the right to hold an opinion or the right to express it."[75]

Mrs. Reagan ended her speech with some advice for future First Ladies. "First, be yourself," she told them, and "do what you're interested in"; next, don't be afraid "to look after your husband" and speak out, when necessary, "to either him or his staff." Above all, she advised, be realistic about what lies ahead. "Once you're in the White House, don't think it's going to be a glamorous, fairy-tale life," she warned. "It's very hard work with high highs and low lows. Since you're under a microscope, everything is magnified, so just keep your perspective and your patience."[76]

✲ ✲ ✲ ✲

Playing a Part

When Nancy Reagan was a little girl, she saw her mother perform a death scene in a play in New York and began crying so loudly her mother had to wave from the stage in order to quiet her. "You can't take this sort of thing seriously," Mrs. Robbins told the little girl back stage after the play ended. "It's not real. It's make-believe. It's a play and I'm playing a part.[77]

On Business

One evening young Nancy called on a retired judge who lived nearby. "Judge," she said, "I've come to see you on business." "What is it, Nancy," he asked. "I'd like to know how to adopt Dr. Davis." "That's a little difficult," said the judge gently. "But I think it can be arranged." After the little girl left, the judge called Dr. Davis to tell him what Nancy had asked him. "I've always wanted that," said Davis, immensely pleased. "But I didn't know how to approach Nancy or her mother." A little later Davis legally adopted Nancy.[78]

Advice for Uncle Walter

One summer, producer Joshua Logan traveled to Arrowhead, where the Davises were visiting their friend Walter Huston to ask the latter to do a musical called *Knickerbocker Holiday*. Young Nancy sat with "Uncle Walter," as she called the actor, and Logan by the pool as Logan read the play aloud. When he finished, Huston told Logan he would think about it. After Logan left, Huston asked Nancy whether she thought he ought to do the part. "Oh, Uncle Walter," said the young woman, thinking he didn't have a good voice, "I don't think it's right for you. I think it would be a big mistake for you to do it." Huston thanked her, but signed up for the part of Peter Stuyvesant anyway, and when the play opened on Broadway it was a smash hit. Huston's singing of Kurt Weill's "September Song" was one of the highlights of the production.

Years later Huston sent Nancy a copy of the book based on the play and inscribed it, "To Nancy, who advised me to do the play!"[79]

Butter

When Nancy was a young actress living in New York, her boy friends frequently took her to dinner at the Stork Club. To economize, she got in the habit of sneaking a roll or two into her pocketbook to take home for breakfast. One evening, to her surprise and embarrassment, Sherman Billingsley, owner of the Stock Club, sent a package of butter over to her table with the note: "I thought you might enjoy some butter on my rolls."[80]

Momentous Call

Shortly after a kinescope of the small part she played in *Ramshackle Inn* was sent to Hollywood, Nancy went to Chicago to visit her family. One day the phone rang; it was Dore Schary, M.G.M. head, on the line. Nancy was filled with excitement; Schary must have thought highly of her performance to call all the way from Hollywood. But it turned out he had a bad back and wanted to talk to her stepfather about a possible operation.[81]

Something Funny

In a scene for *Night into Morning* (1951), Nancy had to walk down a long flight of stairs toward the camera with John Hodiak, playing her fiancé, at her side, and as they approached the camera, Hodiak was supposed to be laughing uproariously. But Hodiak was a serious young man and found it hard to come up with the kind of mirth the director wanted. The director insisted on take after take and still didn't get what he wanted. Finally he took Nancy aside. "You've got to think of something funny to say to him to make him laugh," he told her. As they started down the stairs again on the next take, her mind was a blank. She looked at Hodiak and realized he didn't even have a chuckle left in him. They were almost within microphone range and she was desperate. Then, at the last second, out of the corner of her mouth, the usually prim and proper young actress whispered, "Belly button." They took one more step and suddenly Hodiak began howling with laughter. "Cut and print that one," cried the director gleefully. "I basked in his gratitude for days," Nancy said afterwards.[82]

Has a Baby

Once the Reagans drove to Tijuana with Robert Taylor and his wife, left the car at the border and took a taxi into town to attend a bullfight. As the taxi careened recklessly down the street, Nancy became frightened, but couldn't bring herself to tell the driver to slow down. Then she suddenly relaxed. "Look," she said, pointing to some baby shoes hanging from the rearview mirror. "He must be all right. He must know what he's doing. He has a baby."[83]

Particular Kind of Work

When Nancy was filming *Donovan's Brain* (1954) the schedule called for her to arrive very early in the morning for one sequence and she left home at 4:30 a.m. As she drove through the empty streets of Beverly Hills before sunrise, she suddenly saw a flashing red light in her rearview mirror. She pulled over to the curb and waited as two policemen got out of the squad car and approached her, one on each side of the car. When she gave them her driver's license and asked what she had done, there was no answer. Instead, the policemen questioned her about where she had been and where she was headed and searched her car with a flashlight. Eventually they sent her on her way without explaining why they had stopped her. When she got home that night and told Reagan about it he started laughing. As she looked at him puzzledly, he pointed out that the policemen, seeing a young woman in a nice-looking convertible wheeling through town in the wee hours of the morning, probably thought she was "a girl on her way home from a particular kind of work, not on her way to work."[84]

Easiest

In her last picture, *Hellcats of the Navy* (1957), Nancy played the part of a Navy nurse and her husband played the part of a submarine commander to whom she is engaged. One scene called for her to see him just before he went off to the war zone to risk his life. "I started to say goodbye to him and I, well . . . I started to cry," Mrs. Reagan said years later, and she produced tears not called for in the script. "They had to stop the camera three times and reshoot it. I know it sounds silly." Finally she managed to say the farewell lines without going to pieces. As for other scenes in the film: " . . . I must say the love scenes in this film were the easiest I ever had to do."[85]

My Husband

Once, shortly after Reagan became Governor of California and was having a big fight with the state legislature over the state's budget, Mrs.

Reagan boarded a plane for a flight to San Diego to attend a meeting. As the plane took off, three men seated behind her got into a heated discussion about Reagan's efforts to cut the budget and began denouncing him so vehemently that she finally pushed the button to lower the back of her seat and leaned back to confront them. "That's my husband you're talking about," she cried, "and more than that, you don't know what you're talking about. He's going to be on television tonight, and if you listen you'll learn all the facts." While the Governor's aide sitting next to her sank into his seat in embarrassment, the three anti-Reaganites lapsed into silence.[86]

Nancy's Chocolates

One of Mrs. Reagan's assignments during the 1980 campaign was to cultivate the goodwill of reporters traveling with her and her husband on the campaign plane, *Leadership '80*. To break the ice, she got in the habit of rolling an orange down the aisle at takeoff, and then, during the flight, walking up and down the aisle handing out chocolates and answering questions. But when some newspapers carried a story saying that the reporters were afraid that if they didn't eat her chocolates she would keep them from getting interviews with her husband, she was deeply hurt. Thinking it over, however, she decided to laugh it off. The next time she walked down the aisle with her candy she wore a sign: "TAKE IT OR ELSE."[87]

Wrapped in Towels

Late in the afternoon on election day, November 4, 1980, the Reagans decided to get ready for the long night ahead watching the returns. Mrs. Reagan got into the tub and her husband jumped into the shower, and that's where they were when Jimmy Carter went on television to concede the election. "I had just gotten out of the tub, wearing nothing but a towel," Mrs. Reagan recalled, "when I glanced at the TV set. Then I shouted, 'Ronnie!' He dashed from the shower, also wrapped in a towel, and together we watched in astonishment as President Carter conceded the election to him. There we were, standing there, each clutching our towels, trying to take in the amazing truth that he would be President and I First Lady." As they embraced, Mrs. Reagan murmured, "I don't think this is the way it's supposed to be. . . . I somehow saw it completely different from this." Then the phone rang and it was Carter offering congratulations. "Was that really the way it was supposed to be?" cried Mrs. Reagan, as she and her husband stood there looking at each other in their towels.[88]

Go Into Politics

Mrs. Reagan referred to the attempt on her husband's life as he emerged from the Washington Hilton on March 30, 1981, as "the thing that happened to Ronnie" or as "March thirtieth." When she saw him in the hospital just before surgery to remove the bullet that had lodged close to his heart, he tried to laugh it off. "Honey," he said, "I forgot to duck." After the operation she mentioned the hardships of the Presidency to one of the doctors and he said, "It comes with the job." A little later the doctor told her, "I'm going home, Mrs. Reagan. I haven't been home since Monday and my wife's a little upset." "It comes with the job," Mrs. Reagan said softly with a little smile. By the time the President returned to the White House to recuperate from the surgery, Mrs. Reagan had lost ten pounds. Some of her friends thought she looked fine and asked how she managed to get so slim. "Just have your husband go into politics," she said grimly.[89]

4 More

After Reagan did poorly in his first debate with Walter Mondale during the 1984 campaign, Mrs. Reagan went into a huddle with his campaign crew to discuss what went wrong. They decided he had been too intense and too eager to spout all the facts he had memorized to come across in his usual free and easy way. But when his advisers started rehearsing him for the second debate, they found him apathetic and disspirited. He muffed his lines, forgot things he planned to say, and couldn't seem to concentrate. Hearing about how badly rehearsals were going, Mrs. Reagan decided to intervene. One day, in the midst of a rehearsal, she suddenly walked into the room unexpectedly, went up to her husband, stood in front of him for a moment, and then threw her coat open the way a flasher would. On her sweater underneath were the words: "4 MORE IN '84." Reagan blinked; then his wife flashed again. By this time he was doubled up in laughter. "All right, okay," he finally cried. "Let's all take it from the top." He went back to rehearsing in good humor and his wife slipped happily out of the room. He did much better in the second debate.[90]

Roommate

It was so bitterly cold on Inauguration Day, January 20, 1985, that inaugural officials canceled the big parade planned for the afternoon. This was a big disappointment to all the youngsters who had come to Washington with their school bands and floats to march proudly down Pennsylvania Avenue. But to make up for it, arrangements were made for them to meet in the Capital Centre and entertain each other there instead. There was even a promise that the First Lady might make an appearance. She did. Late in the afternoon she arrived there with the President and made a little speech. She welcomed the kids to Washington, apologized for the weather, expressed regrets that the parade had been called off, and then sat down. Suddenly she put her hands exasperatedly to her face, jumped up, and cried: "I forgot to introduce my roommate!" The crowd rocked with laughter, a reporter noted, and gave her a response "usually reserved for a rock star.[91]

Whispers

In the fall of 1984, President Reagan had a conference with Soviet Foreign Minister Andrei Gromyko in the White House and then Mrs. Reagan entertained him with some fruit juice while she sipped Perrier. After they had exchanged small talk for a minute or two, Gromyko suddenly turned serious. "Does your husband believe in peace or war?" he asked. "Peace," she said. "You're sure?" he said. "Yes," she returned. "Well, then," he said, when it came time to go to lunch, "you whisper peace in his ear every night." "I will," she replied. "I'll also whisper it in your ear."[92]

Unforgettable

One day Mrs. Reagan had a society woman who was a member of the White House Preservation Committee in for tea. When the woman was ready to leave, Mrs. Reagan stood up to shake hands and noticed the woman was staring at her feet. "I looked down," Mrs. Reagan said later, "and there was my skirt around my feet!" But without releasing the woman's hand, she reached down with the other hand and swept her skirt back up where it belonged. "Well," she smiled, "I guess this is *one* meeting you'll never forget."[93]

Livening Things Up

In October 1986, the famous eighty-two-year-old pianist Vladimir Horowitz put on a dazzling performance at the White House, and afterward President Reagan got up to utter some words of appreciation. Suddenly, as his wife listened, the leg of her chair slipped off the edge of the platform and she fell down into a row of potted yellow chrysanthemums. "I'm all right," she cried, to reassure everyone. "I just wanted to liven things up." When she had regained her seat, Horowitz went over, put his arm around her, and, as the guests applauded, gave her hand a gallant Old World kiss. "This is why I did that," she exclaimed. "Honey," interjected the President, "I told you to do it only if I didn't get any applause!"[94]

Occupied by the Poll

In June 1987 Reagan attended a seven-nation economic conference in Venice and his wife left him for a two-day visit to Sweden to campaign against drug abuse. While she was in Sweden, *Fast Lane* magazine announced that a Gallup Poll it had sponsored put Nancy Reagan in first place as the woman American men would most like to spend an evening with. "Of course I'm flattered," said Mrs. Reagan when reporters cornered her in Stockholm. "I talked to my husband last night and he said, "I'm sitting here with a poll in my hand . . . and I think you better get over here soon." She also told reporters, "This may be the last trip we take alone." Asked what the President said about the conference when he called her, she said: "He was pleased. . . . It was late at night, and he was more occupied by the poll."[95]

Married to the President

When Mrs. Reagan was visiting a school in the South, a little boy asked her how she liked being married to the President. "Fine," she smiled, "as long as the President is Ronald Reagan."[96]

NOTES

1. Martha Washington

1. Benson J. Lossing, *Mary and Martha: The Mother and Wife of George Washington* (New York, 1886), 189.
2. Mrs. John A. Logan, *Thirty Years in Washington* (Hartford, Conn., 1901), 45.
3. To Eleanor Parke Custis, Sept. 14, 1794, *Writings of George Washington* (39 vols., ed., John C. Fitzpatrick, Washington, 1931–44), XXXIII:501.
4. Lossing, *Mary and Martha,* 106.
5. Ibid., 129.
6. Ibid.; Elswyth Thane, *Washington's Lady* (New York, 1960), 80.
7. Lossing, *Mary and Martha,* 132; Thane, *Washington's Lady,* 89.
8. Nathanial Hervey, *The Memory of Washington* (Boston, 1852), 13.
9. Lossing, *Mary and Martha,* 168.
10. Marianne Means, *The Woman in the White House* (New York, 1963), 28.
11. Margaret C. Conkling, *Memoirs of the Mother and Wife of Washington* (Auburn, N.Y., 1850), 155.
12. *New Letters of Abigail Adams, 1788–1801* (ed., Stewart Mitchell, Boston, 1947), 13.
13. Ibid., 285, 15.
14. Lossing, *Mary and Martha,* 277; Rufus Wilmot Griswold, *The Republican Court, or American Society in the Days of Washington* (New York, 1854), 216.
15. Means, *Woman in the White House,* 12.
16. Hervey, *Memory of Washington,* 35.

17. Lossing, *Mary and Martha,* 277; Thane, *Washington's Lady,* 287; Meade Minnigerode, *Some American Ladies* (New York, 1926), 26.
18. Lossing, *Mary and Martha,* 279; Hervey, *Memory of Washington,* 37; Griswold, *Republican Court,* 202.
19. Lossing, *Mary and Martha,* 281.
20. Minnigerode, *American Ladies,* 44.
21. *Memoirs of Washington by His Adopted Son George Washington Parke Custis* (ed., Benson J. Lossing, Chicago, 1859), 476–77n.
22. Minnigerode, *American Ladies,* 45.
23. William Parker Cutler and Julia Perkins Cutler, *Life, Journals and Correspondence of Rev. Manasseh Cutler* (2 vols., Cincinnati, 1888), II: 56–58.
24. *Memoirs of Washington,* 514n.
25. Means, *Woman in the White House,* 19.
26. Lossing, *Mary and Martha,* 97–99; Minnigerode, *American Ladies,* 8; *Memoirs of Washington,* 499–501.
27. Means, *Women in the White House,* 34.
28. Hervey, *Memory of Washington,* 85–86; Lossing, *Mary and Martha,* 179–80
29. Conkling, *Mother and Wife of Washington,* 107–8.
30. "An Account of a Visit Made to Washington at Mount Vernon, by an English Gentleman, in 1785,," *Pennsylvania Magazine of History and Biography,* XVII (1893): 81.
31. Anne Hollingsworth Wharton, *Martha Washington* (New York, 1897), 256–57.
32. Ibid., 164.
33. Hervey, *Memory of Washington,* 39; Conkling, *Mother and Wife of Washington,* 158.
34. Girswold, *Republican Court,* 314; *Memoirs of Washington,* 408n.

2. Abigail Adams

1. Dec. 23, 1782, *The Book of Abigail and John: Selected Letters of the Adams Family, 1762–1784* (ed., L.H. Butterfield, Cambridge, Mass., 1975), 333.
2. Ibid., 41.
3. Nov. 2 and 10, 1818, quoted in Lynne Withey, *Dearest Friend: A Life of Abigail Adams* (New York, 1981), 315.
4. Oct. 25, 1782, Withey, *Dearest Friend,* 137–38.
5. March 2, 1780, quoted in Janet Whitney, *Abigail Adams* (Boston, 1947), 148.
6. April 10, 1982, *Book of Abigail and John,* 313.
7. Sept. 14, 1774, *Familiar Letters of John Adams and His Wife Abigail During the Revolution* (ed., Charles Francis Adams, Boston, 1875), 33.
8. Dec. 9, 1781, *Book of Abigail and John,* 302–3.
9. Feb. 1779, quoted in Meade Minnigerode, *Some American Ladies* (New York, 1926), 66.
10. Dec. 2, 1781, *Book of Abigail and John,* 300.
11. Oct. 25, 1778, ibid., 226.
12. Nov. 12–23, 1778, ibid., 229.
13. Ibid.
14. Dec. 2, 1778, *Book of Abigail and John,* 230.
15. Nov. 28, 1800, quoted in Withey, *Dearest Friend,* 276–77.

16. Charles W. Akers, *Abigail Adams: An American Woman* (Boston, 1980), 37.
17. March 31, 1776, *Book of Abigail and John,* 121.
18. April 14, 1776, ibid., 123.
19. May 7, 1776, ibid., 127.
20. Aug. 14, 1776, ibid., 153.
21. Akers, *Abigail Adams,* 100; Withey, *Dearest Friend,* 196.
22. Akers, *Abigail Adams,* 143.
23. Laura E. Richards, *Abigail Adams and Her Times* (New York, 1928), 157.
24. Akers, *Abigail Adams,* 37; Richards, *Abigail Adams,* 268.
25. Akers, *Abigail Adams,* 138.
26. April 26, 1798, *New Letters of Abigail Adams, 1788–1801* (ed., Stewart Mitchell, Boston, 1947), 167.
27. *The Works of John Adams* (10 vols., Boston, 1856), I: 547.
28. Akers, *Abigail Adams,* 161.
29. Whitney, *Abigail Adams,* 305; Akers, *Abigail Adams,* 177; *Letters of Mrs. Adams, the Wife of John Adams* (ed., Charles Francis Adams, Boston, 1848), 386.
30. Akers, *Abigail Adams,* 177.
31. Withey, *Abigail Adams,* 310.
32. *Memoirs of John Quincy Adams* (12 vols., ed., Charles Francis Adams, Philadelphia, 1874–77), IV: 157.
33. *Works of John Adams,* IV:155.
34. *Familiar Letters, Book of Abigail and John,* 42–43.
35. Richards, *Abigail Adams,* 99; Whitney, *Abigail Adams,* 105.
36. Sept. 9, 1776, *Familiar Letters,* 225–26.
37. To Mrs. Cranch, on board ship *Active,* July 6, 1784, *Letters of Mrs. Adams,* 156–73.
38. To Mrs. Cranch, July 24, 1784, *Book of Abigail and John,* 397.
39. To Elizabeth Shaw, Auteuil, near Paris, Dec. 14, 1784, *Letters of Mrs. Adams,* 223.
40. To Lucy Cranch, Auteuil, Sept. 5, 1784, ibid., 199–200.
41. To Mrs. Cranch, Auteuil, Feb. 20, 1785, ibid., 234.
42. To Mrs. Shaw, London, March 4, 1786, ibid., 276; Akers, *Abigail Adams,* 140–42; Withey, *Dearest Friend,* 181–82.
43. Withey, *Dearest Friend,* 194.
44. To Mrs. Cranch, London, June 24, Aug. 14, 1985, *Letters of Mrs. Adams,* 255–60; 270.
45. To Mrs. Cranch, Grosvenor Square, Sept. 15, 1787, ibid., 330–31.
46. Marianne Means, *The Woman in the White House* (New York, 1963), 42.
47. Withey, *Abigail Adams,* 237.
48. To Mary Cranch, May 16, 1797, *New Letters,* 90; March 18, 1800, ibid., 241–42.
49. Nov. 21, 1800, ibid., 256–60.
50. Quincy, May 20, 1804, *Letters of Mrs. Adams,* 389–90; Akers, *Abigail Adams,* 90.
51. Withey, *Dearest Friend,* 300–301.

3. Martha Jefferson

1. Gordon Langley Hall, *Mr. Jefferson's Ladies* (Boston, 1966), xiv.
2. Ibid., 38, 50; Page Smith, *Jefferson: A Revealing Biography* (New York 1976),

121, 140; Sarah N. Randolph, *The Domestic Life of Thomas Jefferson* (New York, 1958), 59.

3. *Autobiography,* in *The Life and Selected Writings of Thomas Jefferson* (ed., Adrienne Koch and Willian Peden, New York, 1944), 53.
4. Dumas Malone, *Jefferson the Virginian* (Boston, 1948), 157.
5. Henry S. Randall, *The Life of Thomas Jefferson* (3 vols., New York, 1858), I: 64; Laura Carter Holloway, *The Ladies of the White House* (New York, 1870), 117; Randolph, *Domestic Life of Jefferson,* 44.
6. To Robert Skipworth, *The Papers of Thomas Jefferson* (ed., Julian Boyd, Princeton, 1950–), I:78.
7. Randall, *Life of Jefferson,* I: 64–65.
8. Virginius Dabney, *The Jefferson Scandals: A Rebuttal* (New York, 1982).
9. Mary Ormsby Whitton, *First First Ladies, 1789–1865* (New York, 1948), 46.
10. To Martha Jefferson, Annapolis, Dec. 22, 1783, Edwin Morris Betts and James Adam Bear, Jr., *The Family Letters of Thomas Jefferson* (Columbia, Mo., 1966), 22; to Martha Jefferson, Aix en Provence, March 28, 1787, ibid., 35; to Nathaniel Burwell, March 14, 1818, *The Writings of Thomas Jefferson* (20 vols., Washington, 1904–1905), XV:165.
11. Whitton, *First First Ladies,* 43; Hall, *Jefferson's Ladies,* 59.
12. *Papers of Jefferson,* VI: 199–200n.
13. Ibid., 196–97.
14. Ibid., 199n.
15. Ibid.

4. Dolley Madison

1. Noel B. Gerson, *The Velvet Glove: A Life of Dolly Madison* (Nashville, Tenn., 1975), 12.
2. Ibid., 195; Katharine Anthony, *Dolly Madison: Her Life and Times* (Garden City, N.Y., 1949), 118.
3. Meade Minnigerode, *Some American Ladies* (New York, 1926), 108; Marianne Means, *The Woman in the White House* (New York, 1963), 65–66; Lucy C. Lillie, "The Mistress of the White House," *Lippincott's Monthly Magazine,* 40 (July 1887): 85.
4. Gerson, *Velvet Glove,* 20.
5. *Memoirs and Letters of Dolly Madison* (ed., L. B. Cutts, Boston, 1886), 14.
6. Ibid., 15.
7. Gerson, *Velvet Glove,* 78.
8. Ibid.
9. Ibid., 77.
10. *Memoirs,* 60.
11. Harriet Martineau, *Retrospect of Western Travel* (2 vols., New York, 1838), I: 193.
12. Gordon Langley Hall, *Mr. Jefferson's Ladies* (Boston, 1966), 167; Means, *Woman in the White House,* 67–68.
13. Minnigerode, *American Ladies,* 108.
14. Gerson, *Velvet Glove,* 80.
15. Ibid., 135; Allen C. Clark, *Life and Letters of Dolly Madison* (Washington, 1914), 60–63.

16. Gerson, *Velvet Glove,* 136.
17. Ibid., 137.
18. *National Intelligencer,* Friday, Feb. 3, 1804, 3.
19. Gerson, *Velvet Glove,* 138.
20. Ibid.
21. Ibid.
22. Laura Carter Holloway, *The Ladies of the White House* (New York, 1870), 188.
23. Ibid.
24. Means, *Woman in the White House,* 66.
25. Gerson, *Velvet Glove,* 203.
26. Ibid., 223.
27. Paul Jennings, *A Colored Man's Reminiscences of James Madison* (Brooklyn, N.Y., 1865), 15.
28. Means, *Woman in the White House,* 72.
29. Gerson, *Velvet Glove,* 234.
30. Ibid., 240.
31. Ibid., 241.
32. Mary Ormsbee Whitton, *First First Ladies, 1789–1865* (New York, 1948), 69.
33. Anthony, *Dolly Madison,* 79–80.
34. *Memoirs,* 15–16.
35. Clark, *Life and Letters,* 475.
36. Gerson, *Velvet Glove,* 146–47.
37. *Memoirs,* 60; Clark, *Life and Letters,* 147–48; Gerson, *Velvet Glove,* 10, 120, 121.
38. Josephine Seaton, *William Winston Seaton of the "National Intelligencer"* (Boston, 1871), 89–91.
39. *Memoirs,* 141–42.
40. Clark, *Life and Letters,* 168–69.
41. *Memoirs,* 140–41.
42. Margaret Bayard Smith, *The First Forty Years of Washington Society* (New York, 1906), 237.
43. Anthony, *Dolly Madison,* 388–89
44. Clark, *Life and Letters,* 411.
45. *Memoirs,* 208–9.
46. Ibid., 209–10.

5. Elizabeth Monroe

1. Laura Carter Holloway, *The Ladies of the White House* (New York, 1870), 237.
2. Margaret Bayard Smith, *The First Forty Years of Washington Society* (New York, 1906), 134, 141.
3. W. P. Cresson, *James Monroe* (Chapel Hill, 1946), 92; Harriet Taylor Upton, *Our Early Presidents: Their Wives and Children* (Boston, 1890), 243.
4. *Memoirs of John Quincy Adams* (12 vols., ed., Charles Francis Adams, Philadelphia, 1874–77), IV: 45–46.

5. Dec. 1819, Josephine Seaton, *William Winston Seaton of the "National Intelligencer"* (Boston, 1871), 144.
6. Anne Hollingsworth Wharton, *Social Life in the Early Republic* (Philadelphia, 1902), 189.
7. Upton, *Our Early Presidents,* 272; Holloway, *Ladies of the White House,* 237.
8. Marian West, "Our Four Year Queens," *Munsey's Magazine,* 25 (Sept. 1901): 891.
9. Quoted, Mary Ormsbee Whitton, *First First Ladies, 1789–1865* (New York, 1948), 89–90.
10. *The Autobiography of James Monroe* (ed., Stuart Gerry Brown, Syracuse, N.Y., 1959), 70–71.

6. Louisa Catherine Adams

1. Jack Shepherd, *Cannibals of the Heart: A Personal Biography of Louisa Catherine and John Quincy Adams* (New York, 1980), 259.
2. Paul C. Nagel, *The Adams Women: Abigail and Louisa Adams, Their Sisters and Daughters* (New York, 1987), 215.
3. Shepherd, *Cannibals of the Heart,* 25.
4. Ibid., 63.
5. "Record of a Life or My Story," begun, July 23, 1824, The Adams Papers, Part III, Reel 265, Louisa Catherine Adams, Miscellany (microfilm, Massachusetts Historical Society, Boston, 1956).
6. To C. F. Adams, Sept. 4, 1850, Adams Papers, Reel 540.
7. July 26, 1811, *Memoirs of John Quincy Adams* (12 vols., ed., Charles Francis Adams, 1874–1877), II:282–83.
8. To C. F. Adams, Sept. 4, 1850, Adams Papers, Rel 540.

Women, 169–70.

9. Shepherd, *Cannibals of the Heart,* 104.
10. Ibid., 116.
11. Ibid., 164.
12. George Dangerfield, *The Era of Good Feelings* (New York, 1952), 7.
13. Anne Royall, *Sketches of History, Life and Manners in the United States* (New Haven, Conn., 1826), 169n.
14. "Mrs. John Quincy Adams's Ball," *Harper's Bazar,* 4 (March 18, 1871): 167.
15. Gerda Lerner, *The Grimké Sisters from South Carolina: Rebels against Slavery* (Boston, 1967), 167.
16. To C. F. Adams, Feb. 4–14, 1838, Adams Papers, Reel 508.
17. Dorothie Bobbé, *Mr. and Mrs. John Quincy Adams* (New York, 1930), 301.
18. *The Education of Henry Adams: An Autobiography* (Boston and New York, 1918), 16–17.
19. Shepherd, *Cannibals of the Heart,* 118–19.
20. Ibid., 227.
21. "Mrs. John Quincy Adams's Narrative of a Journey from St. Petersburg to Paris in February 1815," *Scribner's Magazine,* 34 (July-Dec. 1903): 448–63.
22. Shepherd, *Cannibals of the Heart,* 408–9; to C. F. Adams, March 14, 1831, Adams Papers, Reel 495.

7. Rachel Jackson

1. James Parton, *The Life of Andrew Jackson* (3 vols., New York, 1860), I:153.
2. Marquis James, *Andrew Jackson: Portrait of a President* (New York, 1940), 155.
3. Robert V. Remini, *Andrew Jackson and the Course of American Freedom, 1822–1832* (New York, 1981), 119.
4. James, *Portrait of a President,* 156.
5. *Truth's Advocate and Monthly Anti-Jackson Expositor* (Cincinnati), Jan. 1828, p. 6.
6. James, *Portrait of a President,* 157–58; July 8, 1827, *Correspondence of Andrew Jackson* (7 vols., ed., John S. Bassett, Washington, D.C., 1925–35), III:372.
7. Parton, *Jackson,* I:168.
8. Robert V. Remini, *Andrew Jackson and the Course of American Empire, 1767–1821* (New York, 1977), 43.
9. Meade Minnigerode, *Some American Ladies* (New York, 1926), 202.
10. Parton, *Jackson,* I:146–47.
11. Remini, *Jackson and the Course of American Empire,* 43–67.
12. Ibid., 66–67.
13. Marquis James, *Andrew Jackson: The Border Captain* (New York, 1933), 156; Jan. 18, 1813, *Correspondence,* I:271–73.
14. April 18, 1833, *Correspondence,* V:60.
15. Henry A. Wise, *Seven Decades of the Union* (Philadelphia, 1872), 101.
16. Parton, *Jackson,* I: 163–64.
17. Remini, *Jackson and the Course of American Freedom,* 149.
18. Ibid., 154.
19. Wise, *Seven Decades,* 102–3.
20. Laura Carter Holloway, *The Ladies of the White House* (New York, 1870), 301–3; Parton, *Jackson,* II:595–96, 604.
21. Elizabeth Ellet, *The Court Circles of the Republic* (Hartford, Conn., 1869), 87.

8. Hannah Van Buren

1. *Albany Argus,* Feb. 8, 1819; Laura Carter Holloway, *The Ladies of the White House* (New York, 1870), 346.
2. Ibid., 345–46.
3. Mary Ormsbee Whitton, *First First Ladies, 1789–1865* (New York, 1948), 145.
4. Denis Tilden Lynch, *An Epoch and a Man: Martin Van Buren and His Times* (New York, 1929), 170.

9. Anna Harrison

1. *Cincinnati Daily Gazette,* April 13, 1841, quoted in Freeman Cleaves, *Old Tippecanoe: William Henry Harrison and His Times* (New York, 1939), 328.
2. Ibid., 23.
3. Catherina Y.R. Bonney, *A Legacy of Historical Gleanings* (2 vols., Albany, N.Y., 1875), II:141.

4. Laura Carter Holloway, *The Ladies of the White House* (New York, 1870), 357.
5. Charles S. Todd and Benjamin Drake, *William Henry Harrison* (Cincinnati, 1840), 18.
6. Holloway, *Ladies of White House,* 374.
7. Mary Ormsbee Whitton, *First First Ladies, 1789–1865* (New York, 1948), 170.

10. The Tyler Wives: Letitia Christian Tyler and Julia Gardiner Tyler

1. Laura Carter Holloway, *Ladies of the White House* (New York, 1870), 384.
2. Ibid., 416.
3. Mary Ormsbee Whitton, *First First Ladies, 1789–1865* (New York, 1948), 182.
4. Holloway, *Ladies of the White House,* 380–81.
5. Ibid., 382–83.
6. Lyon G. Tyler, *The Letters and Times of the Tylers* (2 vols., Richmond, Va., 1884), I:276.
7. Elizabeth Tyler Coleman, *Priscilla Cooper Tyler and the American Scene* (University of Alabama, 1955), 73–75.
8. Ibid.
9. Tyler, *Letters and Times,* I:547.
10. Quoted in *Richmond Whig,* Sept. 13, 1842; Oliver P. Chitwood, *John Tyler: Champion of the Old South* (New York, 1939), 396.
11. Whitton, *First First Ladies,* 193.
12. Oliver P. Chitwood, *John Tyler,* 394.
13. Ibid., 406.
14. Ibid., 403.
15. Ibid., 404.
16. George Ticknor Curtis, *Life of James Buchanan* (2 vols., New York, 1883), I:529.
17. Whitton, *First First Ladies,* 188.
18. Ibid., 188–89.
19. Elizabeth F. Ellet, *The Court Circles of the Republic* (Hartford, Conn., 1869), 360.
20. Whitton, *First First Ladies,* 188.
21. Robert Seager II, *And Tyler Too: A Biography of John and Julia Gardiner Tyler* (New York, 1963), 196, 200.
22. Ibid., 207.
23. *Niles' National Register* (Baltimore), 66 (July 6, 1840): 290; *New York Herald,* June 27, 1844, quoted in Seager, *And Tyler Too,* 5–6.
24. Henry A. Wise, *Seven Decades of the Union* (Philadelphia, 1881), 233, 235.
25. Tyler, *Letters and Times,* II:358, 366, 369.
26. Ibid., 369.
27. Chitwood, *Tyler,* 403.
28. Seager, *And Tyler Too,* 243.
29. Ibid., 257.
30. Pierre S.R. Payne, *The Island* (New York, 1958), 221.

31. Tyler, *Letters and Times,* II:513.
32. Seager, *And Tyler Too,* 329.
33. Julia Gardiner Tyler, "To the Duchess of Sutherland and Ladies of England," *Southern Literary Messenger,* 19 (Feb. 1853): 124, 125–26.
34. Chitwood, *Tyler,* 649–51.
35. Payne, *Island,* 202–3.
36. *New York Herald,* June 27, 1844, quoted in Seager, *And Tyler Too,* 6.
37. Payne, *The Island,* 216.
38. Seager, *And Tyler Too,* 196–97.
39. Tyler, *Life and Letters,* 670–72.
40. *New York Herald,* April 17, 1865, quoted in Seager, *And Tyler Too,* 509.

11. Sarah Childress Polk

1. Charles Sellars, *James K. Polk: Continentalist, 1843–1846* (Princeton, 1966), 193.
2. Marianne Means, *The Woman in the White House* (New York, 1963), 79.
3. Ibid.
4. Ibid., 80; Mary Ormsbee Whitton, *The First First Ladies, 1789–1865* (New York, 1948), 203.
5. Charles Sellars, *James K. Polk: Jacksonian, 1795–1843* (Princeton, 1957), 93.
6. Means, *Woman in the White House,* 90.
7. Laura Carter Holloway, *Ladies of the White House* (New York, 1870), 448.
8. *Polk: The Diary of a President, 1845–1849* (ed., Allan Nevins, New York, 1929, 1952), 300; John J. Jenkins, *James Knox Polk and a History of His Administration* (New Orleans, La., 1854), 55–56.
9. Anson and Fanny Nelson, *Memorials of Sarah Childress Polk* (New York, 1892), 92–93.
10. Ibid., 44.
11. Ibid., 93.
12. Means, *Woman in the White House,* 86.
13. Ibid., 86.
14. Whitton, *First First Ladies,* 215–16.
15. "Letters of Mrs. James K. Polk to Her Husband," *Tennessee Historical Quarterly,* 11 (June 1952): 180.
16. Nelson, *Memorials,* 38–40.
17. Means, *Woman in the White House,* 78–79.
18. Nelson, *Memorials,* 78.
19. Ibid., 81.
20. Ibid., 110.
21. Ibid., 39.
22. Ibid., 113.
23. Ibid., 110.
24. Ibid., 257.
25. Means, *Woman in the White House,* 86.

12. Margaret Taylor

1. Laura Carter Holloway, *Ladies of the White House* (New York, 1870), 492.
2. Hamilton Holman, *Zachary Taylor: Soldier in the White House* (Indianapolis, 1951), 25.
3. Hamilton Holman, *Zachary Taylor: Soldier of the Republic* (Indianapolis, 1941), 117.
4. Holman, *Soldier of the Republic,* 110; John H. Bliss, "Reminiscences of Fort Snelling," *Minnesota Historical Society Collections,* 6: 338.
5. Mary Ormsbee Whitton, *First First Ladies, 1789–1865* (New York, 1948), 219.
6. Ibid., 220.
7. Ibid., 232.
8. Holman, *Soldier of the Republic,* 68–69.
9. Brainerd Dyer, *Zachary Taylor* (Baton Rouge, 1946), 400.
10. Irene Gerlinger, *Mistresses of the White House* (Freeport, N.Y., 1970), 43.
11. Holman, *Soldier in the White House,* 316, 396.
12. Bessie Smith, *The Romances of the Presidents* (New York, 1932), 168–71.
13. Fred W. Allsopp, *Folklore of Romantic Arkansas* (2 vols., The Grolier Society, 1931), I:133.
14. Varina Davis, *Jefferson Davis: Ex-President of the Confederate States of America* (2 vols., New York, 1890), I: 95–96, 162.

13. Abigail Fillmore

1. Mary Ormsbee Whitton, *First First Ladies, 1789–1865* (New York, 1948), 328.
2. Ibid., 236.
3. "Millard Fillmore's Youth," *Publications of the Buffalo Historical Society,* Vol. 10, *Millard Fillmore Papers,* I (ed., Frank H. Severance, Buffalo Historical Society, Buffalo, N.Y., 1907): 11.
4. Whitton, *First First Ladies,* 237.
5. *National Intelligencer,* 41 (March 22, 1853): 3.
6. Laura Carter Holloway, *Ladies of the White House* (New York, 1870), 502.

14. Jane Pierce

1. Roy F. Nichols, *Franklin Pierce* (Philadelphia, 1958), 313.
2. Mary Ormsbee Whitton, *First First Ladies, 1789–1865* (New York, 1948), 250.
3. Nichols, *Pierce,* 440.
4. Ibid., 104.
5. Ibid., 77.
6. Ibid., 78.
7. Whitton, *First First Ladies,* 251.
8. Ibid.
9. Nichols, *Pierce,* 94.
10. Ibid., 140.

11. Ibid., 149.
12. Ibid., 205.
13. Ibid, 243; Laura Carter Holloway, *The Ladies of the White House* (New York, 1870), 529.
14. Nichols, *Pierce*, 522.
15. J.R. Irelan, *The Republic* (18 vols., Chicago, 1888), XIV:596; Corn Miley, "Franklin Pierce, the Most Charming Personality of All the Presidents," *Americana*, 24 (April 1935): 180.
16. Ibid., 202; Whitton, *First First Ladies*, 261.
17. Miley, "Franklin Pierce," 173.

15. Mary Todd Lincoln

1. Ruth Painter Randall, *Mary Lincoln: Biography of a Marriage* (Boston, 1953), 308.
2. Marianne Means, *The Woman in the White House* (New York, 1963), 97.
3. Ishbel Ross, *The President's Wife, Mary Todd Lincoln: A Biography* (New York, 1973), 105–6.
4. Ibid., 108.
5. Ibid., 108–9.
6. William O. Stoddard, *Inside the White House in War Times* (New York, 1890), 35.
7. *Lincoln Lore*, January 1975, pp. 1–4.
8. Randall, *Mary Lincoln*, 242–43; 310.
9. Ibid., 317–18.
10. Elizabeth Keckley, *Behind the Scenes: Thirty Years a Slave and Four Years in the White House* (New York, 1968), 103–4.
11. Means, *Woman in the White House*, 105.
12. Ibid., 105.
13. Randall, *Mary Lincoln*, 106.
14. "All About the Domestic Economy in the White House," *Illinois State Register*, 16 (Oct. 30, 1864): 2.
15. F.B. Carpenter, *Six Months at the White House* (New York, 1867), 293.
16. Ross, *President's Wife*, 241; Randall, *Mary Lincoln*, 382.
17. Ross, *President's Wife*, 242; Randall, *Mary Lincoln*, 383.
18. Keckley, *Behind the Scenes*, 199–200.
19. *Herndon's Life of Lincoln* (with an introduction and notes by Paul M. Angle, New York, 1930), 105; David Donald, *Lincoln's Herndon* (New York, 1948), 228.
20. Ross, *President's Wife*, 30.
21. *Herndon's Life of Lincoln*, 347.
22. Keckley, *Behind the Scenes*, 236.
23. Randall, *Mary Lincoln*, 219.
24. Ibid., 112–13.
25. Justin Q. Turner and Linda Levitt Turner, *Mary Todd Lincoln: Her Life and Letters* (New York, 1972), 442.
26. Means, *Woman in the White House*, 110.
27. Turner, *Mary Todd Lincoln*, 267.
28. Ross, *President's Wife*, 298.

29. Ibid., 304.
30. Samuel A. Schreiner, Jr., *The Trials of Mary Lincoln* (New York, 1987), 85–86.
31. Ross, *President's Wife,* 313.
32. Ibid., 318.
33. Emily Todd Helm, "Mary Todd Lincoln," *McClure's Magazine,* 11 (Sept. 1898): 480.
34. Katherine Helm, *The True Story of Mary, Wife of Lincoln* (New York and London, 1928), 42–43.
35. Ibid.
36. Carlos W. Goltz, *Incidents in the Life of Mary Todd Lincoln* (Sioux City, Iowa, 1928), 15–16.
37. Helm, *True Story,* 74.
38. Goltz, *Incidents,* 18–20.
39. Ibid., 20–21.
40. Lord Charnwood, *Abraham Lincoln* (New York, 1917), 86.
41. Helm, "Mary Todd Lincoln," 479; Goltz, *Incidents,* 6.
42. Henry B. Rankin, *Personal Recollections of Abraham Lincoln* (New York and London, 1916), 190–91.
43. Ross, *President's Wife,* 186–87; Randall, *Mary Lincoln,* 88–89.
44. Carl Sandburg, *Mary Lincoln: Wife and Widow* (New York, 1932), 81; Ross, *President's Wife,* 92.
45. Randall, *Mary Lincoln,* 201–2; Means, *Woman in the White House,* 94.
46. Means, *Woman in the White House,* 98.
47. Helm, *True Story,* 179.
48. Carpenter, *Six Months at White House,* 301–302.
49. Keckley, *Behind the Scenes,* 124–25.
50. Ibid., 128–35.
51. Eugene Tripler, *Some Notes of Her Personal Recollections* (New York, 1910), 139–40.
52. Helm, *True Story,* 228–231.
53. Randall, *Mary Lincoln,* 360–61.
54. Adam Badeau, *Grant in Peace* (Hartford, Conn., 1887), 357–60.
55. Turner, *Mary Todd Lincoln,* 438–39.
56. Goltz, *Incidents,* 27–28; Randall, Mary Lincoln, 425–29.
57. Sarah Bernhardt, *Memories of My Life* (New York, 1907), 370.

16. Eliza Johnson

1. W.H. Crook, *Memories of the White House* (Boston, 1911), 67.
2. Robert W. Winston, *Andrew Johson: Plebeian and Patriot* (New York, 1928), 106.
3. James S. Jones, *Life of Andrew Johnson* (Greeneville, Tenn., 1901), 333.
4. Milton Lomask, *Andrew Johnson: President on Trial* (New York, 1960), 110; Lately Thomas, *The First President Johnson* (New York, 1968), 245.
5. Bess Furman, *White House Profile: A Social History of the White House, Its Occupants and Its Festivities* (Indianapolis, 1951), 200.
6. Mary Clemmer Ames, *Ten Years in Washington: Life and Scenes in the National Capital* (Hartford, Conn., 1880), 245.

7. Thomas, *First President Johnson,* 319.
8. Ames, *Ten Years,* 244.
9. Crook, *Memories,* 52–53.
10. Ames, *Ten Years,* 251.
11. Kathleen Prindiville, *First Ladies* (New York, 1964), 165; Thomas, *First President Johnson,* 21; Winston, *Johnson,* 294.

17. Julia Dent Grant

1. Ishbel Ross, *The General's Wife: The Life of Mrs. Ulysses S. Grant* (New York, 1959), 204.
2. *New York Tribune,* Feb. 25, 1867, quoted in William B. Hesseltine, *Ulysses S. Grant: Politician* (New York, 1935), 294.
3. Esther Singleton, *The Story of the White House* (2 vols., New York, 1907), II:129, 131.
4. Ross, *General's Wife,* 210.
5. Ibid., 239.
6. Ibid., 5–6, 65.
7. Ibid., 142.
8. Ibid., 218.
9. Adam Badeau, *Grant in Peace* (Freeport, N.Y., 1887, 1971), 409–10.
10. *The Personal Memoirs of Julia Dent Grant* (ed., John Y. Simon, New York, 1975), 104–5, 110–11, 132–33.
11. Ibid., 111.
12. Ibid., 187, 194.
13. Ibid., 185–86; Badeau, *Grant in Peace,* 243–44.
14. *Memoirs,* 196–97.
15. Badeau, *Grant in Peace,* 288.
16. *Memoirs,* 321–22.
17. Ross, *General's Lady,* 295, 309.
18. *Memoirs,* 331.
19. Hamlin Garland, *Ulysses S. Grant: His Life and Character* (New York, 1898), 63–64.
20. Ross, *General's Lady,* 30–31.
21. Bruce Catton, *Grant Moves South* (Boston, 1960), 11–12.
22. *Memoirs,* 136.
23. Ibid., 137–38.
24. Ibid., 175.
25. W.H. Crook, *Memories of the White House* (Boston, 1911), 98–99.
26. William S. McFeely, *Grant: A Biography* (New York, 1981), 404.
27. Badeau, *Grant in Peace,* 175.
28. *Memoirs,* 325.
29. Bess Furman, *White House Profile: A Social History of the White House, Its Occupants and Its Festivities* (Indianapolis, 1951), 206.
30. *Memoirs,* 178.
31. Ross, *General's Lady,* 228–29.
32. Badeau, *Grant in Peace,* 288.
33. *Memoirs,* 246.
34. Ibid., 211–12.

35. Ibid., 219.
36. Ibid., 239–40.
37. Marie Dressler, *My Own Story* (Boston, 1934), 110–13.

18. Lucy Webb Hayes

1. Mary Clemmer Ames, "A Woman's Letter from Washington: The Inaugural," *Independent*, 29 (March 15, 1877): 2, 29.
2. Charles Richard Williams, *The Life of Rutherford Birchard Hayes* (2 vols., Columbus, Ohio, 1928), II:299.
3. Ibid., 317n.
4. Emily Apt Geer, *First Lady: The Life of Lucy Webb Hayes* (Kent State University, 1984), 147.
5. Ibid., 151.
6. *Puck*, 1 (May 1877): 3.
7. Geer, *First Lady*, 184.
8. Williams, *Hayes*, II:294n.
9. Emily Apt Geer, "Lucy Webb Hayes and Her Influence upon Her Era," *Hayes Historical Journal*, 1 (Spring 1976): 25.
10. H.J. Eckenrode, *Rutherford B. Hayes: Statesman of Reunion* (New York, 1930), 43.
11. Ibid., 44.
12. Ibid., 45.
13. Ibid.
14. Feb. 27, 1853, *Diary and Letters of Rutherford B. Hayes* (5 vols., Columbus, Ohio, 1924), I:444.
15. Geer, *First Lady*, 30.
16. April 27, 1870, *Diary and Letters*, III:105.
17. Geer, *First Lady*, 232.
18. Beverly Beeton, "The Hayes Administration and the Woman Question," *Hayes Historical Journal*, 2 (Spring 1978): 52; see Williams, *Hayes*, II:381.
19. Geer, *First Lady*, 167.
20. Ibid., 169.
21. Ibid., 168.
22. Ibid., 169.
23. Eckenrode, *Hayes*, 312.
24. Ibid., 230.
25. Geer, *First Lady*, 230.
26. Bess Furman, *White House Profile: A Social History of the White House* (Indianapolis, 1951), 224.
27. Williams, Hayes, II:316–17.
28. July 22, 1889, *Diary*, IV:492.
29. Geer, *First Lady*, 273.
30. Mrs. John Davis, *Lucy Webb Hayes: A Memorial* (Cincinnati, 1890), 19–21.
31. Ibid., 31–32.
32. Ibid., 53.
33. Williams, *Hayes*, II:309–10n.

34. Watt P. Marchman, ed., "The 'Memoirs' of Thomas Donaldson," *Hayes Historical Journal,* 2 (Spring-Fall 1979): 235; Geer, *First Lady,* 229–30.
35. Elizabeth Cochrane, "Nellie Bly Visits Spiegel Grove," *Hayes Historical Journal,* 1 (Fall 1976): 144.

19. Lucretia Garfield

1. Extract from a letter to James G. Garfield, undated, in The Papers of Lucretia Garfield: Diaries, Writings, & Memorabilia, Box 89, Manuscript Division, Library of Congress.
2. "Woman's Station," *The Eclectic Star,* Sept. 18, 1851, ibid., Box 88.
3. Allan Pesk, *Garfield* (Kent State University, 1978), 657; letter to Whitelaw Reid, Dec. 9, 1868, in Margaret Leech and Harry Brown, *The Garfield Orbit* (New York, 1978), 265–66.
4. The Diary of James A. Garfield (3 vols., ed., Harry James Brown and Frederick D. Williams, Michigan State University, 1967), I:234–45.
5. Ibid., 251.
6. Pesk, *Garfield,* 41.
7. *Diary,* I:272.
8. Pesk, *Garfield,* 41.
9. Ibid., 54.
10. Theodore Clarke Smith, *The Life and Letters of James Abram Garfield* (2 vols., Yale, 1925), I:111.
11. Pesk, *Garfield,* 55.
12. Ibid.
13. Smith, *Garfield,* I:111.
14. Pesk, *Garfield,* 75.
15. Ibid., 146.
16. Ibid., 160.
17. Ibid., 160–61.
18. Leech and Brown, *Garfield Orbit,* 195.
19. Smith, *Garfield,* II:896.
20. *Diary,* II:171.
21. Pesk, *Garfield,* 347.
22. "Statement for the Family," Feb. 21, 1898, Lucretia Garfield Papers, Box 89, LC; Smith *Garfield,* II:900.
23. *Diary,* II:159, 171; Smith, *Garfield,* II:926; Pesk, *Garfield,* 347.
24. *Diary,* III:181.
25. Pesk, *Garfield,* 348.
26. Ibid.
27. Smith, *Garfield,* II:944. The lines are taken from the fourth stanza of Alfred, Lord Tennyson's "In Memoriam."
28. Smith, *Garfield,* II:1074.
29. Diary of Lucretia R. Garfield from March 1 to April 20, 1881 (typescript), 3, Papers, Box 89, LC.
30. Leech and Brown, *Garfield Orbit,* 242–43.
31. Pesk, *Garfield,* 598.

32. D.W. Bliss, "The Story of President Garfield's Illness," *Century Magazine,* 23 (Dec. 1881): 304.
33. Margaret Bassett, *Profiles & Portraits of American Presidents & Their Wives* (New York, 1969), 198.
34. John M. Taylor, *Garfield of Ohio: The Available Man* (New York, 1970), 288.

20. Ellen Herndon Arthur

1. Thomas C. Reeves, *Gentleman Boss: The Life of Chester Alan Arthur* (New York, 1975), 84.
2. Ibid., 20.
3. Ibid., 21.
4. Ibid., 32.
5. Ibid.
6. Ibid., 33.
7. "Personal Intercourse with the Vice Presidential Candidate," *Boston Herald,* June 13, 1880, quoted, Reeves, *Gentleman Boss,* 85.
8. Chauncey M. Depew, *My Memories of Eighty Years* (New York, 1924), 117.
9. Reeves, *Gentleman Boss,* 159.
10. Allan Pesk, *Garfield* (Kent State University, 1978), 546.
11. Esther Singleton, *The Story of the White House* (2 vols., New York, 1907), II:176.
12. George Frederick Howe, *Chester A. Arthur: A Quarter-Century of Machine Politics* (New York, 1935), 175.

21. Frances Cleveland

1. Mrs. John A. Logan, *Thirty Years in Washington* (Hartford, Conn., 1901), 698–99.
2. Allan Nevins, *Cleveland: A Study in Courage* (New York, 1932), 72.
3. Bess Furman, *White House Profile* (Indianapolis, 1951), 240; Harry Thurston Peck, *Twenty Years of the Republic* (New York, 1907), 60–62.
4. Grover Cleveland, "Woman's Mission and Woman's Clubs," *Ladies' Home Journal,* 22 (May 1905): 3.
5. Rexford G. Tugwell, *Grover Cleveland* (New York, 1968), 45.
6. Horace Samuel Merrill, *Bourbon Leader: Grover Cleveland and the Democratic Party* (Boston, 1957), 89; Nevins, *Cleveland,* 303.
7. Nevins, *Cleveland,* 303.
8. Robert McElroy, *Grover Cleveland: The Man and the Statesman* (2 vols., New York, 1923), I:184.
9. Nevins, *Cleveland,* 304; McElroy, *Cleveland,* I:186.
10. To Mrs. M.E. Hoyt, March 21, 1886, *Letters of Grover Cleveland* (ed., Allan Nevins, Boston, 1933), 103.
11. McElroy, *Cleveland,* I:187; Nevins, *Cleveland,* 307, 309–10; Furman, *White House Profile,* 243.
12. Nevins, *Cleveland,* 311.
13. Irwin Hood (Ike) Hoover, *Forty-two Years in the White House* (Boston, 1934), 12, 14.
14. Nevins, *Cleveland,* 318; McElroy, *Cleveland,* I:188.

15. Nevins, *Cleveland,* 311.
16. Furman, *White House Profile,* 246; Denis Tilden Lynch, *Cleveland: A Man Four-Square* (New York, 1932), 359–60.
17. Nevins, *Cleveland,* 454; Furman, *White House Profile,* 246.
18. To Mary Cleveland Hoyt, *Letters of Grover Cleveland, 1850–1908,* 104.
19. W.H. Crook, *Memories of the White House* (Boston, 1911), 186–87.
20. Furman, *White House Profile,* 258.
21. Hoover, *Forty-two Years,* 15.
22. Nevins, *Cleveland,* 311–12; *The Letters and Friendships of Sir Cecil Spring-Rice* (2 vols., Boston and New York, 1929), I:72–73.
23. Nevins, *Cleveland,* 726.
24. Margaret Bassett, *Profiles & Portraits of American Presidents & Their Wives* (Freeport, Maine, 1969), 218.
25. Cleveland, "Woman's Mission and Woman's Clubs," 3–4.
26. "Mrs. Preston Dies; Wed to Cleveland," *New York Times,* Oct. 30, 1947, p. 25.
27. Hoover, *Forty-two Years,* 13–14; Crook, *Memories of White House,* 179–80; Merrill, *Bourbon Leader,* 91.
28. McElroy, *Cleveland,* I:186.
29. Nevins, *Cleveland,* 311.
30. Crook, *Memories of the White House,* 194–96.
31. Frank G. Carpenter, *Carp's Washington* (New York, 1960), 47–49.
32. Nelle Scanlan, *Boudoir Memoirs of Washington* (Chicago, 1923), 95–96.
33. Crook, *Memories of the White House,* 197–98.
34. *Autobiography of William Allen White* (New York, 1946), 362–63.
35. Margaret Truman, *Souvenir: Margaret Truman's Own Story* (New York, 1956), 172.

22. Caroline Harrison

1. Esther Singleton, *The Story of the White House* (2 vols., New York, 1907), II:221–22.
2. Harry J. Sievers, *Benjamin Harrison: Hoosier President* (Indianapolis, 1968), 52.
3. Margaret Bassett, *Profiles & Portraits of American Presidents & Their Wives* (Freeport, Maine, 1969), 226.
4. Ophia D. Smith, *Old Oxford House* (Oxford, Ohio, 1941), 72–73.
5. Ophia D. Smith, *Fair Oxford* (Oxford, Ohio, 1947), 190.
6. Harry J. Sievers, *Benjamin Harrison: Hoosier Warrior* (New York, 1952), 72.
7. Ibid., 76–77.
8. Ibid., 77.
9. Ibid., 99.
10. Ibid., 167.
11. Ibid., 264.
12. Ibid., 293.
13. Ibid., 296.
14. Ibid., 316.
15. Harry J. Sievers, *Benjamin Harrison: Hoosier Statesman, 1865–1888* (New York, 1959), 354.

16. Sievers, *Hoosier President,* 207.
17. Bess Furman, *White House Profile* (Indianapolis, 1951), 253.
18. Frank G. Carpenter, *Carp's Washington* (New York, 1960), 301.
19. Furman, *White House Profile,* 253.
20. Ibid., 250.
21. Sievers, *Hoosier President,* 156.
22. W.H. Crook, *Memories of the White House* (Boston, 1911), 233–34.
23. Sievers, *Hoosier President,* 256n.
24. Ibid., 257.

23. Ida Saxton McKinley

1. Margaret Leech, *In the Days of McKinley* (New York, 1963), 15.
2. *New York Times,* May 27, 1907, p. 1.
3. Charles S. Olcott, *The Life of William McKinley* (2 vols., Boston, 1916), I:68.
4. Thomas Beer, *Hanna* (New York, 1929), 102–3; Leech, *Days of McKinley,* 14.
5. Charles Willis Thompson, *Presidents I've Known and Two Near Presidents* (Indianapolis, 1929), 17.
6. T.W. Shannon, *Eugenics* (Marietta, Ohio, 1917), 404–5.
7. Leech, *Days of McKinley,* 432.
8. Belle Case and Fola LaFollette, *Robert M. LaFollette* (2 vols., New York, 1953), I:63.
9. Olcott, *McKinley,* II:363.
10. Julia B. Foraker, *I Would Live It Again* (New York, 1932), 257.
11. Mrs. John A. Logan, *Thirty Years in Washington* (Harford, Conn., 1901), 735.
12. Olcott, *McKinley,* I:291.
13. Leech, *Days of McKinley,* 64–65.
14. Hugh Baillie, *High Tension* (New York, 1959), 29.
15. Leech, *Days of McKinley,* 444.
16. Ibid., 445.
17. Ibid.
18. Olcott, *McKinley,* II:363.
19. "Mrs. McKinley, Mrs. Bryan: A Comparison," *Harper's Bazar,* 33 (Aug. 11, 1900), 955–56.
20. Leech, *Days of McKinley,* 434.
21. Olcott, *McKinley,* I:64.
22. Leech, *Days of McKinley,* 439.
23. Ibid., 459.
24. Ibid., 590.
25. Olcott, *McKinley,* II:316; *Literary Digest,* 23 (Sept. 14, 1901): 302.
26. *New York Times,* Sept. 14, 1901, p. 1.
27. Ibid., May 27, 1907, p. 1.
28. H. Wayne Morgan, *William McKinley and His America* (Syracuse, N.Y., 1963), 526.
29. Foraker, *I Would Live It Again,* 259–60.
30. H.H. Kohlsaat, *From McKinley to Harding* (New York, 1923), 66–68.

31. James Barnes, *From Then Till Now* (New York, 1934), 219–20.
32. Foraker, *I Would Live It Again,* 259.

24. The Roosevelt Wives: Alice Lee Roosevelt and Edith Kermit Roosevelt

1. Herman Hagedorn, *The Roosevelt Family of Sagamore Hill* (New York, 1954), 89; *The Letters of Theodore Roosevelt* (8 vols, Harvard, 1951–54), II:1340, 1345.
2. Edmund Morris, *The Rise of Theodore Roosevelt* (New York, 1979), 90.
3. Ibid., 104.
4. Henry F. Pringle, *Theodore Roosevelt: A Biography* (New York, 1931), 42.
5. Morris, *Rise of Roosevelt,* 127.
6. William Henry Harbaugh, *Power and Responsibility: The Life and Times of Theodore Roosevelt* (New York, 1961), 45.
7. Noel F. Busch, *T.R.: The Story of Theodore Roosevelt and His Influence on Our Times* (New York, 1963), 46.
8. Morris, *Rise of Roosevelt,* 241.
9. Harbaugh, *Power and Responsibility,* 46–47.
10. Michael Teague, *Mrs. L: Conversations with Alice Roosevelt Longworth* (Garden City, N.Y., 1981), 5.
11. Carleton Putnam, *Theodore Roosevelt: The Formative Years, 1858–1886* (New York, 1956), 556.
12. Teague, *Mrs. L,* 5.
13. Morris, *Rise of Roosevelt,* 338.
14. Sylvia Jukes Morris, *Edith Kermit Roosevelt: Portrait of a First Lady* (New York, 1980), 90–91.
15. Hagedorn, *Roosevelt Family,* 10.
16. Morris, *Rise of Roosevelt,* 359.
17. Bess Furman, *White House Profile* (Indianapolis, 1951), 266.
18. *Selections from the Correspondence of Theodore Roosevelt and Henry Cabot Lodge, 1884–1918* (2 vols., New York, 1925), I:111.
19. Busch, *T.R.,* 86.
20. Edward Wagenknecht, *The Seven Worlds of Theodore Roosevelt* (New York, 1958), 176.
21. *The Letters of Archie Butt, Personal Aide to President Roosevelt* (ed., Lawrence F. Abbott, Garden City, N.Y., 1924), 103.
22. "Death of a Lady," *Time,* 52 (Oct. 11, 1948):28.
23. Hagedorn, *Roosevelt Family,* 18.
24. Ibid.
25. "Some Great Americans," *Life,* 25 (Dec. 13, 1948):48; Owen Wister, *Roosevelt: The Story of a Friendship* (New York, 1930), 89–90.
26. Busch, *T.R.,* 170.
27. Irwin Hood (Ike) Hoover, *Forty-two Years in the White House* (Boston, 1934), 28.
28. *Letters of Roosevelt,* III:161–62.
29. W.H. Crook, *Memories of the White House* (Boston, 1911), 281.
30. Hagedorn, *Roosevelt Family,* 195–96.

31. Anne O'Hagan, "Women of the Hour: No. 3, Mrs. Roosevelt," *Harper's Bazar,* 39 (May 1905):413.
32. Ibid., 194.
33. Margaret Bassett, *Profiles & Portraits of American Presidents & Their Wives* (Freeport, Maine, 1969), 253.
34. *Letters of Archie Butt,* 30.
35. Hagedorn, *Roosevelt Family,* 193.
36. *Letters of Archie Butt,* 322–23.
37. *Letters of Theodore Roosevelt,* VIII:1301.
38. Harbaugh, *Power and Responsibility,* 512.
39. Ibid., 519.
40. Busch, *T.R.,* 329–30.
41. "Mrs. Theodore Roosevelt Dies at Oyster Bay," *New York Times,* Oct. 1, 1948, p. 25; Karl Schriftgiesser, *The Amazing Roosevelt Family, 1613–1942* (New York, 1942), 244.
42. "President Unlike 'T.R.,' Widow Says," *New York Times,* Oct. 26, 1936, p. 5.
43. "Some Great Americans," 48.
44. Wister, *Roosevelt,* 24.
45. *Theodore Roosevelt: An Autobiography* (New York, 1913), 345.
46. Wagenknecht, *Seven Worlds,* 168.
47. Hagedorn, *Roosevelt Family,* 304.
48. Ibid., 196.
49. *Letters of Archie Butt,* 214–15, 225–26.
50. Busch, *T.R.,* 249.
51. Teague, *Mrs. L,* 128.
52. Hagedorn, *Roosevelt Family,* 283.
53. *Taft and Roosevelt: The Intimate Letters of Archie Butt* (2 vols, Garden City, N.Y., 1930), II:231–32.
54. Busch, *T.R.,* 87.
55. Hagedorn, *Roosevelt Family,* 317–21; Joseph L. Gardner, *Departing Glory: Theodore Roosevelt as Ex-President* (New York, 1973), 276–77; Charles Willis Thompson, *Presidents I've Known and Two Near Presidents* (Indianapolis, 1929), 150–55.

25. Helen Herron Taft

1. Mrs. William Howard Taft, *Recollections of Full Years* (New York, 1914), 333.
2. H.H. Kohlsaat, *From McKinley to Harding* (New York, 1923), 161–62.
3. *Recollections,* 304.
4. Henry F. Pringle, *The Life and Times of William Howard Taft* (2 vols., New York, 1939), I:72–73.
5. *Recollections,* 10.
6. Pringle, *Taft,* I:79.
7. Ibid., 69.
8. *Recollections,* 21.
9. Ibid., 22.

10. Ibid., 24.
11. Ibid., 30.
12. Judith Icke Anderson, *William Howard Taft: An Intimate History* (New York, 1981), 48.
13. Pringle, *Taft,* I:82.
14. Ibid., 149.
15. Ibid., 116.
16. *Recollections,* 33.
17. Ibid., 212.
18. Ibid., 267.
19. Ibid., 269.
20. Ibid.
21. Ibid., 272.
22. Ibid., 276.
23. Margaret Bassett, *Profiles & Portraits of American Presidents & Their Wives* (Freeport, Maine, 1969), 263.
24. Ishbel Ross, *An American Family: The Tafts, 1678 to 1964* (Cleveland and New York, 1964), 188.
25. Pringle, *Taft,* 1:333.
26. Henry L. Stoddard, *As I Knew Them: Presidents and Politics from Grant to Coolidge* (New York, 1927), 338–39.
27. Ona Griffin Jeffries, *In and Out of the White House* (New York, 1960), 286.
28. Marianne Means, *The Woman in the White House* (New York, 1963), 131.
29. Irwin Hood (Ike) Hoover, *Forty-two Years in the White House* (Boston, 1934), 41.
30. *Recollections,* 347.
31. Bess Furman, *White House Profile* (Indianapolis, 1951), 283.
32. *Taft and Roosevelt: The Intimate Letters of Archie Butt* (2 vols., Garden City, N.Y. 1930), I:88.
33. Means, *Woman in the White House,* 130.
34. Pringle, *Taft,* 1:70.
35. Furman, *White House Profile,* 284.
36. Pringle, *Taft,* 2:769; *Intimate Letters of Archie Butt,* II:846, 850.
37. Furman, *White House Profile,* 287.
38. Means, *Woman in the White House,* 134.
39. Anderson, *Taft,* 262.
40. *New York Times,* May 23, 1943, 43.
41. *Recollections,* 19–20.
42. Pringle, *Taft,* I:114.
43. Means, *Woman in the White House,* 121.
44. *Recollections,* 126–27.
45. Anderson, Taft, 31.
46. *The Autobiography of John Hays Hammond* (2 vols., New York, 1935), II:546–47.
47. *Recollections,* 292–94.
48. Alice Roosevelt Longworth, *Crowded Hours* (New York, 1933), 158; *Intimate Letters of Archie Butt,* I:228.
49. Means, *Woman in the White House,* 128.

50. *Intimate Letters of Archie Butt,* I:357.
51. Ibid., II:461–62.

26. The Wilson Wives: Ellen Axson Wilson and Edith Bolling Wilson

1. Edwin A. Weinstein, *Woodrow Wilson: A Medical and Psychological Biography* (Princeton, 1981), 83.
2. Ray Stannard Baker, *Woodrow Wilson: Life and Letters* (8 vols., New York, 1927–39), II:57.
3. Ibid., 58.
4. Ibid., I:162.
5. Ibid., 239.
6. Ibid., 242.
7. Eleanor Wilson McAdoo, *The Woodrow Wilsons* (New York, 1937), 22.
8. Tom Schachtman, *Edith and Woodrow: A Presidential Romance* (New York, 1981), 24–25; Frances Wright Saunders, *Ellen Axson Wilson: First Lady Between Two Worlds* (Chapel Hill, 1985), 201.
9. Baker, *Wilson,* III:210.
10. McAdoo, *Wilsons,* 155.
11. Baker, *Wilson,* III:409.
12. Ibid., II:462.
13. Ibid., 463.
14. McGregor, "The Social Activities of the White House," *Harper's Weekly,* 58 (April 25, 1914): 26–27.
15. Mrs. Ernest P. Bicknell, "The Home-Maker of the White House," *Survey,* 33 (Oct. 3, 1914): 22.
16. Schachtman, *Edith and Woodrow,* 37.
17. McAdoo, *Wilsons,* 300; Arthur S. Link, *Wilson: The New Freedom* (Princeton, 1956), 462.
18. Edith Bolling Wilson, *My Memoir* (New York, 1939), 56.
19. Ibid., 62.
20. Ibid., 76–77.
21. Edmund W. Starling, *Starling of the White House* (Chicago, 1956), 56.
22. Alden Hatch, *Edith Bolling Wilson: First Lady Extraordinary* (New York, 1961), 66.
23. Ibid., 16.
24. *Memoir,* 271.
25. Hatch, *Edith Wilson,* 200.
26. Ibid., 211.
27. *Memoir,* 289–90.
28. Ibid., 290.
29. Ibid., 289.
30. Hatch, *Edith Wilson,* 219; Bess Furman, *White House Profile* (Indianapolis, 1951), 299.
31. Hatch, *Edith Wilson,* 228.
32. *Memoir,* 297.
33. Ibid.
34. Ibid., 299.
35. Ibid., 309.

36. Schachtman, *Edith and Woodrow,* 275.
37. Baker, *Wilson,* II:50; McAdoo, *Wilsons,* 55.
38. McAdoo, *Wilsons,* 19.
39. Ibid., 13–14.
40. Ibid., 54.
41. Ibid., 15.
42. Ibid., 129.
43. Ibid., 124.
44. Ibid., 167.
45. Ibid., 248; Baker, *Wilson,* IV:109.
46. Starling, *Starling of White House,* 44.
47. *Parade,* Jan. 5, 1986, p. 2.
48. *Memoir,* 84–85.
49. Ibid., 63–64.
50. Harry M. Daugherty, *The Inside Story of the Harding Tragedy* (New York, 1932), 237.
51. *Memoir,* 112–13.
52. Ibid., 154–55.
53. Ibid., 153.
54. Ibid., 158–59.
55. Ibid., 140–41; Hatch, *Edith Wilson,* 109.
56. Hatch, *Edith Wilson,* 118.
57. Ibid., 129–30.
58. Margaret Bassett, *Profiles & Portraits of American Presidents & Their Wives* (Freeport, Maine, 1969), 287.
59. *Memoir,* 227–30.
60. Ibid., 237–39; Ishbel Ross, *Power with Grace: The Life Story of Mrs. Woodrow Wilson* (New York, 1975), 155–56.
61. Schachtman, *Edith and Woodrow,* 201–202.
62. Ross, *Power with Grace,* 339.
63. Hatch, *Edith Wilson,* 272.
64. Ibid., 272–73.
65. Ibid., 272.
66. Ross, *Power with Grace,* 373.

27. Florence Kling Harding

1. Samuel Hopkins Adams, *The Incredible Era: The Life and Times of Warren Gamaliel harding* (Boston, 1939), 211.
2. Means, *The Woman in the White House* (New York, 1963), 168.
3. Harry Daugherty, *The Inside Story of the Harding Tragedy* (New York, 1932), 170.
4. "Mrs. Harding Dies after Long Fight," *New York Times,* Nov. 22, 1924, p. 3.
5. Adams, *Incredible Era,* 253.
6. Elizabeth Jaffray, *Secrets of the White House* (New York, 1927), 85, 89.
7. Adams, *Incredible Era,* 215; Means, *Woman in the White House,* 173.
8. Bess Furman, *White House Profile* (Indianapolis, 1951), 301–2; Means, *Woman in the White House,* 169.

9. Adams, *Incredible Era,* 19.
10. Ibid., 22.
11. Francis Russell, *The Shadow of Blooming Grove: Warren G. Harding in His Times* (New York, 1968), 84–85.
12. Adams, *Incredible Era,* 25; Russell, *Shadow of Blooming Grove,* 90.
13. Irene Gerlinger, *Mistresses of the White House* (New York, 1950), 93.
14. John D. Hicks, "Florence Kling Harding," *Notable American Women, 1607–1950* (3 vols., Harvard, 1971), II:132.
15. Means, *Woman in the White House,* 185.
16. Ibid., 180; Daugherty, *Inside Story,* 10.
17. Adams, *Incredible Era,* 27; Russell, *Shadow of Blooming Grove,* 91.
18. Adams, *Incredible Era,* 75.
19. Michael Teague, *Mrs. L: Conversations with Alice Roosevelt Longworth* (Garden City, N.Y., 1981), 170.
20. Evalyn Walsh McLean, *Father Struck It Rich* (Boston, 1936), 220.
21. Daugherty, *Inside Story,* 15.
22. Means, *Woman in the White House,* 185.
23. Russell, *Shadow of Blooming Grove,* 347.
24. Ibid., 353–54; Adams, *Incredible Era,* 100.
25. Russell, *Shadow of Blooming Grove,* 291; Nan Britton, *The President's Daughter* (New York, 1927), 45.
26. Russell, *Shadow of Blooming Grove,* 220; R.W. Apple, Jr., "250 Love Letters," *New York Times,* July 10, 1964, pp. 1, 27.
27. Russell, *Shadow of Blooming Grove,* 216.
28. Ibid., 371; Adams, *Incredible Era,* 163; Charles Willis Thompson, *Presidents I've Known and Two Near Presidents* (Indianapolis, 1929), 334.
29. Means, *Woman in the White House,* 187.
30. Paul F. Boller, Jr., *Presidential Campaigns* (New York, 1984), 216–17.
31. Daugherty, *Inside Story,* 176.
32. Means, *Woman in the White House,* 178.
33. Edmund W. Starling, *Starling of the White House* (Chicago, 1956), 200–201; Daugherty, *Inside Story,* 272.
34. "Mrs. Harding Dies after Long Fight," 3.
35. McLean, *Father Struck It Rich,* 275.
36. Jaffray, *Secrets of White House,* 92.
37. "Mrs. Harding Dies after Long Fight," 3.
38. Anne O'Hagan, "The Woman We Send to the White House," *Delineator,* 94 (Nov. 1920): 56.
39. "The Other Presidents," *Good Housekeeping,* 94 (Feb. 1932): 21.
40. Daugherty, *Inside Story,* 36–38.
41. Ibid., 52–55.
42. Ibid., 58, 227–28.
43. Ishbel Ross, *Grace Coolidge and Her Era* (New York, 1961), 64.
44. Starling, *Starling of the White House,* 195–96.

28. Grace Coolidge

1. M.E. Hennessy, *Calvin Coolidge* (New York, 1924), 53.
2. *The Autobiography of Calvin Coolidge* (New York, 1929), 93.

3. Donald R. McCoy, *Calvin Coolidge: The Quiet President* (New York, 1967), 31; Ishbel Ross, *Grace Coolidge and Her Era* (New York, 1961), 9.
4. McCoy, *Quiet President,* 31; Claude M. Fuess, *Calvin Coolidge: The Man from Vermont* (New York, 1939), 88.
5. Hennessy, *Coolidge,* 52.
6. McCoy, *Quiet President,* 32.
7. Fuess, *Coolidge,* 89; McCoy, *Quiet President,* 33.
8. Ross, *Grace Coolidge,* 28.
9. Fuess, *Coolidge,* 91; Ross, *Grace Coolidge,* 30.
10. Grace Coolidge, "The Real Calvin Coolidge," *Good Housekeeping,* 100 (March 1935): 23.
11. Ross, *Grace Coolidge,* 35.
12. Fuess, *Coolidge,* 91.
13. Mrs. Calvin Coolidge, "When I Became First Lady," *American Magazine,* 108 (Sept. 1929): 12–13.
14. Fuess, *Coolidge,* 91.
15. Hennessy, *Coolidge,* 152.
16. G. Coolidge, "The Real Calvin Coolidge" (March 1935), 218.
17. McCoy, *Quiet President,* 98; Ross, *Grace Coolidge,* 51.
18. Fuess, *Coolidge,* 201.
19. Ross, *Grace Coolidge,* 50.
20. Ibid.
21. Fuess, *Coolidge,* 201.
22. William Allen White, *A Puritan in Babylon: The Story of Calvin Coolidge* (New York, 1938), 214.
23. Fuess, *Coolidge,* 287.
24. McCoy, *Quiet President,* 139.
25. Fuess, *Coolidge,* 300.
26. Ross, *Grace Coolidge,* 65.
27. Hennessy, *Coolidge,* 161.
28. G. Coolidge, "Real Calvin Coolidge" (April 1935):38.
29. Ross, *Grace Coolidge,* 64.
30. Ibid., 62.
31. Ibid., 68.
32. Margaret Bassett, *Profiles & Portraits of American Presidents & Their Wives* (New York, 1969), 312.
33. Fuess, *Coolidge,* 287; Ross, *Grace Coolidge,* 69.
34. White, *Puritan in Babylon,* 241; Ross, *Grace Coolidge,* 88.
35. White, *Puritan in Babylon,* 245.
36. Ross, *Grace Coolidge,* 194.
37. McCoy, *Quiet President,* 160; Vera Bloom, *There's No Place Like Washington* (New York, 1944), 20.
38. Bess Furman, *White House Profile* (Indianapolis, 1951), 306.
39. Ross, *Grace Coolidge,* 96.
40. Ibid., 191
41. Fuess, *Coolidge,* 490; McCoy, *Quiet President,* 161.
42. Ross, *Grace Coolidge,* 99.
43. White, *Puritan in Babylon,* 63.
44. G. Coolidge, "Real Calvin Coolidge" (March 1935), 225.

45. McCoy, *Quiet President,* 161.
46. Mrs. Grace Coolidge, "When I Became First Lady," 106.
47. Ibid., 13.
48. G. Coolidge, "The Real Calvin Coolidge" (March 1935), 217.
49. Lillian Rogers Park, *My Thirty Years Backstairs at the White House* (New York, 1961), 175–76; Mrs. Calvin Coolidge, "How I Spent My Days at the White House," *American Magazine,* 108 (Oct. 1929): 142.
50. G. Coolidge, "Real Calvin Coolidge" (March 1935), 214.
51. Mrs. Calvin Coolidge, "When I Became First Lady," 108.
52. Furman, *White House Profile,* 309.
53. "Making Ourselves at Home in the White House," *American Magazine,* 108 (Nov. 1929): 160.
54. Ross, *Grace Coolidge,* 259.
55. Ibid., 260.
56. Ibid., 140.
57. James M. Cox, *Journey Through My Years* (New York, 1946), 201.
58. White, *Puritan in Babylon,* 438.
59. Ross, *Grace Coolidge,* 29.
60. Ibid, 307.
61. Ibid., 321.
62. *Autobiography of Coolidge,* 93.
63. G. Coolidge, "The Real Calvin Coolidge" (June 1935), 205.
64. Ibid. (Feb. 1935), 18–19.
65. Ibid. (June 1935), 42.
66. *Autobiography of Coolidge,* 93; Fuess, *Coolidge,* 88.
67. White *Puritan in Babylon,* 86–87; Ross, *Grace Coolidge,* 34.
68. Hennessy, *Coolidge,* 148.
69. G. Coolidge, "The Real Calvin Coolidge" (May 1935), 39.
70. Ross, *Grace Coolidge,* 70.
71. Ross, *Grace Coolidge,* 52; McCoy, *Quiet President,* 99.
72. Irwin Hood (Ike) Hoover, *Forty-two Years in the White House* (Boston, 1934), 133.
73. Ross, *Grace Coolidge,* 107–108, 168.
74. Mrs. Calvin Coolidge, "How I Spent My Days at the White House," 138; "The Real Calvin Coolidge" (Feb. 1935), 187.
75. Ibid., 185.
76. Fuess, *Coolidge,* 490; Ross, *Grace Coolidge,* 152.
77. Fuess, *Coolidge,* 351–52.
78. White, *Puritan in Babylon,* 352–55.
79. Ross, *Grace Coolidge,* 305.
80. Ibid., 27.

29. Lou Henry Hoover

1. Richard Norton Smith, *An Uncommon Man: The Triumph of Herbert Hoover* (New York, 1984), 71.
2. "Mrs. Hoover's International Housekeeping," *Literary Digest,* 99 (Nov. 24, 1928): 46.
3. Ibid., 44.

4. Will Irwin, *Herbert Hoover: A Reminiscent Biography* (New York, 1928), 65–66.
5. Frederick Palmer, "Mrs. Hoover Knows," *Ladies' Home Journal,* 46 (March 1929): 6.
6. *The Memoirs of Herbert Hoover: Years of Adventure, 1874–1920* (New York, 1951), 36.
7. Ibid.
8. Irwin, *Hoover,* 86.
9. Helen B. Pryor, *Lou Henry Hoover: Gallant First Lady* (New York, 1969), 52; Palmer, "Mrs. Hoover Knows," 6.
10. Palmer, "Mrs. Hoover Knows," 6.
11. Ibid.
12. "Hoover's Silent Partner," *Literary Digest,* 105 (Sept. 8, 1917): 52.
13. *Memoirs of Hoover: Years of Adventure,* 140.
14. Pryor, *Lou Henry Hoover,* 95.
15. Ibid., 118.
16. Ibid., 101.
17. *The Memoirs of Herbert Hoover: The Cabinet and the Presidency, 1920–1933* (New York, 1952), 187–88; Pryor, *Lou Henry Hoover,* 126.
18. Pryor, *Lou Henry Hoover,* 144.
19. Ibid, 129–30.
20. Ibid., 154–55.
21. Ibid., 184.
22. Ava Long, with Mildred Harrington, "Presidents at Home," *Ladies' Home Journal,* 50 (Sept. 1933): 8.
23. Lillian Rogers Parks, *My Thirty Years Backstairs at the White House* (New York, 1961), 226.
24. Bess Furman, *White House Profile* (Indianapolis, 1951), 311.
25. Long, "Presidents at Home," 8.
26. Parks, *Thirty Years Backstairs,* 229.
27. Alonzo Fields, *My 21 Years in the White House* (New York, 1960), 30.
28. Ava Long, "900,000 Callers a Year!" *American Magazine,* 115 (June 1933): 48.
29. Pryor, *Lou Henry Hoover,* 190.
30. Ibid., 210.
31. Parks, *Thirty Years Backstairs,* 218.
32. Ibid., 232.
33. Smith, *Uncommon Man,* 173.
34. Pryor, *Lou Henry Hoover,* 247.
35. Carol Green Wilson, *Herbert Hoover: A Challenge for Today* (New York, 1968), 22–25.
36. Frederick L. Collings, "Mrs. Hoover in the White House," *Woman's Home Companion,* 55 (April 1928): 64.
37. Ibid.
38. Nelle Scanlan, *Boudoir Memoirs of Washington* (Chicago, 1923), 250–51.
39. Collins, "Mrs. Hoover in White House," 64.
40. *Memoirs of Hoover: Cabinet and Presidency,* 324; Pryor, *Lou Henry Hoover,* 180.

30. Eleanor Roosevelt

1. Joseph P. Lash, *Eleanor and Franklin* (New York, 1971), 356.
2. Bess Furman, *Washington By-Line* (New York, 1949), 151.
3. Eleanor Roosevelt, *This Is My Story* (New York, 1937), 17–18.
4. Ibid., 11.
5. Ibid., 29.
6. Lash, *Eleanor and Franklin,* 94.
7. Elliott Roosevelt and James Brough, *An Untold Story: The Roosevelts of Hyde Park* (New York, 1973), 25.
8. *This Is My Story,* 112.
9. Ibid., 138, 162.
10. Ibid., 181.
11. Ibid., 171.
12. Ibid., 188.
13. Lash, *Eleanor and Franklin,* 210–11.
14. Ibid., 220.
15. *This Is My Story,* 148–49.
16. Lash, *Eleanor and Franklin,* 262.
17. Richard Harrity and Ralph G. Martin, *Eleanor Roosevelt: Her Life in Pictures* (New York, 1958), 89.
18. *This Is My Story,* 352.
19. Lash, *Eleanor and Franklin,* 280.
20. Ibid., 287.
21. J. William T. Youngs, *Eleanor Roosevelt: A Personal and Public Life* (Boston, 1985), 139.
22. Eleanor Roosevelt, *This I Remember* (New York, 1949), 56.
23. Youngs, *Eleanor Roosevelt,* 142.
24. Lash, *Eleanor and Franklin,* 339; Earle Looker, *This Man Roosevelt* (New York, 1932), 140.
25. Lash, *Eleanor and Franklin,* 356.
26. Frank Freidel, *Franklin D. Roosevelt: The Triumph* (Boston, 1956), 226–27.
27. *This I Remember,* 100; Youngs, *Eleanor Roosevelt,* 177.
28. Margaret Bassett, *Profile & Portraits of American Presidents & Their Wives* (Freeport, Maine, 1969), 349.
29. *This I Remember,* 125.
30. Ibid., 172.
31. Youngs, *Eleanor Roosevelt,* 170.
32. *Life,* 8 (Feb. 5, 1940): 72–73.
33. Youngs, *Eleanor Roosevelt,* 175.
34. *This I Remember,* 178.
35. Lash, *Eleanor and Franklin,* 433.
36. *This I Remember,* 231.
37. Ibid., 92.
38. Grace Tully, *F.D.R.: My Boss* (New York, 1949), 107; Harrity and Martin, *Eleanor Roosevelt,* 121.
39. Ibid.
40. *This I Remember,* 4, 162.

41. Youngs, *Eleanor Roosevelt,* 178.
42. Ibid.
43. Nathan Miller, *FDR: An Intimate History* (New York, 1983), 450.
44. Ibid.
45. *This I Remember,* 218.
46. Tamara K. Hareven, *Eleanor Roosevelt: An American Conscience* (New York, 1975), 176.
47. Lash, *Eleanor and Franklin,* 685, 691.
48. *The Journals of David E. Lilienthal* (5 vols., New York, 1964–71); III:364.
49. Bernard Asbell, *When F.D.R. Died* (New York, 1961), 53.
50. Harry S. Truman, *Memoirs* (Garden City, N.Y., 1955), I:5.
51. Elliott Roosevelt and James Brough, *A Rendezvous with Destiny: The Roosevelts in the White House* (New York, 1975), 285.
52. *This I Remember,* 348–49.
53. Roosevelt and Brough, *Rendezvous,* 94.
54. James Roosevelt, *My Parents: A Differing View* (Chicago, 1976), 96–98, 103–5.
55. Joseph P. Lash, *Eleanor: The Years Alone* (New York, 1972), 15.
56. Ibid., 159.
57. Hareven, *Eleanor Roosevelt,* 260; Eleanor Roosevelt, "If You Ask Me," *McCall's,* 82 (Dec. 1959): 56; "How To Take Criticism," *Ladies' Home Journal,* 15 (Nov. 1944): 155.
58. Harrity and Martin, *Eleanor Roosevelt,* 207–8.
59. Youngs, *Eleanor Roosevelt,* 215.
60. Eleanor Roosevelt, *On My Own* (New York, 1958), 230.
61. "Publisher's Note," Eleanor Roosevelt, *Tomorrow Is Now* (New York, 1963), x.
62. Ibid.
63. Lash, *Years Alone,* 332; Edward P. Morgan, *This I Believe* (New York, 1953), 155–56.
64. *Tomorrow Is Now,* 135.
65. Lash, *Eleanor and Franklin,* 278; William Phillips, *Ventures in Diplomacy* (Boston, 1953), 70.
66. Jean Vanden Heuvel, "The Sharpest Wit in Washington," *Saturday Evening Post,* 238 (Dec. 4, 1965): 33; Eleanor Roosevelt, *You Learn by Living* (New York, 1961), 80–81.
67. Harrity and Martin, *Eleanor Roosevelt,* 91.
68. Heuvel, "Sharpest Wit," 33.
69. "Mrs. Roosevelt Bans Police Guard," *New York Times,* March 16, 1933, p. 19; Lash, *Eleanor and Franklin,* 367–68.
70. James Roosevelt, *My Parents,* 110, 216; Joan Hoff-Wilson and Marjorie Lightman, eds., *Without Precedent: The Life and Career of Eleanor Roosevelt* (Bloomington, 1984), 15.
71. Lash, *Eleanor and Franklin,* 378.
72. Geoffrey T. Hellman, "Mrs. Roosevelt," *Life,* 8 (Feb. 5, 1940): 78.
73. Ibid., 73–74.
74. *This I Remember,* 112.
75. Marianne Means, *The Woman in the White House* (New York, 1963), 206.

76. Ibid., 202.
77. Lash, *Eleanor and Franklin,* 453; Helene Huntington Smith, "The First Lady," *McCall's,* July 1935.
78. Lash, *Eleanor and Franklin,* 424.
79. Ibid., 457.
80. Ibid.
81. Ibid., 390.
82. *This I Remember,* 275.
83. Ibid., 283–84.
84. Ibid., 343; Lash, *Eleanor and Franklin,* 715–16.
85. Hareven, *Eleanor Roosevelt,* 235.
86. Jeanette Eaton, *The Story of Eleanor Roosevelt* (New York, 1956), 257–58.
87. *Tomorrow Is Now,* 15.
88. Ibid., 138.

31. Bess W. Truman

1. "Mrs. Harry S. Truman," *Current Biography, 1947* (New York, 1948), 646.
2. "First Lady," *Newsweek,* 27 (Jan. 7, 1946): 26.
3. "Behind Mrs. Truman's Social Curtain: No Comment," *Newsweek,* 30 (Nov. 10, 1947): 16.
4. Margaret Truman, *Bess W. Truman* (New York, 1986), 276.
5. Ibid., 265.
6. Albin Krebs, "Bess Truman Is Dead at 97; Was President's 'Full Partner,' " *New York Times,* Oct. 19, 1982, p. 1.
7. Margaret Truman, *Souvenir—Margaret Truman's Own Story* (New York, 1956), 12.
8. Truman, *Bess Truman,* 11.
9. Merle Miller, *Plain Speaking: An Oral Biography of Harry S. Truman* (New York, 1973), 104.
10. Truman, *Bess Truman,* 234.
11. Ibid., 20.
12. Margaret Truman, *Harry Truman* (New York, 1973), 55.
13. Alfred Steinberg, *The Man from Missouri: The Life and Times of Harry S. Truman* (New York, 1962), 37.
14. Robert H. Ferrell, *Truman: A Centenary Remembrance* (New York, 1984), 45.
15. Ibid., 47.
16. Ibid.
17. Truman, *Bess Truman,* 49, 51.
18. *Souvenir,* 21.
19. Truman, *Bess Truman,* 127.
20. Ibid., 132.
21. Ibid., 133.
22. Marianne Means, *The Woman in the White House* (New York, 1964), 239.
23. Truman, *Souvenir,* 64.
24. Ferrell, *Truman,* 108–9.
25. Truman, *Bess Truman,* 231; Jonathan Daniels, *The Man of Independence* (Philadelphia, 1950), 253.
26. Ibid., 232; Truman, *Bess Truman,* 231.

27. Steinberg, *Man from Missouri,* 219.
28. Truman, *Bess Truman,* 236.
29. Steinberg, *Man from Missouri,* 231; Truman, *Bess Truman,* 245.
30. Steinberg, *Man from Missouri,* 234; Truman, *Souvenir,* 83–84.
31. Daniels, *Man of Independence,* 258.
32. Means, *Woman in the White House,* 227.
33. "First Lady," *Newsweek,* 27 (Jan. 7, 1946): 26.
34. Truman, *Bess Truman,* 295.
35. Means, *Woman in the White House,* 226.
36. Ibid., 218.
37. Ferrell, *Truman,* 154.
38. Truman, *Bess Truman,* 258.
39. Ibid., 297–98.
40. Robert J. Donovan, *Conflict and Crisis: The Presidency of Harry S Truman, 1945–1948* (New York, 1977), 147; Carl Solberg, *Riding High: America in the Cold War* (New York, 1973), 12.
41. Truman, *Bess Truman,* 266–67.
42. Ibid., 281.
43. Monte M. Poen, *Strictly Personal and Confidential: The Letters Harry Truman Never Mailed* (Boston, 1982), 172–73.
44. Krebs, "Bess Truman Is Dead," 1.
45. Means, *Woman in the White House,* 217.
46. Ibid.
47. Truman, *Bess Truman,* 229.
48. Ibid., 303.
49. Means, *Woman in the White House,* 231.
50. Steinberg, *Man from Missouri,* 324–25.
51. Ibid., 328.
52. Truman, *Bess Truman,* 335.
53. J.B. West, *Upstairs at the White House: My Life with the First Ladies* (New York, 1973), 101–2.
54. Truman, *Bess Truman,* 383.
55. Ferrell, *Truman,* 222; Truman, *Bess Truman,* 381.
56. Harry S. Truman, *Mr. Citizen* (New York, 1960), 24.
57. Miller, *Plain Speaking,* 107.
58. Ferrell, *Truman,* 246.
59. Means, *Woman in the White House,* 231.
60. Truman, *Bess Truman,* 424.
61. Ibid., 426.
62. Charles Robbins, *Last of His Kind: An Informal Portrait of Harry S. Truman* (New York, 1979), 138.
63. Truman, *Bess Truman,* 429.
64. Krebs, "Bess Truman Is Dead," 1.
65. Truman, *Bess Truman,* 278–79.
66. Steinberg, *Man from Missouri,* 243.
67. Daniels, *Man of Independence,* 273.
68. Truman, *Bess Truman,* 265.
69. Truman, *Souvenir,* 130.
70. Ibid., 359.

71. Ibid., 163; Truman, *Bess Truman,* 300.
72. Krebs, "Bess Truman Is Dead," 1; Steinberg, *Man from Missouri,* 249.
73. Truman, *Souvenir,* 118.
74. West, *Upstairs at the White House,* 74.
75. Truman, *Bess Truman,* 282–83; West, *Upstairs at the White House,* 93–94.
76. Means, *Woman in the White House,* 219.
77. Steinberg, *Man from Missouri,* 291.
78. Robert J. Donovan, *The Tumultuous Years: The Presidency of Harry S Truman, 1949–1953* (New York, 1982), 333.
79. West, *Upstairs at the White House,* 82–83.
80. Ibid., 122.
81. Truman, *Mr. Citizen,* 102.
82. Means, *Woman in the White House,* 232.
83. Truman, *Mr. Citizen,* 64; Truman, *Bess Truman,* 402.
84. Miller, *Plain Speaking,* 107–108.
85. Truman, *Bess Truman,* 430.

32. Mamie Doud Eisenhower

1. Mamie Doud Eisenhower, "My Memories of Ike," *Reader's Digest,* 96 (Feb. 1970): 70.
2. Ibid.
3. Alden Hatch, *Red Carpet for Mamie* (New York, 1954), 129.
4. Lester David and Irene David, *Ike and Mamie: The Story of the General and His Lady* (New York, 1981), 66.
5. Hatch, *Red Carpet,* 2.
6. Ibid., 57.
7. Dorothy Brandon, *Mamie Doud Eisenhower* (New York, 1954), 47.
8. Hatch, *Red Carpet,* 66.
9. Mamie Doud Eisenhower, "My Memories of Ike," 71; Dwight D. Eisenhower, *At Ease: Stories I Tell to Friends* (Garden City, N.Y., 1967), 113.
10. Hatch, *Red Carpet,* 73.
11. Ibid.
12. Ibid., 89.
13. Eisenhower, "Memories of Ike," 71; Eisenhower, *At Ease,* 123.
14. Julie Eisenhower, *Special People* (New York, 1977), 199.
15. David, *Ike and Mamie,* 16.
16. Brandon, *Mamie,* 77.
17. David, *Ike and Mamie,* 65.
18. Brandon, *Mamie,* 37, 246.
19. Hatch, *Red Carpet,* 92.
20. Steve Neal, *The Eisenhowers: Reluctant Dynasty* (Garden City, N.Y., 1978), 38; Andrew F. Tully, "Ike and Mamie at Home," *Collier's,* 131 (June 20, 1953): 17.
21. Brandon, *Mamie,* 89–90; Hatch, *Red Carpet,* 111.
22. David, *Ike and Mamie,* 73–74; Hatch, *Red Carpet,* 119.
23. David, *Ike and Mamie,* 82.
24. Neal, *Eisenhowers,* 67.
25. David, *Ike and Mamie,* 92.

26. Eisenhower, *At Ease,* 194.
27. Brandon, *Red Carpet,* 141.
28. Stephen E. Ambrose, *Eisenhower* (2 vols., 1983–1984), I:107–8.
29. Brandon, *Mamie,* 179.
30. Hatch, *Red Carpet,* 167.
31. Brandon, *Mamie,* 209–10; David, *Ike and Mamie,* 100.
32. David, *Ike and Mamie,* 112.
33. Ibid., 114.
34. Ibid., 111.
35. Hatch, *Red Carpet,* 192.
36. David, *Ike and Mamie,* 121.
37. Dwight D. Eisenhower, *Letters to Mamie* (Garden City, N.Y., 1978), 76, 99.
38. David, *Ike and Mamie,* 158.
39. Merle Miller, *Plain Speaking: An Oral Biography of Harry Truman* (New York, 1973), 340.
40. Kay Summersby, *Ike Was My Boss* (New York, 1948); *Past Forgetting: My Love Affair with Dwight D. Eisenhower* (New York, 1976), 194.
41. Dwight D. Eisenhower, *Letters to Mamie.*
42. Ibid., 132.
43. David, *Ike and Mamie,* 151.
44. Ibid., 152.
45. Brandon, *Mamie,* 229.
46. Hatch, *Red Carpet,* 202.
47. Ibid., 222; David, *Ike and Mamie,* 13–14.
48. "The General's Lady," *Newsweek,* 40 (Oct. 13, 1952): 30.
49. Margaret Bassett, *Profiles & Portraits of American Presidents & Their Wives* (Freeport, Maine, 1969), 381.
50. Hatch, *Red Carpet,* 238.
51. Ibid., 243.
52. Ibid., 250.
53. Ibid., 250–51; Ambrose, *Eisenhower,* I:552; David, *Ike and Mamie,* 178–79; Nanette Kuttner, "I Remember Mamie," *Woman's Home Companion,* 80 (Aug. 1953): 88.
54. Nick Thimmesch, "Mamie Eisenhower at 80," *McCall's,* 104 (Oct. 1976): 214.
55. David, *Ike and Mamie,* 181.
56. Ibid., 20–21.
57. Brandon, *Mamie,* 296.
58. "General's Lady," 29.
59. David, *Ike and Mamie,* 179.
60. "General's Lady," 29.
61. Ibid., 20–30.
62. Brandon, *Mamie,* 300.
63. "First Lady's First Days," *Newsweek,* 41 (Feb. 2, 1953): 18.
64. Andrew F. Tully, "Ike and Mamie at Home," 15.
65. Ibid., 16.
66. Eisenhower, *Special People,* 203; J.B. West, *Upstairs at the White House: My Life with the First Ladies* (New York, 1973), 141.
67. West, *Upstairs at the White House,* 131,132.

68. Ibid., 130–31.
69. Elaine Shepard, "Why Mamie Will Be Glad to Leave the White House," *Better Homes and Gardens,* 38 (Aug. 1960): 78.
70. West, *Upstairs at the White House,* 155.
71. Ibid., 155–56.
72. Neal, *Eisenhower,* 401.
73. Ibid.
74. Ibid, 400.
75. Dwight D. Eisenhower, *Mandate for Change* (New York, 1963), 571.
76. David, *Ike and Mamie,* 237–40.
77. Ambrose, *Eisenhower,* II:281; John Eisenhower, *Strictly Personal* (New York, 1974), 183–84; Eisenhower, *Special People,* 204.
78. Ellis D. Slater, *The Ike I Knew* (n.p., 1980), 171.
79. Eisenhower, *Special People,* 209.
80. Ambrose, *Eisenhower,* II:630.
81. David, *Ike and Mamie,* 23; Ambrose, *Eisenhower,* II:674.
82. Eisenhower, *Special People,* 209.
83. David, *Ike and Mamie,* 261.
84. Neal, *Eisenhower,* 460.
85. Ibid., 461.
86. "A Letter from Mamie Eisenhower," *McCall's,* 100 (Feb. 1973): 30.
87. "Mamie Eisenhower Dies at 82," *Time,* 114 (Nov. 12, 1979): 26.
88. Neal, *Eisenhower,* 461.
89. "The Ike I Remember," *Ladies' Home Journal,* 91 (april 1974): 64.
90. David, *Ike and Mamie,* 257.
91. Ibid.
92. Ibid., 258.
93. Hatch, *Red Carpet,* 56–57.
94. Eisenhower, *At Ease,* 124–25.
95. Ibid., 127–28.
96. Hatch, *Red Carpet,* 159.
97. Brandon, *Mamie,* 135–36; Hatch, *Red Carpet,* 128–29; Eisenhower, *At Ease,* 184.
98. Hatch, *Red Carpet,* 186–87.
99. Ibid, 206.
100. Virgil Pinckly, *Eisenhower Declassified* (Old Tappan, N.Y., 1979), 269.
101. Brandon, *Mamie,* 3.
102. Pinckly, *Eisenhower Declassified,* 373.
103. West, *Upstairs at the White House,* 134–36.
104. David, *Ike and Mamie,* 14.
105. Ibid., 259.

33. Jacqueline Bouvier Kennedy

1. Margaret Mead, "A New Kind of First Lady," *Redbook Magazine,* 118 (Feb. 1962): 9.
2. Ibid.
3. Marianne Means, *The Woman in the White House* (New York, 1963), 267.

4. Ibid.
5. Ibid.
6. Ibid., 269.
7. Stephen Birmingham, *Jacqueline Bouvier Kennedy Onassis* (New York, 1978), 7.
8. John H. David, *The Kennedys: Dynasty and Disaster, 1848–1984* (New York, 1984), 174.
9. Gordon Hall and Ann Pinchot, *Jacqueline Kennedy: A Biography* (New York, 1964), 10, 64.
10. Ibid., 68; Mary Van Rensselaer Thayer, *Jacqueline Bouvier Kennedy* (New York, 1961), 20, 21.
11. Hall and Pinchot, *Jacqueline Kennedy,* 67.
12. Ibid., 83; Thayer, *Jacqueline Kennedy,* 73; Igor Cassini, "How the Kennedy Marriage Has Fared," *Good Housekeeping,* 155 (Sept. 1962): 69.
13. Robert Harding and A.L. Holmes, *Jacqueline Kennedy* (New York, 1966), 13.
14. Hall and Pinchot, *Jacqueline Kennedy,* 90.
15. Jacqueline Kennedy Onassis, *One Special Summer* (New York, 1974).
16. Thayer, *Jacqueline Kennedy,* 83–84.
17. Ibid., 86.
18. Ralph G. Martin, *A Hero for Our Time: An Intimate Story of the Kennedy Years* (New York, 1983), 79; Davis, *Kennedys.* 246; Harding and Holmes, *Jacqueline Kennedy,* 200
19. Peter Collier and David Horowitz, *The Kennedys: An American Drama* (New York, 1984), 194; Hall and Pinchot, *Jacqueline Kennedy,* 112.
20. Davis, *Kennedys,* 157; Martin, *Hero,* 81.
21. Collier and Horowitz, *Kennedys,* 197; Martin, *Hero,* 127.
22. Cassini, "How Kennedy Marriage Has Fared," 183; Martin, *Hero,* 80; Hall and Pinchot, *Jacqueline Kennedy,* 115.
23. Thayer, *Jacqueline Kennedy,* 92.
24. Ibid., 95.
25. Martin, *Hero,* 88.
26. Thayer, *Jacqueline Kennedy,* 106; Hall and Pinchot, *Jacqueline Kennedy,* 131.
27. Birmingham, *Jacqueline Onassis,* 77.
28. Martin, *Hero,* 101.
29. Ibid., 193.
30. Collier and Horowitz, *Kennedys,* 196–97.
31. Ibid., 205; Thayer, *Jacqueline Kennedy,* 112.
32. Ibid.; Harding and Holmes, *Jacqueline Kennedy,* 56.
33. Martin, *Hero,* 123.
34. Hall and Pinchot, *Jacqueline Kennedy,* 142.
35. Martin, *Hero,* 144.
36. Ibid.
37. Davis, *Kennedys,* 236.
38. Martin, *Hero,* 151.
39. Ibid., 230.
40. Ibid., 209.
41. Robert J. Levin, "Senator Kennedy's Wife — If He Wins, How Much Does She Lose?" *Redbook,* 114 (April 1960): 82.

42. Ibid., 36.
43. Martin, *Hero,* 227.
44. Collier and Horowitz, *Kennedys,* 251.
45. Martin, *Hero,* 225; Hall and Pinchot, *Jacqueline Kennedy,* 151.
46. Marianne Means, *Woman in the White House,* 274; Hall and Pinchot, *Jacqueline Kennedy,* 151.
47. Martin, *Hero,* 279; Collier and Horowitz, *Kennedys,* 281; J.B. West, *Upstairs at the White House: My Life with the First Ladies* (New York, 1973), 268.
48. Mary Van Rensselaer Thayer, *Jacqueline Kennedy: The White House Years* (Boston, 1971), 181.
49. Hall and Pinchot, *Jacqueline Kennedy,* 230.
50. Cassini, "How Kennedy Marriage Has Fared," 188.
51. Martin, *Hero,* 251; Hall and Pinchot, *Jacqueline Kennedy,* 214.
52. Ibid., 215.
53. Norman Mailer, *The Presidential Papers* (New York, 1963), 95.
54. "Toward the Ideal," *Time,* 82 (Sept. 6, 1963): 67.
55. Martin, *Hero,* 510.
56. Harding and Holmes, *Jacqueline Kennedy,* 72.
57. Ibid., 74.
58. Martin, *Hero,* 349; Mary Barelli Gallagher, *My Life with Jacqueline Kennedy* (New York, 1969), 177.
59. Martin, *Hero,* 354.
60. "Queen of America," *Time,* 79 (March 23, 1962): 13.
61. Martin, *Hero,* 295.
62. Joan Braden, "An Exclusive Chat with Jackie Kennedy," *Saturday Evening Post,* 235 (May 12, 1962): 85.
63. Collier and Horowitz, *Kennedys,* 197.
64. Martin, *Hero,* 382.
65. Ibid., 383.
66. Hall and Pinchot, *Jacqueline Kennedy,* 194.
67. Martin, *Hero,* 520.
68. Ibid., 7.
69. Collier and Horowitz, *Kennedys,* 283.
70. Martin, *Hero,* 474.
71. Ibid., 528.
72. Ibid., 534.
73. Ibid.
74. Thayer, *White House Years,* 268.
75. Kenneth P. O'Donnell and David F. Powers, *"Johnny, We Hardly Knew Ye"* (New York, 1970), 23.
76. Ibid., 23; Martin, *Hero,* 550.
77. Ibid.; O'Donnell and Powers, *Johnny,* 23.
78. Martin, *Hero,* 551.
79. O'Donnell and Powers, *Johnny,* 24.
80. Ibid.; Davis, *Kennedys,* 432; Harding and Holmes, *Jacqueline Kennedy,* 98.
81. O'Donnell and Powers, *Johnny,* 25.
82. Ibid., 24.
83. Davis, *Kennedys,* 431; Birmingham, *Jacqueline Onassis,* 121.
84. O'Donnell and Powers, *Johnny,* 26.

85. Davis, *Kennedys,* 433.
86. Ibid.
87. Ibid., 434, Hall and Pinchot, *Jacqueline Kennedy,* 23.
88. Lady Bird Johnson, *A White House Diary* (New York, 1970), 6.
89. Davis, *Kennedys,* 535, 539.
90. Harding and Holmes, *Jacqueline Kennedy,* 125.
91. Ibid., 117.
92. Collier and Horowitz, *Kennedys,* 318.
93. The JFK Memorial Issue, *Look,* 28 (Nov. 17, 1964): 36.
94. Birmingham, *Jacqueline Onassis,* 146.
95. Davis, *Kennedys,* 509.
96. Collier and Horowitz, *Kennedys,* 367.
97. Martin, *Hero,* 573; Birmingham, *Jacqueline Onassis,* 151.
98. Davis, *Kennedys,* 569; Collier and Horowitz, *Kennedys,* 367; Birmingham, *Jacqueline Onassis,* 153.
99. Davis, *Kennedys,* 571.
100. Martin, *Hero,* 573.
101. Birmingham, *Jacqueline Onassis,* 198.
102. Ibid., 216.
103. "Confessions of a Public Son," *Time,* 127 (Jan. 20, 1986): 68.
104. Thayer, *Jacqueline Kennedy,* 14–16.
105. Ibid., 21.
106. Ibid., 1 8–20.
107. Martin, *Hero,* 76–77.
108. Ibid., 84.
109. Collier and Horowitz, *Kennedys,* 196.
110. Martin, *Hero,* 102,127.
111. Ibid., 131–32; O'Donnell and Powers, *Johnny,* 26.
112. Martin, *Hero,* 132.
113. Lester David, *The Lonely Lady of San Clemente: The Story of Pat Nixon* (New York, 1978), 117.
114. Ibid.
115. Gallagher, *My Life,* 44.
116. Thayer, *White House Years,* 21.
117. Davis, *Kennedys,* 264–70; Birmingham, *Jacqueline Onassis,* 95.
118. Gallagher, *My Life,* 143.
119. Thayer, *White House Years,* 152–53.
120. Birmingham, *Jacqueline Onassis,* 123.
121. Martin, *Hero,* 349.
122. Hall and Pinchot, *Jacqueline Kennedy,* 189.
123. Braden, "An Exclusive Chat with Jackie Kennedy," 85.
124. Harding and Holmes, *Jacqueline Kennedy,* 61.
125. Davis, *Kennedys,* 389.
126. Martin, *Hero,* 317.
127. Gallagher, *My Life,* 223.
128. David, *Kennedys,* 343–44.
129. Gallagher, *My Life,* 159.
130. Martin, *Hero,* 534.
131. Ibid., 293.

132. Ibid., 349.
133. Frank Saunders, *Torn Lace Curtain* (New York, 1984), 298.

34. Lady Bird Johnson

1. Ruth Montgomery, *Mrs. LBJ* (New York, 1964), 38; Gordon Hall, *Lady Bird and Her Daughters* (Philadelphia, 1967), 121.
2. Montgomery, *Mrs. LBJ,* 101; Frances Spatz Leighton and Helen Baldwin, *They Call Her Lady Bird* (New York, 1964), 182–83.
3. Marie Smith, *The President's Lady: An Intimate Biography of Mrs. Lyndon B. Johnson* (New York, 1964), 91; Carole Bannett, *Partners to the President* (New York, 1966), 72–73.
4. Montgomery, *Mrs. LBJ,* 180.
5. Ibid., 186.
6. Leighton and Baldwin, *They Call Her Lady Bird,* 102.
7. Liz Carpenter, *Ruffles and Flourishes* (Garden City, N.Y., 1970), 14.
8. Hall, *Lady Bird,* 245.
9. Montgomery, *Mrs. LBJ,* 101; Hall, *Lady Bird,* 247.
10. Smith, *President's Lady,* 26.
11. Montgomery, *Mrs. LBJ,* 186.
12. Margaret Bassett, *Profiles & Portraits of American Presidents & Their Wives* (Freeport, Maine, 1969), 411.
13. Smith, *President's Lady,* 27; Leighton and Baldwin, *They Call Her Lady Bird,* 11.
14. Smith, *President's Lady,* 35.
15. Hall, *Lady Bird,* 108.
16. Ronnie Dugger, *The Politician: The Life and Times of Lyndon Johnson* (New York, 1982), 176.
17. Smith, *President's Lady,* 40; Hall, *Lady Bird,* 109.
18. Smith, *President's Lady,* 41; Hall, *Lady Bird,* 109; Dugger, *Politician,* 176.
19. Hall, *Lady Bird,* 111.
20. Smith, *President's Lady,* 40–41.
21. Leighton and Baldwin, *They Call Her Lady Bird,* 35; Montgomery, *Mrs. LBJ,* 26.
22. Smith, *President's Lady,* 42.
23. Ibid.
24. Ibid., 43–44; Montgomery, *Mrs. LBJ,* 11–12.
25. Leighton and Baldwin, *They Call Her Lady Bird,* 39.
26. Smith, *President's Lady,* 45.
27. Ibid., 42.
28. Robert A. Caro, *The Years of Lyndon Johnson: The Path to Power* (New York, 1981), 302.
29. Montgomery, *Mrs. LBJ,* 37.
30. Caro, *Years of Johnson,* 302–3.
31. Montgomery, *Mrs. LBJ,* 60–61.
32. Ibid., 34.
33. Dugger, *Politician,* 177.
34. Alfred Steinberg, *Sam Johnson's Boy: A Close-Up of the President from Texas* (New York, 1968), 90–91.

35. Smith, *President's Lady,* 182–83; Montgomery, *Mrs. LBJ,* 93; Leighton and Baldwin, *They Call Her Lady Bird,* 86, 92.
36. Hall, *Lady Bird,* 121.
37. Dugger, *Politician,* 185.
38. Leighton and Baldwin, *They Call Her Lady Bird,* 46.
39. Smith, *President's Lady,* 4; Montgomery, *Mrs. LBJ,* 31.
40. Hall, *Lady Bird,* 132.
41. Ibid., 133; Smith, *President's Lady,* 140.
42. Hall, *Lady Bird,* 153.
43. Montgomery, *Mrs. LBJ,* 181.
44. Ibid., 51–52.
45. Ibid., 54–55; Leighton and Baldwin, *They Call Her Lady Bird,* 94–95; Hall, *Lady Bird,* 148–49.
46. Leighton and Baldwin, *They Call Her Lady Bird,* 131.
47. Smith, *President's Lady,* 25–26.
48. Hall, *Lady Bird,* 160.
49. Ibid., 161; Smith, *President's Lady,* 22.
50. Smith, *President's Lady,* 23–24.
51. Ibid., 8.
52. Ibid, 19; Leighton and Baldwin, *They Call Her Lady Bird,* 216; Lady Bird Johnson, *A White House Diary* (New York, 1970), 16.
53. Montgomery, *Mrs. LBJ,* 182.
54. J.B. West, *Upstairs at the White House: My Life with the First Ladies* (New York, 1973), 283–85, 291.
55. Carpenter, *Ruffles and Flourishes,* 192.
56. Hall, *Lady Bird,* 247.
57. West, *Upstairs at the White House,* 338.
58. Smith, *President's Lady,* 235.
59. Hall, *Lady Bird,* 249–50.
60. West, *Upstairs at the White House,* 306.
61. Carpenter, *Ruffles and Flourishes,* 75.
62. Ibid., 115.
63. Smith, *President's Lady,* 241.
64. George Christian, *The President Steps Down: A Personal Memoir of the Transfer of Power* (New York, 1970), 83.
65. Jack Valenti, *A Very Human President* (New York, 1975), 39.
66. Christian, *President Steps Down,* 260.
67. Valenti, *Very Human President,* 389.
68. Carpenter, *Ruffles and Flourishes,* 145–60; Hall, *Lady Bird,* 220–21.
69. Carpenter, *Ruffles and Flourishes,* 146; Hall, *Lady Bird,* 220.
70. Ibid., 234.
71. Carpenter, *Ruffles and Flourishes,* 246; Hall, *Lady Bird,* 233.
72. *Diary,* 103.
73. Ibid., 576.
74. Ibid., 588–89.
75. Ibid., 556.
76. "The LBJ Nobody Knew," *U.S. News and World Report,* 75 (Dec. 24, 1973): 34.
77. *Diary,* 646.

78. Ibid., 783.
79. "Six Former First Ladies," *U.S. News and World Report,* 82 (June 20, 1977): 53.
80. Ibid.
81. Liz Carpenter, "Lady Bird's Long Walk Back from Grief," *Good Housekeeping,* 184 (Jan. 1977): 129.
82. "Lady Bird Johnson Remembers," *American Heritage,* 32 (Dec. 1980): 13.
83. "The Painful Price Our First Ladies Pay," *Ladies' Home Journal,* 95 (July 1978): 138; *Fort Worth Star-Telegram,* Feb. 1, 1987, E5.
84. Carpenter, *Ruffles and Flourishes,* 191.
85. Dugger, *Politician,* 320.
86. Hall, *Lady Bird,* 166.
87. Bannett, *Partners to President,* 54.
88. "Lady Bird Johnson Remembers," 8; Ralph G. Martin, *A Hero for Our Time: An Intimate Story of the Kennedy Years* (New York, 1983), 223.
89. *Diary,* 332; West, *Upstairs at the White House,* 307.
90. Leighton and Baldwin, *They Call Her Lady Bird,* 169.
91. Smith, *President's Lady,* 198.
92. "Lyndon Baines Johnson: An American original," *People,* 27 (Feb. 2, 1987): 38.
93. Hall, *Lady Bird,* 210; Carpenter, *Ruffles and Flourishes,* 214–15.
94. "Lady Bird Remembers," 9; *Diary,* 159–60.
95. Traphes L. Bryant, "My Life in the White House Doghouse," *Ladies' Home Journal,* 89 (Nov. 1972): 186.
96. *Diary,* 503.
97. West, *Upstairs at the White House,* 348–49.
98. Ibid., 240–41.
99. "Research, Cast Make 'LBJ' Work," *Fort Worth Star-Telegram,* Feb. 1, 1987, E5.

35. Pat Nixon

1. Fawn Brodie, *Richard Nixon: The Shaping of His Character* (New York, 1981), 466; Julie Nixon Eisenhower, "My Mother," *Newsweek,* 87 (May 24, 1976): 13; Lester David, *The Lonely Lady of San Clemente* (New York, 1978), 18.
2. "Mrs. Pat—A British View," *New Republic,* 139 (Dec. 22, 1958): 5.
3. David, *Lonely Lady,* 19.
4. Jean Lippiatt, "Pat Nixon Was My Typing Teacher," *Saturday Evening Post,* 243 (Summer 1971): 29.
5. David, *Lonely Lady,* 40.
6. "The Silent Partner," *Time,* 75 (Feb. 29, 1960): 25.
7. Brodie, *Nixon;* Earl Mazo, *Richard Nixon: A Political and Personal Portrait* (New York, 1959), 31–32.
8. "The Silent Partner," 25.
9. "The Girl He Chased—Candidate for First Lady," *Life,* 65 (Oct. 11, 1968): 40.
10. Julie Nixon Eisenhower, *Pat Nixon: The Untold Story* (New York, 1986), 27.
11. Brodie, *Nixon,* 151.

12. John Brady, "Freelancer with No Time to Write" (interview with Gloria Steinem), *Writer's Digest,* 54 (Feb. 1974): 17.
13. "Silent Partner," 26.
14. David, *Lonely Lady,* 33.
15. Brodie, *Nixon,* 152; David, *Lonely Lady,* 34; J.B. West, *Upstairs at the White House* (New York, 1973), 127.
16. Eisenhower, *Pat Nixon,* 40.
17. "Silent Partner," 26.
18. David, *Lonely Lady,* 39.
19. Ibid.
20. Ibid., 48; Brodie, *Nixon,* 147; *The Memoirs of Richard Nixon* (New York, 1978), 23.
21. David, *Lonely Lady,* 47; Mazo, *Nixon,* 31; Eisenhower, *Pat Nixon,* 55.
22. "The Girl He Chased," 40.
23. Mazo, *Nixon,* 31.
24. Eisenhower, *Pat Nixon,* 58.
25. Ibid., 79.
26. "Silent Partner," 26.
27. David, *Lonely Lady,* 65.
28. Ibid., 67; Brodie, *Nixon,* 178.
29. Eisenhower, *Pat Nixon,* 88.
30. David, *Lonely Lady,* 73–74.
31. Brodie, *Nixon,* 178; *Memoirs of Nixon,* 36.
32. "Dick about Pat," *Life,* 65 (Oct. 11, 1968): 44.
33. Eisenhower, *Pat Nixon,* 97.
34. Brodie, *Nixon,* 255; *Memoirs of Nixon,* 86; Mazo, *Nixon,* 95; David, *Lonely Lady,* 82.
35. Henry D. Spalding, *The Nixon Nobody Knows* (Middle Village, N.Y., 1972), 300; Brodie, *Nixon,* 256.
36. *New York Times,* July 12, 1952, p.5.
37. Eisenhower, *Pat Nixon,* 128.
38. Ibid.
39. Richard M. Nixon, *Six Crises* (New York, 1962), 87.
40. Eisenhower, *Pat Nixon,* 120.
41. David, *Lonely Lady,* 93.
42. Brodie, *Nixon,* 282; David, *Lonely Lady,* 94.
43. "Radio Address of Senator Nixon, Sept. 23, 1952," *U.S. News and World Report,* 33 (Oct. 3, 1952), 66–70.
44. Eisenhower, *Pat Nixon,* 124.
45. David, *Lonely Lady,* 91.
46. Eisenhower, *Pat Nixon,* 125.
47. *Memoirs of Nixon,* 108.
48. David, *Lonely Lady,* 98.
49. Brodie, *Nixon,* 336.
50. David, *Lonely Lady,* 100.
51. Robert L. Riggs, "No Trouble in Our Home," *New Republic,* 131 (Oct. 25, 1954): 8.
52. David, *Lonely Lady,* 101.
53. Brodie, *Nixon,* 337.

54. Eisenhower, *Pat Nixon,* 142.
55. Ibid., 143.
56. David, *Lonely Lady,* 112; Brodie, *Nixon,* 371.
57. Eisenhower, *Pat Nixon,* 175.
58. David, *Lonely Lady,* 112.
59. Brodie, *Nixon,* 371; Nixon, *Six Crises,* 220.
60. Eisenhower, *Pat Nixon,* 189.
61. Ibid., 190.
62. Ibid., 189.
63. Ibid., 197.
64. Brodie, *Nixon,* 451.
65. Ibid., 455–56; *Memoirs of Nixon,* 240.
66. David, *Lonely Lady,* 124; Brodie, *Nixon,* 480.
67. Eisenhower, *Pat Nixon,* 232.
68. Ibid., 243.
69. Ibid., 238, 243.
70. Ibid., 247.
71. David, *Lonely Lady,* 129.
72. Ibid., 140.
73. Allen Drury, *Courage and Hesitation: Notes and Photographs of the Nixon Administration* (Garden City, N.Y., 1971), 43.
74. *Newsday,* July 8, 1969.
75. David, *Lonely Lady,* 136.
76. Ibid., 152.
77. Eisenhower, *Pat Nixon,* 264.
78. David, *Lonely Lady,* 136.
79. Eisenhower, *Pat Nixon,* 293.
80. Ibid., 333.
81. Ibid., 281.
82. Lenore Hershey, "The 'New' Pat Nixon," *Ladies' Home Journal,* 89 (Feb. 1972): 126.
83. Ibid.
84. David, *Lonely Lady,* 155.
85. Ibid., 153; Eisenhower, *Pat Nixon,* 296.
86. David, *Lonely Lady,* 161.
87. Ibid., 167.
88. Ibid, 168.
89. Helen McCain Smith, "Ordeal! Pat Nixon's Final Days in the White House," *Good Housekeeping,* 83 (July 1976): 127.
90. Eisenhower, *Pat Nixon,* 424.
91. David, *Lonely Lady,* 185.
92. Brodie, *Nixon,* 514; *Memoirs of Nixon,* 1053.
93. Kandy Stroud, "Pat Nixon Today," *Ladies Home Journal,* 91 (March 1975): 132.
94. David, *Lonely Lady,* 6.
95. Stroud, "Pat Nixon Today," 133.
96. Smith, "Ordeal!", 133.
97. David, *Lonely Lady,* 7.
98. Ibid., 6.

99. "It's Morally Wrong," *Time,* 128 (Oct. 6, 1986): 22.
100. Carol Lawson, "The Eisenhowers: 2 Writers Look Ahead," *New York Times,* Oct. 13, 1986, p. 20.
101. David, *Lonely Lady,* 54–55.
102. Eisenhower, *Pat Nixon,* 160.
103. David, *Lonely Lady,* 100–101.
104. Ibid., 101–2.
105. William Safire, *Before the Fall* (Garden City, N.Y., 1975), 530.
106. Eisenhower, *Pat Nixon,* 221–22.
107. West, *Upstairs at the White House,* 356.
108. Eisenhower, *Pat Nixon,* 404; David, *Lonely Lady,* 187, 191; "There Goes the Presidency," *Newsweek,* 103 (April 16, 1984): 37.

36. Betty Ford

1. Elizabeth Pope Frank, "Betty Ford's Secret Strength," *Good Housekeeping,* 187 (Sept. 1978): 86.
2. "A Talk with Betty Ford," *Newsweek,* 86 (Dec. 29, 1975): 21; Frank, "Secret Strength," 84.
3. Ibid., 89.
4. Ibid., 84.
5. Ibid., 89.
6. Ron Nessen, *It Sure Looks Different from the Inside* (New York, 1978), 28.
7. Betty Ford, *The Times of My Life* (New York, 1978), 225.
8. Ibid., 223.
9. Ibid., 220.
10. Nessen, *Sure Looks Different,* 28.
11. *Times of My Life,* 195.
12. Myra MacPherson, "Betty Ford at 60: 'My Life Is Just Beginning,' " *McCall's,* 106 (March 1979): 139.
13. Ibid., 138.
14. *Times of My Life,* 6.
15. Ibid., 9.
16. Ibid., 18.
17. Ibid., 14, 22.
18. MacPherson, "Betty Ford at 60," 139.
19. Lynn Minton, "Betty Ford Talks about Her Mother," *McCall's,* 103 (May 1976): 74.
20. Ibid., 76.
21. *Times of My Life,* 24.
22. Ibid., 32.
23. Ibid., 34.
24. MacPherson, "Betty Ford at 60," 138.
25. Ibid.
26. Gerald R. Ford, *A Time to Heal: The Autobiography of Gerald R. Ford* (New York, 1979), 65.
27. *Times of My Life,* 49; Ford, *Time to Heal,* 63.
28. Ford, *Time to Heal,* 65.
29. Ibid., 65.

30. *Times of My Life,* 60.
31. Ibid., 61.
32. Ford, *Time to Heal,* 67.
33. *Times of My Life,* 132.
34. Ibid., 135–36; Frank, "Betty Ford's Secret Strength," 88.
35. Phyllis Battelle, "Betty Ford: Finding Courage in Pain," *Ladies' Home Journal,* 98 (Jan. 1981): 43; *Times of My Life,* 133.
36. *Times of My Life,* 137.
37. MacPherson, "Betty Ford at 60," 139.
38. "Betty Ford's Operation," *Newsweek,* 84 (Oct. 7, 1974): 32.
39. *Times of My Life,* 3; Ford, *Time to Heal,* 39.
40. Trude B. Feldman, "Gerald and Betty Ford," *McCall's,* 28 (Jan. 1977): 28; *Times of My Life,* 176; "Betty Ford's Operation," *Newsweek,* 84 (Oct. 7, 1974): 32.
41. *Times of My Life,* 178; "Chatting with Betty and Susan, *Time,* 104 (Aug. 26, 1974): 11.
42. *Times of My Life,* 141.
43. "Betty and Jerry Are at Home," *Time,* 104 (Dec. 30, 1977): 9.
44. Hugh Sidey, *Portrait of a President* (New York, 1975), 119.
45. *Times of My Life,* 226.
46. Sheila Rabb Weidenfeld, *First Lady's Lady: With the Fords at the White house* (New York, 1979), 217.
47. *Times of My Life,* 220.
48. Ibid.; Weidenfeld, *First Lady's Lady,* 84–93.
49. *Times of My life,* 220.
50. Ibid., 221.
51. Ibid., 224–25; Weidenfeld, *First Lady's Lady,* 170–74; "Betty Ford Would Accept 'An Affair' by Daughter," *New York Times,* Aug 11, 1975, p. 16.
52. *Times of My Life,* 224–26; Ford, *Time to Heal,* 307; "Woman of the Year," *Newsweek,* 86 (Dec. 29, 1975): 19.
53. *Times of My Life,* 226; Weidenfeld, *First Lady's Lady,* 201, 216–18; "On Being Normal," *Time,* 106 (Aug. 25, 1975): 15.
54. Weidenfeld, *First Lady's Lady,* 215; Nick Thimmesch, "Ten-Four First Mama," *Saturday Evening Post,* 248 (Sept. 1976): 63.
55. Weidenfeld, *First Lady's Lady,* 274.
56. *Times of My Life,* 229.
57. Weidenfeld, *First Lady's Lady,* 258; Feldman, "Gerald and Betty Ford," 30.
58. "Betty vs. Rosalynn: Life on the Campaign Trail," *U.S. News & World Report,* 81 (Oct. 18, 1976): 23.
59. Weidenfeld, *First Lady's Lady,* 295.
60. Ibid., 389.
61. Feldman, "Gerald and Betty Ford," 130; Nessen, *Sure Looks Different,* 277.
62. *Times of My Life,* 295; Weidenfeld, *First Lady's Lady,* 389.
63. *Times of My Life,* 302.
64. Andrea Chambers, "Frank as Ever, Former First Lady Betty Ford Describes Her Harrowing Years of Addiction," *People,* 27 (March 9, 1987): 91.
65. *Times of My Life,* 306.
66. Ibid., 307.

67. Ibid, 309–10; Betty ford, *Betty: A Glad Awakening* (Garden City, N.Y., 1987), 55.
68. Phyllis Battelle, "Betty Ford Finding Courage in Pain," *Ladies' Home Journal*, 98 (Jan. 1981): 46.
69. "TV Taking a Frank Look at Betty Ford's Drama," *New York Times*, Feb. 25, 1987, p. 21; Chambers, "Frank as Ever," 89.
70. *Newsweek*, 107 (Aug. 11, 1986): 13.
71. *Times of My Life*, 47.
72. Weidenfeld, *First Lady's Lady*, 38.
73. *Times of My Life*, 183.
74. Ibid., 222.
75. Ibid., 72.
76. Ford, *Time to Heal*, 207.
77. Candice Bergen, "An Intimate Look at the Fords," *Ladies' Home Journal*, 92 (May 1975): 131.
78. *Times of My Life*, 289–90; Weidenfeld, *First Lady's Lady*, 318.
79. Weidenfeld, *First Lady's Lady*, 417.
80. *Betty: A Glad Awakening*, 124.
81. MacPherson, "Betty Ford at 60," 142.

37. Rosalynn Carter

1. Rosalynn Carter, *First Lady from Plains* (New York, 1984), 176, 178.
2. "Of Many Things," *America*, 138 (June 11, 1977): 512; "Rosalynn's Turn at Diplomacy 'Family Style,' " *U.S. News & World Report*, 82 (June 6, 1977): 36; *First Lady from Plains*, 191.
3. "Some Surprises as First Lady Sets a Style of Her Own," *U.S. News & World Report*, 83 (Nov. 11, 1977): 40.
4. Ralph G. Martin, "When Jimmy Carter Married Her, He Married Magic," *Ladies' Home Journal*, 96 (March 1979): 99.
5. Betty Glad, *Jimmy Carter: In Search of the Great White House* (New York, 1980), 500; Kandy Stroud, "Growing Up with Rosalynn Carter," *Good Housekeeping*, 185 (Aug. 1977): 173, 176.
6. "Some Surprises as First Lady Sets a Style of Her Own," 41.
7. Ibid., 40.
8. Bruce Mazlish and Edwin Diamond, *Jimmy Carter: A Character Portrait* (New York, 1979), 104.
9. Stroud, "Growing Up with Rosalynn," 176; Martin, "When Carter Married Her," 99.
10. John Osborne, "Rosalynn at Work," *New Republic*, 179 (Aug. 1978): 14.
11. Martin, "When Jimmy Carter Married Her," 168.
12. Ibid., 99.
13. *First Lady From Plains*, 9.
14. Ibid., 11.
15. Glad, *Carter*, 54–55.
16. Ibid., 55.
17. *First Lady from Plains*, 14.
18. Ibid., 9.
19. Ibid., 15.

20. Ibid.; Glad, *Carter,* 55; Stroud, "Growing Up with Rosalynn," 174.
21. *First Lady from Plains,* 17.
22. Ibid., 18.
23. Ibid., 17.
24. Ibid., 20–21.
25. Ibid. 21.
26. Glad, *Carter,* 56; Martin, "When Jimmy Carter Married Her," 182.
27. *First Lady from Plains,* 22.
28. Ibid., 23.
29. Ibid.
30. Jimmy Carter, *Why Not the Best?* (Nashville, Tenn., 1975), 63; Phyllis Battelle, "The Jimmy Carters' Untold Love Story," *Good Housekeeping,* 183 (Oct. 1976): 186.
31. Martin, "When Jimmy Carter Married Her," 101.
32. Stroud, "Growing Up with Rosalynn," 174.
33. Trude B. Feldman, "Rosalynn Carter at 50," *McCall's,* 104 (Aug. 1977), 126.
34. Martin, "When Jimmy Carter Married Her," 101.
35. *First Lady from Plains,* 36.
36. Ibid., 37.
37. Glad, *Carter,* 70.
38. Stroud, "Growing Up with Rosalynn," 174; *First Lady from Plains,* 45.
39. *First Lady from Plains,* 41.
40. Ibid.
41. Ibid., 50.
42. Ibid.
43. Battelle, "Carters' Untold Love Story," 187.
44. *First Lady from Plains,* 92.
45. Ibid., 89.
46. Ibid., 99–100.
47. Feldman, "Rosalynn at 50," 198.
48. Stroud, "Growing Up with Rosalynn," 174.
49. Ibid., 173–74; Battelle, "Carters' Untold Love Story," 186.
50. *First Lady from Plains,* 113, 132.
51. Martin, "When Jimmy Carter Married Her," 174.
52. Jimmy Carter, *Keeping Faith: Memoirs of a President* (New York, 1982), 19.
53. Ibid., 32.
54. "The New Women: Tough and Gracious," *Christian Century,* 100 (May 11, 1983): 443.
55. Martin, "When Jimmy Carter Married Her," 168.
56. Ibid.
57. Ibid.
58. *First Lady from Plains,* 173–74.
59. Ibid., 211.
60. Ibid., 305.
61. Ibid., 322.
62. Ibid., 323.
63. Ibid., 325.
64. Ibid., 338.

65. Barbara Gamarekian, "A Former First Lady Returns to the City She Loves," *New York Times*, April 13, 1984, p. 8.
66. Jean Saunders Wixon, "Saga of an 'Ordinary Woman,' " *Modern Maturity*, 30 (Feb.-March 1985): 70
67. William V. Shannon, "The Other Carter in the Running," *New York Times*, Sept. 15, 1976, p. 45; Rosalynn Carter, "We Have to Stop Running," *McCall's*, 105 (June 1978): 198.
68. Allethea Wall, "My Sister, Rosalynn," *Good Housekeeping*, 186 (Jan. 1978): 76.
69. *First Lady from Plains*, 96; Glad, *Carter*, 184.
70. Stroud, "Growing Up with Rosalynn," 102; *First Lady from Plains*, 100.
71. Ibid., 119–20.
72. Ibid., 115–16.
73. Ibid., 118–19.
74. Ibid., 132.
75. Dennis Farney, "Rosalynn Carter: Complex, Driven, Striving to Improve," *Wall Street Journal*, Oct. 26, 1977, p. 1.
76. *First Lady from Plains*, 317.
77. "The Carters Make the Most of a New Life," *New York Times*, May 18, 1987, C20; "Some Things You Don't Do with Spouses," *Fort Worth Star-Telegram*, June 12, 1987, Section 1, p. 3; "On the Road with the Carters," *Time*, 129 (June 15, 1987): 78–79; Samuel Hudson, "The Final Chapter," *Fort Worth Star-Telegram*, June 15, 1987, Section 2, p. 1.

38. Nancy Reagan

1. Lawrence Leamer, "The First Couple," *Ladies' Home Journal*, 100 (April 1983): 112.
2. "The World of Nancy Reagan," *Newsweek*, 98 (Dec 21, 1981): 26.
3. Ibid., 22.
4. Emily Yoffe, "Puttin' on the Ritz," *New Republic*, 184 (May 23, 1981): 18.
5. "Hail, Hail to the Chief," *Newsweek*, 97 (May 25, 1981): 44.
6. Nancy Reagan, with Bill Libby, *Nancy* (New York, 1980), 220.
7. Frances Spatz Leighton, *The Search for the Real Nancy Reagan* (New York, 1987), 192.
8. Laurence Leamer, *Make-Believe: The Story of Nancy and Ronald Reagan* (New York, 1983), 334.
9. "The World of Nancy Reagan," 22.
10. "Co-Starring at the White House," *Time*, 125 (Jan. 14, 1985): 26.
11. Ibid.
12. Leighton, *Search for Nancy Reagan*, 266.
13. Bill Adler, *Ronnie and Nancy: A Very Special Love Story* (New York, 1985), 186.
14. Leighton, *Search for Nancy Reagan*, 161–62; "Co-Starring at the White House," 26.
15. "The World of Nancy Reagan," 24.
16. Adler, *Ronnie and Nancy*, 186.
17. Leamer, *Make-Believe*, 300.
18. *Nancy*, 1.

19. Ibid., 12.
20. Ibid., 16
21. Ibid., 20.
22. Leamer, *Make-Believe,* 46.
23. Ibid., 49.
24. *Nancy,* 56–57.
25. Ibid., 79.
26. "Co-Starring at the White House," 34.
27. Leamer, *Make-Believe,* 70; Adler, *Ronnie and Nancy,* 111.
28. Bill Adler, "The Reagans' Untold Love Story," *Good Housekeeping,* 200 (June 1985): 239.
29. Leamer, *Make-Believe,* 139.
30. *Nancy,* 101.
31. Ibid., 102; Anne Edwards, *Early Reagan: The Rise to Power* (New York, 1987), 394, 400–403.
32. Ibid., 113–14.
33. Ibid., 114.
34. Ibid., 113.
35. Ibid., 117.
36. Nancy Reagan, "My Life in the White House," as told to Jean Libman Block, *Good Housekeeping,* 193 (Sept. 1981): 188.
37. Adler, *Ronnie and Nancy,* 115.
38. *Nancy,* 144.
39. "The World of Nancy Reagan," 23; Leighton, *Search for Nancy Reagan,* 95; Eleanor Harris, "What is Nancy Reagan Really, Really Like?" *Look,* 31 (Oct. 31, 1967): 40.
40. *Nancy,* 148; Susan Granger, "Nancy Reagan: A Private Visit in the White House," *Redbook,* 156 (July 1981): 64.
41. "California's Leading Lady," *Look,* 31 (Oct. 31, 1967): 40.
42. Joan Didion, "Pretty Nancy," *Saturday Evening Post,* 241 (June 1, 1968): 20.
43. Leamer, *Make-Believe,* 215.
44. "The Woman behind Reagan," *Newsweek,* 95 (April 28, 1980): 33.
45. Leighton, *Search for Nancy Reagan,* 165.
46. "Woman behind Reagan," 33.
47. Leighton, *Search for Nancy Reagan,* 266.
48. Ibid., 267; Fred Barnes, "Nancy's Total Makeover," *New Republic,* 93 (Sept. 16 and 23, 1985): 16.
49. "Co-Starring at the White House," 26.
50. Barnes, "Nancy's Total Makeover," 18.
51. "World of Nancy Reagan," 25; Leighton, *Search for Nancy Reagan,* 269.
52. Barnes, "Nancy's Total Makeover," 21; Cory SerVaas, "Nancy Reagan: A Love Story," *Saturday Evening Post,* 257 (Oct. 1985): 56; Leighton, *Search for Nancy Reagan,* 277–78.
53. Leighton, *Search for Nancy Reagan,* 365.
54. Barnes, "Nancy's Total Makeover," 16.
55. Ibid., 18.
56. Hugh Sidey, "It's Morally Wrong," *Time,* 128 (Oct. 6, 1986): 22.
57. "The First Couple," 112.
58. Adler, *Ronnie and Nancy,* 158.
59. Ibid.

60. Charlotte Curtis, "A Private Talk with Nancy Reagan," *Ladies' Home Journal,* 98 (June 1981): 133; Leighton, *Search for Nancy Reagan,* 160.
61. Adler, *Ronnie and Nancy,* 156.
62. Leamer, *Make-Believe,* 268.
63. *Parade,* Feb. 9, 1986, p. 2.
64. William Safire, "113 Days Is Enough," *New York Times,* March 12, 1987, p. 27.
65. "Who's in the Kitchen with Nancy?" *Newsweek,* 110 (March 16, 1987): 22.
66. Bernard Weinraub, "Nancy Reagan's Power is Greater than Ever, Aides and Friends Say," *New York Times,* March 3, 1987, p. 7.
67. Fred Barnes, "President Nancy," *New Republic,* 196 (March 23, 1987): 12; "Just Say Goodbye, Don," *Time,* 129 (March 9, 1987): 28.
68. Bernard Weinraub, "Regan Days Were Numbered after Clash with First Lady," *New York Times,* March 1, 1987, p. 1; "The First Lady: A Hang-Up about Don Regan," *Newsweek,* 110 (March 2, 1987): 22.
69. "President Can't Recall Arms Order," *Fort Worth Star-Telegram,* Feb. 24, 1987, Section 2A.
70. "Under Heavy Fire," *Time,* 128 (Dec. 15, 1986): 19; Susanne M. Schafer, "Ex-Aides Deceived Reagan, Nancy Says," *Fort Worth Star-Telegram,* Dec. 18, 1986, A17.
71. "The Week of the Dragon," *Time,* 129 (March 16, 1987): 24; Barnes, "President Nancy," 12; William Safire, "The First Lady Stages a Coup," *New York Times,* March 2, 1987, p. 19.
72. Leighton, *Search for Nancy Reagan,* 385.
73. Bernard Weinraub, "Angry Reagan Calls Reports of Wife's Power 'a Fiction,'" *New York Times,* March 5, 1987, A19; "Who's in the Kitchen with Nancy?" 22.
74. "Poll Finds Most Back Role of Nancy Reagan," *New York Times,* May 10, 1987, p. 21.
75. Frank Lombardi, "Nancy Takes the Pulpit," *New York Daily News,* May 5, 1987, p. 3; "Poll Finds Most Back Nancy Reagan," 21; "Nancy Reagan Defends Her Rights in Advising the President," *New York Times,* May 5, 1987, A26.
76. Ibid.
77. Leamer, *Make-Believe,* 35.
78. Ibid., 39.
79. *Nancy,* 50.
80. Leighton, *Search for Nancy Reagan,* 25; Leamer, *Make Believe,* 64.
81. Adler, *Ronnie and Nancy,* 86.
82. *Nancy,* 87–88.
83. Leamer, *Make-Believe,* 173.
84. *Nancy,* 89–90.
85. Adler, *Ronnie and Nancy,* 97; *Nancy,* 90.
86. *Nancy,* 151–52.
87. Leighton, *Search for Nancy Reagan,* 161–62.
88. "Barely a Footnote," *New York Times,* Aug. 11, 1981, B10; Leamer, *Make-Believe,* 285; Leighton, *Search for Nancy Reagan,* 171–72.
89. Adler, *Ronnie and Nancy,* 131; Leighton, *Search for Nancy Reagan,* 248–50; Leamer, *Make-Believe,* 326.
90. Leighton, *Search for Nancy Reagan,* 301–302.

91. Ibid., 308.
92. "Co-Starring at White House," 25.
93. Leighton, *Search for Nancy Reagan,* 344.
94. "People," *Time,* 128 (Oct. 20, 1986): 83.
95. "Poll Pleases Nancy, But Ron's Her No. 1," *Fort Worth Star-Telegram,* June 11, 1987, Section 1, p. 3.
96. Adler, *Ronnie and Nancy,* 202.

INDEX